# THE EXISTENTIAL COORDINATES OF THE HUMAN CONDITION: POETIC – EPIC – TRAGIC

ANALECTA HUSSERLIANA

THE YEARBOOK OF PHENOMENOLOGICAL RESEARCH

VOLUME XVIII

*Editor-in-Chief:*

ANNA-TERESA TYMIENIECKA

*The World Institute for Advanced Phenomenological Research and Learning*
*Belmont, Massachusetts*

# THE EXISTENTIAL COORDINATES OF THE HUMAN CONDITION: POETIC – EPIC – TRAGIC

## *The Literary Genre*

*Edited by*

ANNA-TERESA TYMIENIECKA

Published under the auspices of
*The World Institute for Advanced Phenomenological Research and Learning*
A-T. Tymieniecka, President

D. REIDEL PUBLISHING COMPANY

A MEMBER OF THE KLUWER ACADEMIC PUBLISHERS GROUP

DORDRECHT / BOSTON / LANCASTER

**Library of Congress Cataloging in Publication Data**

CIP

Main entry under title:

The Existential coordinates of the human condition, poetic – epic – tragic.

(Analecta Husserliana ; v. 18)
A selection of studies presented at three annual seminars of the International Society for Phenomenology and Literature, 1980–1982.
"Published under the auspices of the World Institute for Advanced Phenomenological Research and Learning."
Includes bibliographical references and index.
1. Literature—Addresses, essays, lectures. I. Tymieniecka, Anna-Teresa. II. International Society for Phenomenology and Literature. III. World Institute for Advanced Phenomenological Research and Learning. IV. Series.
B3279.H94A129 vol. 18. [PN45] 142'.7s [809] 84–1960
ISBN 90–277–1702–8

---

Published by D. Reidel Publishing Company,
P.O. Box 17, 3300 AA Dordrecht, Holland.

Sold and distributed in the U.S.A. and Canada
by Kluwer Academic Publishers
190 Old Derby Street, Hingham, MA 02043, U.S.A.

In all other countries, sold and distributed
by Kluwer Academic Publishers Group,
P.O. Box 322, 3300 AH Dordrecht, Holland.



Printed in The Netherlands

# TABLE OF CONTENTS


ANNA-TERESA TYMIENIECKA / The Theme: The Poetic, Epic and Tragic Genres as the Existential Coordinates of the Human Condition — ix

ACKNOWLEDGEMENTS — xvii

## PART ONE: POETRY AND ITS ROLE IN LITERATURE

### A. POETRY IN THE HUMAN CONDITION

ANNA-TERESA TYMIENIECKA / Aesthetic Enjoyment and Poetic Sense. Poetic Sense: The Irreducible in Literature — 3

WOLFGANG WITTKOWSKI / Movement in German Poems — 23

MARIA-TERESA BERTELLONI / Why Be a Poet? — 37

LOIS OPPENHEIM / The Field of Poetic Constitution — 47

### B. THE NATURE OF THE POEM: PHILOSOPHICAL AND LITERARY ANALYSIS

CYNTHIA A. MILLER / The Poet in the Poem: A Phenomenological Analysis of Anne Sexton's 'Briar Rose (Sleeping Beauty)' — 61

JEANNE RUPPERT / Nature, Feeling, and Disclosure in the Poetry of Wallace Stevens — 75

### C. METAPHYSICS OF THE POETIC

MEENA ALEXANDER / "Fallings from Us, Vanishings . . .": Composition and the Structure of Loss — 91

ZYGMUNT ADAMCZEWSKI / Poetic Thinking to Be — 99

### D. THE POET AND THE WORLD OF POETRY

L. M. FINDLAY / From Helikon to Aetna: The Precinct of Poetry in Hesiod, Empedokles, Hölderlin, and Arnold — 119

CHRISTOPH EYKMAN / What Can the Poem Do Today? The Self-Evaluation of Western Poets after 1945 — 141

TERESA GELLA / Poetry as Essential Graphs — 157

## PART TWO: THE EPIC COORDINATE IN HUMAN HISTORICITY

### A. HISTORICITY OF HUMAN EXISTENCE "CIPHERED" IN THE GREAT EPOS

L. M. FINDLAY / The Shield and the Horizon: Homeric Ekphrasis and History 163
GILA RAMRAS-RAUCH / The Myth of Man in the Hebraic Epic 175
JESSE GELLRICH / On Medieval Interpretation and Mythology 185
VALDO H. VIGLIELMO / The Epic Element in Japanese Literature 195
BEVERLY ANN SCHLACK / A Long Day's Journey into Night: The Historicity of Human Existence Unfolding in Virginia Woolf's Fiction 209

### B. THE EXISTENTIAL SIGNIFICANCE OF RHETORIC

ANGEL MEDINA / The Existential Sources of Rhetoric: A Comparison Between Traditional Epic and Modern Narrative 227
PATRICIA M. LAWLOR / Metaphor and the Flux of Human Experience 241
CHRISTOPH EYKMAN / The Literary Diary as a Witness of Man's Historicity: Heinrich Böll, Karl Krolow, Günter Grass, and Peter Handke 249
VICTOR CARRABINO / The French *Nouveau Roman*: The Ultimate Expression of Impressionism 261

## PART THREE: TRAGEDY – THE TRAGIC FEELING AND THE DRAMA IN THEIR EXISTENTIAL SIGNIFICANCE

### A. THE EVERLASTING TRAGIC ELEMENT OF HUMAN EXISTENCE AND OF DRAMATIC ART

MARLIES KRONEGGER / The Birth of Tragedy out of the Spirit of Music: Claudel, Milhaud and the *Oresteia* 273
ANNA-TERESA TYMIENIECKA / Tragedy and the Completion of Freedom 295
SHERLYN ABDOO / Hardy's Jude: The Pursuit of the Ideal as Tragedy 307

WOLFGANG WITTKOWSKI / Values and German Tragedy 1770–1840 319
A. BEN CHEHIDA / La Destinée de la tragédie dans la culture Islamique 333
EUGENE KAELIN / Toward a Theory of Contemporary Tragedy 341

B. SOME RECURRING MOMENTS OF TRAGIC SIGNIFICANCE IN THE DRAMA

BEVERLY KENNEDY / The Re-emergence of Tragedy in Late Medieval England: Sir Thomas Malory's *Morte Darthur* 363
MICHAEL PLATT / Tragical, Comical, Historical 379
FRANÇOISE RAVAUX / The Denial of Tragedy: The Self-Reflexive Process of the Creative Activity and the French New Novel 401

C. THE MODES OF THE TRAGIC

JOHN LYONS / Tragic Closure and the Cornelian Wager 409
BARBARA WOSHINSKY / Intuition in *Britannicus* 417
MARILYN STEWART / Myth and Tragic Action in *La Celestina* and *Romeo and Juliet* 425
BERNADETTE LINTZ MURPHY / Du désordre à l'ordre: le rôle de la violence dans *Horace* 435

PART FOUR: LITERARY SIGNIFICANCE AND MAN'S SELF-INTERPRETATION WITHIN THE LIFE-WORLD – LITERARY THEORY AND PRACTICE

A. LITERARY EXPRESSION AND MAN'S BEING-IN-THE-WORLD

JACQUES GARELLI / The Act of Writing as an Apprehension of the Enigma of Being-in-the-World 451
MICHAEL D. RILEY / The Truth of the Body: Merleau-Ponty on Perception, Language, and Literature 479

B. LITERARY INTERPRETATION OF "REALITY"

FÉLIX MARTÍNEZ-BONATI / Fiction and the Transposition of Presence 495
JESSE GELLRICH / The Structure of Allegory 505

MARLIES KRONEGGER / Literary Impressionism and Phenomenology: Affinities and Contrasts 521
PETER STOWELL / Phenomenology and Literary Impressionism: The Prismatic Sensibility 535

C. LITERARY "READING" OF THE HUMAN DRAMA

ADNAN MOUSSALLY / Un modèle d'analyse dy texte dramatique 547
JOSEPH MARGOLIS / The Problem of Reading, Phenomenologically or Otherwise 559

INDEX OF NAMES 569

ANNA-TERESA TYMIENIECKA

# THE THEME: THE POETIC, EPIC AND TRAGIC GENRES AS THE EXISTENTIAL COORDINATES OF THE HUMAN CONDITION

The investigation of the literary genres – the poetic, epic and tragi-comic – as the existential coordinates of the human condition has a double relevance: to the *life significance* of literature, and to the *metaphysics of the human condition*. This enquiry calls for a context that includes them both. The literary genres are, in fact, in need of a new, *philosophical* investigation: they require philosophical reflection to go to their ultimate source in order to have their *irreducible* residuum, which persists through innumerable fluctuations of their historical genesis, retrieved and estimated with respect to their artistic and aesthetic role and validity. In turn, they offer to philosophy a thread to follow in assessing the significance of art and its role in man's *meaning-inventing*. To begin with, let us affirm that philosophy and literature – art at large – stem from a common source: the *invention of the sense of experience*. "Inventing the sense" means for the living individual crossing the borderline of the strictly *vital significance* of the "external" life-elements, as well as his "inward" promptings, propulsions, reactions, etc., to that of the *specifically human significance of life*. It means orchestrating his functions of sensing, intellecting and imagining in a mode of reflection upon himself, his life and its meaning; it means the origin of culture and the establishment of the specifically human life-world.[1]

In this spectrum of studies we continue in an individually diversified fashion, proper to our common work, to probe into this ultimate source common to philosophy and literature.[2]

We may here follow our initial plan, in which we proposed a comprehensive investigation of the source-relationship between philosophy and literature.[3] This study should, on the one hand, yield profound and innovative perspectives for the understanding and interpretation of literature in its role in *man's self-interpretation in existence*; on the other hand, it will lead us to the hitherto inaccessible germinal phase of this self-interpretation itself: the factors, criteria, aims of its expansive élan, and its limits.

We have begun this twofold investigation through the descent to the "creative crucibles" of man's *sense-inventing* for the sake of his own *cultural*



*A-T. Tymieniecka (ed.), Analecta Husserliana, Vol. XVIII*, ix–xvi.

*individualization* – as well as for that correlated with it – of his *human life-world*. Literary analysis leads us through the labyrinth of the creative processes, with their structural schema and their pulp of experiential complexes; philosophical reflection scrutinizes the many-sided relevances of the creative endeavor and distills from their interfaces moments pregnant with various kinds of significance for man's self-interpretative effort and progress. Both meet ultimately at their *common source*: in the *conjectural inquiry into the Human Condition*, where the *subliminal virtualities* of the human inventive power and its shaping proficiencies both meet: the promptings of nostalgias, hopes, and Visionary dreams, as well as their limitative measure of their fulfillment. In this *poetics* we have outlined a meeting ground, which may be progressively unraveled in a metaphysical inquiry into man's creative forge.[4]

Indeed, the striking phenomenon – when we think of it – from which the literary analysis and theory usually begin, as from a "natural" state of affairs, is the most intimate intermingling and interpenetration between the "literary language," "aesthetic" or "poetic" language – having its established ground-work in literary works – and our thinking, evaluating, and feeling of "everyday life." It is as if poetry, fiction, drama, and comedy, with their distinctive ways of viewing and approaching human life and the human world, would not only penetrate our actual "pedestrian" thinking of the world and ourselves, but even fuse with the sense of enactment of our actual existence. It is not enough to attribute to the aesthetic element in life the role of giving it a "taste." It is true that the aesthetic aspect, with its vast range of nuances along various spectres (e.g., of the beautiful, grotesque, sublime, nostalgic, infinite, ominous, and so on) gives "flavor" to life. Like various condiments added to plain nourishing substance they lift it in innumerable mixtures from the simply palatable quality to those of "tasty" (delicious, . . .); thus the advent of the aesthetic element transposes the narrow pedestrian "quality of life" into a different key. Yet unlike the condiments, the aesthetic element does not limit itself to the quality of taste. On the contrary, it is merely its secondary function, itself the result of the primary one: *sense-giving*.

The inverse emphasis, namely, upon the ingredient or even substantial reference of literature to "reality" of life, is the ever-recurring subject of puzzlement and investigation on the side of literary analysis and theory. To distinguish sharply between the *vital significance* of the elements of the human language – as expressed by individual life-proceedings, demeanor, stand on life-matters, reactions, etc. comprising all types of life-manifestation – and the *aesthetic* or more specifically "poetic" significance of the innumerable means of our communication with the Other, is a vain task, so much are they at the borderline fused together in infinite nuances into the common task of weaving various segments of *man's self-interpretation-in-existence*. In this basic task of all human functioning, the *vital* and the *aesthetic* senses – which are differentiated into various degrees of complexity – are, albeit in different registers, equally instrumental in promoting individual human existence. Yet in the creative works, works of man's art brought to their complete accomplishment, the aesthetic element in its full measure distinguishes itself radically from the concrete, survival-subservient, vital significance.

It is of crucial importance for the searching mind to ask – and to know – what is the source, the validity, and the varying conditions of the significance he himself gives to the innumerable moments of his life-enactment. Indeed, it is not enough for man to enact his life in establishing passively its significant course by implementing the postulates of Nature's interest in his frame; he seeks to individualize himself beyond them in order to give it its specific, *its very own significance*; he has to attempt to "understand" it. He has to ask, What am I *really after* in my evaluations, stands on matters, manifestations of my feelings and self-appreciation as I establish the meaningfulness of my life-course? From what do I make my satisfactions, defeats, ideals, aims – as well as the criteria of their evaluation? That is, What kinds of meanings do I choose to give to the events, efforts and struggles of my life and how do they emerge and whence? The human being undeniably draws significant elements for the apparatus, by means of which he establishes the network of his meaning-bestowing upon everything in his concern, not only from the vital factors but also from the aesthetic ones. Stated differently, we may say that man feeds himself not exclusively on bread but concurrently on art, and especially on literature. Beginning with the invention of allegories, in which men and events appear larger than life, musing about one's feeling by improvising a song, etc., up the ladder to the most elaborate and skillful literary forms, all of these artistic efforts belong to life. They bring into life's narrow frame perspectives of significance that correspond to the innermost longings of man.

It is no wonder that literary analysis attempts to pry, as inquisitively as its respective tools allow it, into linguistic forms of the literary text in which the author, like a sorcerer, has enclosed a unique "key" to the enigmatic meaning of life. In order to find this key both the critic and the philosopher scrutinize the entire range of significant elements in the author's expression – language, values, action, etc. – and seek their relevances in life, human consciousness, the world, and society. Yet in no single line of research, nor within one schema established by intuitive conjectures, may we reach conclusive answers. All the attempts, however revealing, open either a sector or give us an "overview" – one out of many – of the genesis of this meaningfulness; but none alone bring us to its source. No doubt we have to seek a point of crossing of all ways, indicating the poles upon which their mutual distantiating as well as intergenerative fusion is suspended: a context within which the significance of various life-functions – the vital-intellectual-aesthetic among them – is projected, starting from the germinal virtualities and proceeding upward into various segments of the self-interpretative effort of man. Such a context must be established in order to treat these questions plausibly.

Such a context has been offered by Moritz Geiger, the founder of phenomenological aesthetics, and others before him in the so-called "aesthetic situation." No doubt, the aesthetic significance is suspended upon three poles: the author, the work, the recipient/reader/spectator. Within this tri-polar situation the "meaning" of the literary work, established by the author within it, is scrutinized by the recipient with a view to being received and "understood" by him. These three poles each present a selection of constants and variables which seem to account for this process. The situation of each – the author, work, reader – in the historical period of time with its sets of significant clues to interpret life elements, the tastes and artistic conventions of the period, cultural trends and their aspirations, are the constants which allow room for varying historical circumstances within the author, the work and the recipient, all three. However, in spite of the unique singularity of the set of historically conditioned variables which the author as a human being – who unfolds his individual beingness out of a specific set of circumstances, himself endowed necessarily with some specific bents, modes of sensibility, passional strivings, likes and idiosyncracies, tendencies and ideals – brings into his creative works, whatever "message" or "vision" he conveys through them, they are set in the form of a ***universal schema***. His "message" or vision, as the key to the significance of existence, springs, indeed, from the most particular and specific ground; and yet, by entering into the work of art, it becomes purified of concrete trivia and imponderably subjective features. A work of art incarnates aesthetic significance only insofar as it transposes that significance from suggestive, uncrystallized, evanescent concreteness into the pluri-vocal and yet universalized key; the work of art thus crystallizes the concrete complex into one predominant, significant form, and thereby accomplishes the metamorphosis. The work thus suspended in an intentional mode in midair, pregnant with the meaningful content of life, but emptied of its juices, draws its existential sustenance from its intrinsic appeal to the infinitely complex singularity of the recipient. The reader or spectator is called upon to lend its lived substance which she/he draws from the concrete entanglement within her/his life-world. To come to its own in the exfoliation in full of its virtual endowment, in which its clue is ensorcelled, it needs again the full-fledged singular concreteness of the reader or spectator to fill its empty constants with palpitating variables of life-significance.

It is to be granted to the proponents of structural aesthetics that the intentional context puts literary analysis upon a firm ground and is only too well known to the phenomenologically oriented critic or philosopher. However, it must be restated. But in fact it has to be observed that this "aesthetic situation," in spite of its indubitable merits, exhibits fundamental shortcomings. No doubt the tri-polar approach to the understanding of the work of art is helpful in establishing the interpretation of the artwork as a specific singular text within the radius of meaningfulness which it encircles. However, this type of interpretation has only the text as its objective and its guideline and does not focus upon the condition of its origin. The interpretation within this context draws upon the "last" stage of the genesis of meaningfulness as acquired by the author-work-reader within their respective life-worlds. It varies with respect to the historical situation of each. In this way it uses and takes for granted the entire apparatus of "understanding" acquired within the historical process, which, at the one extreme, is incarnated within the work by the author partaking of it and, at the other extreme, is to be brought in by the reader or spectator as also pertinent to it in his "education."

Consequently, this type of approach, which the intentional "situation" yields, is itself fleeting on the tides of history and of the cultural progress of human kind taking its very condition for granted. Thus, it can only offer us answers relative to the specific phases of culture; our search, in contrast, is for the *ultimate origin of the meaningfulness of the aesthetic significance* as such and its relation to the vital significance.

We cannot find this origin by scrutinizing an intentional model; moreover, we cannot find the answer to a state of affairs involving the *entire human beingness*, assuming human functioning as reduced to intentionality. Lastly, in seeking for an "understanding" of the significance of art – its source, genesis, dependencies – we cannot expect to find it in the already-established cultural life of either the author or the reader. Their "cultural development" might be decisive for the fixating, on the one hand, and grasping, on the other hand, of the concrete, contextual meaning of the work. Nevertheless, in creating a specific work the author is already virtually in possession of culturally relative skills to incarnate his innermost vision in the text; the reader, to understand it, has in turn to pertinently develop his aesthetic sense on the one hand, and his life-reflection on the other hand, to be capable of endowing the skeleton of the work with the life-substance of his experiential scheme.

If we seek for a context within which the origin of the aesthetic sense and of the vital sense differentiate, as well as fuse into each other, develop and then interlace, so that we may clarify their respective *source-dependencies* and *proficiencies* within man's self-interpretative life-course, we must search for it not in its already-made product. The completed work of art itself is the result of innumerable sedimentations of the meaning-bestowing (and meaning-inventing) processes. We must seek it in the *context establishing the means of this meaning-inventing*, first, within their specific significance-proficiency with respect to the significant role of each (e.g., bringing into source-experience the significant streak of the aesthetic sentiment, or that of projecting objectifying patterns in which the dispersed elements of functioning would come together – the role of the intellect – or of the moral sentiment infusing benevolence into the reactions toward the Other, etc.); second, by bringing these heterogeneous *sense-proficiencies* into common molds of experience and initiating the specifically human phase of self-interpretation. I have emphasized already that this phase of man's self-interpretation resides beyond the response to vital needs; it responds to man's innermost propensity in endowing his life-course with his own invented meaning. Thus, in order to understand the significance of art within life we have to reach the *creative context of the Human Condition*, from which the antennae spring, fuse, or intertwine, assuming complementary roles, and with which the existential script of man is spun. Consequently, it is within the creative context of the Human Condition that we situate our probings into the aesthetic significance of art, art-sentiment, and art-language.

To bring back the literary genres into the focus of attention when, on the one hand, extensive studies of them abound, and, on the other hand, the distinction between genres is no longer clear in contemporary literature, would have seemed obsolete were it not for two factors. One is that, together with some recent literary critics, we cannot fail to observe their persistent residual vestiges in contemporary literature; and this in spite of the ever-evolving transformations of the literary styles, conventions, and forms which, after classic standards have long since been broken, have in contemporary literature reached the point of almost total rejection. Having abandoned the formal precepts of excellence, with their standards and criteria of structure, expression, language, etc. (e.g., in Western literature, sonnet in poetry, three unities in drama, etc.), contemporary literature seems, in fact, after centuries of probings into the "ideal" models and then, the quest after perfection of their implementation, to have gained complete freedom of expression. Nevertheless, the basic distinctions, which spontaneously differentiate the literary art from its origin in cultural history, remain valid. The epic, lyrical, tragic, comic – poetic and prosaic – distinctions play here, as always, a forming/differentiating role in literary art, making it "art," and making it "literary." First, the lyrical, tragic, comic, or epic sentiment pervades the literary expression of man. It pervades the feeling and its cadences, language with its vocabulary, and syntax as well as the formal arrangements of elements of the literary work, and make, from what otherwise would be a "matter of fact" mode of information, an aesthetic (literary) one. Second, each of them finds its own occasion to gain prominence and an appropriate way to exercise it. In fact, the lines or forms of the final embodiment in literary work may largely differ from the clasic ones. Nevertheless, the basic differentiation into epic, dramatic, or poetic genres through an emphasis upon one or other type of significant experience that is inventively ciphered, on the one hand, and upon the creative form which the conveying of its significance demands, on the other hand, affirm their residual but seemingly unavoidable role. The aims and means to assume this role are adjusted in a complex and selectively extended span of the creative process. Stretching further, it appears that inventively projecting a network of aesthetic significance – that causes the straightjacket of the strictly *vital* significance to burst open and expand – is postulated by the *virtualities* of the Human Condition. They are the means for the expansion of human life. On these analyses, the Human Condition actualizes itself, beginning by life's germinal state, and progressively through the individual's self-interpretative effort, and is revealed by the forms of human existence which it has projected.

The probings into the ways in which what we call the "literary genre" appears in the most interesting contrast between its function with respect to classical and contemporary literature when we investigate this function within the creative context, reveal its unique task in this meaning inventing and projecting effort. The epic, poetic (lyric), tragic genres serve as *coordinates between the virtualities of the human condition and man's specific self-interpretative endeavor through the aesthetic significance of life*. Through the infinite range of possible fluctuations in the structural crystallization of aesthetic significance, in which the elements of signifying complexes either loosen their expressive poignancy due to the contrary emphasis, or intensify it to the full measure by becoming the focus of concentration of all expressive means – length, temporal spread, experiential purity, alternation of cadences, there persists an "objective/subjective" coordination between the "feeling" of aesthetic experience and its "significance." The differentiated genres may be considered the essential antennae which organize, transfuse, and transmit the aesthetic experience by means of the literary work. They thereby lead to the actualizing of the virtualities of the Human Condition into a full-fledged expansion of existential significance.

Within the *creative context* of the Human Condition we bring all the seemingly disparate issues which literary work in the perspective of its diversification into epic, tragic and poetic (lyric) genres raises to an ultimate common reference schema. The epic, lyric, tragic moments of the aesthetic significance of life run by their ramified threads through the very heart of literary endeavor, literary work and its interpretative understanding. The creative context serves as a pipeline for the philosophical findings drawn from their singular investigations about man's actualizing his Human Condition by way of art, and also offers the literary analysis an access to "understanding" its own final significance.

## NOTES

[1] Cf. *The Philosophical Reflection of Man in Literature, Analecta Husserliana*, Vol. XII, 1982.

[2] Cf. A-T. Tymieniecka, '*Poetica Nova*: The Creative Crucibles of Human Existence and of Art', *Analecta Husserliana*, Vol. XII, 1982, pp. 1–93.

[3] Cf. *The Philosophical Reflection*, op. cit.

[4] Cf. '*Poetica Nova*', op. cit.

# ACKNOWLEDGEMENTS

This volume is a selection of studies presented at three annual seminars and conferences of the International Society of Phenomenology and Literature (1980–1982). Warm thanks are due to our collaborators and to the dedicated staff of the World Phenomenology Institute for their care in preparing the details of these events. Mrs. Marie Lynch, the executive secretary of the Institute, who directs its day to day activities and keeps track of the various concerns of our collaborators, deserves our thanks in the first place. My assistant, Miss Rebecca Ramsay, is to be thanked not only for taking charge of the practical problems of carrying out the conferences but also for her expert typing of some of the material in this volume. Mr. Webb Dordick, our senior editor, deserves our continuing appreciation. Thanks are also due to Mr. Louis T. Houthakker for his editorial help.

ANNA-TERESA TYMIENIECKA

# PART ONE

# POETRY AND ITS ROLE IN LITERATURE

# A. POETRY IN THE HUMAN CONDITION

ANNA-TERESA TYMIENIECKA

# AESTHETIC ENJOYMENT AND POETIC SENSE

## *Poetic Sense: The Irreducible in Literature*

### THE SHIFT FROM STRUCTURE TO ENJOYMENT

In this study I propose to reinvestigate from a new perspective the relation between literary theory and aesthetics. This perspective opens up when, upon taking into consideration the vast field of theories proposed recently on the subject, we still – in appearance – ask, What makes a poem poetic? or, What makes a poem a poem? These questions have been asked from time immemorial. Yet I understand by them something particular. Instead of understanding these questions in the currently established "structural" way, namely, as asking, How is the poem as a literary work constructed that it can be differentiated, first, from other types of literary works (e.g., fiction), and second, from nonliterary writing? my above formulation emphasizes the "What" over the "How." This is a significant shift from the contemporary stress, especially in phenomenological literary theory and aesthetics, upon the structure of the literary work, to "that" which this structure is "structurizing" and ultimately "what" the structurizing is working with. The superficial impression which this explanation seems to give, namely, that by this "what" is meant anything objective (e.g., the content of a work), if we see it in erroneous terms (in contrast to the "form/matter" distinction) as a message, may make us want to reduce the work to its author's intentions – or we may even be misled by supposing the "what" to be the structure itself. I understand here, by contrast, a shift away from the "objective" means of structure (including the expressed or represented objectified reality) toward the *source* of the human experience from which the author drew it and into which the structural means and ways lead.

In what follows I will redirect attention from the forms in which the poetic work dwells – like in a cocoon – to the *aesthetic enjoyment* in which it frees itself from the structural bonds and flourishes, and also, from which it originally surges. This shift has a particular significance with respect to the current tendencies in literary criticism, on the one hand, and with respect to the present issues of philosophy, in particular phenomenology, on the other hand.

First, in both phenomenology and literary criticism – partly interdependently – the mistrust of the sovereignty of reason, which I have voiced



*A-T. Tymieniecka (ed.), Analecta Husserliana, Vol. XVIII*, 3–21.

myself – has reached its peak. The "de-constructive" approach in literary interpretation – in which the furthest consequences of the mistrust of reason are reached – either leaves nothing of significance after having cleared the field (and we must ask whether the literary work retains an aesthetic value, and if so, in what it consists) or resorts to purely metaphysical realities, which the work, structurized on a "de-construction" basis, makes appear, prompting the reader to constitute them (e.g. the Theater of the Absurd of Ionesco, Mrozek, etc.).

If the structure of the literary work can be de-composed down to the meanings and tendencies of the language, what – if anything – sustains its aesthetic impact? What, ultimately, makes it a work of literature, in contradistinction to a piece of pedestrian writing? No matter how we may distort or decompose it, it is always still with respect to "understanding," "feeling," "meaning" that we treat and receive it. That is, in a most general way we can say that it is with respect to the aesthetic *sense* that we handle the literary work.

I submit that the mistrust of reason and hence the following "de-construction" of essences, *eidoi*, meanings, objects represented, situations and life-references of the literary text, has instead of dissolving it, indeed, emphasized more strongly than ever the specific nature of literature. Moreover, as we will show, it has helped us to formulate the perspective which this questioning of the prerogatives of reason and rational structure opens upon the primeval sources of the literary creation in the Human Condition. Indeed it allows us to raise the question of the *sense* which it ultimately embodies.

Second, the proposed shift in the approach to the literary work from the ontological structure of the literary work (Ingarden and his followers), from the structures of the subjective experience exemplified by the text (the Geneva Critics) and from the structure of the text as a basis for hermeneutics (Heidegger, Gadamer, Ricoeur and their followers), and from the general relatedness of the work of literature to the theory of meaning – from that in which the work reposes and by which it is embodied – to the strictly aesthetic *enjoyment* which it is to warrant, to embody and to release in the receptor, is also the shift from the traditional object/subject – the work/reader-approach. The "subject/object" issue, strongly emphasized at the heart of phenomenology, will find in this shift a sort of dissolution. The work of art, and especially of literature, reveals a perspective into the recesses of the virtualities of the human being from which it springs. The distinctions between subject and object, reason and emotion, belong to a certain level at which these virtualities are already actualized. With the aesthetic enjoyment we *actually reach* the pre-structural and pre-rational condition of literature.

It appears that it is the condition of aesthetics as well. I may anticipate the unfolding of the argument by pointing out that, in the face of reductionistic and de-constructing attempts, the *poetic sense* of the aesthetic enjoyment will emerge as the "irreducible" – or to respond to the now fashionable de-constructive interpretations and philosophies – as the "indestructible" nerve of literature as well as of aesthetics at large. At the same time, the relationship between the irreducibly aesthetic and the irreducibly artistic factors in literature – as well as in art in general – will be re-inforced in its *existential significance*.

# *PART I*

## *THE AESTHETIC ENJOYMENT*

### *The Poetic Sense as the Key to Literary Theory*

### 1. THE PREDICAMENT OF VALUE-AESTHETICS AND OF LITERARY TEXT-BOUND THEORIES

What is literary work? In contrast to the ontological stress upon the nature of the literary work itself, let us say that the literary work is *a specific type of result obtained by the creative activity of the human being*. It is "literary" first, insofar as it is "embodied" in the media of written (and spoken) language; it is a "literary creative result" insofar as this embodiment concerns the *poetic sense* brought to fruition through the *aesthetic enjoyment*. In the exfoliation of this definition we must concern ourselves first with aesthetic enjoyment.

In every conception of the literary work we have to take into consideration its essential polarity: (a) there is the aesthetic enjoyment by the reader which makes him acquainted with the literary work, but this enjoyment on the side of the experiencing subject is strictly bound up with (b) the nature of the literary work as an object, that is its availability in its own right. All depends now upon how we conceive of the *subjective experience of "enjoyment"* and of the *objective nature of the work*. Moritz Geiger who stressed this polarity initiated phenomenological aesthetics; but this polarity, first elaborated by Geiger and running through the entire line of phenomenological aesthetics, has yet to be developed so far as to yield crucially important results. The polarity between the objective and experiential dimensions of the approach to the literary work was accepted by Roman Ingarden – a student of Geiger in Göttingen – who pursued it to its furthest limit. Ingarden's aesthetic theory of the literary work follows two distinctive lines. Let us recall that, first, having stressed the ontological question, What type of object is the literary work? Ingarden developed an ontology of the literary work. Second, faced with the question, How do we experience the literary work? he proposed a systematic analysis of the ways of "cognition of the literary work." Last, he gave a culminating touch to this twofold investigation in the conception of the specifically "aesthetic object" understood as (1) the constitutive result of the cognitive process, which, however, (2) is supposed to be found in the aesthetic values inherent in the literary work itself. With these sharp differentiations, Ingarden's theory remains incapable of accounting for what makes the literary *work* aesthetic, or for resolving what makes the *cognition* of the literary work aesthetic. We are referred from the cognition to the values within the object and conversely.

Neither the origin of supposedly aesthetic values, nor that of the retrieving (or crystallizing) in cognition of the aestheticity of these values, is accounted for. If Ingarden is brought to the fore in the present argument it is not merely for an historical reason. I suggest in fact, that this predicament is unsolvable in any aesthetic theory that refers ultimately to aesthetic values. Yet how may one overcome the value approach?

These basic shortcomings of phenomenological aesthetics bring us back to its source: the Geigerian initial distinction between aesthetic enjoyment and the literary object. To investigate this distinction might reveal some important and hitherto unobserved features. Does not the original approach to the polarity of the literary work, as devised by Geiger, contain some basic hints missed by Ingarden, who accepted Geiger's views in a simplified, approximate way? If Geiger himself did not exploit his conception fully, what can we find in it that may perhaps put us on the right track?

## 2. THE CONCEPTION OF THE "AESTHETIC ENJOYMENT" IN MORITZ GEIGER'S AESTHETICS

It should be emphasized that, unlike Ingarden, it is from the reader's side that Geiger approaches the issue of literature: in discussing access to a literary work he begins with aesthetic enjoyment rather than ontological structure. In fact, he proposes aesthetic enjoyment as the Archimedean point about which revolve questions concerning both the literary work's nature and the cognitive experiences in terms of which it is to be crystallized. To begin with, we have to distinguish between enjoyment as experience and enjoyment as a function. It is the experience of enjoyment that enters here into consideration. Experience as a function will be considered in the second part of this study.

(a) *Enjoyment Differentiated from Cognition*. Following Geiger, who focuses upon experience, we may differentiate from enjoyment in general the experience of enjoyment as a peculiar category understood as (1) an affective reverberation of our natural conscious acts – in contradistinction we may say to the "pure acts" of the intellectual (transcendental) consciousness. (2) He distinguishes it from enjoyment seen as a conscious dwelling upon experience itself, pleasurable or painful, or as a meditative inquiry. (3) He differentiates it from enjoyment seen in a more general sense as an emotive (affective) sensuous "thickness" of experiences of the natural conscious acts. (4) Being different from pleasure, enjoyment is also something other than *joy about* something. Joy about something is motivated objectively, whereas enjoyment does not for motivation extend "beyond" its object to something else.[1] Yet enjoyment of a work of art has its *foundation* and a *cause*. What can be an object of enjoyment? Objects of enjoyment are numerous. We may enjoy our experiences, moods, etc.; a great many psychic experiences may become objects of enjoyment; situations may also be objects of enjoyment. But enjoyment must have an individual object (whether real or imaginary) and this object must have a "certain fullness."[2] So much for the differentiation of enjoyment with reference to its object.

(b) *Enjoyment Distinguished from Experiences*. But enjoyment should also be differentiated from cognitive and volitional experiences with respect to the way in which the experience of enjoyment constitutes its object. That question Geiger formulates by asking, How does the experiencing self stand in relation to the experience of enjoyment? first, with respect to the act of enjoyment, second with respect to the act by which the object of enjoyment is constituted for me. We will here pursue the specific nature of the experience of enjoyment as a category of an intentional act.

Here, of course, it is the relation of the act of enjoyment to the object of enjoyment that is at stake. In contradistinction to the cognitive acts "striving for understanding of something," the act of enjoyment has a different direction. In cognitive instances the "movement" of the intentional act is to be grasped going from the self to the object; in the act of enjoyment, Geiger emphasizes, the movement of the intention proceeds inversely, namely, *from the object to the self*. In addition, the act of joy consists in the movement of intention proceeding from the self, which "sends out a wave of joy." In contrast the act of enjoyment – even the most "active" in itself – is not a trap into which the object should fall, but is a "taking in" of what comes from the object. This latter is not grasped, structured, or constituted; it is "drunk up and absorbed."[3] No matter how active, or intensely pronounced the enjoying as a function is, the experience of enjoyment itself remains a "listening to the object."[4] It should be emphasized that the intentional activity concerns only the self-to-enjoyment but not the attitude of the self *in* the experience of enjoyment.

(c) *Enjoyment Differentiated from the Constitutive Features of the Will.* If we consider all manifestations of will, either direct as desiring, willing, wishing, preferring, or indirect as present in every feeling (anger, satisfaction, etc.) in the turning of the subject "toward" or "against" an object – even in joy, if it is not the joyful state of the subject but as "joy about" – a *position* is adopted by the self. In contrast, although desire, wishing, contentment may precede or follow the experience of enjoyment, the experience of enjoyment of a poem, music, or play does not leave to the enjoying subject the *distance* indispensable to taking a stand of any sort toward what he enjoys. In fact, he is not only absorbed by, but even merged in, what he experiences: poem, musical melody, drama. He no longer may hold a position toward it; even not that of pleasure or displeasure.

Hence it follows that enjoyment has no relation to will. On the contrary, enjoyment is an unreserved acceptance of, and an unconditional surrender to, its object.[5]

At the one extreme we see, then, the experience of enjoyment utterly suspended and drawing upon its object; at the other extreme, the experiencing act is conceived as free from all features of the constitutive/cognitive intention as well as of all strictly "subjective" volitive features. As such it appears as utterly "free" from all subjective/empirical/constitutive contingencies, a type of "purist feeling" or "nothing but feeling."[6] But it is not an "unqualified" feeling: it is permeated by its object.

(d) *Enjoyment and the Self.* Having thus freed the nature of the act of enjoyment from all dependency upon the constitutive functions of consciousness, both cognitive and volitive on the one hand, and having previously distilled it from the empirical dependencies upon senses, emotions, and empirical feelings (even of pleasure and displeasure type), on the other, we must now try to discover what this thus "purified" experience depends upon.

There is only the self which we may refer it to.[7] We have, however, first to distinguish the self from the system of consciousness. Acts of consciousness may, according to Geiger's analysis, either just "happen in consciousness" as "something which happens to me" and impinges upon my conscious field even to the point of "taking possession of me" or be "self-centered," that is, arise from the self as the center of consciousness. In the first case, acts which just happen to me have not engaged my self. Anger, desire, or joy, for instance, may just happen in my consciousness whether I myself like it or not; they might be alien to myself; its source is the sentient context of my functions but not myself.[8] But my enjoyment never "happens" to me. It does not take possession of me – I am in it from the very beginning. While feelings, desires, etc. may happen in consciousness without or even against my consent, the consent of the self, I take the urge as *mine*. In the first case, the self does not participate in the "happenings" of consciousness, which burst out without my self-experience being engaged in them. Or the enjoyment is taken by myself as mine; as something which belongs to my self. In contradistinction to the experiences which appear at random within consciousness as superficial – even alien to the self – the enjoyment belongs to those experiences which proceed from the depth of the self.

We may then conclude about the features of enjoyment with respect to the self: (1) it is always self-centered; (2) it is an experience of taking in; (3) the experience of surrender to the object. These features presuppose that (4) the self-participation of enjoyment means the experiencing by the self. In enjoyment we take what streams from the object but we do the enjoying (in contrast to all the experiences as well as simple feelings which do not include the self-participation).

To conclude, the self might be more or less absorbed in enjoyment, which may be deeper or more superficial. Enjoyment even has the tendency to become our master. And yet enjoyment itself does not consist in these three features: self-centeredness, surrender, and acceptance. "Enjoyment itself is that which is conjoined with all those factors; it is a special way of being affected," in itself "it is an arousing of the self in which the self reacts to what flows into it."[9]

With Geiger we find a "series of properties which were shown to be characteristic of the essence of the phenomenon of enjoyment." They emphasize the kernel of experience which is neither structurizable nor qualifiable.

Geiger concludes:

1. Enjoyment is motiveless.
2. The object of enjoyment has "fullness."

3. All enjoyment has "participation" by the self.
4. Enjoyment is an experience of taking in.
5. Enjoyment contains no position toward the object, but a surrender to the object.
6. Enjoyment is centered upon the self.
7. Enjoyment fills the self.
8. Enjoyment is an affect upon the self.
9. Enjoyment displays certain qualities and colorings, such as serious, light, deep, which determine it more closely.[10]

(e) *The Aesthetic Specificity of Aesthetic Enjoyment*. Aesthetic enjoyment is surrendering to the object of art in contemplation/observation. In contemplation we accept the object – and yet hold ourselves open; we still keep a distance from it. In the enjoyment of observation the object takes possession of us, for it is neither the enjoyment of the *act* of observing nor the enjoyment of the act of painting an object, but the *object* itself, which is being painted.

However, it must be emphasized, first, that following other thinkers who treated aesthetic enjoyment (Kant, Külpe, Edith Landmann, Lipps), Geiger's contemplation/observation of the work of art in enjoyment is not meant as an isolated or a particularized descriptive intention, but as the "designation of the total situation." Second, it is also significant – especially for what I intend to develop in the second part of my inquiry – that although strictly correlated with its object, the analysis of enjoyment does not begin from the object, but *from the experience of enjoyment itself*.

Geiger himself insists that what is understood by aesthetic object is narrower than the field of objects which can be aesthetically enjoyed. Since such strong emphasis is laid upon the fact that it is the object which is the occasion of the "arousing of the self" and the "reaction" of the self with respect of what flows in it, we may now venture the question whether in principle there is a restriction on objects which can be aesthetically enjoyed. Is it the object-side that determines the release of the subject's enjoyment experience of an aesthetic type? Last, whence does the specifically aesthetic nature of aesthetic enjoyment arise? These questions gain all their interrogative significance when we consider two important points. The specific way of being aroused could be caused either (1) by specific features or intrinsic virtualities of the object which (a) may be *already* "aesthetic" and affect the self in a direct impact upon his functioning, or (b) are *virtually* aesthetic, that is, within the contemplating/observing encounter between the object and the subject, these virtually aesthetic elements provoke the enjoyment as aesthetic and they themselves assume an aesthetic form within the object as concretized in the enjoyment. Or (2) the aestheticity of the enjoyment stems from the *specific affectivity of the self* to a certain type of the object's features – this affectivity making the experience aesthetic and the object aesthetic.

Geiger's own answer to these questions is developed in a double line. On the one hand, he stresses the specific attitude of the subject: observation holding a distance, concentration (outer or inner), looking through the fullness of the object, excluding the self from it, etc. Furthermore, the aesthetic interest is characterized by the attitude of the subject; it is disinterested, when the inner aesthetically enjoyed sympathy exceeds a certain minimum.[11] The attitude of the subject makes the aesthetic enjoyment the *enjoyment of the disinterested contemplation of the fullness of the object.*[12] Yet, on the other hand, there is a difference in the experience of enjoyment depending upon whether "it arises from aesthetic values or merely from objects which can be aesthetically enjoyable." The above-mentioned characteristics of the aesthetic enjoyment depend upon the subject's *attitude*, whereas the aesthetic enjoyment is still *more specifically determined* when the objects embody *aesthetic values*. That is, the aesthetic enjoyment is not only the surrender to the object but to what Geiger calls "deep aesthetic values." Furthermore, it depends upon the "value" of the personality which is capable of experiencing such enjoyment.[13]

On this analysis there is only one step to be made, namely, to appreciate the value of the aesthetic enjoyment itself in its depth, seriousness, intensity, etc. with reference to the aesthetic values of the object. Thus, even granting to the subjective attitude the capacity to be aesthetically aroused, the entire weight of aesthetic enjoyment may lean upon the aesthetic values as supposedly intrinsic to the object. With this we naturally branch off into the aesthetics of values and objective structures that were first investigated by Ingarden, Hartmann, Bense, Dufrenne, and Wellek, and by others subsequently.

Second, including the factor of the author, Geiger has set the universal pattern for contemporary phenomenological aesthetics in its different variations. The pattern of the "aesthetic situation" is suspended upon three poles: the aesthetic enjoyment of the recipient (the reader or spectator), the work of art in its objective structure as embodying aesthetic values, and the writer who devised the structure of the work of art. Variants of this situation pattern can be detected in every phenomenologically inspired literary theory. Whether "aestheticity" as a quality of experience appears in the enjoyment itself or whether it is seen in the entire set-up of the aesthetic object as experientially concretized, it is brought down to values. Aesthetic values are assumed by its foundation. But the aestheticity of the values remains a well-sealed secret. No key to it is offered. Consequently within phenomenological aesthetics at large the aesthetic sensibility of the subject and his relation to values remains an open question.

Yet, Geiger's deepest intuitions have been ignored. I suggest that these crucial shortcomings do not compromise the question of aesthetic enjoyment, which he strongly emphasized as lying at the origin of phenomenological interest in aesthetics and containing most precious indications worth pursuing further. I have indicated their direction by the above questions.

Now we should ask concretely, what is the specific affectivity of the self to a certain type of objective feature – whether of inner or outer concentration (objects or our moods) – that makes the experience *aesthetic*, or that makes us retrieve the *aesthetic* virtuality of the objects? Where does the aesthetic quality of values come from? What is the origin of the aesthetic values? What, if any, is the relation of aesthetic values at their origin to this aesthetic affectivity of the self?

To these questions we will turn in what follows.

## *PART II*

## *THE AESTHETIC ENJOYMENT AND THE POETIC SENSE*

I propose to go back to the seminal intuition which resides concealed in Geiger's emphasis upon the enjoyment by exfoliating it myself in the line of my thought. First, I propose to approach it as a unique moment present in all experiences: (a) insofar as they are concrete experiences; and (b) insofar as they are experiences in which the self "is living." This means that I will shift the emphasis from the experience of enjoyment to *enjoyment as a function*. Thus, the function of enjoying will appear first as expressing itself in the essential way of enacting the life of experiences and of the self and as the function which – to use Hölderlin's expression – manifests itself as a *vis viva* which brings to manifestation everything insofar as it is alive. Second, I raise the issue of the life-significance which this function entails. I propose to distinguish between the *vital* significance and the *poetic* significance of the enjoying function. Then, I submit, that the function of enjoying brings into our meaning-bestowing the *poetic sense*. The *poetic sentiment, poetic feelings, poetic imagination,* etc., account within our language-forming for the "poetic language," which is the *basis for the aesthetic essence of all arts.*

## 1. ENJOYING IN CONSCIOUS EXPERIENCES. CONSCIOUS ACT AS AN "ACT" VERSUS CONSCIOUS ACT AS "OPERATION"

In the attempt to isolate in phenomenology the "pure" transcendental acts of intentionality from the lived experiences, we simultaneously bring to light what we isolate it from; indeed, the strictly intentional skeleton of all our experiences (*Erlebnisse*) – that is, of our natural, empirical consciousness, to use Husserl's terms – are clad in a lived experiential pulp. While they concern the structurizing proficiency and order suspended upon this intentional schema, they are simultaneously, however, themselves manifestations of life consciousness. Intentional acts, which are "purified" in the phenomenological perspective from the empirical embodiment, manifest intentional operational order; in contrast, the experiential pulp, although far from being "formless," manifests in its complex net of intermotivations various types of human functioning which it establishes and by which it carries on the *vis viva*: the continuity of the living self.

This concrete experiential pulp of sensing, feeling, and emotive stirrings is "conscious" not in virtue of its rational order, but in virtue of the "enjoyment" of its acts within which the self appears, first, as the one "who" enjoys them. Second, it is in virtue of the *enjoyment in the acts* that the subject, as their center to which all of them – even the remotest simplest feelings have some sort of relevant relation – are *my* experiences; not only are they "experiences" but *eo ipso* "mine" – as experiences of this specific self, uniquely his own.

## 2. THE NATURE OF ENJOYING

It is in the acts that the self lives. Without the acts, which it either performs, or which are performed at the peripheries of the territory which the self unifies, the self neither would have a role of an ordering reference system with itself as the vortex, nor would it be "alive." The self does not control the acts in its spread of references. It *dwells* in what we could call a "fountain of enjoyment," from which the self-centered acts stream – or burst forth – and which is the core of the conscious network. It is not responsible for the quality or determination of the acts, nor for their emergence; but, on the one hand, it participates in the life of the acts (directly or remotely); on the other hand, as the representative of their unity it carries their virtual performance. Last, in the circumference of its direct reach and as the central factor of this ever-evolving/intermotivating unity it is "clad" by the qualitative nature of the actual, as well as by the virtual content of the acts, as a person.

Yet, basically, it is not the qualitative feature of the content, nor the ordering proficiency, nor anything other than the enjoyment of the acts, which makes the self "alive." In the enjoyment of the acts the *self unites into one living system the singular acts past, present, and virtual.* No matter how diversified, and in themselves isolated, the singular acts may be, how close or remote from the self as the center they may emerge, disregarding the divergent ordering networks of intentionality which bring them together, the crucial factor of the unity of this dispersed and discrete spread of acts, a factor which fills out all the crevasses of structure and all the discontinuities, is the enjoyment of the acts. The enjoyment continues through all of them in various degrees of intensity, vigor, etc., and yet carries on a continuing line of life; a line sustained by the self coming with it to life. The self thus opens upon a *vis viva*, a living power, with which it "feels itself alive."

## 3. THE ENJOYING FUNCTION DIFFERENTIATED FROM COGNITIVE, VOLITIVE, OR MORAL FUNCTION

Returning to our analysis of enjoyment, I must reaffirm that enjoyment (1) as the thread of *life-experience* of *being alive*, (2) as the reverberation of the acts' *life-reverberation*, and (3) as the *filum Ariadne* of the labyrinth of the life of the self, is totally different from the *nature* of the acts. Although their very life sparkles, it is entirely "purified" of the willing, desiring, striving, and wishing features of the *willful acts*; nor do we see in enjoyment the features of the cognitive acts: empirical or purely intellectual. Nor do we mean by "enjoyment" anything that we could identify with the moral features of valuation or judgment. Last, inasmuch as we might be tempted to identify the life-dynamism of enjoyment with the exalting, elevating, energizing, aesthetic experience, we may not attach to enjoyment any of these energizing features. That is, leaving the experience of enjoyment aside, we move into the analysis of the *function of enjoying* which manifests itself in the experience of enjoyment. Here we definitely part from Geiger's conception of enjoyment, who considers it essentially as object-bound. By the "self enjoying itself in the acts", and by the "acts' reverberations in enjoyment", I mean the uniquely *subjective* nature of the enjoying function. In contrast to the intentional reaching-out of the volitive, cognitive, and aesthetic functions as such, *the enjoying function is self-consuming*. Enjoying as the function of the subject is an inside-drawn, self-reposing state of the subject, who quivers *in* and *with enjoyment*. As such, it permeates all subjective functions or rather carries on their manifestation. The subject, albeit in fractions of instants – if we may speak in temporal terms – reposes in itself in the "enjoyment" of the acts which *eo ipso* become his. Last, in this self-consuming nature, enjoying draws all the experiential tentacles of the subject which are spread within the life-world unto the subject itself. In the enjoyment the subject exults in all its operations; it comes into its own. Although stretching through intentional directions outside, in enjoyment it is self-absorbed; in the enjoyment the subject becomes the innermost foothold for himself; he is the one who lives.

Does this mean, however, that it "energizes" so to speak the subject and is self-consumed in performing it, disappearing without a trace, without expanding in an expressive way? Or, to put it differently, is self-consuming without – as Geiger insists it to be – being necessarily object-related? In what follows, I will try to show how the enjoyment of our acts – as the root-factor of our constructive self-interpretative progress – does not depend on the already constituted objectivity nor upon the qualitative meaningfulness borrowed elsewhere. On the contrary, in disagreement with Geiger, I propose that the *function of enjoyment is to bring its own unique sense.*

## 4. THE POETIC SENSE OF THE ENJOYING FUNCTION: THE "VITAL SENSE" AND THE "AESTHETIC" SENSE

Enjoyment of our acts varies along a widely differentiated ladder of intensities and degrees of self-absorption as well as extension within the circuits of our functioning. Thus, it varies in the nature of its *life-significance*. In fact, although a sharp demarcation line between the one and the other cannot be drawn because they are partly fused with each other, nevertheless in their respectively pronounced visibility they stand out as clearly different – as if we could switch one of them on or off – and we distinguish one as the vital sense of our act-enjoyment function and the other as the aesthetic sense. Both of them are carried by the *root-factor of the enjoyment function*: the self living in his acts; and yet there is a special "qualitative" (still "pre-objective") *sense* which differentiates them. We may either emphatically experience our self as vibrating in the richness of our life-potency – that is, in our *vital sense*, which endows all our individual enjoyments with a special vigorous sense of power – or we may overshadow it by emphasizing in our experience an imaginative "aesthetic" sense of our enjoyment within which the self finds something akin to a new dimension.

In fact, the innumerable acts which we perform and which carry our vital progress onward (for example, acts of pulsations, instinctual acts, acts of sensations, feelings, desires, volitions), which express our vital or as it is usually said "animal" phase of existential progress, as well as our specifically human circuit of experience, which begins with the entrance into play of cognitive, valuative, aesthetic, etc. – that is, fully developed human acts – are tempered in their respective intensities by the entire circuit within which they participate. To a certain degree they appear self-absorbing and subdued to specific objectives (e.g., either as reactions "to . . . ," or as acts "of . . . ," directed toward objects). Consequently, they are extinguished or muted in their *own* "voice". The spans of their extension upon the entire circuit of functioning differ from each other, but their singular voices are coordinated with those of other acts within the constructive design which they unfold together, unless a particularly strong intensity disrupts the harmony of the entire circuit, resulting in one act – or a complex of acts – putting the entire being on an alert of pain, anguish, pleasure, joy, expectation, etc. Nevertheless, their mode of performance, which is muted in the fulfilling of their specific role within a vast and self-ordering pattern, is not devoid of a specific *sense*. In fact, neither is it self-consumed without a trace in the performance of the act, nor is it "neutral" with respect to the specific and further extending role which it either performs or within which it at least participates. On the contrary, it endows this role with a specific *life-significance.*

It must be emphasized again that acts are neither neutral with respect to the individual progress which they carry, nor with respect to the self, of whose "life" they are: they carry this progress as "ours"; they make this specific self "alive" inasmuch as they "live within him." That is, our acts, are neither neutrally mechanical, nor empty of a sense.

Here I must describe what I understand by "sense." In the first place, *sense is to be seen as prior to all structuration of meaning as well as to the rules and principles of meaningfulness*. Consequently, it is also prior to quality, "quality" understood as the simplest structure unit (e.g., "pure red," "pure joyfulness," etc.). Hence, although sense acquires its meaningfulness from objectifying interrelations into which it fuses, yet within these relations the sense plays the determining role: it infuses the virtual formal/qualitative moments with a "qualitative reverberation" of a certain modality. That is, although the sense in itself cannot be defined – it can be described only in its interrelations into which it spreads – yet we may say that it is essentially contributing to the modality of the experiences of the self. The function of enjoyment insofar as it carries the sense, bubbles with qualitative reverberations to be infused into, to spread, to activate, to color, to tonalise, etc., the "contents of experience" which emerge. In this capacity the *function of enjoyment carries through its acts a life-significance*. It is with respect to the life-spread of the self and with respect to its spread of experiences that the qualitative reverberations of the enjoyment function live within its acts. There may be detected in every one of our acts a germinal qualitative complexity under the eye of attention, in a virtual state and only in the vaguest of forms, while virtually extending its tentacles into our entire living system. However, the qualitative virtuality of the enjoying function does not come forth, does not stand out in any qualitative fashion: does not come to take voice. In their function of enjoyment, acts carry in their first, elementary role muted together *in unison, the vital significance of the life-progress.*

However, the vital significance, which finds its manifestation in our vital vigor, and which constantly vacillates between enthusiasm and indifference, even apathy, for our undertakings and spreads through an infinitely differentiated gamut in between the two extremes, is not the only sense that the vibration of enjoyment carries. The vital sense of life, if alone, would confine our existence strictly to life-survival meaningfulness in its restricted scope had our enjoying function not also carried still another sense. It must be pointed out that both of them are so closely interlaced within the Human Condition that we may draw a line between them only at the level of abstract thinking. On the one hand, we experience our being in the vibrations of force, vigor, "faith in ourselves," and a feeling of completeness, fullness to the brim. On the other hand, those same vibrations are loaded with expectation of the unforeseeable, with longing for far-off, open, and limitless horizons, with foreboding of the unknown, the unseen, the marvellous as well as a subsequent compulsion to "leave it all" and go, undertake, reach the extraordinary and marvellous. This new sense "raises" our "vision" above the *hic et nunc* of life-concerns; they appear in contrast "pedestrian," "limited," etc. The new sphere of significance lifts us on the wings of the "elevated," the "sublime," the "beautiful."

Flowing into the source-experience from which human life originates, this new modality of our vibration stirs our inventive function which expands into the *Imaginatio Creatrix*. Within the workings of their orchestration the new sense of life, the aesthetic sense, crystallizes. It creates its own, new vision within life but reaching beyond the life realities; it silences the hitherto captivating murmur of the life-necessities, bringing forth its own voice; this voice makes our entire frame and our experiential compass reverberate with aesthetic significance.

## 5. IMAGINATIO CREATRIX, "HOMO LUDENS" AND "HOMO CREATOR"

Indeed, its basic role consisting in energizing our entire system and prompting toward its manifestation, the vibration of enjoyment reaches a state of special *life-exuberance* of our being. It does not need to reach any extravagant gestures, outcries, acts with which we often associate exuberance. We are exuberant in an overall prompting to "get out of ourself" at every phase of our "healthy" that means "equilibrated" functioning. Life exuberance plays manifold, important roles in human existence, and in its progress, development, learning, capacity to overcome obstacles, and, above all, in maintaining an equipoise in the struggles of life. Its meaningfulness is particularly strong at the crucial developmental phase of childhood. In fact, the childhood development/learning phase is most vividly expressed by the activity of "play." The learning, experimentation with the surrounding world, and communicative role of play is too often pointed out to require much attention. Play as learning about the world, as the means of unfolding skills and learning about our capacities, is common also to other living beings, especially in the same initial phase of learning/experimenting with the world and forming skills to handle the elements of life. Yet it seems that the exuberance of life that prompts it does not lead much further than to fulfill these above mentioned functions. As a matter of fact, as ingenious, complex, and fanciful as playing might be, in its strict nature it is merely *experimental, not inventive*. This distinction is important: invention presupposes experimentation and play, but the converse does not hold.

The experimental moment of play does not entail an inventive turning point; it is restricted to the discovering of the rules of direct manipulation and to acquiring skills to manipulate. Young animals, and children, share this learning and acquiring of skills as the role of the exuberance of the enjoying function. They maintain it beyond this phase and throughout life in a narrower sense for the sake of alternating "work," which demands concentration, rigor, and purposeful orientation with "play," which is meant as a spontaneous and therefore relaxing exercise of our vital forces in games (which refresh the psyche), in sports (which enable us to exercise our muscles and limbs), and in social or communal artistic activities (where we reorient our social relations affected by the tensions of work). It is meant as an outlet for our whimsical/spontaneous energies, in contrast to controlled, imposed work – as an outlet for our exuberance and a means of life-enjoyment.

Yet the same exuberance of our enjoyment-vibrations entering into the source-experience – which like a watershed gathers all our vital operations and functions, and, then like a sieve, after having mixed them, filters them in a special proportion – stir all of them at once; doing so it releases the *inventive imaginary function*. It is here that from a mere exercise of fancy – which in agreement with Aristotle and with Husserl I believe to be one of the simplest vital-sense phenomena, one which allows for the most elementary functions of perception, flexibility, etc. – we move to a radically different level of *Imaginatio Creatrix.*[14]

The same vibrations of enjoyment of our acts, which accompanied by fancy lead us to play, now, having released the inventive function, stir and carry the workings of the creative imagination. Imaginatio Creatrix proposes to us to "leave it all" behind and to create for ourselves wings and devices to explore the extraordinary and the marvellous, the unprecedented and the infinite. The individual armed with them becomes *homo creator.*

## 6. THE AESTHETIC SENSE: ITS VOICE AND THE AESTHETIC LANGUAGE

As mentioned above, the exuberance of enjoyment releases its "voice" while stirring the inventive imagination within the *source experience,*[15] within which the specifically human faculties crystallize. By "voice" I do not mean an appeal to, e.g., the "call of the wild." Exuberance prompts a "voice" expressing its state – expression, which at the same time responds to the tendency of the prompting to crystallize definitively in some or other "mood" for our own completion, as well as for its communication to the others. Under its promptings man hums, whistles, dances, jumps for joy, embraces others. In these gestures there vibrates a specifically "sounding" rhythm, pitch, cadence, "warmth" or "coolness," appeal or repulsion, "vibration with," or "distanciation"; finally, it culminates by conveying a *sense*.

How can this sense be "read" from the gesture, the tune, the rhythm, the movement? Had they remained as expressions of pure exuberance given a form by fancy, they would express a marginal outburst of undirected, spontaneous, whimsical, playful moods. As such they would take the place of a simply "pragmatically" oriented, life-carrying operation. However, once the entire system of source-experience brings about, under the direction of the creative imagination, the creative orchestration of all human functioning, the galvanizing and transforming of the entire experiential system, man may go beyond the strictly life-subservient operations.[16] The function of enjoyment crystallizes indeed, within this orchestration, into the *poetic sentiment*, which infiltrates the "neutral" voice of the enjoyment-vibrations. "Poetic" and "aesthetic" mean at this stage two interchangeable notions. However, they direct different aspects of our sense-bestowing.

The *"aesthetic" significance of life* which enters into the *existential self-interpretation of the human being is at its roots poetic.*[17] The poetic sentiment calls upon all our noisy routine vital operations to pause and to silence their beat; to turn our attention from their busy-ness to the *inward spectacle*. It reveals within our entire experiential system a dimension of the *sublime*, the *beautiful*, and opens it out to the *infinite*. The reverberation-virtualities of our experiences respond within the source experience to these three major lines of poetic sentiment. They bring forth life-significant moments – which the vital current of life left alone would not bring forth, but which are lying there in waiting. They do it in innumerable spectra of significance, encompassing entire ranges of virtual qualitative evaluations inherent to our experience. In contrast to the vital principle, the principle of life-significance given by the poetic sense is not meant to promote life operations and growth but to make it pleasurable. Thus emerges an infinitude of moments of the pleasurableness of life: its aesthetic significance, its aesthetic enjoyment which lifts us from pedestrian life-concerns to an elevated plane.

The aesthetic sense invents significance and forms its own meanings through the qualitative moments: aesthetic qualities, principles, values.

It brings forth its system of interconnectedness according to which it functions as an *aesthetic language*; the language of elevated enjoyment which – again contrasted with the vitally oriented language – is not directly life-action oriented but contemplative.

The formation of an aesthetic language culminates in artistic creation. From there it pervades human communication. But the poetic language lives on its own at the foothold of the artistic reaction proper, within the orbit of the inventive imagination, in relation to the elements of Nature, life, and the inward experience. In this phase it is, in fact, the foundation of artistic creativity. The poetic sentiment endowed with voice and sense, uniquely its own, inventive and infinitely expanding, forms the horizon of human life.

## NOTES

[1] Moritz Geiger, *Beiträge zur Phänomenologie des ästhetischen Genusses*, 2, unveränderte Auflage, Tübingen: Max Niemeyer, 1974, pp. 21–31.

[2] Ibid., pp. 35–47.

[3] Ibid., pp. 40–45.
[4] Ibid., p. 44.
[5] Ibid., pp. 39–40.
[6] Ibid., p. 45.
[7] Geiger means not the "phenomenal self" which accounts for the fact that all contents of consciousness are to be characterized as "mine": "That everything that I experience belongs to my individual consciousness and never to another's". He speaks about the experience of the self as the self (Icherleben). Ibid., p. 45.
[8] Ibid., p. 175.
[9] Ibid., p. 180.
[10] Ibid., p. 61.
[11] Ibid., pp. 95–96.
[12] Ibid., p. 97.
[13] Ibid., pp. 103–104.
[14] Cf. the present writer's 'Imaginatio Creatrix', *Analecta Husserliana*, Vol. III, 1974, pp. 3–41.
[15] Cf. the present writer's 'Natural Spontaneity in the Translacing Continuity of Beingness', *Analecta Husserliana*, Vol. XIV, 1983, pp. 125–151.
[16] Cf. The above-cited 'Imaginatio Creatrix'.
[17] Cf. the present writer's '*Poetica Nova*', *Analecta Husserliana*, Vol. XII, 1982, pp. 1–93.

WOLFGANG WITTKOWSKI

# MOVEMENT IN GERMAN POEMS

When asked to contribute observations on the phenomenology of poetry, I chose the aspect which has struck me more than any other: movement. I sought translations of the examples I planned to use and two of my American students hurried to rescue my own English versions from the impossible – and I made a surprising discovery: both translators as well as students showed little feeling for poetic rhythm. Evidently, I was going to explore new territory alone, explaining poetic rhythm and movement to more readers than I had anticipated. Consequently, my intention to use poems in which movement in form demonstrated movement in content seemed all the more justified. However, there will also be more subtle examples.

What, then, are movement and rhythm all about? Looking for help, I found a book with the promising title *Was alles hat Platz in einem Gedicht*?[1] I encountered the very term 'Bewegung' ('movement') on the opening pages. They described "the way a poem moves" ("seine Art sich zu bewegen") in terms of its sentences and lines as possibly the single most crucial characteristic of a work. More than anyone else in recent years, the American poet Charles Olson emphasized this central importance of movement in sentence and line. The line, he says (and vows, according to that book), comes from the breath of the writer in the moment he writes. Sentence and line cause the movement from the subject to the object; they provide a bridge and unity between these two. They make the poem singable (cantable).

So much for Olson and his German commentators.[2] Other terms, like "Gedichtablauf" ("course of the poem") or "the energy within the form" (from William Carlos Williams), address the same phenomenon, but in a fashion much too general and abstract for a phenomenological approach. More importantly, not everyone will emphasize (or even recognize) the aspect of movement I mentioned at the outset. The book quoted above, however, claims to return to just such a point of view, one which indeed seems to apply to most poetry. For a brief period in recent years, there were "poems" created without words and meaning and therefore without movement. This may produce interesting experiments but certainly not masterful or even satisfying lyrics. They seem to have become tiresome to writers as well as readers.



*A-T. Tymieniecka (ed.), Analecta Husserliana, Vol. XVIII,* 23–34.

But why in the world would some of us write, read and love poems with words, meaning and movement? What is the function of poetry among literary genres?

In some way, all literature invites or even requires us to share in its perspective, especially when the effect is one of alienation (Verfremdung). In this sense, we identify with dramatic and narrative characters, with narrative subjects (narrators), but most intensely with what in German is termed the "lyrical I." The fictional persona provides the only perspective with which a reader may identify. This perspective differs from those in other genres. In dramatic and narrative works, the overall view appears to be broader and more complex; it seems to embrace more of life and the world. We do recognize, however, that it does not embrace the entire world, but only a greater or lesser segment of it. By contrast, poetry considers life, earth, universe and eternity easily and with special preference since that perspective is usually not limited and relativised by other views within the poem. It may, and probably will, be contested by the perspectives of other poems by the same author. Each poem, even if it focuses on a limited subject, implies the totality of the real or of a possible world. That is because the persona presents itself in its potential totality during the "lyrical moment" of that poem – which may be consistent throughout the author's lifetime or may change from year to year, from day to day, from minute to minute.

Certain poems, particularly modern ones, emphasize the fragmentation of life. Reading them, we feel alienated, separated from the world but not from the poem and its perspective, whether or not the literary device of alienation (Verfremdung) is being used. There *ex negativo* we feel confronted with the concept of the (here missing) unity of life. Thus, the implied totality of the lyrical perspective rests on two pillars: the implied totality of or behind the subject matter and the implied totality of the persona. Both pillars are linked by the all-embracing totality of the lyrical moment in time (and beyond or above time). A poem, like any work of art, can become alive in the imagination of readers and listeners at any time. The totality of time and scope is, of course, just the other side of an exclusive limitation in time and topic. It therefore brings with itself a single and distinct style and mood or, at the very least, considerable limits on style and mood (again the implication of all that also *ex negativo*). These limitations aim toward one single moment, topic, style, and mood. This, in turn, enhances the totality of the perspective and the intensity, totality of our involvement, that is, of our identification with the poem's perspective.

Total involvement in a work of art grants the delight of feeling all one's faculties at play and in display. Total involvement in a poem, therefore, consists of more than intellectual sharing, of more than, for instance, analyzing the meaning of poetic imagery. Total involvement is intellectual and emotional involvement. It is, at its basis, the very process of becoming involved. This process, again, overlaps with the involvement in the perspective as it develops in the process of the poem. Like all literature (and music) poetry is experienced as a process which begins and proceeds to its end. Dramatic and narrative works move a relatively long and complex way. Poems move within a short moment, homogeneous in topic, mood, tone and style, allowing them to be grasped in their totality and intensely felt in their impact *if* – I would maintain – *if* they are read the right way, that is, if they are read the way they move.

The importance of movement can be seen in that fact that poets pay particular attention to it. As though to compensate for the brevity of a poem's process, they order and intensify the language with great care through verse, stanza, meter, and rhyme. All of these serve the movement. Therefore, the poem is almost destroyed if, as I mentioned at the beginning, translators sacrifice the rhythm in order to save the rhyme and eventually present an understandable syntax and meaning where the original is far from clear in this respect. This reverses the way in which the reader experiences the original as a progressing structure in which every word falls into place, that is, into its place within the movement of the language. The reader understands immediately the process or after he has overcome certain obstacles, surprises. Poetry works in a way similar to music. The beginning states a theme or, more fundamentally, sets a pattern of rhymes, especially rhythm, movement, and meter. The pattern raises expectations which are fulfilled in the course of its repetitions, variations, countermovements, gradations, climaxes, cadenzas, etc. Identification with a poem's perspective means, not least, riding with and on its movement.

My examples, German poems from around 1800 to the present, demonstrate syntactic movement as a carrier of meaning. At the same time, however, in most of the examples, the meaning is that of movement (or of its absence), of movement in the world as perceived and sung by the poet, the "lyrical I." These poems may actually have too much movement to be good examples; poems which do not address movement as a topic alone would prove the thesis that verbal movement is at the very heart of poetry. Therefore I shall present some poems of this kind as well. Nevertheless, I find it all the more important to concentrate on examples in which movement as topic is mirrored by movement of verbal structure since American translators, professionals and amateurs alike, do not seem to recognize or appreciate such a relationship. This will be demonstrated by some excerpts added to my own translations. Finally, there is another reason that I concentrate on poems with a threefold movement (in syntax, fictional world, and fictional persona). In fact, it is the primary reason. It is my impression that in such poems verbal movement and even lyrical language itself rise to their greatest heights – an impression which may be debatable.

I shall begin with three short pieces by Joseph von Eichendorff (1788–1857) whose songlike poems were frequently set to music. For several generations they were more popular with Germans than any other poetry, *Volkslieder* included. The first example may well serve as a motto for this study; in fact, its final line served as the title of a book on poetry (*Triffst du nur das Zauberwort*).

The poem's topic is the romantic process of transforming the entire world of objects into poetry in a magic and mysterious way. It consists of only four lines. Note the change between the state of rest in the first two lines and the movement of life and art suggested in the second two.

*Wünschelrute*

Schläft ein Lied in allen Dingen,
Die da träumen fort und fort,
Und die Welt hebt an zu singen,
Triffst du nur das Zauberwort.

*Divining Rod*

Sleeps a song in all the things
Which are dreaming on and on,
And the world sets off to sing
Once you touch the magic word.

The main difficulty in translating is to find an equivalent for the position of "Triffst" at the beginning of the last line. This position endows "Triffst" with the dynamic energy which carries over the rest of the line, literally and emphatically bringing it to touch the "magic word" at the end of the poem. The syntactical function of "Triffst" is beyond doubt; not so the "Schläft" at the outset of the poem. It suggests the beginning of a conditional clause (as "Triffst" initiates a conditional and temporal clause). This impression, however, fades once it is realized that the third line does not properly connect this way. Conclusion: "Schläft" lacks the grammatical subject "Es"

("There"). Therefore, syntactically as well as semantically, "Schläft" wavers between "If there sleeps" and "There sleeps," shifting – moving – the emphasis from an "As if" impression to the conviction "The mystery is true, the miracle is real."

The first two lines establish the (dynamic) potential for just such a typically romantic development which becomes clear in the third line. At the beginning of this line, the potential is ignited with soft but irresistable force by the unspectacular "Und," and is entirely unleashed by the accelerating "Triffst" at the beginning of the last line. Logic would ask to interchange the sequence of lines three and four; the author may have decided otherwise in order to continue intensification of the language movement. This intensifying movement and its climax are totally lost if the "Und" is translated by "But." The delicate structure of gradual, mysterious development changes into a sudden dramatic turn-about: a miraculous potential being realized becomes, instead, the contrast between sleep and "will start to sing" as the translation progresses, frittering away its noisy, superficial violence before the beginning of the last line, the true climax of the poem's movement. Certainly no English version seems able to equate the poem's accelerating finish from "Triffst" to "magic word."

Like the first poem, the second includes not only an awareness of the external world but also of past times. This overall openness allows the "lyrical I" as well as the reader to be absorbed by everything. The process occurs in breaths or movements:

*Der Abend*

Schweigt der Menschen laute Lust:
Rauscht die Erde wie in Träumen
Wunderbar mit allen Bäumen,
Was dem Herzen kaum bewußt,
Alte Zeiten, linde Trauer,
Und es schweifen leise Schauer
Wetterleuchtend durch die Brust.

*Evening*

When man's noisy pleasures rest:
Earth is rushing as in dreams
Wondrously with all its trees
What the heart does scarcely know,
Times that passed, griefs that softened,

And there strays a summer lightning,
Shivering gently through the breast.

The first line goes far back in order gently to release a soft movement which seems to be without end but eventually does end in line five. A similar though shorter movement occurs in the last two lines. The world, time past and time present, are conceived of in two breaths; the second fulfilling and crowning the entire movement, intertwining world and soul swiftly, though without visible effort.

*Mondnacht*

Es war als hätt der Himmel
Die Erde still geküsst,
Da sie im Blütenschimmer
Von ihm nun träumen müsst.

Die Luft ging durch die Felder,
Die Ähren wogten sacht,
Es rauschten leis die Wälder,
So sternklar war die Nacht.

Und meine Seele spannte
Weit ihre Flügel aus,
Flog durch die stillen Lande
Als flöge sie nach Haus.

*Moon Light*

It was as if the heavens
Kissed silently the earth
So that in all her blossoms' shimmer
She had to dream of him.

A breeze blew through the meadows,
The long stalks softly swayed,
The forests gently rustled,
So star-clear was the night.

And there my soul outstretched,
It spread wide out its wings,
Flew over silent countries
As though returning home.

The overall movement proceeds from heaven down to earth and back up to heaven, or, if you wish, from sky to earth to heaven. But this "What" draws its artistic and human impact only from its "How," from the "physical" movements as they breathe in stanzas and lines. In the first stanza, the movement is a very gentle one: a kiss causes a dream, and that in the mode of the "As if." Each of the two "breaths" passes over to the following line but clearly pauses briefly at the end of the first and third respectively. The pause at the end of the fourth line is a final stop, while the pause at the end of the second establishes the hovering which characterizes the entire stanza. This element is not conveyed in D. G. Wright's translation "The sky had kissed the earth to sleep," etc.

The second stanza picks up speed. The movement of breeze and stalks is still gentle but distinct. However, the forward movement of the breeze diminishes from line to line – "swayed," "rustled," until "was" marks a standstill in the last line. Here, the universe of the stars provides the stage for the triumphant movement which characterizes the last stanza. As in poems one and two, the movement is virtually ignited by an "And" which is excluded by all translations I have seen. The soul, having undertaken considerable preparations for "takeoff," achieves the impossible: it flies over countries, that is, the earth, into the infinite. The movement into that last dimension may have enough momentum to perpetuate the flight even beyond the poem. At the same time, however, the drive becomes less distinct and strong, hovering between earth and heaven in the mode of the "As if" – like the first stanza with its movement from heaven. A return to heaven takes place, indeed – but, to be sure, only in the "As if" mode of the unimaginable as projected by the poem.

Movement is a matter of style. It is different in the song-like poems of Eichendorff from that of Hölderlin (1770–1843). Identifying with Hölderlin's sense of movement demands, for the most part, that we divide ourselves into the "I" which identifies with the movement and the other which is watching, listening with awe and admiration. The poem "Menschenbeifall" or "The World's Acclaim" moves with the majestic steps of an antique form, like a ballet dancer constantly changing his pace, the length and form of his energetic and decorative steps. The subdued or fierce violence dictated by the prescribed scheme also makes the inverted word order seem forced and artificial – an essential element of the poem's constrained movement. As far as the topic is concerned, as in the last Eichendorff poem, we move in a circle or spiral, this time from the "holy I" of the present to the not so "holy I" of the past, from the unchanging crowd back to the few who are like – or almost like – God Himself.

*Menschenbeifall*

Ist nicht heilig mein Herz, schöneren Lebens voll,
Seit ich liebe? warum achtetet ihr mich mehr,
Da ich stolzer und wilder
Wortereicher und leerer war?

Ach! der Menge gefällt, was auf dem Marktplatz taugt,
Und es ehret der Knecht nur den Gewaltsamen;
An das Göttliche glauben
Die allein, die es selber sind.

*The World's Acclaim*

Is not holy my heart, full of lovelier life
Since I love? Why then did you respect me more
When I was prouder and fiercer
Fuller of words, and emptier yet?

Oh, the world likes that which fits on the marketplace,
And the slave does pay homage only to stronghanded masters;
In the godlike believe
Those only who are so themselves.

Also Rilke speaks on an equal footing with God. In *Herbsttag* he advises God on the orders He should give. The tone matches the ambition: sentences move with majestic calmness. They begin short, then become longer as the stanzas also become longer. The first two are similar in content and structure. Each accelerates by way of enjambement between the two last lines, the second stanza proceeding much more forcefully than the first by using a transitive imperative – "jage," "force" – at the beginning of the enjambement.

*Herbsttag*

Herr, es ist Zeit. Des Sommer war sehr groß.
Leg deinen Schatten auf die Sonnenuhren
Und auf den Fluren Lass die Winde los.

Befiehl den letzten Früchten, voll zu sein.
Gib ihnen moch zwei südlichere Tage,
Dränge sie zur Vollendung him und jage
Die letzte Süsse in den schweren Wein.

Wer jetzt kein Haus hat, baut sich keines mehr.
Wer jetzt allein ist, wird es lange bleiben,
Wird wachen, lesen, lange Briefe schreiben
Und wird in den Alleen hin und her
Unruhig wandern, wenn die Blätter treiben.

*Autumn Day*

Lord, it is time. The summer was most great.
Set down your shadow there upon the sundials,
And on the plains let loose the winds.

Command the last of fruits that they be full.
Give them yet two warm southern days
Urge them towards perfection and force
Their final sweetness into heavy wine.

Who has no home now never builds another.
Who now is lonely will long remain alone,
Will stay awake, will read, will write long letters
And will wander up and down the boulevards
Restlessly as the leaves are blowing.

The third stanza is totally different in form and content: no longer is God given any advice. Nature serves only as the setting for man's isolation. The five lines consist of two sentences. The first one sets the scheme which the second follows, enlarging it. However, line two repeats the scheme of line one; line three continues the second half of line two, shortening the syntactical scheme and thereby accelerating the pace. This is further increased through the enjambement – which is surpassed in speed by the following and again by the last enjambement (as the enjambement of stanza one was surpassed by that of stanza two). The syntactical scheme is extended beyond line four. This causes the restlessness indicated by "hin und her/Unruhig wandern" and reinforced by the pause and the final construction: the "wenn" clause of the last line.

Thus, the stanzas establish the steps of an ever growing movement. Like a piece of music, it develops by repetition, variation and gradation, proceeding from God and the world to man and back again – not quite as with Eichendorff's poems. Hölderlin stands between the two in this respect also.

The Swiss C. F. Meyer (1825–1898) provided the German language with probably its most famous poetic example of movement and the halting counterforce. The poem symbolizes both by rhythm as well as by its very theme: the interplay between the streaming of water, its acceleration in descent, its slowing down, coming to a halt and being shaped by a solid container. The writer obviously tries to equate the "What" with the "How."

*Der römische Brunnen*

Aufsteigt der Strahl und fallend gießt
Er voll der Marmorschale Rund,
Die, sich verschleiernd, überfließt
In einer zweiten Schale Grund;
Die zweite gibt, sie wird zu reich,
Der dritten wallend ihre Flut,
Und jede nimmt und gibt zugleich
    Und strömt und ruht.

*The Roman Fountain*

Up shoots the stream and falling fills
The marble basin to its rim
Which, veiling its surface, overflows
Into a second basin's ground;
The second gives, it grows too rich,
Its surging flood on to the third,
And each receives and gives at once,
    And streams and rests.

(To still a third its surplus store
The second gives and rolling grows;
Each gives and takes forevermore
And rests and flows . . .
    Isabel S. MacInnes)

The spouting of the fountain is expressed at the outset by two emphatic syllables (*Aùfstèigt*). The pause in the middle of the first line indicates the moment of balance at which the water stops and then begins to fall. The enjambement here (and later together with the one syllable cadenzas of all lines) represents the water pressing against and over the opposing pressure of the basin's rim. This same end is served by the present participles "sich verschleiernd," "wallend," and the pauses at the commas. Like the lines which precede it, but much more directly, the final two lines attempt to create a balance between movement and rest, the two polar forces central to life, by form and definition as well as rhythmical movement.

All of the poems mentioned to this point can and should be read in a distinct meter and rhythm which corresponds to the movement present in the work. The following two examples by the contemporary poet Sarah Kirsch (b. 1935) are in "free rhythms" ("freie Rhythmen"), popular in Germany since the 18th century. Her poems generally lack commas, thereby forcing the reader to discover the right movement. The first poem represents the monotonous staggering and hesitating movement of that melancholic mood which depressed the author, especially during her last years in East Germany. The second reflects her high-flying exuberance when she finally crosses the border to the public freedom of the West and to the private happiness she was too long denied.

*Schwarze Bohnen*

Nachmittags mahle ich Kaffee
Nachmittags setze ich den zermahlenen Kaffee
Rückwärts zusammen schöne
Schwarze Bohnen
Nachmittags ziehe ich mich aus mich an
Erst schminke dann wasche ich mich
Singe bin stumm.

*Black Beans*

Afternoons I grind coffee
Afternoons I put the ground coffee
Backwards together beautiful
Black Beans
Afternoons I undress I dress
First I make up than I wash myself
Sing am silent.[3]

*Raubvogel*

Raubvogel süss ist die Luft
So kreiste ich nie über Menschen und Bäumen
So stürz ich nicht noch einmal durch die Sonne
Und zieh was ich raubte ins Licht
Und flieg durch den Sommer!

*Bird of prey*

Bird of prey sweet is the air
Never before cruised I thus over people and trees
No other time will I dive thus through the sun
And pull what I grasped into the light
And wing away through the summer![4]

Do you feel the up and down, the power of the dive? Or do you still believe it does not matter how you read a poem and whether or not you ride on its movement?

*State University of New York at Albany*

## NOTES

[1] Ed. Hans Bender and Michael Krüger (Münich: Hanser, 1977).
[2] Ibid., pp. 7 f., 20–25.
[3] From *Zaubersprüche* (Ebenhausen: Langewiesche-Brandt, 1976).
[4] From *Rückenwind* (E. Berlin: Aufbau Verlag, 1976).

From left to right: Marcellino de C̊isneros, Anna-Teresa Tymieniecka, Eugene Kaelin, María-Teresa Bertelloni, and Robert Magliola at the Phenomenology and Literature Seminar organised by Professors Bertelloni and de C̊isneros at the University of Puerto Rico, Mayagüez, March, 1982.

MARIA-TERESA BERTELLONI

# WHY BE A POET?

"What is the use of being a poet in times of need?" asks Hölderlin in his elegy "Brod und Wein" quoted by Heidegger in his lecture on R. M. Rilke whose title is precisely "What is the use of being a poet?"

I believe the real question is not what is the use of being, but rather why be a poet, and not only in times of historical need but at all, at any time in the very course of existence.

Actually, need is not a historical situation of man, but rather the existential situation of the human being. For that reason it is difficult, perhaps impossible but existentially necessary to answer the question "Why be a poet?" To answer it implies finding the key to all human endeavors since poetry encompasses in its winged word the other acts carried out by man in his attempt to reconcile phenomenological fragmentation with primigenial inspirational unity. To poetry belongs the magical-symbolical dimension of religion as well as the intuitive dimension, means of all philosophical activity, and in poetry the diverse ways of reading from the "book of nature," as Galilo called the universe, find their difficult, otherwise inconceivable, equilibrium.

But the answer to the question is possible only by searching, seeking out in the gloom of prehistory, the birth of the human phenomenon, the *time-place* of that birth which opened a fissure, a crack in the unity of the magic through which knowledge of the fragmentation, and therefore the fragmentation itself, seeped out.

We find that *time-place* suggested in one the symbols legend has attributed to Dionysus: the mirror symbol. Thus, and Plotinus clearly points it out, the god's figure acquires its creative character and at the same time faithfully embodies the essential problem of Greek thought: the contrast between the unique and the multiple which, since then, has been the central problem of Western thought.

But the dionysian myth has elements of oriental origin which explain even better the presence of the mirror among the toys of the boy god. In oriental thought, the Whole, the Unity, is the only reality accepted without anguish and without dismay. The phenomenological world is a world of appearances and is lived as such; in oriental man's attiude there doesn't



*A-T. Tymieniecka (ed.), Analecta Husserliana, Vol. XVIII*, 37–45.

exist the attachment to worldly things so characteristic of the Western world which has at times led to consider platonic thought as idealistic because it places intutive reality in the *topos uranós*.

Oriental thought easily accepts the idea of the phenomenological world as a degradation of the original Unity and that degradation as the cause of all the evil and pain man suffers. And in Orphism, whose oriental roots are clear, there exists the concept of the *σῶμα σημα*, the body as the tomb of the soul, the only element which ties us to that Unity lost in the embodiment and to which that Unity will return after death.

Yet to the Western world, death is a mystery which not even religions have been able to fathom and all the poetic thought which appears in tragic poetry has its roots in that mystery, in that mysterious fact which destroys and annuls the sense of existence. But to talk of poetic thought is to fix our gaze upon a moment posterior to that of the primigenial birth, when the tasks have already been assigned and human activity has been differentiated. Actually, poetic thought springs forth when man knows his mortal destiny and, in the attempt at fulfilling it, carves out a path in the wilderness of the intricate forest of existence using different tools: philosophy, religion or poetry. At that moment poetry is already a solution, partially or totally liberating, to the enormous burden which living places upon the shoulders of "that one day's creature" to whom Prometheus, in his heroic madness, gave the fire of intelligence.

But the myth of Dionysus's mirror takes us back to the very moment of the birth when the god, while looking at himself in the mirror, was fragmented into the phenomenological multiplicity whose essence is precisely that of the reflected image: pure appearance. And all reality, including man, is *ἐίδο λον*, ghost, shadow, illusion. All perceptible reality is only *ἐίδο λον*, but of all the *ἐίδολα* only one embodies that fleeting, ephemeral quality of the image in the mirror: man, because only in him does cosmic conscience become reflected conscience, life which lives itself, person.

The concept of time is clearly explained in Marcelino De Cisneros's thought as that which determines the human, since "the consciousness of time, that is "being aware of," allows us to settle back into the present retreating into ourselves, being able to have a perspective of the past and foreseeing the possibilities in the horizon of the future."[1]

This conception allows us to place the birth of the human phenomenon precisely at the beginning of time when the harmony of the magic circle, of which Cassirer speaks, has been broken, pierced, and the human being finds himself torn asunder by the encounter with himself, with his me-ness, which in order to exist opposes otherness, an otherness which is not only constituted by the other εἴδολα but also by the conditions and the limits of existence. For that reason the birth of the human phenomenon is a birth "in times of need."

Need, in its metaphysical sense, is the helplessness of the man who hasn't yet installed himself in the logos, from there to attempt the reconciliation of his me and the other. Need is also his finding himself imprisoned in the ferreous logical structure of reason while life blooms indefatigable around him, sparkling and contradictory. Need is, therefore, the existential situation of man and in it he must live and watch himself live.

It is not the death of the gods which produces eras of need, as Hölderlin affirms; on the contrary, the existence of the gods is the most forcible testimony of human need because religion, as a phenomenon of reunity, is also an answer, a solution, a path in the forest. If a logical, universally valid, answer were possible, the gods wouldn't be necessary.

Thus, we have found the human phenomenon to be a εἴδο λον which feels the inevitable need to transform the imaginary reality which it must live into a body of meaningful facts, with a cause and a goal, But how? Ghosts are slippery, they cannot easliy be grasped; they appear and disappear like the phantasmagoric tricks of a divine juggler whose sense of the comic is, without a doubt, frightening.

Therefore, existence is sorrowful; the human being inevitably condemned to his fragile, imaginary existence is placed before a wall which separates him from the final, radical and definitive explanation. And, in his first encounter with existence, he tries to penetrate it by creating or perhaps finding suspended in the mirage of surrounding reality the magic words capable of clarifying the mystery, of defeating death and fate, even crueler than death itself. And thus poetry is born.

Isn't Orpheus perhaps the first human being? Isn't the worst need that which derives from the loss of the *alter ego,* the other being which has become me? Isn't love perhaps the only sublime, divine, way of reconciliating me with the other? Therefore, there is no worse need than the loss of love. And Orpheus is the first man to live that hardship and to try to defeat fate not knowing any other path but the word.

Orpheus, more than any other man, deserves to be called *deinóteron*, and the canto in Sophocles's *Antigone* may refer to him since it indicates precisely the marvelous character of one who is capable of everything except defeating death.

In his *Introduction to Metaphysics* Heidegger points out a particular meaning of the term *δεινόν*; that of fearful. Indeed, man is not only fearful to the world when he tries to capture it, manipulate it or live it as a function of himself; he is, above all, fearful unto himself, a destructive impostor who, in the end, comes face to face with a shadow like Orpheus on his return from Hades. His magic word charms mature, even charms the gods of death, but in the end reveals his human frailty and falls, impotent, before the wall of mystery.

With Orpheus man embodies himself in poetry and through the word penetrates the *σοφία*, the poetic and pre-logical wisdom which moves in what Husserl calls the *Lebenswelt*, the world of life. Thus, poetry accomplishes an intuitive ἐποχή, since it is not a matter of adjourning a trial but rather of living the impossibility of the trial as the only path towards getting to be.

If life "is always temporality, the surpassing of oneself and the clarification of an inexhaustible, obscure reality" as Enzo Paci affirms in his penetrating study on Husserl,[2] poetry is situated on the threshold of that reality and doesn't attempt to shed light upon its dark folds but rather to brighten the path towards it with brief and intense bursts of lightning.

The poetic epoch reveals humanity's innermost core to us, the terrible anguish of hanging over the abyss of being or nothingness; it matters very little whether it is one or the other, for both are forbidden to man since they lie beyond his comprehension and constitute the insurmountable limits of logical activity itself.

Thus placed in the face of anguish, man has but two alternatives: he may destroy himself or he may conquer that anguish by accepting it as the cause of all human endeavors. Poetry is found in the second of these alternatives.

Being a poet, it seems, is man's first response to existential anguish; it is an attempt at climbing over the wall of mystery by intuitively guessing its secret codes in the words spoken whose suggestiveness creates a complex and articulate world. For that reason Gusdorf states that "to name is to call forth existence; it is to create from nothingness"[3] and, we may add, it is to call forth and unveil being.

The poet speaks and things come into being and they are his. This is a world in which nothing has any sense other than that suggested by the poetic word, but it is a first world from whose chaos the universe begins to spring forth. Thus Gusdorf states that "the first language (which to us is the poetic language) is essential language; it possesses a magical and relious value.[4]

Now then, poetry is not man's only response to existence, but it is the first and from it, delineated by the fields of pathos and logos, spring forth the other two responses: religion and philosophy.

Poetry, religion and philosophy all undertake tasks which are strictly and specifically human since the subject of their activity is so implied in them that, as Heidegger would say, it is risked. Being risked means, we believe, that the active *me* is at the same time the point of departure and the point of arrival of the activity undertaken in such a way that success as much as failure affirms that *me*. These activities are existential because they are necessary, because in order to be the *me* cannot do without them even if he believes that the goal of that search is someone other than himself.

Actually, if in a hypothetical way we affirm that the goal of the search, the result of the activity, is to encounter something else, even when that something else is Totality, the result of the activity will always be the *me*, since the *me* will result fortified from that encounter. In the reconciliation with Totality, it is the *me* that does the searching, it is the *me* that is in need and it is the *me* which takes the risk. If the result is negative, risking oneself will result in radical solitude, helplessness, anguish or self-destruction. If it is positive, risking oneself will result in self-exaltation and victory, as Pascal affirmed in his *Pensée* when he spoke of faith as a risk and not just an ordinary daily risk but a vital risk.

The Attic tragic poets clearly understood it this way.

As an example, let us consider Eurypedes's *Hippolytus* and the tragic character of Phedra. Her vital action is presented dramatically; first of all in the internal conflict between loving passion and they and then in her conflict with Hippolytus, the hero. No one would dare say that the conclusion is positive, and yet Phedra's *me* finds a place in history through the *iter* of destruction.

We find the opposite case in Sophocles's *Antigone*. The apparent result is the same, physical death, but Antigone's *me*, which is presented in dramatic conflict whth the world of sensible men, installs itself in time as the embodiment of divine Justice, thus transcending temporal and spatial conditions.

Nevertheless, in both cases, whether destroyed or triumphant, it is the *me* which results strengthened.

Now then, poetry is not only the primigenial existential response, root of all strictly human activity, *ποιεῖν* by antonomasia which manifests a hermeneutic side of a constitutive character. Poetic hermeneutics is justly and precisely creation because it is not undertaken as a function of transcendency, as in the case of religion, or as a function of transcendental metaphysics, as in the case of philosophy. Poetic hermenutics is always creation, even when dealing with religious poetry – as in the case of mystical poetry – or with the so-called metaphysical poetry.

This is so because the hermeneutic side of poetry is a function of something more radical and more decisive to poetry itself; that is, its linguistic nature. By paraphrasing Heidegger we may state that poetry is "the home of language" insofar as it is essential language constitutive, at the same time, of the *me* and the *other*.

At the same time, poetry's linguistic nature is a phenomenon, in the strict sense of a manifestation of the ontological, and it is in that respect that Heidegger affirms, in the lecture previously mentioned, that "language is the home of being." The fact that the philosopher makes this affirmation when referring precisely to a poet and to poetry is a determinating factor for its comprehension.

Poetic language, however, is different from all other types of language because it is not situated in the sphere of the entity but rather in that of the valent and, specifically, in that of the beautiful.

It is precisely this living in the sphere of values which makes poetry something which transcends the spatial and the temporal in order to install itself in a perennial history which is developed by constantly affirming the primigenial.

In it there are no improved beliefs or systems; only different ways of affirming beauty, something which is not only intangible but also indestructible. Even in historical periods such as ours where the "esthetic of the ugly" reigns, it is not beauty which has died but rather man's capacity to appreciate it.

Undoubtedly, poetry as a primingenial artistic manifestation encompasses within it, in the bud, the other literary and artistic manifestations and that is why Croce stated that "artistic intuition is always lyrical intuition"[5] and why noun and adjective are synonymous. This is equivalent to saying that all art is poetry. This sure judgement clearly situates poetry at essential heart of literature; indeed, poetry is the structural axle around which all other literary activity revolves. We can affirm that poetry is to literature what metaphysics is to philosophy – that which justifies them, gives them life and bestows value upon them.

But what is lyrical poetry?

The Spanish scholar Pilar Gómez Bedate points out that the goal of poetry is "to achieve poetic emotion"[6] and that emotion must be understood "as an excitement which affects the mood in its totality, on the intellectual as much as on the sentimental plane."[7] This statement agrees with our point of view that poetry, being a primigenial and primary activity, contains within it the seed if the other existential activities: the intellectual seed of philosophy and the sentimental seed of religion.

Pilar Gómez Bedate also states that "the factors which cause the unleashing of poetic emotion are basically two: surprise and suggestion. These two elements belong to the plane of the signified but are expressed through the significant."[8] Those two characteristics are precisely what allows poetry to be analyzed from two different angles: the literary and the metaphysical angles, insofar as they establish a relationship of an immediate character between the word and the object.

Actually, in non-lyrical literary activity "there generally exists a mediation between the author and his poetic world; the rhetorical forms selected have a mediary nature insofar as they become weighed down with meaning and lead to reflexion and the search for that which they hide. Words, in all their expressive possibilities, mediatize the immediate: intuition, perception, contemplation . . ."[9] But in lyricism, where the perfect adequacy between the signified and the significant is really achieved, the intuited being offers itself in the suggestive and suggesting nakedness of the poetic word.

To speak as Croce does of the cosmos *sub specie poesiae* is equivalent to saying that the being reveals itself in the poetic word as the Christian god does through the Verb. It is the coincidence expressed in the gospel according to Saint John: "In principio erat Verbum, et Verbum erat apud Deum, et Deus erat Verbum."

This brief outline of what the poetic word encloses no doubt explains the statement that poetry is the essence of literature and that without it non-lyrical literature would never attain the nature of artistic manifestation. Beauty can only exist in intuition, not in logical discourse, and it is precisely in the poetic flash that beauty is revealed to us.

If we return to Montale's metaphor about the wall which closes man's path towards the truth and constitutes a mystery, we find ourselves before three different ways of overcoming it: poetry, which delves into the mystery and which, in its suggesting character, tries to bore into the wall; philosophy, which annuls the mystery through the logical explanation of the structure and function of the wall; and religion, which lives with and in the mystery thanks to the sentimental leap of faith which allows it to go over the wall.

But there is no doubt that in every human being there exists a minimum of fantasy which allows him, even if just for a fleeting moment, to feel he is a poet and thus find liberty.

Why, then, be a poet? In order to be free and so that literature may exist, since without poetry nothing can be.

Poets embody language; they have eliminated its functional aspect and speak from language.

If "language is the home of being," as Heidegger affirms, the poet is its herald, its prophet. Through him the being reveals itself in the suggesting suggestion of language; through him the being finds the way to direct the efforts of man's integration with Totality; through him truth becomes *alètheia*, the unveiled, which resides in the darkness of the word. Thus states E. Paci in the previously cited study: "poetry is not the word of man, but rather the word of the being. However, though necessary, the being has within itself its own hiding place: it is revelation, but it is also occultation. The being is the truth insofar as it is a truth anterior to man which necessarily reveals itself and hides in the word."[10] In the poet, the miracle of creation which is possible only through the word is verified.

The poet will always be more Orpheus than Prometheus; he can't foresee the future; he can only make it possible by finding the word.

## NOTES

[1] De Cisneros, Marcelino, Paper read at the Tenth Interamerican Congress of Philosophy in Tallahassee, Florida, in October 1981.

[2] Paci, Ezio, *La filosofía contemporánea* (Garzanti, 1984), p. 175.

[3] Gusdorf, G., '*La palabra*,' trans. Horacio Crespo (Buenos Aires: Nueva Visión, 1971), p. 35.

[4] Ibid., p. 15.

[5] Croce, Benedetto, *Breviario di estetica* (Bari: Laterza, 1972), p. 35.

[6] Gomez Bedate, Pilar, *Introducción a la poesía lírica* (Puerto Rico: Editorial Universitaria, 1977), p. 19.

[7] Ibid.

[8] Ibid.

[9] Bertelloni-Bazzichi, María Teresa, *El mundo poético de Angel Crespo*, p. 66.

[10] Paci, Enzo, p. 211.

## BIBLIOGRAPHY

Aristotle. *Poetica*. Aguilar, 1966.

Casares, Angel Jorge. *Sobre la esencia del hombre*. Puerto Rico: Universitaria, 1979.

Cassirer, Ernst. *Antropología filosófica*. Mexico City: Fondo de Cultura Económica, 1968.

——. *Filosofía de las formas simbólicas*. Mexico City: Fondo de Cultura Económica, 1971.

Croce, Benedetto. *Breviario di Estetica*. Bari: Laterza, 1972.
——. *Estética*. Buenos Aires: Nueva Visión, 1969.
Capelle, Wilhelm. *Historia de la filosofía griega*. Madrid: Gredos, 1972.
Ferrater Mora, José. *Diccionario de filosofía*. 2 vols. Buenos Aires: Sudamericana, 1969.
Gomez Bedate, Pilar. *Introducción a la poesía lírica*. Puerto Rico: Universitaria, 1977.
Heidegger, Martin. *Introducción a la metafísica*. Buenos Aires: Nova, 1966.
——. *Sendas perdidas*. Buenos Aires: Lozada, 1960.
Hölderlin, Friedrich. *Poesie*. Einaudi, 1963.
Husserl, Edmund. *Ideas*. Mexico City: Fondo de Cultura Económica, 1962.
Mondolfo, Rodolfo. *La comprensión del sujeto humano en la cultura antigua*. Buenos Aires: Universitaria de Buenos Aires, 1968.
Paci, Enzo. *La filosofía contemporánea*. Garzanti, 1974.
Sartre, Jean-Paul. *¿Qué es la literatura?* Buenos Aires: Lozada, 1969.
Schajowicz, Ludwig. *Mito y existencia*. Puerto Rico: Universitaria, 1962.
Vossler, Karl. *Filosofía del lenguage*. Buenos Aires: Lozada, 1968.

LOIS OPPENHEIM

# THE FIELD OF POETIC CONSTITUTION

Any effort toward a theoretical delineation of the role of poetry in literature implies a more general consideration of the role literature can and should play in the evolution of a postmodern social, political and esthetic consciousness. Literature that not only reflects change, but inspires it must be investigated, moreover, within the perspective of its fundamental constitution and poetry, as a literary event, reveals its function precisely to that degree that it reveals the conditions of its very existence.

Current interpretive studies in both art and literature reflect an increasing awareness of the limitations of representational thinking in the analysis and appreciation of the work of art or literary text. Despite the movement away from the imperialistic hold of the notion of sign over contemporary criticism, however, the view of creative expression as an arbitrary, symbolic re-production of reality persists. It remains up to the phenomenological esthetician to prove that the creative act is at once entirely referential, that is to say, perceptual in its relation to a world already there, and auto-constitutional. Thus, what is required is the demarcation of that field in which the work of art opens onto ontological regions unapproached by thinking founded on the traditional dichotomies of mind and matter and self and world. Within the non-presuppositional perspective of phenomenological methodology, post-Husserlian attempts to re-situate inquiry into the creative process on the level of an absolute reciprocity between the creative imagination and Being have contributed significantly to the weakening of the idealistic and positivistic interpretive tendencies of Occidental thought. A continued emphasis on the inaugural power of art and its transcendence over the materiality of its expression is needed, however, to further restrict the exploitation and distortion of art viewed within the cognitive perspective of objective representation. It is with this in mind that the present study of the field of poetic constitution is offered.

## I

For the purpose of revealing the interrelation of presuppositional and representational thinking in poetic theory as the *source* of the dispute over



*A-T. Tymieniecka (ed.), Analecta Husserliana, Vol. XVIII*, 47–59.

the foundation of meaning in the poem, we will limit our focus to precisely that aspect of poetic constitution which gives the verbal articulation its *poetic value*: the operation of perceptual experience, To describe the primary function of this experience in the constitution of poetic meaning implies demonstrating that inner cohesion of poetic expression directly relating to a primordial, pre-cognitive awareness of the poet, a cohesive integration of self and world which cannot be reduced either to a given reality or to those abstractions from reality, sign and symbol, which, in recent years, have been affirmed by the semiotician as necessarily non-perceptual.

In affirming the non-perceptual foundation of linguistic reference, the critic of semiotic inspiration has overlooked one important component of the word-world relation which appears in the study of poetic theory to contradict the assumption that language, relating concepts, is fundamentally unrelated to the world of *things*. If, in the tradition of structural linguistics, we understand the primary operation of the sign as the substitution of a linguistic phenomenon – the totality of a sonorous, imagistic expression (the signifier) and a concept (the signified) – for a non-linguistic phenomenon, we are obliged to view language solely with respect to the limits and constraints of an assumed linguistic autonomy. By the very nature of communicaton, however, language as *meaningful* expression cannot justifiably be conceived as an absolute, independent of that which appears *beyond* the immediate linguistic situation. Rather, it demands, by virtue of its intersubjectivity, to be considered in terms of a *particular* implicit or explicit reference to a pre-conceptual comprehension of the self in relation to the world, a relation described by Calvin O. Schrag in an essay on Heidegger, as a pre-reflexive understanding "in which the intentional structure of experience is already operative."[1] The communicability of language, in other words, is specifically located within the projection of this apprehension of Being and it *is* necessarily perceptual in so far as this awareness reveals to the poet the specificity of his affective mode of Being-in-the-world. Thus, while the poetic utterance distinguishes itself from the discursive expression of a thetic or ideational level of thinking, both reveal a perceptual referentiality within the not necessarily overt manifestation of the primacy of a pre-cognitive knowledge. And it is within the context of this perceptual, as inherently intentional, referentiality that the *experiential* mode of the poetic articulation is revealed. When we consider, therefore, the concept of textuality in the terms of many of our recent critics for whom the text is everywhere revealed and yet, in the words of Roland Barthes, "'nowhere *fully* present,'"[2] we see that it is the very notion of *reproduction* which, in supplanting the perceptual and primordial experience of a pre-cognitive *constitution* of meaning, *de-constructs* the referential operation of communication into the order of representation: to cite Jacques Derrida,

There is no present text in general, and there is not even a past present text, a text which is past as having been present. The text is not thinkable in an originary or modified form of presence . . . . Everything begins with reproduction.[3]

In substituting for a representational thinking on the poetic text the order of intentionality whereby meaning is determined solely through a unique orientation of the mind toward the world, it becomes clear that the act of writing, as a projection of a pre-conceptual awareness of the "In-der-Welt-sein," is a constituive, as opposed to a re-productive, process. This is to say that the creative imagination in the act of poetic utterance is the bringing into being of that which has never before been, and this precisely to the degree that the verbal situation is a primary experience of language. In the words of Merleau-Ponty, "Pour le sujst qui parle . . . , qui n'est pas *obervateur* en face de la langue prise comme *objet*, mais qui la pratique, qui l'assume, il y a incontestablement une realité de la langue . . . ."[4] In so far, therefore, as the constitutive process embodies both the poet's immediate apprehension of the world as already there and his preliminary recognition of himself as, within the context of any of the multiple possible affective dimensions, always within it, situating the analysis of the poetic text on the level of this co-determination serves to eliminate the erroneous representatonal framework and allows the verbal experience to reveal itself – not as the *translation* of a non-verbal phenomenon – in which the transcendental subject would be divorced from the experiential situation of language – but as the projection of the transcendental *movement* of a perceptual, intentionalizing consciousness.

In his book, *La Gravitation poëtique*, Jacques Garelli has noted that "There is a paradox of poetic activity which is the measure of its very condition. Opening onto the world, the poem, in the act of unveiling, is that which is unveiled. Looking, it makes itself seen. Enlightening, it illuminates itself. Revealing, it reveals only in being itself revealed."[5] It is within this paradox that the enigmatic poetic value of the poem – the emergence of meaning through the syntactic and paradigmatic disposition of rhythm and sound – resides. Within the context of this ambiguity, the field of poetic constitution unfolds not only as a *constitutional* process, but as an *auto*-constitutional process, the presencing of that which opens onto a presence. To determine *how* the articulation of the poet functions as poetry– as, in other words, the revelation of that which in the process of revealing, is itself revealed – it is the notion of the function of poetic expression as the translation into language of some ineffable zone of reality that must first be eliminated. This necessity, to be sure, has already been outlined by Garelli in *La Gravitation poétique* where all conceptualization of a so-called silent reality awaiting expression through the magic of the poetic word was automatically rejected within the substitution of the intentional order for objective and representational thinking in the analysis of the poetic act. It is only in maintaining Garelli's initial effort to exorcize this notion of language as translation, however, that the full weight of Merleau-Ponty's achievement in *Le Visible et l'Invisible* can be felt, for in acknowledging with Garelli the paradox of poetic constitution as fundamentally *auto*constitutional, it becomes apparent that the progressive surrealization described by Merleau-Ponty as emerging from within the defunctionalization of quotidian reality through the linguistic experience is a function of the *Logos* conceived not as a *property* of man, but as a *visibility* emanating from within him. It is in this sense that the site or place of poetic activity – the *locus* of the field of poetic constitution – is to be located within the metamorphosing movement of the creative imagination which, by a transcendental and *invisible* power, effects on the world a transformation such that a particular ontological domain becomes *visible*.

If the field of poetic constitution, moreover, may be described as *auto* constitutional, the originality and specificity of the poetic text may be said to reside within the fundamental structures of meaning which are apprehended only through the *actuality* of the uttered word. To locate both the *site* and *origin* of a particular experience of langauge, therefore, is to situate that *place* or *locus* from which the linguistic experience emerges, from which the cognitive apprehension of meaning arises, and this, paradoxically, is to delineate the very sphere of the poetic act itself. This is to say that the realm of poetic creation, the field of poetic constitution, is, of and by itself, the origin of its own accomplishment and thus the poetic utterance is originary precisely in so far as the poetic *value* emerges through the temporal structure of its own enunciation. That which the poem achieves, in other words, its very realization of a newly constituted "world" or "sur-reality," is inseparable from the corresponding temporalizing movement of the text and, phenomenologically speaking, any investigation into the constitution of the poetic field is necessarily an inquiry into the transcendental movement of *Logos* into *Topos*,[6] the movement from word to place, whose *temporal* immateriality offers proof of the irreducibility of poetic expression to the objectivations of representational thinking.

## II

Two aspects of the temporalization process at work within the poem become evident when one considers, within the theoretical perspective of semantic analysis, the notion of poetic reference *not* according to the realistic conception of representational thinking whereby the word "refers back" to the thing in the world, but rather, according to the ontological dimension of perceptual and intentional referentiality. Understood in the context of the relation between thought and language conceived *within* the unity of their relation to Being, the ontology of poetic reference resides in the pre-cognitive recording of the world as already constituted and yet forever open to modification and modulation. This, of course, serves to eliminate any conceptualization of poetic referentiality as a logical designation of some level of experience existing outside of, and prior to, the poetic utterance. The temporalizing movement of poetic reference, in other words, may be described according to a dialectical system of opaque and dispersive modes of signification – what Garelli calls the "recel" and the "dispersion" of meaning – an integration of the pre-reflexive apprehension of reality with the fundamental *interrogation* of the modes or structures of Being which is the text functioning as an "opening onto" the world.

The first temporal element revealed within the ontological dimension of poetic referentiality is the verbal irruption of a field of *sonorous* meaning. Into the silence preceding the linguistic expression, the presence of the word is thrown and yet, joined syntactically and paradigmatically with the surrounding words of the poem, the initial utterance, as a verbal presence, stands detached from the background of silence which is also that of a familiar or known reality. At this point, the temporal dimension is felt by the reader as a *rupture*, for he is torn from his habitual orientation toward the world by the temporalizing movement of the verbalization. And the preliminary effect of the reading activity is perceived, therefore, as a risk: that inherent in the possible *loss*, as opposed to the acquistion, of meaning.

A second temporal element revealed within the ontological dimension of poetic referentiality may be described as that which gives the poem its poetic *value*: the process of 'opening onto' the fundamental structures of the world and the constitution of a sur-reality. Gianni Vattimo, in a paper entitled 'On the Way to Silence (Heidegger and the Poetic Word),' has explained this process as follows: "A resounding of the authentic word can only spring out of silence. The authentic word is the inaugural word, the word which lets the truth happen, i.e., which opens new historic horizons."[7] Specifically, it is the antepredicative imagery which combines with the sonority of the expression to create that which is representationally irreducible to a cognitive a priori, a pre-existing experience of reality. And it is through these two components of the temporalization process that is the poem – first, the explosion, *in time*, of sound into silence, and second, the revelation, recapitulation and reflection of the spatio-temporal constitution of the world – that the horizons of meaning emerge.

Given the temporalizing movement of the poetic semantic sphere, by which the poem transcends the objectivity of its expression, we can agree with Vattimo that "Analyzed more deeply, the horizon context shows itself not as a structure of connections between things, but as a system of meanings,"[8] a system displayed as *time* unfolding an experience of language whose meaning is both reactivated and variable from one act of reading to the next. For this reason, meaning resists being *established*. Rather, it *occurs*, and this through the poet's intersubjective communication with a reader posted simultaneously as an object and as a subject in an intentionally constituted world. Occurring and re-occurring with each and every act of reading, meaning is vivified through the temporal continuity operating within the *retentive* and *protentive* movements of a perceptual referentiality inherent within the flow of words.

Writing of the text as an "object" which assumes meaning through a particular historical context, Thomas E. Lewis, in his essay 'Notes Toward a Theory of the Referent,' defines the literary referent as a "cultural ideological unit,"[9] a definition derived from his correlation of the semiotic and Marxist conceptions of meaning. While it connot be denied, even within the ontological perspective, that the sphere of poetic meaning presupposes a certain cultural conditioning, the coming together of psychological, sociological and historical determinants, the semantic sphere of the poetic text cannot be reduced to such causative factors any more than it can be reduced to the objective a prioris of representational thinking. The signifying poetic universe unequivocally *transcends* its material expression and despite what may be regarded as the cultural transposing of signs into other signs, the limitless chain of semiosis, textuality is not *the same as* the organization of cultural units which stimulates it. In so far as the poetic text, which does not *relate* experience but *is* experience, is an adventure in expression, it must be recognized as unfolding a world of meaning fully present and irreducible to any preceding it.

In limiting the referent's horizon of meaning to cultural units and in refusing to recognize the ontology of literary reference, reductionist theory fails to take into account the fundamental operation of that temporality constituting the semantic space. Lewis, for example, clearly distinguishes between the reality of the textual meaning and that of the cultural elements transformed through the referentiality of literary language: "Crucially, we should never underestimate this difference between literary raw material and the literary signification that emerges through textualizing that raw material."[10] And yet, in locating the referential horizon of the textual semantic sphere in the "cultural universe *preceding* textual production,"[11] he ignores the temporal functioning of the newly constituted reality of the text. Viewing only the ideological determination of signification, Lewis, like so many of his semiotically oriented contemporaries, disregards the entire topography of the temporal metamorphosis. Establishing, nevertheless, an interesting connection between the Marxist and semiotic conceptions of referentiality, Lewis, like his colleagues in post-structural criticism, neglects that very process which both allows for the inception or *inauguration* of textual meaning and gives rise to the poetic value of the text. And thus Lewis' and related theories of the referent neglect as well the fundamental site or topos which, in the ontological perspective of a perceptual and pre-reflexive apprehension of the "In-der-Welt-sein," appears as the most crucial determinant of the specificity and originality of any poetic text.[12]

## III

To fully understand how it is that the operative intentional structure of experience discloses within the semantic sphere of the poetic text the primacy of perception and the ego's relation to his surroundings, we might consider both the theoretical and practical examples of two poet-estheticians, Ezra Pound and Garelli, for whom perceptual reference, rather than reproduction, is necessarily the site and source of poetic meaning.

In *Gaudier-Brzeska*, a memoir of the painter written by Pound in 1916, Pound defined his poetry as a "sort of poetry where painting or sculpture seems as if it were 'just coming over into speech.'"[13] Pound's focus here is on the interplay between two zones of creativity: the verbal and the non-verbal. Well aware of that dialectical movement between the objective world and a perceiving consciousness which provides the ontological foundation for creative (verbal and non-verbal) expression, Pound effectively disassociated the word "art" from the word "object" through a theory of the image which redefined art and, specifically, the written text as a non-objective phenomenon: his notion of the image as a vortex, a *mobile* nexus receiving and transmitting the flow of ideas, eliminated any possibility of the assumption of the poem as fixed in its state of objectivity. The vorticist movement, in fact, resided in the effort to reveal the analogous situation of the artist working in one medium with that of an artist working in another. And the analogy disclosed that creative activity sets into motion a spatio-temporal phenomenon constitutive of a *new* reality which is re-inaugurated with each act of viewing or reading:

> The arts have indeed "some sort of common bond, some inter-recognition." Yet certain emotions or subjects find their most appropriate expression in some one particular art. The work of art which is most "worth while" is the work which would need a hundred works of any other kind of art to explain it. A fine statue is the core of a hundred poems. A fine poem is a score of symphonies. There is music which would need a hundred paintings to express it.[14]

If we find in vorticism a decisive repudiation of the tradition of presuppositional and representational thinking on art, it is that within this esthetic Pound locates precisely that perceptual referentiality which prevents art from being both mimetic and arbitrary:

> in writing poems, the author must use his *image* because he sees it or feels it, *not* because he thinks he can use it to break up some creed or some system of ethics or economics.
>
> An image, in our sense, is real because we know it directly . . . It is our affair to render the *image* as we have perceived or conceived it.[15]

Pound could not have expressed any more clearly the fundamental intentionally determined structure of the creative imagination.

Pound's well-documented preoccupation with the interrelatedness of one art with another points to his appreciation of the function and value of primary or non-thetic perception as the referential foundation of creative activity. Similarly, Garelli's efforts to reveal the perceptual and experiential foundation of poetic activity within a temporally explosive structure defined as a sort of "Intentional Nucleus in Sonorous Articulation of Variable Tensions, *these tensions being none other than the explosive manifestations of time*,"[16] is indicative of a shared view that the image has that value of the vortex as described by Gaudier-Brzeska as "The point of maximum energy,"[17] a value which prevents its conceptualization as a non-mobile, objective phenomenon. The poetic value of the image for both Pound and Garelli, in other words, is not only intentional, and thereby perceptually referential, but temporal, its organic-like temporality being implicated within the power and intensity of its energetic constitutional movement.

Not only theoretically in their writing on the image, but creatively within the content of many of their poems, both Pound and Garelli reveal a preoccupation with the interrelation of poetic expression and the contours and surfaces of non-verbal art. As Donald Davie, certainly the authority on Pound in this area, points out in his excellent study, *Ezra Pound: Poet as Sculptor*, Pound's concern is with "the analogy between poetry and sculpture" and "Since sculpture, or at least one aspect of sculpture, can be expressed in terms of a relationship between plane surfaces, Pound speaks in these terms of images of a poem":[18]

The pine-tree in mist upon the hill looks like a fragment of Japanese armour.

The beauty of this pine-tree in the mist is not caused by its resemblance to the plates of the armour.

The armour, if it be beautiful at all, is not beautiful *because* of its resemblance to the pine in the mist.

In either case the beauty, in so far as it is beauty of form, is the result of "planes in relation."

The tree and the armour are beautiful because their diverse planes overlie in a certain manner.[19]

What is of particular interest here in regard to our consideration of Pound's poetic referentiality is that in disclosing – *through the poetic act* – the interrelation of sculptural *planes* and their projected *image*, Pound demonstrates how the correspondence resides precisely within the *self-revealing* dimension of creativity. It was also in writing of this dimension that Garelli, in a paper cited earlier, described the self-illuminating process as the paradoxical determination of the primary poetic (as opposed to both ontic and representational) value.

Furthermore, Garelli has written that "The poetic act arises at the intersection of two reigns."[20] Here again we have evidence of a profound similarity between Pound's and Garelli's esthetic viewpoints for the "two reigns" to which Garelli refers are not the unintelligible and its magical translation by the poet into signifying expression, but those of the, to quote Pound, "diverse planes" of a poetically constituted world *analogous* to those of chiseled stone which assume an esthetic form and meaningful constitution as, like the "real" and the "surreal," they "overlie in a certain manner." The presupposition of art as translation necessarily implies a representational or reproductive value and that "intersection" from which the poetic act arises can be perceived, analogously, as that which promotes the beauty of the sculpted image, as "planes in relation," *only* when such assumptions are abandoned in favor of the non-presuppositional perspective of phenomenological thinking.

A recent poem written by Garelli in conjunction with the monumental stone sculptures of the Uruguayan Gonzolo Fonseca focuses on this question of poetic reference and considers, metaphorically, the analogous esthetic situation of poetic and sculptural creation. The follwing excerpts provide an indication to the *intracreative movement* explored by both poets, that movement between stone and word which constitutes a unique correspondence whose meaning can be understood not in terms of either a transference or a translation of a system of signifiers, but in terms of the primacy of perceptual experience:

I

Death joined with the word against these stones will cause the eye beyond ashes and sleep on marble night fire burst to scratch the field of a life for little which breaks into bits like man ejected in jerks on the motionless surface of death

IV

He who blends the night with the sound of the toll who seizes the look in a block of cold let him construct as a system the terror of the stone walls

VI

Order Block Night Time

. . . . . . . . . . . . . . . . . . . . . . . . . . . . . . . . . . . . . . . . . . . . . . . . . .

Closer to the block To its sovereign life Something invisible tells me that this is within Something stable compact wan . . . Thus have I journeyed curled several decades in the scarcely explored grains of stone

Garelli's effort in writing this poem must not be confused with an attempt to "reproduce" Fonseca's sculpture through poetry. Neither must it be thought that it was undertaken to enlighten the stone pieces with the power and magic of the poetic word (again, presupposing translation as the function of creative expression). Rather, it is on a level of ontological *investigation* that the referentiality must be considered, a level of *inquiry* very closely related to that of Pound who, in fact, wrote that "All poetic language is the language of exploration."[21]

To consider Garelli's text in terms of an ontological inquiry into the relation between self and world and, specifically, poet and sculpture, it must be understood that the focus of the inquiry is not the *object* of perception, but the *fusion* of subject and object in the referential *movement* of the poetic articulation. This fusion operates within the appreciation of the poet for both the *energy* inherent within the related sculptural planes and masses and what Pound calls the "sense of the stone."[22] It is in these terms that we can understand Pound's feeling that Gaudier-Brzeska's success resided precisely in his ability to *reveal* the stone, for "The real sculptor 'sees' or is aware of, not only all the sides of his work, but of the 'through,' that is, the diameters that can be passed through it from any angle."[23] For the poet, it is the perceptual *word*, as a transcendental projection of the mind's eye, which comes up against the ontic reality, the mass of stone, causing, in the words of Garelli's poem, "the eye beyond ashes and sleep on marble" to "burst to scratch the field of a life for little which breaks into bits like man ejected in jerks. . . ." This confrontation of word and marble is nothing other than the "discovery," the unveiling, of that primordial movement of consciousness manifesting itself within the temporality of poetic expression as the forever renewable vivification of the sculptural energy. Topologically described by Garelli above, it is located in that "Interplace Interworld which inscribes its calls on the paroxysm of a stone wall." *That focus of the poet's eye on the stone opens – through his utterance – onto precisely those "diameters" which constitute for Pound the "'through'" and in that visual perception, which "scratches" the sculptured mass as a "field of a life," the field of sculptual presence is revealed in the realm of its specificity as the "sense of [that] stone."*

## IV

My aim in this paper has been to delineate the field of poetic constitution with a particular focus on the ontology of perceptual reference in the poetic text and to locate the horizon of linguistic meaning within the context of the textual 'Weltlichkeit'. An existentially dynamic process comprehensible on the level of the intentionality of consciousness, we have attempted to demonstrate that the poetic text is fundamentally a bringing into being of a world whose noetico-noematic structure directly reflects that of the creative imagination. Thus it may be interpreted only in the light of that which it reveals: the world and the individual's pre-conceptual apprehension of his existence therein. If, for us, neither Pound's nor Garelli's poetic activity can be categorized either stylistically or semiologically, this is due to an inherent *cohesiveness* of the poetic act – founded in the subliminal apprehension of the ontological "Being-in-the-world" – a cohesiveness which defies the non-experiential value too easily assigned to the poetic text by contemporary critics. Only within the actuality and intimacy of the poet's perceptual contact with his surroundings can we locate the experiential and poetic value and "discover" that space, as described by Garelli, "where the ear listens, the eye inspects, the gesture conveys the near and far . . . ." Thus if the dynamic form of the poetic act may be shown to represent *nothing other than itself*, it is precisely that the intentional, world-constitutional and temporalization processes interrelate in the act of poetic writing to effect a field of meaning whose horizon is perceptual, referential and ontological.

Finally, it must be noted that the current practise of applying non-literary theory to the analysis of literature is proving to be, more often than not, restrictive and distorting, for in confining the literary text to a level of interpretation which objectifies the connection between language and thought *outside* of their relation to Being, the specificity of the text, and the actuality of its experience as well, remain hidden. That inaugural criticism, on the other hand, appears valid to us in the light of that very relation of thought and language to Being is due to our recognition of the literary event as, what Merleau-Ponty calls an "inscription of Being." Localizing the *invisible* power of the poetic utterance to make *visible* a particular realm of ontological experience allows for an authentic demaration of the topology of the poetic field, for it is only within the context of the metamorphosing interplay of the individual and the world as disclosed within the ego's perceptual endowment of reality that the foundation of poetic constitution can ultimately be revealed. And it is for this reason that the integration of time and space and word and place reinforces neither an *external* objectivation of experience nor an *internal* egoistic enclosure of subjectivety within itself, but rather, a *field* in which the movement of temporality opens through the Word onto a World to unveil the ubiquitous nature of Being.

*New School for Social Research*

NOTES

[1] Calvin O. Schrag, 'Phenomenology, Ontology and History in the Philosophy of Heidegger,' in *Phenomenology: The Philosophy of Edmund Husserl and Its Interpretation*, ed. Joseph J. Kockelmans (New York: Doubleday, 1967), p. 281.

[2] Roland Barthes, cited in 'Critical Factions/Critical Fictions,' in *Textual Strategies: Perpectives in Post-Structuralist Criticism*, ed. Josué V. Harari (Ithaca: Cornell University Press, 1979), p. 37.

[3] Jacques Derrida, 'Freud and the Scene of Writing,' trans. Jeffrey Mehlman, *Yale French Studies* 48 (1972): 92; cited in Harari, p. 37.

[4] Maurice Merleau-Ponty, *Les Sciences de l'Homme et la Phénomenologie* (Paris, n.d.), p. 42.

[5] Jacques Garelli, *La Gravitation Poétique*, trans. Lois Oppenheim (Paris: Mercure de France, 1966), p. 9.

[6] For a most interesting consideration of this question of the movement of *Logos* into *Topos*, see Garelli, *Artaud et la Question du Lieu* (Paris: José Corti, 1982).

[7] Gianni Vattimo, 'On the Way to Silence (Heidegger and the Poetic Word),' trans. Daniel Scanlon, paper presented at the 'Favorite Malice,' an Italian-Ameican symposium on poetry and poetic theory held at New York University in March 1979, pp. 7–8.

[8] Ibid.

[9] Thomas E. Lewis, 'Notes toward a Theory of the Referent,' *PMLA*, 94 (May 1979): 459.

[10] Ibid., p. 472.

[11] Ibid. (my italics).

[12] Brief passages concerning the Lewis article are excerpted from a more complete analysis of the question in my essay 'Ontological Reference and the Horizon of Meaning,' *Criticism*, fall 1982.

[13] Ezra Pound, *Gaudier-Brzeska: A Memoir* (1916; reprint, London: New Directions, 1960), p. 82.

[14] Ibid., p. 84.

[15] Ibid., p. 86.

[16] Garelli, *La Gravitation poétique*, p. 188, trans. Lois Oppenheim; concluding chapter published as 'Temporality, Imagery, and Unity', in *New Literary History* 12 (Spring 1981):453–473.

[17] Pound, *Gaudier-Brzeska*, p. 81.

[18] Donald Davie, *Ezra Pound: Poet as Sculptor* (New York: Oxford University Press, 1964), p. 55.

[19] Cited in Davie, p. 15.

[20] Garelli, *La Gravitation poétique*, p. 9.

[21] Pound, *Gaudier-Brzeska*, p. 88.

[22] Ibid., p. 137.

[23] Ibid., p. 145.

# B. THE NATURE OF THE POEM: PHILOSOPHICAL AND LITERARY ANALYSIS

CYNTHIA A. MILLER

# THE POET IN THE POEM: A PHENOMENOLOGICAL ANALYSIS OF ANNE SEXTON'S 'BRIAR ROSE (SLEEPING BEAUTY)'

Confessional poetry differs in content and style from other types of poetry. Confessional poets present in their works intensely personal life experiences of their own suffering and crises; thus, the content of confessional poetry is a projection and reflection of the author. Ralph J. Mills notes that confessional poetry concerns the "more intimate aspects of life, areas of experience that most of us would instinctively keep from public sight."[1] M. L. Rosenthal indicates that the "private life of the poet himself, especially under stress of psychological crisis, becomes a major theme."[2] The authors of confessional poetry expose highly subjective world-views in the attempt to come to terms with themselves and their reality. The style of confessional poetry allows the author to deal with the subject matter in an unrestrained manner. Robert Phillips comments that "openness of language leads to openness of emotion .... The confessional poets, at the risk of all else, return that which is uniquely human to poetry. Rather than fearing emotion. they make it their stock-in-trade."[3] Confessional poetry reveals the author's inner-most existential struggles. This unique classification of literature offers an opportunity to uncover and understand the author's experience within the work.

Anne Sexton, a contemporary confessional poet, uses poetry to express her painful and extremely personal experiences. On some level, which is not always clear to her, Sexton is able to grasp her private life through poetry. She confesses,

> Sometimes the doctors tell me that I understand something in a poem that I haven't integrated into my life. In fact, I may be concealing it from myself, while I was revealing it to the readers. The poetry is often more advanced in terms of my unconscious, than I am. Poetry, after all, milks the unconscious. The unconscious is there to feed it little images, little symbols, the answers, the insights I know not of.[4]

From the motifs of guilt, mental illness, fear and suffering, Sexton constructs a poetic world of reality with the persona at the center. A. R. Jones states that Sexton's poems, "create largely the world of her *persona*. the 'I' of the poems which undergoes continuing development and is clearly related, intimately and painfully to the poet's auto-biography."[5] In Sexton's



*A-T. Tymieniecka (ed.), Analecta Husserliana, Vol. XVIII*, 61–73.

works, her personae convey aspects of her own existence or being-in-the-world. No research has yet explored, phenomenologically, how the experience of confessional writers is conveyed in their works, and specifically, how the experience of Anne Sexton emerges in her poetry.[6]

Phenomenological criticism, a criticism of the creating consciousness in literature, offers a way to explicate Anne Sexton's experience as conveyed in her works through her personae. To discover the existential experience of an author, the Geneva School of phenomenology[7] employs this intrinsic approach to literature by treating the poem as "a self-contained world where human experience takes shape as literature."[8] The Geneva School contends that every piece of literature offers traces of the author's experience because the work cannot be cut off from the author's intentionality of the act of consciousness which creates it. Robert Magliola explains this point of view, stating that,

> Precisely because language is gestural, is expressive, the literary work bears within itself the unique imprint of the author's own consciousness. The Geneva critic considers the author's unique imprint immanent in the literary work and critically available there.[9]

Georges Poulet, of the Geneva School, asserts, "To understand a literary work then is to let the individual who wrote it reveal himself to us in the work. It is not the biography which explicates the work, but rather the work which sometimes enables us to understand the biography."[10]

The confessional poetry of Anne Sexton offers a rich source from which to discover more about the author's experience in her work. *Transformations*,[11] a collection of Grimm's fairy tales written by Sexton and recast into contemporary and shocking parables, lends itself to phenomenological criticism. Fairy tales, in their original form, confront the human existential predicament. Bruno Bettelheim posits that "the fairy tale . . . takes these existential anxieties and dilemmas very seriously and addresses itself directly to them: the need to be loved and the fear that one is thought worthless; the love of life and the fear of death."[12] And, he adds,

> As with all great art, the fairy tale's deepest meaning will be different for each person at various moments in his life. . . When given a chance, he will reurn to the same tale when he is ready to enlarge on old meanings or replace them with new ones.[13]

In *Transformations*, Sexton revisits the fairy tale and recreates the story to confront problems in her own life. In the narrative mode through the voice of the persona (speaker) telling the story, the author is able to examine her own experience by distancing her emotions. Poulet suggests that painful experiences become more manageable when they are buffered by time, and there is a "new chance which is given us to enter once more upon the existence of our lived moments, but, this time, at a distance, at an interval remote enough to allow us to rectify them."[14] Sexton acknowledges in a letter to Kurt Vonnegut, dated November 17, 1970, that the story form allows her to explore her consciousness: "I think they [poems in *Transformations*] end up being as wholly personal as my most intimate poems, in a different language, a different rhythm, but coming strangely, for all their story sound, from as deep a place."[15] The modernized fairy tales in *Transformations* offer Sexton a new perspective from which to contemplate her way of being-in-the world.

This study examines one poem from *Transformations*, 'Briar Rose (Sleeping Beauty)', and focuses on one facet of Sexton's consciousness, the experience of time. 'Briar Rose' is especially appropriate for phenomenological analysis because the poem contends with the basic issues of life and death – a struggle interwoven with the author's experience of time. This analysis describes the persona's experience of time in relation to the character Briar Rose, reduces the description to the persona's experience of time, and interprets the meaning of the experience of the persona in relationship to the author's life.[16]

## DESCRIPTION

The first section of 'Briar Rose' is a lyric prologue. The persona objectifies her experience by narrating in third person. As the poem opens, the persona asks the reader to reflect upon a girl's emotional experience. "Consider," the persona implores, "a girl who keeps slipping off into the hypnotist's trance." A state of suspended animation, frozen time, allows the girl to drift into a "spirit world speaking with the gift of tongues." The religious allusion to prophecy implies that the girl can see into the future while in the trance. The persona observes that, during this experience, the girl feels "stuck in the time machine," as if she were wedged in time, unable to pull free.

The time frame then shifts "suddenly" and the persona reports that the girl retreats uneasily into the past, "two years old . . . as inward as a snail." Time, for the girl, is not only a diachronic experience of moving chronologically backward, but also a synchronic experience of turning inward, as if for protection. The "voyage" is overlaid with conflict and the persona states that the girl must swim "further and further back, up like a salmon."

The journey into memory is exhausting as the girl confronts a strong flow of feelings and must pull herself through and over emotional tides. The metaphor involves the future as well as the past. The reference to fish fighting the current to spawn is prophetic because, like the salmon, one must inevitably face death. The girl regresses into the past and draws inward to peer at the future.

The persona relates that the girl, now firmly in her memory, dwells on a past experience with her father which carries the imaginary insinuation of death. The girl's father becomes a death-figure as he beckons seductively, "Little doll child, come here to Papa." Death proffers the appeal: "I have kisses for the back of your neck." "A penny for your thoughts, Princess," is a further tendering of consultation and support. His entreaty, "I will give you root," is enticing, promising stability and fulfillment. The solicitation is assuring, yet malodorous and corrupt. The persona acidly describes the trip into the girl's past as "that kind of voyage, rank as honey-suckle." The advent of death compels an inviting and overpowering fragrance while simultaneously exuding the putrid smell of evil.

In the second section of the poem, the persona moves into the narrative mode: "Once a king had a christening for his daughter Briar Rose." The fairy tale exists in a state of timelessness that belongs to all stories that begin, "Once upon a time," and is thereby distanced from the present moment. The baptismal name portends the double nature of the character, which is beautiful and alluring and, at the same time, repulsive and hurtful. The ceremony intimates protection, sanctioning the future of the child. But, an unexpected and intrusive guest appears at the "grand event," the thirteenth fairy. The univited and forboding fairy arrives with an "evil gift" of doom, prophecying that Briar Rose will make the choice to die when she shall "Prick herself on a spinning wheel" and "then fall down dead." The king is aghast; he "looked like Munch's *Scream*," a hideous shrieking face surrounded by darkness. Then, tongue-in-cheek, the persona remarks, "Fairies' prohecies, in times like those, held water," understating the certainty of death. However, the twelfth fairy mitigates the curse, changing "that death into a hundred-year sleep." The promise of life seems as powerful as the prophecy of death, and the ogre is temporarily assuaged.

The persona tells that the king then words off the evil spirits to protect Briar Rose by ordering "every spinning wheel exterminated and exorcized." The persona remarks sarcastically that "he forced every male in the court to scour his tongue with Bab-o." The King's desire to give her a secure and pure life only results in a sterile, imprisoning environment. The persona comments, "she dwelt in his odor. Rank as honeysuckle." The reference to the king's odor suggests again, a smell that is stimulating but offensive; the king's presence is comforting yet repugnant.

The persona continues that on Briar Roses's fifteenth birthday, the prophecy is fulfilled. The princess "pricked her finger," choosing to live in a sleep-like death. In this time state, all activity halts as, "the clocks stopped. . . . The fire in the hearth grew still and the roast meat stopped crackling." Objects and people become petrified non-entities as "trees turned into metal and the dog became china," and, "they all lay in a trance, each a catatonic." The suspended existence is like being "stuck in a time machine" from which there is no escape. The persona states that "only a bunch of briar roses grew forming a great wall of tacks around the castle." The princess's wall of safety attracts with blossoms yet repels with thorns. In sleep, the princess is protected from intruders who "had not scoured their tongues."

The persona's statement that "In *due* time a hundred years passed and a prince got through" [emphasis mine], suggests that the princess must pay her debt to the prophecy of death. The only event that arouses Briar Rose is the prince's brief kiss and "She woke up crying: Daddy! Daddy!" Unnerved and frightened by her new ability to feel, Briar Rose harks back to the life-giving protection of her father. The excitement of the prince instills a new freedom from her deadened existence and "Presto! She's out of prison!"

The persona reports that the princess now lives in dreaded anticipation of the future which may bring back the numbness of sleep. The persona avers that "all went well except for the fear – the fear of sleep." The princess is anxiously aware of the dangers of inertness. The persona contends, "Briar Rose was an insomniac. . . . She could not nap or lie in sleep without the court chemist mixing her some knock-out drops." The princess realizes that sleep is an existence too close to death, and in this state she becomes powerless to really live. The princess also dreads sleep because there she foresees her death. The princess states, "I must not dream for when I do I see the table set and a faltering crone at my place, her eyes burnt by cigarettes as she eats betrayal like a slice of meat." The withering old woman resembles the evil fairy of death and sits ready, waiting to consume.

In the third section of the poem, as the fairy tale ends, the persona's lyric voice emerges from the dialogue of Briar Rose. The persona, using "I," describes a sleep-like existence equivalent to death: "I must not sleep for while asleep I'm dying." The persona feels petrified and weighted down in this condition, stating, "I wear tubes like earrings, I lie still as a bar of iron." The line, "I'm all shot up with Novocain," expresses the persona's diffident and stuporous emotional state. In time of sleep, "this trance girl" is completely submissive and "yours to do with." She feels as if she were dead already and reflects in disgust, "you could lay her in a grave and shovel dirt in her face." The persona feels self-loathing, characterizing herself as an "awful package." The persona refers to herself in third person to objectify the realization that intense sensations of the moment alone set her free from her passive state of imprisonment: "But if you kiss her on the mouth her eyes would spring open and she'd call out: Daddy! Daddy! Presto! She's out of prison." Although the persona believes temporary stimulation will free her from a life of resigned confinement, she cries for support and supervision from her father.

The persona uses the past tense in her lyric voice to delve into her memories. The character feels at some point in her past, "there was a theft," wherein part of her personhood was stolen. She states, "I was forced backward. I was forced forward." As a passive participant, she was pushed into being submissive and childlike, and thrust into being self-sufficient and adult. In the first case, she feels her growth was stunted; in the latter, her maturation was too quickly advanced. In either event, the persona believes she was treated as a non-person or object, stating, "I was passed hand to hand like a bowl of fruit." Due to these incidents, the persona states, "Each night I am nailed into place and I forget who I am." Her past influences her passive resignation in the present, where she is suspended in time. Her allusion to the crucifixion reveals her feeling of deadened existence.

The persona reflects on the father image, stating, "Daddy? That's another kind of prison." The life-giving protection of her father, though safe, is recognized as a loss of freedom to live as an individual. This vision of her father represents a type of death. She posits that her father is "not the prince at all," but a figure "drunkenly bent over my bed, circling the abyss like a shark, my father thick upon me like some sleeping jellyfish." The father image is not the prince of sustenance, but a dangerous, death-like predator. However, this kind of death does not strike quickly; it is a dormant, enveloping force which may smother her.

In the final lines of the poem, the persona examines the moment of writing as she asks herself rhetorically, "What voyage this, little girl? This coming out of prison?" The persona accepts that writing is a journey which provides stimulation and escape from her life's routine.

The persona works through her private experiences by referring to herself in the third person in the prologue, by transferring her experiences to Briar

Rose in the fairy tale, and by speaking openly in the lyric "I" in the final section. The persona recognizes the poem will live into the future, and "God help," she pleads, this part of me as well, that is immortalized in the poem – "this life after death?"

## 2. REDUCTION

The major theme for the entire poem is the persona's voyage through time. First, the persona journeys into her past, regresses to childhood, and, through memory, predicts her future. Next, the persona travels into the timelessness of the fairy tale. The persona then navigates from the fairy tale to the present and dwells on her current and past life. Her trip through time brings her to the poem, "What voyage this, little girl?" The moment of writing is recorded now and propels into the future. Writing is the harbor of safety from which the persona travels through time. The entire voyage, winding into past, present and future, permits the persona to examine the tensions between life and death.

The motif of the father-king is significant in that the image explicates the struggle of living and dying in the persona's experience of time. In the past, the father-king offers sustenance and life. Yet, in this sterile existence, the persona escapes from living fully. There is a strong pull towards this safer condition, but revulsion from its captivity. In the present, a sleep-like passive state simulates the secure life with the father king. Moments of genuine feeling provoke the persona to call his name for protection. In the future, the father-king symbolizes ultimate death: a provocative, magnetic force and, concomitantly, a repulsive, dreaded power. Death's double nature is concretized in the sexual metaphor. The persona is drawn toward the sensual experience but senses the danger therein. Death, lurking in the future, is an enticing escape, but the stultifying end to all life.

The prison theme throughout the poem is a graphic metaphor of the persona's experience of time. In the present, the persona feels captive to time. Weighted down and reluctant to move through time, she restrains herself from feeling and living fully. Present time is a confinement from which she tries to escape. The persona lives for moments of excitement when, "Presto! She's out of prison." Only those moments of special stimulation, represented by a kiss, liberate the persona from her inert life. Writing is also acknowledged as a stimulating experience, a "coming out of prison," which offers escape. The past is yet "another kind of prison." Memories of her father are correlated with smothering overprotection. This time dimension represents her loss of freedom to be an autonomous self. The future is also implicated when the persona remembers her father. Her memories project the message of her death. A prisoner of time in the present and past, she views her death as the final escape; thus, there is an impelling attraction towards future.

Grammatical structure further illuminates the persona's experience of time in the past. The persona employs "ing" verbs, such as "learning," "swimming," "struggling," as she describes her retreat into memory. The verb forms depict the persona's active participation in attempting to understand her past. The choice of verbs reveals that her efforts require mental and physical exertion. Later, the persona describes her past using such passive verbs as "was forced" and "was passed." The verb tenses show her feelings of being acted upon and being ineffectual. The persona desires to understand her past, but her memories are repelling. The persona's conflicting sensations about this time period make resolution difficult.

Other themes from the poem shed light on the persona's experience of time in the present. The persona is mired in the present, "stuck in the time machine." A time machine usually implies access to past or future events; however, the persona is held fast in the present, unable to travel easily in time. Trapped in her existence, she chooses to escape from time through "trances" and "sleep." In these synonymous passive states, the persona confronts life. In the fairy tale, the elaboration of images shows the persona's fascination with cessated time. Activity slows down; animals are immobilized; people are depersonalized. The persona, like an object, becomes dull to all feeling. Although frozen time appears safe, the persona likens the experience to death. She begins to fear sleep, her catatonic prevalence. Suspended in time, the persona chooses to exist in passivity, a state of non-being, to cope with reality.

The motif of prophecy explicates the persona's experience of time in the future. Although the twelfth fairy changes the curse of death to sleep, death remains inevitable. The persona dwells on her future and predicts her own death as she refers metaphorically to the salmon and the enticement of her father. In her dreams, she envisions the withered crone, death. The persona is obsessed with sleep as death. Sleep, which is likened to death, becomes intolerable. Fear causes her to live in anxious anticipation of time, and yet, she recoils from the future.

The metaphors of time in the poem related by the persona reveal her conflicting impulses. The persona longs to journey freely in time. She desires to visit the safe memories in her past and to move with excitement through the moments of the present. However, the persona is bound in her past by feelings of throttling restraint and is a captive in her present by feelings of non-being existence. A prisoner of time, the persona seeks escape in the future.

### 3. INTERPRETATION

In the poem, 'Briar Rose,' as narrated by the persona, Anne Sexton's experience concerning her self-image emerges. The persona feels as if she faced reality as an impotent and incapable child. Sexton expressed her own feelings of inadequacy in a letter, dated November 5, 1963: "As the salmon fight upstream to spawn, I fought to stay in Europe. . . . But I began to feel smaller and smaller, unreal."[17] The persona assumes the role of princess which allows her to romanticize her need to be safeguarded. Sexton's battle between her roles and her own identity was reflected in a letter to her husband, dated October 6, 1963:

> I want to become MORE than Princess Anne (much as I want to be her) . . . I want to become a woman who lives side by side with Kayo. Not a princess, not a queen, but a friend. . . . I must learn to be a woman, not a child. Yes, I'm your princess and want to have you say it . . . also your dearest friend, your companion, your mistress, your mother. I'm me.[18]

Sexton struggled with the process of realizing her full potential in the adult world. She was pulled between the desire to remain in the secure child-like role of princess and the need to become an adult person. The persona in 'Briar Rose' possesses such a poor self image that she feels self-abhorrence. In the same letter to her husband, Sexton disclosed her own self-aversion stating, "Kayo, sometimes I *am* ugly."[19] In the poem the persona becomes so distressed with the existential struggle that she chooses sleep. Anne Sexton also opted for a passive condition when life became too stressful. She poured out her feelings of alienation and inertia to a friend, writing, "I'm like a Jew who ends up in the wrong country. I'm not a part. I'm not a member. I'm frozen."[20] The author's wish to grow and cope with life was complicated by the exigence to learn to like herself. Sexton's conflicting self-image and the strife of everyday life resulted in the inability to deal effectively with her existential predicament.

Sexton's life, as related to the poem, is revealed in the writer's statement, "I always have this desperate feeling about time."[21] In a recent collection of letters, there exists corroborating evidence that Anne Sexton, like the persona, experienced time in the past with conflict and anger. The persona is unable to reconcile past protection with wanting safekeeping in the present. While the events in Sexton's past are different from incidents in the poem, the prevailing mood are similar. Linda Gray Sexton and Lois Ames write that Anne Sexton's

> memories were of a childhood studded with mental and physical abuse. Yet her older sister Blanche saw Anne as a "much-loved child, over-indulged – the center of attention." Whatever the reality, an early sense of rejection was to haunt Anne throughout her life and shaped much of her poetry.[22]

Although Sexton did not believe she was over-insulated in her past, she and the persona felt as if they were treated as objects and thereby lost some of their self-identity. The author and persona experience their memories with tension, and their remembrances are bitter. The persona goes into a hypnotic trance to delve into her past. Sexton's biographers reveal that "Dr. Martin had used hypnosis with some success."[23] Sexton used psychiatric aid to heal her past experiences, and complained to her husband in a letter that "the trouble with therapy is that it makes life go backwards . . . and I am so tired of that old suffering."[24] During her entire life, Sexton struggled with her feelings toward time in her past.

Like the persona in the poem, Sexton also experienced time in the present as unbearable. She, as does the persona, searched for different avenues of escape. In a letter Sexton disclosed her obsession with a sleep-like existence: I want to sleep all day and night (i. e., not live but not quite die)."[25] Her biographers contend that Sexton withdrew from the present, "not only by repeated suicide attempts but by various forms of hysteria, including trances, as well."[26] In correspondance to a friend Sexton admitted,

> I have black outs. I don't mean I pass out. I mean everything goes black. . . . I was having them everyday, sometimes seventeen a morning. . . . I've had these blackouts and this heart condition for four years. . . . I attributed it to anxiety and took seven 50 mg. Librium.[27]

The idea of death is frightening, yet Sexton toyed with the prospect through suicide. In a letter to someone close, Sexton voiced her wish to die: "I didn't want to live. I'm only writing to tell you about it, not warn you. . . . Only me do I threaten. Only me do I kill. No one else. Now listen, life is lovely but I CAN'T LIVE IT."[28] Sexton felt mired in present time and was drawn toward death for relief; yet, she could not quite forsake life. Sexton and her persona chose to cope with time in the present by retreating from life.

Sexton was made to feel alive during moments of intense sensation, like the persona who is stimulated only by a kiss. Sexton found ways to wrench herself from a non-feeling existence for brief spans of time. She wrote of her passive condition and her attempts to arouse herself: "I am like a stone that lives . . . locked outside of all that's real. . . . And because I'm frozen I drink to simulate [stimulate?] it; a drug up too [*sic*]."[29] One source of animation for the persona in 'Briar Rose' is the experience of writing, which provides her a way to work out her stress. Sexton understood the value of writing in maintaining her own sanity. She wrote, "All is an emotional chaos. Poetry and poetry alone has saved my life."[30]

Sexton, as well as the persona, experienced time in the future as both provocative and horrifying. The future summoned both with promises of death, the ultimate flight from time. As Sexton and the persona lived in a death-like state and dwelt on their final death, future time became a persistent and disturbing preoccupation. In an interview, Sexton discussed her obsession with her future and death in frequent talks with Sylvia Plath:

> Sylvia and I would talk in length about our first suicide, in detail and in depth. . . . We talked death with burned-up intensity, both of us drawn to it like moths to an electric light bulb, sucking on it . . . as if death made each of us a little more real at the moment.[31]

Sexton was equally frightened over the prospect of death. As the persona wishes sleep would take her unaware, Sexton wanted to relinquish control over her future. The author confessed, "I wish stupidly that someone but me would kill me and take the responsibility away."[32]

Thus, the fairy tale, 'Briar Rose,' is essentially the story of Anne Sexton's self-fulfilling prophecy. The child-hood tale is the nexus of Sexton's own modernized abult script. Flirtation with death and death-like images pervade both. The memories of her past are suffocating and intolerable. In the present, Sexton withdraws into trance-like states in which she is numb to all feeling, a waking death. Stimulation from drugs allows her to function and to feel as if alive. Contemplation of suicide proves exhilarating and demonstrates her preoccupation with death. Unable to cope with past, present, or future, as she is pulled between life and death, Sexton becomes a prisoner of time.

Anne Sexton took her own like on October 4, 1974.

## 4. CONCLUSION

By phenomenologically analyzing 'Briar Rose (Sleeping Beauty)' through the subjective perspective of the persona, a simple poem has taken on deeper meanings. The themes embodied in the language of the poem become the vehicle which enables us to project ourselves into the persona's consciousness and to understand her experience of time. The motifs have not only given us information about the persona's experience in the world of the poem but also told us the import and meaning of that experience for author. Sexton's ambivalence toward time as reported through the persona, is not a mysterious happening, but a meaningful reaction based on her existential struggle with life and death.

Confessional poetry becomes a significant genre when analyzed phenomenologically. This approach to literature returns the emphasis to the authors' unconscious experiences in the poems and turns attintion away from the lives of the authors outside the poems. Therefore, the authors' experiences embedded in the works becomes primary and biographical information which may elucidate the works becomes secondary. Because confessional writers are openly struggling with their existential dilemmas, their poetry is a rich source for understanding the deeper meaning of their lives. Uncovering the authors' ways-of-being-in-the-world in confessional poetry is a starting point for understanding our own existential pleght.

*Austin College*

## NOTES

[1] Ralph J. Mills, Jr., *Contemporary American Poetry* (New York: Random House, 1965), p. 156.
[2] M. L. Rosenthal, *The New Poets: American and British Poetry Since World War II* (New York: Oxford University Press, 1967), p. 15.
[3] Robert Phillips, *The Confessional Poets* (Carbondale: Southern Illinois University Press, 1973), p. 10.
[4] 'The Art of Poetry XV,' *Paris Review* **52** (Summer 1971):162.
[5] A. R. Jones, 'On Necessity and Freedom: The Poetry of Robert Lowell, Sylvia Plath, and Anne Sexton,' *Critical Quarterly* 7 (Spring 1965):15.
[6] Various analyses of confessional poetry are available. Ralph J. Mills in *Contemporary American Poetry* employs biographical criticism; M. L. Rosenthal in *The New Poets: American and British Poetry Since World War II* and Robert Phillips in *The Confessional Poets* feature formal analysis. Phenomenological analyses of poetry, other than confessional poetry, do exist. Janet Larsen McHughes discusses 'A Phenomenological Analysis of Literary Time in the Poetry of James Dickey' (Ph.D. diss., Northwestern University, 1972). David Halliburton focuses on poems by Edgar Allen Poe in *Edgar Allen Poe: A Phenomological View* (Princeton: Princeton University Press, 1973).

[7] Sarah N. Lawall's book on the Geneva School, *Critics of Consciousness* (Cambridge: Harvard University Press, 1968) offers a summary of Genevan practice. Robert R. Magliola in *Phenomenology and Literature* (West Lafayette, Ind.: Purdue University Press, 1977) expands the list of Geneva critics and focuses on the importance of phenomenological philosophy and aesthetics to this practice.

[8] Lawall, p. 6.

[9] Magliola, p. 28.

[10] Georges Poulet, 'The Phenomenology of Reading,' *New Literary History* 1 (October 1969):58.

[11] Anne Sexton, *Transformations* (Boston: Houghton Mifflin, 1971).

[12] Bruno Bettelheim, *The Uses of Enchantment: The Meaning and Importance of Fairy Tales* (New York: Random House, 1977), p. 10.

[13] Ibid., p. 12.

[14] Georges Poulet, *The Interior Distance*, trans. Elliott Coleman (Baltimore: Johns Hopkins Press, 1959), p. 80.

[15] Linda Gray Sexton and Lois Ames, *Anne Sexton: A Portrait in Letters* (Boston: Houghton Mifflin, 1977), p. 367.

[16] Chapter 2 of *Phenomenology and Literature* by Robert R. Magliola presents an exhaustive account of the phenomenological methodology of the Geneva School which closely follows the early teachings of Edmund Husserl. A summary is presented on pp. 45–46.

[17] Sexton and Ames, p. 226.

[18] Ibid., pp. 213–14.

[19] Ibid., p. 213.

[20] Ibid., p.252.

[21] Ibid., p. 60.

[22] Ibid., p. 1.

[23] Ibid., p. 225.

[24] Ibid., p. 191.

[25] Ibid., p. 252.

[26] Ibid., p. 225.

[27] Ibid., pp. 364–65.

[28] Ibid., p. 251.

[29] Ibid., pp. 251–52.

[30] Ibid., p. 81.

[31] 'The Art of Poetry XV,' p. 170.

[32] Sexton and Ames, p. 252.

JEANNE RUPPERT

# NATURE, FEELING, AND DISCLOSURE IN THE POETRY OF WALLACE STEVENS

In Wallace Stevens' phenomenological poetry, Nature and the human response to Nature are inseparable components of a single experience of a "world." It is human feeling for the things of the world that makes them "things," but that feeling has its basis in what is already there. This fundamental phenomenological idea is expressed with particular vividness in one of Stevens' last poems, entitled 'An Old Man Asleep':

> The two worlds are asleep, are sleeping now.
> A dumb sense possesses them in a kind of solemnity.
>
> The self and the earth – your thoughts, your feelings,
> Your beliefs and disbeliefs, your whole peculiar plot;
>
> The redness of your reddish chestnut trees,
> The river motion, the drowsy motion of the river R.[1]

As the developing images of the poem make clear, there are not "two worlds" but one, and that world *is* only in the being-together of consciousness and brute being. We recognize this in Stevens' shading of imagery from aspects of the self ("your thoughts, your feelings, / . . . your whole peculiar plot") to aspects of the "outer" world – the "redness" of chestnut trees that *are* "reddish," but that become fully "red" in being "[yours]," yours like "the drowsy motion of the River R" that goes unnamed if you fall off to sleep before naming it. This anonymous "River R" also suggests the brute being of the river, the fact that it *is* (in some way or another) beyond the human idea of it or feeling for it. We must recognize it as a pole of the "world" we inhabit, understandable that way only because that world has another pole, which is consciousness itself.

It is easy to understand the basic phenomenological idea of reality as a "compresence" or being-together of consciousness and brute being (or, let us say, "nature"), but it is not easy to account for the feeling response that the conscious human being makes to nature. Does it arise from what one finds in nature, or from what one brings to nature, or from the fact that one both "finds" and "brings" (that is, from a sense of an ontological truth that is present in every real presence to being)? Stevens poses this problem in poems



*A-T. Tymieniecka (ed.), Analecta Husserliana, Vol. XVIII,* 75–88.

written throughout his career. One of the earliest of these is a poem published in *Harmonium*, his first volume of poetry, entitled 'Frogs Eat Butterflies. Snakes Eat Frogs. Hogs Eat Snakes. Men Eat Hogs.'

It is true that the rivers went nosing like swine,
Tugging at banks, until they seemed
Bland belly-sounds in somnolent troughs,

That the air was heavy with the breath of these swine,
The breath of turgid summer, and
Heavy with thunder's rattapallax,

That the man who erected this cabin, planted
This field, and tended it awhile,
Knew not the quirks of imagery,

That the hours of his indolent, arid days,
Grotesque with this nosing in banks,
This somnolence and rattapallax,

Seemed to suckle themselves on his arid being,
As the swine-like rivers suckled themselves
While they went seaward to the sea-mouths. (*CP*, 78)

For the reader, the ugliness and torpor of this scene is only slightly relieved by the thunder's "rattapallax" (a sound perhaps too whimsical for this environment, and then perhaps purposely inserted to suggest the possibility of another perspective, another way of seeing things). The way things are seen here is the way the man who lived here saw them. As he was a man who "knew not the quirks of imagery," he saw his surroundings in elemental and ultimately stultifying terms. The atmosphere of the place is one of unrelieved aridity despite the rivers "tugging at banks" and the approaching rainstorm. The whole landscape is asleep, its "somnolence" unbroken by the thunder's "rattapallax."

We begin with the terrain itself, where the rivers are described as swine-like in their motion, in their sound, and in their smell. As "bland belly-sounds," these rivers express, perhaps, a zestless hunger and become in turn troughs at which the rest of nature feeds. The air itself is heavy with them and weighted down by the thunder that ought to bring fresh rain. The man who lived here tended this field for only "awhile," the "hours of his indolent, arid days" apparently oppressed by the grotesquerie of the place. But no, these arid hours and days "seemed to suckle themselves on his arid being," their aridity paradoxically nourished by his own inner aridity. The imagery gives us the sense of an insular, self-devouring emptiness that is like the self-devouring of the rivers themselves as they move toward the "sea-mouths" that consume them.

This insularity is in a sense paradoxical, for the poem clearly suggests a relationship between the aridity of the terrain and that of the man who lived there. Indeed, the question the poem asks is where we ought to look for the source of this desolation: is the human life lived there destitute because the place is? or is the place destitute because that is the way it was seen? We have observed that the man's "arid days" seemed "to suckle themselves on his arid being," which suggests that he is to blame for the emptiness of his existence. At the same time, we are told at the outset that "It is true that the rivers went nosing like swine," suggesting that this is how anyone would have seen them. Still, the structure of the imagery suggests that the man is in some way responsible for the continuing destitution of the scene: the poem begins with the imagery of the swine-like rivers, moves through the "heavy air" and "rattapallax" to the man who, at the center of the poem, has tended his land awhile and them left it; from this point, the imagery moves back outward through the "somnolence and rattapallax" to the rivers now seen carrying their desolation to the sea. If the place presented a grotesque surface to the man, it is also true, apparently, that he in some sense perpetuated its grotesqueness, failed to redeem it from its somnolence. But it is virtually impossible to distinguish here between the effects of the imagination and the effects of the "given." There is simply an experience of the world.

It is instructive to compare the poem with a canto from *Notes Toward a Supreme Fiction* that might have been written as a companion piece.

On a blue island in a sky-wide water
The wild orange trees continued to bloom and to bear,
Long after the planter's death. A few limes remained,

Where his house had fallen, three scraggy trees weighted
With garbled green. These were the planter's turquoise
And his orange blotches, these were his zero green,

A green baked greener in the greenest sun.
These were his beaches, his sea-myrtle in
White sand, his patter of the long sea-slushes.

There was an island beyond him on which rested,
An island to the South, on which rested like
A mountain, a pineapple pungent as Cuban summer.

And là-bas, là-bas, the cool bananas grew,
Hung heavily on the great banana tree,
Which pierces clouds and bends on half the world.

He thought often of the land from which he came,
How that whole country was a melon, pink
If seen rightly and yet a possible red.

An unaffected man in a negative light
Could not have borne his labor nor have died
Sighing that he should leave the banjo's twang. (*CP*, 393)

The vivid colors of the planter's island suggest immediately the quality of the life he lived there: one harmoniously close to the natural things the island provided (oranges, limes, sea-myrtle, sand). So intimately did the planter live among these things that he knew them virtually as colors (as "orange blotches" and "zero green"). These colors were "his," like "his beaches, his sea-myrtle . . . / his patter of the long sea-slushes." Nearby another island captured his gaze – and his imagination. It presented to him a chromatic change in almost a musical sense, the oranges and greens modulating into the bright yellows of pungent pineapples and towering bananas. Crowned with a mountainous green, it suggested to him a pineapple mountain, which he transformed into a "great banana tree." And he remembered another place, that from which he came, and turned it over and over in his memory, seeing it now in a pink light, now in a possible red, depending on the quality of his feeling, but expressing always the constant quality of the place, which he remembered as like a "melon." As Ronald Sukenick has pointed out, the planter's pineapple and melon imagery is emblematic of his seeing things in terms of his own experience as a planter, a shaper of nature.[2] We can also see in this imagery Stevens' careful balancing of the effects of the imagination with the effects of the "given." For the natural object (the pineapple that, in a later poem, Stevens will show being "put together" by "someone")[3] is here transformed by the planter's feeling into a banana tree; his feeling as he remembers "the land from which he came" qualifies the light and the consequent color in which he remembers the constant qualities of the place figured in the image of the melon.

The final stanza confirms what we already know from the foregoing imagery, that this man was "affected" by the world in which he lived and remembered, looked and dreamed, and that his involvement in the physical (and imaginary) beauty of this world sustained him in his labor. Because he lived in a "[positive] light," he regretted the loss of life: the loss of the sound, as of the color, of things. The "banjo's twang" is a suggestively neutral term; like the "three scraggy [lime] trees" in the second stanza, it qualifies the impression that all here was perfectly "beautiful." The planter could look at the scraggly trees and enjoy the "zero green" of the lime, could listen to the banjo's "twang" and take pleasure in the mere actuality, the mere being, of the sound.

Unlike this planter, the man in 'Frogs Eat Butterflies . . .' could find no stimulation in the scene in which he lived, not even in the thunder's "rattapallax." He was a man who lived in a "negative light," the negative light of his own insularity, his own aridity. But he also lived in the comparatively bleak light of a monotonous landscape, a land of fields, of sluggish rivers and approaching rain. He too was "affected" by what he saw, but in a wholly negative way that was the result of both the effects of the given and the effects of his imagination (or rather, lack of it). While we cannot distinguish between these effects, it is clear that the experience evoked in the poem is a product of both the aridity of the landscape and the aridity of the man's response to it. Of all the possible responses one might make, his is merely to duplicate what is around him and thus to extend it. The man virtually gives himself to the inertia of the scene, virtually becomes a part of it. He becomes the last link in an absurdity like that of the food chain evoked in the title: 'Frogs Eat Butterflies. Snakes Eat Frogs. Hogs Eat Snakes. Men Eat Hogs.' Human beings must, of course, do more than this if they are to redeem their existence from absurdity.

"The Snow Man" sheds additional light on this theme:

One must have a mind of winter
To regard the frost and the boughs
Of the pine-trees crusted with snow;

And have been cold a long time
To behold the junipers shagged with ice,
The spruces rough in the distant glitter

Of the January sun; and not to think
Of any misery in the sound of the wind,
In the sound of a few leaves,

Which is the sound of the land
Full of the same wind
That is blowing in the same bare place

For the listener, who listens in the snow,
And, nothing himself, beholds
Nothing that is not there and the nothing that is. (*CP*, 9–10)

Here the persona speculates on the condition of a consciousness that could observe the chilling surfaces of winter and not feel a sense of misery. Such a consciousness would have to be like winter itself, would have to "have been cold a long time," like a snowman, would indeed be no consciousness at all. The poem suggests that only a literal snowman could stand within sight of such a barren scene, within hearing of the wind and the scraping of a few leaves, and feel "nothing." For the human being, on the other hand (figured in the poem as the "listener"), to see and to hear *is* to feel, for seeing and hearing bring the human into the "otherness" of the landscape. One observes evergreens after a snowstorm and sees pine trees "crusted," junipers "shagged," spruces "rough" in a distant glitter: all terms that convey various tactile qualities (the touch of hardened snow, of ice, as the human being senses it).

Similarly, for a human being the sound of winter wind can hardly be a neutral sound. As the "sound of a few leaves," it conveys the emptiness of the trees (part of "the nothing that is" there). Or perhaps the sound is "nothing" in the sense of its strangeness, its enigmatic other side. At the same time, however, as sound it exists for the ear that hears it, thus as *something heard* (something, then, as well as nothing). This sound, which is a sound "for the listener," is also "the sound of the land" (thus it is a thing that binds them to one another, a thing that discloses at once their being there together and their difference). And the difference is the difference the listener makes in the scene: unlike the snowman, who sees nothing and thus feels nothing, the listener (a human awareness) sees something as well as nothing, hears something in the sound of the wind.

He is able to do so, paradoxically, because he is "nothing himself." The "nothing" of the listener suggests a correlation with the existential-phenomenological concept of the "nothingness" of consciousness – its being open-ended, futural, full of possibility, thus distinct from the brute being of things that simply are, and consequently capable, as openness, as possibility, of "nihilating" what is, of seeing it as other than it is, even of imagining it as not being at all. Thus, as nothing himself, as consciousness, the listener can imagine even this scene away, can imagine a more total emptiness (the "Nothing that is not there").

Yet by being able to imagine this scene away he also discloses its being (beholds "the nothing that is"). The phenomenological significance of this can be understood through Merleau-Ponty's explanation of what he sees as the only "[solid] meaning of the *cogito*":

> [It is] the act of doubting in which I put in question all possible objects of my experience. This act grasps itself in its own operation . . . and thus cannot doubt itself. The very fact of doubting obdurates doubt. The certitude I have of myself is here a veritable perception: I grasp myself, not as a constituting subject which is transparent to itself, and which constitutes the totality of every possible object of thought and experience, but as a particular thought, as a thought engaged with certain objects, as a thought in act. . . . This thought . . . *feels* itself rather than *sees* itself, . . . searches after clarity rather than possesses it, and . . . creates truth rather than finds it. . . .[4]

This helps us to understand how this present scene is to be conceived as "nothing" – not in its bleakness, as such, but in its ephemerality as something that is revealed only by consciousness. The bleakness of the wintry scene is a metaphor for this ephemerality, and in a sense so is the snowman. For the snowman is merely part of the scene and will disappear with it. By contrast, the listener (who "stands out" from the scene even as he stands in it) will go on (as long as he goes on). And in going on, he sees, hears, and feels his existence and the existence of things, and makes them (as he makes himself) "ek-sist," i.e., stand out.

If we return for a moment to 'Frogs Eat Butterflies,' we can now understand perhaps the strange insularity of the images of "suckling" in the poem (the man's hours and days "suckling themselves on his arid being," the "swine-like rivers suckling themselves as they move seaward to the seamouths"). This imagery expresses the man's failure to forge a vital contact with the place in which he lived, thereby disclosing both his standing in the place and his standing out from it. Like the snowman to a certain degree, the man in this poem thus fails to see the scene in terms of the possibilities of being and of nothingness disclosed by a vital consciousness. He mirrors the outer rather than interacting with it, and in doing so he gives way to its inhumanity, rather than taking upon himself his full humanity as a feeling existence that is part of, and yet significantly not part of, the scene. By contrast, the planter in the "blue island" canto (like the listener in 'The Snow Man') stood out from the happier surfaces of his scene by carrying their qualities further in his imagination.

These poems argue that only by standing out from brute being in the full consciousness of one's difference is one able to stand in the scene, to disclose its being. Feeling, as a power or property of consciousness, unites us with nature only to the extent that consciousness recognizes and maintains in that feeling the difference that, when examined, it discloses. For Stevens, the role of the artwork is to make this clear, to bring the reader into the realization of his own role, as feeling consciousness, in opening a place in which being can be revealed. 'The Idea of Order at Key West' expresses all of this.

> She sang beyond the genius of the sea.
> The water never formed to mind or voice,
> Like a body wholly body, fluttering
> Its empty sleeves; and yet its mimic motion
> Made constant cry, caused constantly a cry,
> That was not ours although we understood,
> Inhuman, of the veritable ocean.
>
> The sea was not a mask. No more was she.
> The song and water were not medleyed sound
> Even if what she sang was what she heard,
> Since what she sang was uttered word by word.
> It may be that in all her phrases stirred
> The grinding water and the gasping wind.
> But it was she and not the sea we heard. (*CP*, 128–29)

The first two stanzas open a meditation on a song sung by a woman walking along the shore and pose the problem for the reader that this singing poses for the persona: does this song express the singer or the sea?

We are told at the outset that the woman sings "beyond the genius of the sea," that her song expresses her own genius if any. But, we note, the sea has a "genius" as well. It never "forms to mind or voice" (never speaks humanly or articulates thought); it never acquires a human substance, but flutters "empty sleeves" (a metaphor wonderfully expressive of the enigmatic quality of a nature that at once stimulates a human response and withdraws into the non-human). Yet its motion seems to "mimic" human feeling, making a "constant cry" that is "not ours" but which we nevertheless understand (and which is part of the woman's song), a cry "of the veritable ocean" (ambiguously both "true" or "actual" and able to be verified, disclosed).

Being "veritable," the ocean is "not a mask." It is itself; it is what it appears to be and not an appearance behind which lurks some other reality. Similarly, the woman is not a mask but a veritable woman. They are two beings, yet apparently become part of a single reality, for "the song and water" are "not medleyed sound" ("medleyed" carrying the connotation not only of "mixed" but of "incongruent"). Thus the woman's song does not incorporate the water's sound as such, "even if what she sang was what she heard," for the reason that the song is "uttered word by word." The song apparently combines the sea and the woman in a single reality, perhaps the woman's reality or a reality of her making. The last few lines of the second stanza recapitulate: it may be that "the grinding water and the gasping wind" stir within this song, but what the persona hears is the woman singing, not the sea. Yet even in the cadences and images of the persona's meditation about the woman's song we hear and see the grinding water, the gasping wind, and thus we are not entirely satisfied by the claim that follows.

> For she was the maker of the song she sang.
> The ever-hooded, tragic-gestured sea
> Was merely a place by which she walked to sing.
> Whose spirit is this? we said, because we knew
> It was the spirit that we sought and knew
> That we should ask this often as she sang.

The ambiguity is apparently unresolved by the persona's identification of the woman as the "maker of the song," for the role of the sea is still clouded in ambiguity, expressed in the incommensurability of the next images of the sea. As "ever-hooded, tragic-gestured," it is mysterious and deeply evocative and cannot be "merely a place by which she walked to sing." Our initial question remains to be answered, and the persona now asks it in this form: "Whose spirit is this?" recognizing that it is this "spirit" or source that he seeks.

> If it was only the dark voice of the sea
> That rose, or even colored by many waves;
> If it was only the outer voice of sky
> And cloud, of the sunken coral water-walled,
> However clear, it would have been deep air,
> The heaving speech of air, a summer sound
> Repeated in a summer without end
> And sound alone. But it was more than that,
> More even than her voice, and ours, among
> The meaningless plungings of water and wind,
> Theatrical distances, bronze shadows heaped
> On high horizons, mountainous atmospheres
> Of sky and sea.

The "spirit" behind or within the song cannot be "only the dark voice of the sea," the persona reasons, although here (as heretofore) in the poem it is the sea (and sky) whose presence is most vivid, most "real." We dwell on the colors of the "voice" of the waves, the "outer voice of sky / And cloud" and of the "sunken coral water-walled." Yet all this, the persona tells us, is "sound alone," and the spirit he seeks is "more than that," more than mere sound. But it is also "More even than her voice, and ours, among / The meaningless plungings of water and the wind . . . ," and more than the "theatrical distances," "high horizons," and "mountainous atmospheres" evoked in the last few lines of the section.

As the persona works through these speculations (culminating in images of distance and perspective), he apparently finds his answer, signalled by the line break in mid-stanza:

> It was her voice that made
> The sky acutest at its vanishing.
> She measured to the hour its solitude.
> She was the single artificer of the world
> In which she sang. And when she sang, the sea,
> Whatever self it had, became the self
> That was her song, for she was the maker. Then we,
> As we beheld her striding there alone,
> Knew that there never was a world for her
> Except the one she sang and, singing, made.

This appears to throw the emphasis back on the woman and to signify that, after all, it is she alone whose "spirit" is expressed in the song. But this would contradict the equally emphatic claim six lines above that the "spirit" was "More even than her voice, and ours. . . ." How, then, can we understand the woman's role here, her "[making] the sky acutest at its vanishing"? Precisely in the persona's recognition that, by virtue of her singing, of her "mortal speech," the woman brings order to that which has a being beyond her. He has come to this recognition phenomenologically, describing what appears in the song: a gesture of signification made in the midst of brute being that discloses that being in its measuring of it. It is the woman who measures out a horizon that defines the "sky" as a "sky," that gives meaning to its "vanishing." But in fact she virtually cannot help but do so. The "ever-hooded, tragic-gestured sea," with all its "meaningless plungings," calls forth from the feeling human witness the need to so measure out a world. The "theatrical distances" of that world exist because of the human perspective that cannot but make these distances significant.

Similarly, the "meaningless plungings" of the sea become more than "sound alone" when heard by the human ear. Apart from such human participation in its being, the sound of the sea is, the persona conjectures, "a summer sound / Repeated in a summer without end," a sound limitless, measureless, and therefore meaningless in itself. But the woman who "measure[s] to the hour its solitude" measures to the sound of the sea a human significance. And thus the sea, "Whatever self it ha[s]," becomes "the self that [is] her song." Indeed, whatever the sea is "in-itself" cannot be known, and yet the sea *is* known – *as one is capable of knowing it*. "Things seen are things as seen," Stevens tells us;[5] the world known is *the world as known*, and it is in this sense that "there never was a world for her / Except the one she sang and, singing, made."

Without a phenomenological orientation to the problem posed by this poem, it is difficult to work out an interpretation of the work that accounts for the various claims made within it. Thus Harold Bloom concludes that the poem "remains equivocal and perhaps impossible to interpret fully in at least two respects, and these are nearly antithetical to each other. The poem affirms a transcendental poetic spirit yet cannot locate it, and the poem also remains uneasily wary about the veritable ocean, which will rise up against Stevens yet again. . . ."[6] Working from a phenomenological perspective we are able to heal this presumed breach between the "spirit" (whether of "Art" or of man) and the physical world. For Stevens there is no "transcendental poetic spirit"; there is only the constantly recurring phenomenon of a world measured out in terms of a human response to brute being.

That being (figured here as the "veritable ocean") is both "real" in the ordinary sense of the word (actual) and "real" in the phenomenological sense (real as revealed or disclosed by *Dasein*). And the "spirit" whose identity the persona seeks – that is, the source of the inspiration motivating both the singer and the persona-listener – comes neither from the singer nor from the sea (nor indeed from any "transcendental" source), but from the "being-together" of human consciousness and the physical "world." In Heidegger's terms, this being-together is the "mutual appropriation" of Being and human be-ing.[7] In be-ing authentically (in being temporal, in measuring the time and place of one's world), the human being responds to the call of Being as the singer responds to the "constant cry" of the sea at Key West, bringing it to realization in her song. In fact, the human being is appropriated by Being to be the point of disclosure of the whole (or the possibility of the whole), and at the same time the human being appropriates the knowable being at hand in his/her own ex-isting (being there). In short, when one lives authentically, one responds to the particular "world" in which one finds oneself in the present moment, spinning out of this "being-there" a known in the midst of the unknown, bringing Being to be within the temporal horizon of one's own lived world.

The "spirit" of this mutual appropriation infects the persona as he listens to the song and reads its significance phenomenologically. At the song's end he asks a question that he goes on to answer:

> Ramon Fernandez, tell me if you know
> Why, when the singing ended and we turned
> Toward the town, tell why the glassy lights,
> The lights in the fishing boats at anchor there,
> As the night descended, tilting in the air,
> Mastered the night and portioned out the sea,
> Fixing emblazoned zones and fiery poles,
> Arranging, deepening, enchanting night.
>
> Oh! Blessed rage for order, pale Ramon,
> The maker's rage to order words of the sea,
> Words of the fragrant portals, dimly-starred,
> And of ourselves and of our origins,
> In ghostlier demarcations, keener sounds.

Stevens claimed that Ramon Fernandez "was not intended to be anyone at all. I chose two everyday Spanish names. I knew of Ramon Fernandez, the critic, and had read some of his criticisms but I did not have him in mind."[8] But as Joseph Riddel says, if this is so "then we are witness to the most astounding instance of critical telepathy in literature. . . ." Riddel points out that Fernandez' aesthetic theory was "a precursor of recent phenomenological criticism."[9] And Stevens admits to having read Fernandez' weekly columns in a Paris newspaper.[10] There is no accounting for Stevens' reluctance to admit his direct reference to this aesthetic theory and its ontological foundations. No doubt Ramon is "pale" because he has recognized the fundamental ontological insight disclosed by the woman's song.

Having recognized this ontological significance, the persona turns to see the immediate scene taking form moment-by-moment against the deepness of night and finally "deepening" and "enchanting" night with its "[arrangements]." And he recognizes that this happens *because* he turns toward this scene with an awakened sense of his own be-ing as a participant in the being of all that is. This sense has been awakened by the woman's song (and by his own meditations upon it), for that song disclosed, like every work of art or artifice, that the world is always a project of the human rage for order – that project of meaning by which *Dasein* opens a clearing in the depths of Being. The lights of the fishing boats that portion out the night and the sea, "Fixing emblazoned zones and fiery poles," are (remarkable coincidence) a vivid figure for this clearing. Further, as the night "[descends], tilting in the air" and the lights of the fishing boats "Master the night," we recognize an astonishing parallel to Heidegger's analysis of the artwork as a place of the conflict (and unity) of "earth" and "world," the known and the unknown.[11]

We also sense the temporality and even the ephemerality of this sense of order, which occurs as one temporal phenomenon (nightfall) touches another (the lights of the fishing boats) as a particular individual turns toward the scene, composing it as he beholds it. What he sees is significantly not a work of art but a matter of perception, yet it is a "making," albeit on a humbler level. (We recall that in Merleau-Ponty's phenomenological philosophy, "the same creative capacity which is at work in imagination and in ideation is present, in germ, in the first human perception.")[12] And this "making" at the level of perception has the same ontological significance as the woman's song.

This ontological significance is expressed again in the imagery of the concluding lines, where the persona's awakened sense of order (a sense developed phenomenologically as he sought to discover the role of the sea in the woman's song) leads him to recognize that "words of the sea" (human descriptions of the sea, of the near-at-hand) *are* "words of the fragrant portals, dimly-starred" (of the gates – the horizon – of the knowable and thus of the imaginable) and thus are words of "our origins" (since we know, by virtue of the knowable and the not-knowable, our condition as those beings that clear a place in which a limited knowledge of Being can arise). In short, having recognized that order is a matter of human perspective, growing out of human need, the persona describes the expression of this human "rage to order" as a "making" that must begin with the near-at-hand (in this case the sea), that necessarily implicates the horizons of knowable being, and that thereby discloses how we ourselves come to be: by virtue of the things of the world, the things at hand, and our perspective upon them.

In 'The Idea of Order at Key West' and in Stevens' poetry in general, it is the sense of our fundamental ontological condition that is the source of our feeling confronting nature. Examining that feeling and the works (of perception, of artifice, of art) that it inspires leads us, as it leads the persona of 'Idea of Order at Key West,' to an understanding of this condition and a deeper commitment to the openness of our own be-ing.

*Florida State University*

## NOTES

[1] Wallace Stevens, *The Collected Poems of Wallace Stevens* (New York: Alfred A. Knopf, 1955), p. 349; hereafter cited in the text as *CP*. Poems reprinted from this collection are by permission of the publisher.

[2] Ronald Sukenick, *Wallace Stevens: Musing the Obscure* (New York: New York University Press, 1967), p. 148.

[3] 'Someone Puts a Pineapple Together,' part of 'Three Academic Pieces,' in *The Necessary Angel: Essays on Reality and the Imagination*, by Wallace Stevens (New York: Vintage, 1951), pp. 83–89.

[4] Maurice Merleau-Ponty, 'The Primacy of Perception and Its Philosophical Consequences,' in *The Primacy of Perception*, ed. James M. Edie (Evanston, III.: Northwestern University Press, 1964), p. 22.

[5] Wallace Stevens, *Opus Posthumous*, ed. Samuel French Morse (New York: Alfred A. Knopf, 1977), p. 162.

[6] Harold Bloom, *Wallace Stevens: The Poems of Our Climate* (Ithaca, N.Y.: Cornell University Press, 1976), p. 104.

[7] See Albert Hofstadter's explanation of this concept and his translation in his introduction to Martin Heidegger, *Poetry, Language, Thought*, trans. Albert Hofstadter (New York: Harper Colophon, 1971), pp. xix–xxi.

[8] *The Letters of Wallace Stevens*, ed. Holly Stevens (New York: Alfred A. Knopf, 1966), p. 798.

[9] Joseph Riddel, *The Clairvovant Eye: The Poetry and Poetics of Wallace Stevens* (Baton Rouge: Louisiana State University Press, 1965), p. 117.

[10] *Letters of Wallace Stevens*, p. 823.

[11] Heidegger, *Poetry, Language, Thought*, pp. 55–57.

[12] Merleau-Ponty, 'The Primacy of Perception,' p. 40.

# C. METAPHYSICS OF THE POETIC

MEENA ALEXANDER

# "FALLINGS FROM US, VANISHINGS . . . ": COMPOSITION AND THE STRUCTURE OF LOSS

What follows is elliptical, often hesitant, searching out a track of thought: the pure naming of poetry and the loss which inscribes it. Reflection strays beyond the borders of the known, identifiable self, into a half-lit realm exquisitely evoked by Milosz as the place where "The clothes of my name fall away and disappear."[1]

I would like to start with Rilke, with the well worn Ninth Elegy. His lines repeat in the mind, echo in the inner ear. The syllables translated from another tongue take on the texture of this. The words are precise in the manner of taut speech yet filled with multiplicities of meaning. They remake themselves with each utterance:

> For the wanderer doesn't bring a handful of that unutterable earth from the mountain-side down to the valley, but only some word he's earned, a pure word, the yellow and blue gentian. Maybe we're here only to say: *house, bridge, well, gate, jug, olive tree, window* – at most, *pillar, tower* . . . but to say them, remember, oh to say them in a way that the things themselves never dreamed of existing so intensely.[2]

How clear Rilke's tones are, lending us words, isolate words of our everyday imaginings: "*house*," "*bridge*," "*well*." But what does it mean to utter them as poetry?

The wanderer, he or she in search of a source, cannot bring back the soil of the mountain slope. No, you cannot carry back that clump of soil with you, earth dark and fragrant. It would mean nothing.

Strange as it may seem, the stuff of the place you were in, the substance of your mountain, lacks meaning. The structure of loss is woven deep into life.

To be human is to place one's trust in words, in the fragile multiplicities of words whose meanings spring from loss. To earn the pure word, in Rilke's terms, is to pluck it, in fear and trembling, out of the perishing world.

Both self and world are in the grip of a relentless mutability. The great pillars of Eternity Edmund Spenser invoked in his '*Mutability Cantos*' are across the black waters. All that moveth, doth not in change delight. Now, not only does the soil of revelation vanish, but the mountain itself and the self that saw the mountain are both fleeting. The only permanence is that of



*A-T. Tymieniecka (ed.), Analecta Husserliana, Vol. XVIII*, 91–97.

rock and root perpetually transfigured out of the meaning they evoked in us.

And so it is, that later in the same poem, Rilke tells us that all things whose "lives are lived in leaving" bear a great tenderness for us. They turn to us, the most perishable of all, for help. "They understand when you praise them."[3]

Words, numinous precisely because their sense springs from loss. A structure of being composed of significations that have their source in what is no longer here.

Yet grace comes and with it a difficult truth of which the listener must be reminded: "but to say them, *remember*. . . . " The tone of the verb is heightened in the need to comprehend. The intensity of understanding ("aber zu *sagen*, verstehs," says Rilke) can come only by plunging into the depths of a mutable self whose utterance paradoxically gives life: a vitality so intense that the lost substance could barely have longed for it.

Almost two decades later, Wallace Stevens struck this very chord. "The poem of the mind" must speak so purely that its syllables will touch off those "sudden rightnesses."[4] Exquisite balances in which desire is perfectly disclosed.

Are we so very far from Husserl who held that "All reality exists through the dispensing of meaning."[5] Aren't we reflecting on the most intimate sense in which meaning may be bequethed?

But what does it mean to bequeath meaning? After the Romantic revolution this has become the essential question of poetry and our ordinary lives are not immune from it.

In paradise, there is no need of language. Spirit is not cleft from body, the senses touch and taste and delight. Being is full. The erosions of meaning have not touched the landscape. To compose poetry, to need to compose, is to touch the gash of a vanished bliss. Meaning-making, the fragile forays of art, are a function of loss.

To identify poetry as pure speech is to recognize that the name you utter is not the thing. Naming is the finest negativity known to us. Composition is bodied forth by the structure of loss. The more perfect the perception of loss, the finer the composition. We live in falling.

To speak of joy at this point seems presumptuous, yet joy comes, unbidden. A fragrance of light, an incipient vision. It is the horizon of the world that greets us, at the very edge of our trembling.

I turn to Rilke. His starting point is our precariousness in this interpreted world. Yet there's "some tree on a hillside we can look at / day after day, one of yesterday's streets. . . . "[6]

If to compose is to live with loss, to chisel meanimgs in the half-darkness of what is no longer there, it is also in that very act to recognize the tree on the hillside, a street, the wall of an abandoned house. In short, the world that is already there.

The world of depth. The world that exists outside us, beyoud the vacancies of loss, which the voice must greet however timidly, if perception is to survive.

This recognition is like grace, it grants the voice a tenuous light. At the threshold of an achieved composition it may even situate the self.

We are speaking now of the world that comes after the suspension of faith in solidities, the world after the epoché of the creative act. Without it, the voice cannot truly exist. All great poets have realized it and Rilke was no exception. He knew that this source, already there, could only be touched in the perilous realm of loss. It came lit by an acute emptiness: "You *still* don't understand? Throw the emptiness in your arms out into that space we breathe; maybe birds will feel the air thinning as they fly deeper into themselves."[7]

It's not simple. One takes time to understand. But when our vacancies touch the outer air, even birds those creatures of a final freedom will fly more purely into themselves, recovering what they had not known they were.

The already there, if only for an instant, is touched and transfigured in the knowledge of loss.

For Rilke, naming house, bridge, well, gate, is to touch our loss and out of it, build the fluent architecture of poetry. But though the treasured objects will rise more intensely within it, the world that lies around, beyond us, cannot be foresworn: "The lost are always searching / for the earth again."[8] When the poet cries for a lover, he cannot limit his call and many girls rise out of crumbling graves. Underfoot, however high one leaps, there is always the plenitude of ground, trembling. Without it the voice could not leap. Nor without it, could it fall, ceaselessly, in the curious joy of falling with which he ends the 'Elegies.'

Wordsworth, it seems to me, understood this ground, this realm of depth, better than many. In touching this depth, forever displaced by our longings, we quicken what is lost; the wound becomes visible and the voice spells out the coiled, hidden self and its moments of vision.

> But there's a Tree, of many, one
> A single Field which I have looked upon,
> Both of them speak of something that is gone. . . .[9]

We understand the details of desire. It is this particular tree and no other; this single field that I have known and loved, light flashing on the soil, bent grass, tiny runnels of water. To say tree, field, is to evoke what has vanished. The task of the voice is to track the course of loss, to render luminous what is still held in memory. But the treasure house of memory is not enough. And this is crucial.

Beyond the *Pothos*, the lingering nostalgia for what is lost, the pain of a past, we find a *Himeros*, a burning intensity of desire for present being. But this desire is not for anything, tree, field rock or person. Rather, it is directed towards the darkest refinements of our knowledge, the groping, pulsing edges of what is given; the irreducible awareness of being human in a world which is neither solid nor all here:

> . . . those obstinate questionings
> Of sense and outward things,
> Fallings from us, vanishings. . . .[10]

The final figuration is death. "Fallings from us, vanishings" are intimations of perishing; stoic, visionary sensations. They are Wordsworth's greatest discovery.

In 'The Prelude' he turns to the sources of his power, those "spots of time" when vision was most luminous. Childhood was full of them.

Layer after layer of feeling evaporates. The voice, tracking itself, cries out: "Oh! mystery of man from what a depth / Proceed thy honours. I am lost. . . ."[11]

The landscape, the ground, instead of being a simple guarantee of the self, as might have been imagined, comes to stand for that infinitely more complex condition: an anonymity, a falling. The voice hesitates, turns, glimpsing the arches through which its being pours.

It's a world without clear directives. There is only the wind blowing its difficult way through a woman's garments; a naked pool; a beacon on a hill. Icons, their sense as yet hidden, their names fringed by darkness.

Beyond this "visionary dreariness" there is nothing. Past and future are vapourised in this perception of our poverty. The voice that names, elaborates and rejoices, is also the voice that strips and peels, that brings us face to face with "blank misgivings . . . worlds not realised."[12]

I have deliberately set aside the various visions on the mountain side that Wordsworth was blest with; recognitions of the soul's glory. Set them aside to show a more difficult truth that came to the poet: his voice as a selfless, quivering impermanence, transfixed by the depths of unknowing, by the dark invisible that seeks us out. The rarest of beginnings.

To begin is to relinquish the known self. It is to fall through "sad perplexity" as Wordsworth calls it in 'Tintern Abbey,' through dim recognitions, "gleams of half extinguished thought," until the "picture of the mind revives."[13]

And the voice trembles, pitched beyond itself, till at the very end it is transfigured into the site of bliss, becoming addressed to an intimate other, a "thou."

We glimpse this in Rilke too, the poignance of address in a realm of perishing: "*Love*, the World exists nowhere but within. . . ."[14] The voice has passed the ordinary limits of identity, a great and clarifying emptiness is touched.

But we cannot forget Wordsworth's "picture of the mind" so easily. What was it? Beyond the real solid world of images, the world of depth that nurtures perception, lay loss. Loss precisely of the ground of those images, loss of a self empowered to taste its own past and rejoice. A reduction so acute that the voice was pitched to a nameless, unutterable region.

There (we are once again on a mountain) time gathers itself together. "Characters of the great Apocalypse"[15] are revealed. But what do we hear? The winds thwart themselves the poet tells us, "bewildered and forlorn," the stream is sick and raves and the prospect giddy.

To plunge to the depths of naming is to purge oneself of time's constancies: name, place, station and the clarities of recollection. The imagination is radically unhoused, it becomes an "unfathered vapour."[16]

Milosz the poet of exile, who longs for a spot beyond Lethe where memory might be healed, recognizes this desolation. The voice has trespassed its name; the clothes, the syllables of ordinary identity drop away and disappear: "Again the other, unnamed one, speaks for me."[17]

Who is this unnamed one? There's no name anymore and in wide waters the stars dwindle.

In the working notes he made before his death Merleau-Ponty speaks of the unnamed one, the 'I' disclosed at the finale of all reductions:

> The I, really, is nobody, is the anonymous; it must be so, prior to all objectification . . . the unknown *to whom* all is given to see or think, to whom everything appeals. . . . It is therefore negativity. . . .[18]

When all that has meaning is stripped away, we stand face to face with an ancient anonymity which phenomenology cannot afford to forget. It's the source of sense Merleau-Ponty tells us, the "invisible meaning" and even further, the "inner framework of speech."[19] He is Husserl's heir: we are given the dark imprint, the obverse of the Transcendental Ego. But how will sense hold together? How is this vital anonymity displayed? The answer is not so very different from that of the poets.

For Merleau-Ponty there is a world af depth, a wild-flowering world born of the "original ecstasy."[20] It cannot be subsumed under any categories, it is ceaselessly displaced in our acts of meaning-making and regroups, ceaselessly, forging the bounds of the visible. He calls it the "flesh" of the world: depth, the ground of imaging.

The task of genius is a stripping and peeling, an entry into that radical and anonymous freedom where alone Being is inscribed.[21] The artist must therefore decreate the self: invoke power for the image. Step by step, in patience, the text is forged and it is only at the very end that the miracle of sense is disclosed, the "crowned" by light.

There's a poignant tale embedded in the *Phenomenology of Perception*. The young Cézanne was haunted for many years ("throughout my youth" he says) by the image of a glorious dinner table. It came to him from the pages of Balzac's *Peau de Chagrin*: "a white tablecloth like a covering of snow newly fallen, from which rose symmetrically the plates and napkins crowned with light coloured rolls."[22] The young painter longed to capture the image, encode it with paints, on canvas. It took him many years of struggle to realize that the symmetrical arrangement of plates, the napkins crowned by rolls, could not be trapped through mere will." If I paint 'crowned,' I'm finished you see," the older Cézanne finally realized. It was only when he balanced and grouped his images in the invisible architectonic of sense, mixing pigments, perfecting tints, that the barely longed for illumination occured, and the cipher was perfected.

What is Merleau-Ponty's comment? The problem of the world he tells us, and of our body, is that its all already there. But the already there, as the older phenomenologist was to recognize, must be lost, cast aside, renounced, struck by lightening till finally in the peril of vision we touch the mind "quiet as water in the fissure of Being."[23]

What is the "fissure of Being"? What does it hold for the frail ciphers of our poetry?

It is the dark underside of all naming, a vital negativity and it clasps the poet to a perpetual groundlessness. Through it, the structures of loss transfix composition. The architectonic of sense is lit by a nothingness.

In the Romantic-Symbolist tradition, in Wordsworth and Rilke as we have glimpsed, nothingness comes to stand for a perilous truth, Stevens' "vital assumption."[24] Or it is the fragrance of lost hawthorn that irradiates Proust's *A La Recherche du Temps Perdu*. As Merleau-Ponty reminds us, for Proust "the *true* hawthorns are the hawthorns of the past." And this past is inexistent, mythic, "farther than India and China."[25]

*Fordham University*

NOTES

1 Czeslaw Milosz, *Bells in Winter*, trans Milosz and Lillian Vallee (New York: Ecco Press, 1978), p. 38.
2 Rainer Maria Rilke, *Duino Elegies*, trans. A. Poulin Jr. (Boston: Houghton Mifflin, 1977), p. 63.
3 Ibid. p. 65.
4 Wallace Stevens, 'Of Modern Poetry,' in *Collected Poems* (London: Faber & Faber, 1966), p. 240.
5 Edmund Husserl, *Ideas, General Introduction to Pure Phenomenology*, trans. W. R. Boyce-Gibson (London: Allen & Unwin, 1952), p. 168.
6 *Duino Elegies*, p. 5.
7 Ibid, pp. 5–6.
8 Ibid. p. 49.
9 William Wordsworth, 'Ode, Intimations of Immortality from recollections of early childhood,' in *Poetical Works* ed. E. de Selincourt (London: Oxford University Press, 1969), p. 460.
10 Ibid., p. 461.
11 Ibid., p. 578.
12 Ibid., p. 461.
13 Ibid., p. 164.
14 *Duino Elegies*, p. 51 (my italics).
15 William Wordsworth, 'The Prelude,' in *Poetical Works*, p. 536.
16 Ibid., p. 535.
17 Ibid.
18 Maurice Merleau-Ponty, *The Visible and the Invisible* trans. Alphonso Lingis, ed. Claude Lefort (Evanston: Northwestern University Press, 1968), p. 246.
19 Ibid., p. 224.
20 Maurice Merleau-Ponty, 'The Philosopheı and his Shadow,' in *Signs*, trans. Richard C. McCleary (Evanston: Northwestern University Press, 1964), p. 174.
21 *The Visible and the Invisible*, p. 197: "Make an analysis of literature in this sense: as *inscription* of Being."
22 Maurice Merleau-Ponty, *Phenomenology of Perception*, trans. Colin Smith (London: Routledge & Kegan Paul, 1962), p. 198.
23 *The Visible and the Invisible*, p. 235.
24 Wallace Stevens, 'The Rock,' in *Collected Poems,* p. 526.
25 *The Visible and the Invisible*, p. 243.

ZYGMUNT ADAMCZEWSKI

# POETIC THINKING TO BE

> Only the poet will never abandon you, insult you, shy away from what you dream or do – faithfully!
> (Stefan Żeromski)

The following phrases, paragraphs, pages, are intended to approach that amazing landscape which is yet no maze for the open-hearted, that unexplainable vision which is but too plain in its truth, that impenetrable terrain which yields with such ease to penetration with feeling: the charmed land of poetry. Approaches to it are ever invitingly free to all but hopelessly prosaic thinkers. But to attempt any such approach demands care, forbearing and foresight, all the way.

How to seek access to what is poetic, in words of our time? Simple paths seem obstructively barred through complex constructions of a bewildered technological age: computing mechanization, robotic data processing, leveling mass media, stultifying ad propaganda, obsessive ideologies, irrelevant quantifications, petty linguistic analyses, impotent pseudo-philosophizing. It may well be that to pass the darkly thickening thickets of this century's neurotic verbal promiscuity the whole aggrandized and aggravating Gordian tangle has to be cut through to its roots – in the ground of the human being, thought and speech. This needs readiness to break in the attempt. Qui vivera verra.

What is poetic?

A gateway for the present approach opens on Martin Heidegger's call for this difference in human awareness: ontic against ontological. To ward off straight away any misapprehension, no abstruse technical jargon will result. Heidegger's difference – which makes a difference – is to be taken not as simple-minded but as open-minded; it is a wide difference, and there is no wider one. Two words are offered, with a corresponding pair referring to existence not available in English; they both address all human awareness, displaying for it not two opposed directions but rather two interlinked dimensions, conflicting in life no more than male and female. Heidegger appreciates as well as Ryle that real dualities dilute when "taken with a pinch of salt." They are two, but confusion between them is not more likely, or



*A-T. Tymieniecka (ed.), Analecta Husserliana, Vol. XVIII*, 99–116.

unlikely, than between the sexes. In a less playful but not less fruitful vein, the difference may be exposed on these planes: ontic awareness grasps and disposes things, facts, affairs, events, entities, ever plural, ever specific, ever determinate for the human being and possibly for other living beings; while under-lying ontological awareness, possibly distinctive for humanity, unifies and holds together its living, existing, dwelling in the world at one time, at one as such with being. The two planes are not separable, any more than the sexes for life; since clearly to be humanly aware is to tend to the manifold of entities – anything other would be less than a dream – and yet at the same time it is to be one human being: alive, conscious, experiencing as only he or she can. In its reaches ontological awareness aims at a sense, humanly yielding, of what it means for a human to be; or, to give it more ground, how you or I or we stand forth toward being. Being: if the word takes you off your feet or turns you off your bearings, you need not gasp; in what you will hear, that word will be put off, though not out.

How on earth can there be seen an opening for the attempted approach? Still on earth, safely on earth. The guiding ray from Heidegger, however dimly it may shine in your eyes, alights on this guessing try: it is human ontological awareness that weaves the threads of poetry most poetically. Discounted here can be jingles, greeting card verses, jabberwocks, and the like offshoot trifles. The cut attempted has to be clean, not a faint jab. Whether the dim shine tried onward will expand ahead into a larger clearing, remains to be seen.

The approach path now has to unwind, slide or climb on; the trying will hardly be proved, yet improved, perhaps approved; the guard needs to be clear-sighted against objective obstructions, hidden hurdles, terminological traps, and many of these have to be bypassed, leaped over or disarmed, while the ray lights the way.

With regard to an ontological grounding of poetry, the first objection may be that any such linkage would be too abstract: that it would amount to a philosophical intrusion, a conceptual extrapolation, a theoretical superfluity. As much as the difficulties to come, this attitude harbors the suspicion that the thinker and the poet are alien to each other, and their confrontation turns mutually disabling. To gentle down such tension, it may help to raise instead the image of those dwellers apart on neighboring mountains, whose friendly appointments point out efforts of descent and ascent, but whose high air in-between extends not only distance but closeness, expecially if each may try on wings of a kind. With that perspective, it may be possible to take another look at the notion of abstraction. It need not stray to vanish in other-wordly clouds; as a Latinate word, it can come to signify a drawing out of ideas, a pulling over of intent, an imaginative teasing on, an unveiling promise, a clear-eyed development. If such a view of abstraction might be plausible at all, then it could entice you to gaze with such awareness toward what is indeed poetic. An illustration would add some lustre. Here is the lonely musing of Hugo von Hofmannsthal's Marie-Therese:

"How can it really be so that I was once the little Resi and that I too shall become an old woman? . . . Just how can that happen? Dear God, just how can you do it? While I am, ever, the same one. And if you must do it this way, why let me look on, myself, with such clear sense? Why not hide it from me? All this is mystery, so much mystery. And one is there to bear it. And in the 'how' lies the whole difference. . . . Time – on its own time alters nothing in things. Time is a remarkable matter. When one lives on so, it means just nothing at all. But then, suddenly, one feels nothing but time. It is all around us, and it is within us, too. It moistens faces, it pours through a mirror, it flows in my temples. It flows between you and me, soundless like a sand trickle. . ., Sometimes I hear it flow – without a halt. Sometimes I get up in the middle of the night and stop all the clocks, all. And still, one must not fear time. It, too, is created by the father of us all."

This appealing Viennese lady would not know anything ontological, if it bit her suddenly. And yet here she is, bitten and smitten by the tempting terrors of time, rather than any temporal things, in an all too human awareness of flowering and aging, of living and dying. Poetic? With striking simplicity. Abstract? If you please. . . . The question dawns: have any theories of time drawn you on this far, pulled you so intently, touched you as tenderly as this girl's insight? For time, remarkably, can moisten your face, too, as you live on.

The second treacherous trap, threatening in the specialized scientific subservience of this age, can be the reservation against an ontological origin for poetry as directing it in terms inapplicably strict, exact, rigorous. Again, it is possible to weigh in thought this terminological burden. On rigidity, even short of rigor mortis, more strictures might be stated; clearly, it has to be alien to the permissive pliancy of the poets' passions; nor need it straightjacket what can be exactly strict in a richer sense. Strictness rewards better not when it is constricting: rather when it gives free reign to the strain of tending efforts in genuine human striving; when it steers the stringing of forces, of fragile nerves, of excited aspirations, of quickened heart-beats; when it is not tamely technical, logical, or technological, but instead strenuously alive, awake, alert, fond of its feelings and finding them full. Such stringency is the strong hold of awareness, that exacts it exorbitantly to its own exposition in the light of existing; therein the poets steep themselves, stripping their heart to "found that which can remain," human and humane, with Hölderlin. Again, to invoke the aid of poetic speech, from a ballad by Adam Mickiewicz:

Feeling and faith can yield and expose
More than glass eyes of a man who 'knows.'
People ignore the dead truths you ponder,
This world you split as dust does stars part;
You have no live truths, will see no wonder:
Have the heart live – look into the heart!

Unless the anglicizing diminished for you the demanding inspiration here, its aim is transparent. The poet who never viewed tele-photos, micro-dots or laser beams, visualizes acutely the exacting perspicacity of a felt and filled awareness where the heart fittingly follows the living fullness of the world. The heart, if you will have faith in it, is manly as well as womanly: it is all-human. The question, in the stronger light: does smug scientific technocracy let you glimpse the faults, failures and fooleries in its ontic splintering and separating of things, as you can see this with the saner sensitivity of poetic vision? Should you wonder – you have but one heart and one life.

A third obstruction on the path pursuing the poetic possibilities ontologically: such an allotment would construe the poets' utterances as universal, uni-form, un-individuated. The Platonic skeleton of that prohibitive structure has been pre-formed and pre-vailing for centuries, with a pauperizing imposition of sameness, of likeness, of indifference. And yet what is and can be universal need not be rigid repetition, anonymous monotony, numbered or numberless stagnation; by contrast, it may be: spreading and sailing, winding and wandering throughout the wholeness of the world – the only universum. The world may not stand still, need not petrify, will not bore human awareness; not if the poets can help it. Listen to the greatest:

"I could be bounded in a nutshell, and count myself a king of infinite space. . . ."

The multi-hued mosaic of a poet's imagination proudly appropriates private shells and shaded valleys and silent angles with equally pulsing penchant as roaring ranges, desperate deserts, and siren seas: the world is anything but poor when prized in song.

Now beauteous Ophelia: "We know what we are, but know not what we may be. . . ."

Not only is the world enticingly vast – even so inexhaustible are prospects, permutations, privileges, potencies and pains of its human habitants: the like-ly, the love-ly, the love-lost, the lost-found, the founding fermentations of the possible – pleasured, plumbed, played poetically.

Hamlet again: "This goodly frame, the earth, seems to me a sterile promontory; this most excellent canopy, the air, look you, this brave o'erhanging firmament, this majestical roof fretted with golden fire – why, it appears no other thing to me than a foul and pestilent congregation of vapours. What a piece of work is a man! . . . the beauty of the world! the paragon of animals! And yet, to me, what is this quintessence of dust?"

Poetic awareness uni-form, un-individuated, unequivocal? Only those sworn to and cursed by prose could so limit their short sight, unrescuable by any glasses. Because for the per-forming, many-voiced sentience of the poet's heart there are hardly enough colors in the rainbow, shining through air, on earth, for the human animal, whether radiant or sombre, rising, or setting be the light of the world.

But – the reticent while rarely inarticulate voice of Portia: "O me, the word choose!" .

As universal appeal, the poetic call is not for discoloring effacement, chain-linking equalization, sterile eradication of spontaneity, but on the contrary, for decisive distillation of the wealth of the world, so abundant that it can overwhelm differing desires, ambivalent attractions, fighting fascinations by its embarras de richesse. Male or female, ambitious or submissive, pedestrian or perverse, all humans may choose, with temerity or with trembling, not as puppets played on but as players of their choice.

And the magic wish of Prospero's adieux: "Gentle breath of yours my sails must fill, or else my project fails, which was to please. . . . "

The poet: no demonic despot, no pedantic pedagogue, no manipulative manager; the poet pleads to please, labors for love, evokes enthusiasm, induces intercourse of breaths and sighs, of cries and smiles, to fill and own and claim the world. The lightning question: is poetic awareness ontologically universal? In truth it is, and multiversal, and magniversal, and beneversal, for you as human beings may all feel good in this great world, all in one and thereby not alone: the heights of your habitation staked out, safeguarded, and signified poetically.

For this attempted approach to the abode of poetry, in circumventing the previous obstacle, the risk is near of stumbling into the next one, which lies in quite a divergent direction. It threatens with this: an ontological derivation of that which speaks in poetry would determime it as historically dependent, traditionally bound, heteronomously fixed in its nature by doctrines, definitions or dogmas invasive, foreign, remote. Falling into this treacherous trench could be prevented by an initial regard lucidly reflecting history. History is not the surpassed past, haunting spectrally with stiff hands to settle human habits in a heteronomous set, ossifying dead tradition as determinate nature; it is nothing if not happening. But letting happen, having it happen, caring that it happen – could that be monopoly of monarchs, legitimization of laws, supremacy of schools, closure of codes, standardizing of sciences, corrupting conformity of crowds? A Nietzschean perspective can see through the surface of these strata: higher history is humanly given forth in saints, thinkers, and poets. Dynasties are degenerative, ordinances disorderly, technicalities tedious, dictates depraving, mobs immobilized. Historic appropriation of what happens can only be solved, dissolved, or resolved by autonomous, art-full, agent human beings.

To confirm such historical concourse – Albert Camus: "Do not wait for the last judgment; it happens every day."

Eschatologically expecting an end – beyond move and act and life – may not contend with the content of moving and acting and living; whether contenting or discontenting: judged. Happening is judged here and now; heaven as well as hell found in other humans; the heart in its felt immediacy can intimate the good and the bad, the needed and the wasted, the graceful and the hideous. So it happens: c'est la vie.

To live and not to lie on life – Charles Baudelaire: "The true travellers are those who part only to depart."

There is a way to go, to travel, to traverse, long or short, devoted or reluctant, arduous or glorious; for a human this is the way to go, go on living for the happening of life; to part is to die a little, yet to stop is wholly to die; to know where the way leads is to pretend prophecy but not to know is to have it happen for you.

To seek no excuses in de-naturing nature – Voltaire: "This is a mockery; there is no nature – all is art."

Nature is where it is sought and found, nature is as much within as without; nature is not just law and force and mechanism, but for a human play and fancy and adventure: it is art. Humanly existing is to take part in the emergence of nature, to co-create it, to let it happen naturally, genuinely, exuberantly, to care for it all as long as you shall live, because however it has come to be, it is now your heart's way to be, your art swaying history. The question, by the caressing soft light of spring: in defiance of any states, statutes, or statements, does not poetic awareness on its own move with resolve to favor the way of human flesh as it happens to come and go – still on earth; is it not your flesh, your blood, your sex, your thirst, your phantasy, your joie de vivre? Unless you happen to come elsewhere.

Going still, moving on, catching breath, sighting far the land of promise, hoping its lay is plainer, planed down, explained, but not away; almost there perhaps? Here is another catch, another twist, another strait; out of precedent reflections, the consequent deflection might still distance a clearing discoveredness. If not faulted as abstract, strict, universal, historicized – poetry springing from subtly shaping ontological sources might be suspect as unpredictably subjective, confounding classifications, comparisons, criteria for comprehension; that would license indulgences, inconsistencies, insecurities. Providing some light to soothe such suspicions involves a flexible rethinking of subjectivity: that uncertifiably certain center of so many modern modulations in conceiving consciousness, concealingly Cartesian. Is the human being a set solid substance, staying safe like stars or stones or sediment? Is it a fleeting fickle flux, failing firmness like falls or fairies or foam? Or is it something else, not at all subject to such surreptitious subordination among extraneous settings of entities? "The proper study of mankind is man"; Alexander Pope's sensible suggestion has been rather improperly and steadily subverted, save perhaps by the poets. On the present path, does some light avail already to charm and dispel such dispropriation, with an ontological ray? Try and try agin.

Being human, if poetically spoken out: that is to under-stand a world, under-go a time, under-take a happening, to be responsive to and for – there being the over-flow, over-coming, over-sight of existence. To be subjective is to be projective, adjective, injective toward being there at all. To dwell between heaven and earth is to turn round in-between the graceful and the grateful. To be a human subject is to live not as a slave nor a master, not in black shadow and not in full sun, neither a solitude nor a plenitude. Subjectivity is ever being temporally fore-cast, heavenly fore-given, worldly fore-seeing, earthly foregone, and lively fore-playing.

For such being in-between under and over, encompassed and endowed, shamefully covered amd barely discovered, what does the poet appropriate, with his full heart, his tender touch, his impassioned voice? Through poetic awareness, this challenge to the sphinx has to be amazingly pushed into speech; poetic speech has to exclaim in sound the expressiveness of existence so as to echo it eclectically; poetic embrace for humanity has to surround it, behind as before and beyond; thus to re-call, unveil, and dis-play, with fore-feeling. Poetry is pre-senting: sentient pre-paring of passing sensation, pre-shaping of potential sense, pre-viewing of prevailing sensitivity. Thought ontologically, the poets are pre-cursors of people's being there.

Can any such potent purposes presumingly pervade a poetic presentation? Here, proclaiming a pride of poetic points for speech is Juliusz Słowacki; his words' fluidity may not be caught in an English netting:

The aim is this: that the pliant tongue should say all that the head might ever think; sometimes blind brightly like a lightening flash, and then cry sadly like a prairie song, sometimes lie softly like a nymph's lament, then reveal beauty of the angels' speech; it should soar over all on spirit's wings; a stanza must show tact, not holding strings. To draw all from it – shades of longing dusk, or a serene shine with a silent glow, then bring it forward in a radiant gold, or in archaic structures' overblow, or weave on subtly like a spider web, then shape it roundly in an earthen mold, like sparrow's nest to a tree branch glued on, that sings each morning to the rising sun. . . .

The speech of a poet is of one who under-stands the over-flow of the world, and subjects it to the yield of an im-pressive sentience, who is cast fore into the clearing of existence and fore-casts gratefully a response of choice words, who fore-plays a pre-lude of people's awareness and pre-sents for them a grateful welcome to the grace of sun-rise. The question, on landing now in the time of light: is not the poetic gift to play round with all there is and for all who humanly are, to sing from a fully sub-jecting heart about the super-abundant wonders of life giving a measure of plenty to the variety of human senses, to touch each and fore-feel any and speak out for every human awareness? There is a hope that your responsive silence may be compliant, contemplative, concurring . . . .

What is your pleasure on arrival, if it be arrival? Not un-like-ly, a pause. What land could have been reached, would it show an easy lay, or swampy or iced or rocky? Exploring onward may invite you, at your convenience and your risk. The way thus far never projected a proof, rather a provisional probe, not a promulgation, only a proposal; but perhaps not without promise. If you find the Heideggerian ground of ontological awareness not entirely arid for poetic seeding, then your own flowers may yet blossom in this soil still assayable. Anything possible is ever so more exciting than all things necessary.

With your enduring patience, the present perspective may be extended into some further reconnaissance, aligning a brief glance at literature with a rather neglected vista of language and then a sweep of the cloudy horizon of thinking; the latter, as you may have expected, has been kept in view from the outset of this adventure, aiming for some airy passage between the high summits of poetry and philosophy. But still on earth, kindly follow.

If this might be a proper poetic way, what place does it find in literature? The question is vaguely perspectival. So many directions can issue from the placement of literature itself. How literal or literary? For the literate or for the literati? In lettering, in letters, or in belles lettres? The placing is so amorphous that some displacing is almost mandatory. Is literature in "readers" or is it for readers? Is literature fiction or is it a fixation? Is literature filmed propagating or prostituting? Does literature offer men mediation, mass mediocrity, mind's meditation, or a mad medley? Hardly an accommodating territory to settle on.

In such a shapelessly pluribund field, one plausible prospect for localizing a goal of discourse appears to emerge from the previously pertinent path of convergence: one human being, one world, one awareness. With this orientation, it is perhaps not preposterous to point tentatively toward one sense for literature, to be construed on some such provisional lines: literature is the cumulative written record of the events and experiences, fortunes and misfortunes, actions and passions, in the lives of men and women, pursued in many languages, by many individuals, on many levels, at various times, through various cultures, with various intentions – yet to be understood by all humanity. If this lay-out does not offend as preclusively narrow or wide or vacuous, then bridging it with the preceding point of departure may come into sight. So to reconnect it was suggested early that on ontological ground human awareness acts, attends to, acquires a grasp of all possible entities, facts, affairs, available in the world, in multiple, mutable, manifold ontic arrangements. Out of this eventuates the recording of literature. Such devolution may well hold fast. The field of literature can be surveyed as available, accessible and arable with the vitalizing spread of the poetic element current throughout it. This differentiation is by no means intended as a bifurcation: literature lives with poetry. Nor is it, though, an identification: literature does not share all the aims and ambitions of poetry. If the notion of symbiosis applies, then it would indicate reciprocally that to the large scope of literature poetry owes possible venting toward the unmapped zones of prose.

In alternative structural terms, the correlation promoted here is modelled on the guidelines of Heidegger's thought about the human being. Without affirming, altering, or abstracting from the traditional concept of an "animal rationale," he introduces a determination intending to penetrate deeper, and discover a substratum enabling and elucidating human rationality which suffers not so much being disputed as disregarded, distorted and disgraced. His unique regard for, and rethinking of what it means to be human would, however, expectably open so many reconsiderations, reflections, responses, that it could not be delved into at present.

The restrained outline offered now intends only to point out how the bright-blooded poetic arteries carry a vital pulsation through the whole body of literature – as long as it is not a corpse – from the heart of existence. This image is inadequate in one respect: it would be intolerable to prohibit the influx of the poetic element into the life of any art other than that of words; but what influence it may contribute there cannot be viewed with the present scope. After this allowance, the sympathetic tie signifies that all poetic speech eo ipso belongs to literature, though what literature may amount to without being poetic is not quite in the light. The work of words is poetized in the measure of its arousing concern with any existing humanity of yesterday, today or tomorrow. It may be hard to imagine literary expression utterly void of such concern, but there is no question of drawing sharp demarcation lines, the display showing only degrees. Still, the obverse can be admitted: the more exclusive a text of human import as such, the less it would be poetic. To symbolize it rather absurdly: *A* does *B* at time *C* in place *D* with effect *E*; if that were the whole of the story, perhaps it would be one fine story not for boys or girls but for computers. Writing, whether by goose quill, ball pen or type keys, has not yet got stuck in such straits; may it never get there. Human beings live, love and laugh together in the world; of such awareness the poet may serve as ontological witness. At the same time, or at any time, each stays a he or a she, doubtfully an it; of this in some bared isolation, various ontic reports become produced – let it remain open which qualify as literature. Must anything ontic be bad? Not at all; nor need it be good to fancy writing about.

After this view on a straight theoretical line, it might be refreshing to move into the teasing shades of paradox. At least one such alley, out of the path travelled so far, inclines that way. In the many puzzle games of language invented to date, which pieces appear most neglected by philosophy, science, logic, mathematics, business, administration, technology, cybernetics, et al.? The unnoticeable answer, really at hand: pronouns – personal, possessive, reflexive pronouns. But these are indispensable poetically. Unless this very linguistic neglect prevents you from reacting to such a provocation, you may well retort: does this not conflict with the trend of previous considerations? Are not pronouns more restrictive, less participatory, more blank, less interesting, more singling out, less inviting on to a meaningful discourse dealing with human affairs? The overemphatic "me," the enigmatic "you," the dismissive "them," the peremptory "we," the tedious "she" or "he"? Is this a paradox or a puzzle box? Obfuscation, opposition, objection? None of this gets off the track, off the pedestrian track. And still: the poets dwell with these little creatures of language and play joyfully with their shapes; it is through their tempting moves that seeming insignificance assigns itself existentially, embraces the world shared openly, enters not only into humanity but even masculinity and femininity. This marvelous dress change is quite possible and surely worth screening for your eyes.

What are pronouns? The uniform grammar teaching classifies them as replacements or substitutes, for nouns, for substantive or adjective building blocks of sentences, particularly declarative, informative, assertive; with interrogative, imperative, and more elusive ones their dependence may be less confining: "Who are you?", "let us", "would that. . . ." etc. The stress in all such defining rules falls, of course, on "nouns": nomina or names. This fact might just hide a hint for some reversal, since singleminded attending on names is by no means inescapable for human thought about the world, but certainly directive for ontic awareness. And how could a reversal be entertained? In all simplicity, the accent might be placed on "pro"; and thus bring into relief its otherness from "contra," from "versus," from "ob." The Latin grammatical formation would then incline word intent to turn: toward, for, on; and thus: over, beyond, away from the various contrarieties, adversities, subversions, obstacles, and even objects. Now the end stage for the cover-play may become visible: uncovering pronouns tears their veils "against" and allows looking "into": compliant content where concealments concede a clearing.

But with sceptical surprise: unless they put on mere fancy dress, how could these bare pronouns move, denuded of their contractually versatile objectivity? Another veil must fall, for a fascinating view. They would pronounce, pro-pose, pro-test, pro-long, pro-create: the primal if not primitive being human in the world. The negligé of pronouns has, prudently or prudishly, to betray the negligence of obsessive ontic objectives. Below the surface supervision of a surfeit of things, existence shown naked shamelessly provokes this quest: to be or not to be, when to be, how to be, with whom to be, what to be for? Dispossession from possession of acquisitions and attributes, articles and artefacts, attachments and affiliations, undresses the living display of ontic awareness, ambivalently attracted among them; that awareness which on its ontological plane aims but toward the all-rounding discoveredness of human being and time.

It is appreciated that this light touching glance may not discover a cleaning expanse for your consent. It may be sensed as only an invocation of a pro-turn toward, rather than a contra-turn against, the poetic play of pronouns turning on your sensitive thought. Some ease might come with fresh source light.

Quintus Horatius Flaccus: "Non omnis moriar: I shall not wholly die, and a great part of me will pass far from burial."

This I given and giving to poetry, is it the I enshrined in the beliefs and customs of imperial Rome, the I sliding only on turns of the Latin tongue, the I soon named on a marble headstone, the dying I? It may happen not. . . .

Francesco Petrarca: "When in those times I was in parts a man other than I am."

Those times, are they parts of the I, is the I an adding or subtracting, then by what count? If that other is I, then am I still an other, ever to be I? Yet this time is, not they: I am.

Juliusz Słowacki:

My first days' love! I am yours again. Look how I come back, with no heart, no name, at your feet falling like a bird astray. . . . For your tears only my voice was a wait, only you know the pain I must bring with me to have that voice want and sing. . . . Think on me – are more rare souls above than mine – who through the world found only one love?

I: ever thought on alone? I, with pain and love, to ask and to sing, to leave and to come, carried and carrying through, and through a whole vast world? Can an I be without a you, will I be mine when not yours? There is you when I fall, only you, my love.

Heinrich Heine:

One night I cried in my dream, I dreamt of your being dead;
That morning I waked and wept, so many tears have I shed.
One night I cried in my dream, I dreamt – from me your were gone;
That morning I waked and wept, bitterly wept, on and on.
Last night I cried in my dream, I dreamt – your heart was mine still;
This morning I wake and weep, wanted to stop, never will.

You and I, love and loss, dream and wake. You dead and I stay, you gone and only I here, love lost yet live? What loss is most worth crying, what heart is most worth having, what world is most worth living or loving or leaving? For wonder of nights no stilling in mornings, but questing all days, all-ways. Perhaps not always losing.

Adam Mickiewicz:

In all those places, ever, but anywhere
I did sigh with you, where you with me would play,
Ever and always, I shall be with you there,
For there I scattered parts of my soul to stay.

With you, I can take possession of all places, I can appropriate the world, I can properly hold my stay in it, not to leave but to dwell. Yet if you leave: I shall be with you, will you be with me, ever? The world will stay, in it happily my soul, but may it be mine if you are not, may the world? Maybe if it is ours . . .

Rupert Brooke:

> Breathless, we flung us on the windy hill,
> Laughed in the sun, and kissed the lovely grass,
> You said, "Through glory and ecstasy we pass;
> Wind, sun, and earth remain, the birds sing still,
> When we are old, are old..." "And when we die
> All's over that is ours; and life burns on
> Through other lovers, other lips," said I
> "Heart of my heart, our heaven is now, is won!"
> "We are earth's best, that learnt her lesson here.
> Life is our cry. We have kept the faith!" we said;
> "We shall go down with unreluctant tread
> Rose-crowned into the darkness!" . . . Proud we were,
> And laughed, that had such brave true things to say,
> – And then you suddenly cried, and turned away.

This poet – who at twenty-eight made a corner of a foreign field forever England – appropriates the world not as his, nor calls it hers. The world is ours, ours the life, ours the ecstasy, ours the cry. Or is anything ours, if not all that? When we go down, it is we who shall . . . let the earth remain. It is we who laugh and love and live; then away, away, or on, burn on, turn on? Our earth, our sun, our heaven is won!

Clearing claiming climax: over-coming, over-timing, over-taking life, earth, world, in the name of human being. Here is not the question of "only me" or "only you," there is over all "we, us, ours," here is an over-flow of existence. Such brave true things are said, sung, spoken poetically for the sake of humanity in time, at any given time, for all our time. Poetic awareness, while in no way absolute, infinite or superhuman, but ever human-bound, end-bound, time-bound, is yet wonderfully enough to pro-ject and pro-gress, toward heights over-looking the world as a whole plane. There is no contra worthy of contention, no adversary relation worth relating, no objects to obtrude, obligate, obsess. The poet plunges in fore-feeling through the screen of ontic disguises and distractions into an open clearing of ontological sense. Are any things, facts or names pertinent for this pro-nounced ecstasy? Hardly. The pro-creation, pro-longing, pro-posing of all human appropriation through the world is amazingly, paradoxically, but clearly due to pro-nouns. Their turn over comes on: while not particularized they allow partaking, while not conceptual they expand conceiving, while not predicative they play for pre-paring and pre-senting humanity with poetic wonder.

If the illumination above has incited you up the avenue of rapprochement between poetic and ontological attestations of man's being in the world, you may be more acquiescent with regard to the luminosity of pronouns. To let them shine on further would exceed the present perspective. But before leave-taking it may be proper to note that their electric excitement is by no means without difference equally available: some are visibly more equal than others: the prior lit passage was conducted from a quasi-natural start and arrived apparently but mid-way. How? Beginning with "I," it moved irrepressibly to face "you," then climbed to a joining with "we." And yet perhaps mid-way comes here all the way. To hazard a glimpse, other pronouns are transparently more feeble in the light of their sense, giving just lunar reflections not of their own radiation. Alternatively speaking, they are no more than alternates. And still differently pro-posed: I-you-we turn in toward high over-view of the existentially shared world, are in-turning; it-one-they turn outward from this panorama to obviate its property, are thus off-turning; and the intermediate he-she turn the focus of awareness sideward to provide no more help than hindrance, are con-turning. Still other projections to shape the fits and folds of pronouns are possible, once their pliable play applies to your feeling. Is anything on earth impossible for poetic possession?

The time approaches, though, to turn back. If it please you at all, so approaching the charms of poetic speech, if this probe were poeticized appropriatingly, if its promise might have been filled as felt, then perhaps turning around is now proper. The prospect remains to retrace the trail and regain its source: to return to thinking. The fresh wonders, sighted at least distantly and instantly, yield some refreshment for the return way, Their light still shines on, the vital current still flows, the poetic speech still sings; this should prevent a finish stuck in the dim, dry stocks of prose.

What thinking lies back and ahead? Need it be prosaic? Why does thinking part from poetry?

Most assuredly, there is no monopoly in human thinking, none whatever. None in asserting, in demonstrating, in calculating; none for teachers or preachers, for leaders or readers; none with logicians or statisticians, with scientists or economists, with judges or critics. Philosophy has long claimed privileged access to thought; brought about many a painful abscess on the living body of humanity; and recently entered into a recess of recondite recriminations, without act or verdict or appeal in its process for sentient and sensible men or women. This does not mean that the thinking animal is dead, even if already subject to dismembering dissection; but it shows that disqualifications, exclusions, strictures, are deadening, unless sensitized by sweet saving sympathy. Naturally, there can be conscientious individuation favoring this or that mode of thought, yet this may have a thin flavoring of caprice, if it is not agreeable to a founding properly broad: homo sum et nihil humanum mihi alienum esse puto.

Under this wide canopy of tolerance, where to look on the temporal horizon to locate thought? On these pages, a favoring inclination was initially acknowledged, by now perhaps approved as not unfounded: Martin Heidegger. And in retrospect, some of those whose acclaim as philosophers may sound with no dissonance: Plato with his successor, Marcus Aurelius apart from his empire, Augustine through his conversions, Descartes on his pilgrimage, and even in his symmetrical tower Immanuel Kant. Still, on its own, this may not ring sound: such naming is restrictive and not freely unbound to the winds of humanity, which can fly elsewhere for you. Allowing for that, in search of genuine human thought, perhaps one entrance needs opening widely, the entrance passed already but not bypassed. Whose thinking will enter? Quite a choice. If thinking of man's life and time was not enriched by Dante, Cervantes, Molière, Goethe, Blake, Flaubert, Ibsen, Hemingway – or for that matter by Beethoven, Chopin, Wagner, Debussy, and Michelangelo, Rembrandt, Rodin, Picasso – then thinking is not what is thought on here. But these men were all: poets, no more and no less. You may breathe queries about the letters of their professions, the format of their pronouncing, the sound of their pronouns. Yet such fluttering breezes need not slow the air bursting through their in-spiration as poets: because this entrance opens into another elevation in-habited poetically. From the charmed dwelling of poetry issues not only a pulsing element to en-liven literature, and not only a light display of re-dress for language, but also, with no less power, an in-fluence to shape and beautify thought. How can this happen? Con amore – stimmungsgemäss – par raison du coeur: moving.

To be humanly moving – a host of philosophers to witness – thought does hardly rest with intellect, understanding, reason alone. So well forgotten is this: human motion is e-motion e-jected from deep seated rest; it is not in-ert but art-full; it is passion not of cool passivity but of hot pathos en-acted; it is sentiment, sentient mentality both sensible and sensitive for the sense of life, not e-vitable but in-viting toward vital movement: it is feeling alive. To live thinking and not live feeling, can only forget, forgo, forbid to live. It needs to happen that humanity re-member, re-collect, re-assemble, re-unite itself feeling alive. To make that happen it may become af-fluent with the in-fluence of poets, because this is who they have been from their Hellenic origins: makers of happening human, sayers of logos as human speech, seers of lumen as human light on life. Is it properly thinkable that proper thinking should be moving ahead without being moving, that it should happen to speak away from poetic happening of speech, that it should live on and not say much on feeling alive? May it be? Perhaps – by some mishaps. But. . . .

Perhaps the parting of thinking and feeling departs from human life as lethally as the Cartesian partition of mind and body at the start, or false start, of what is referred to as the modern age. Perhaps the author of that partition was not quite so disoriented as sometimes painted, when he visualized feeling as inherent in thinking, much as believing is inherent in it for faith, knowing is inherent in it for science, acting is inherent in it for actuality. Perhaps thoughtless neglect of feeling in what passes for modern thought is on a par with negligence in regard to, or without regard for, pro-nouns as shaping how humans speak. Perhaps we can still have the dearly departed, thinking and feeling, reconvene as dearly gathered together. Perhaps you may well retreat from hastily merging them in indifferent sameness, yet not maltreat their alienation so as to treat them as inhumanely alien to each other. Perhaps I might endeavor to look for a way toward not just a rapprochement but a rassemblement in the clearing, suffused by the charms fore-sighted above.

Thinking poetic: that is to comply with the contours of the path previously pursued to reach poetic property; that is to respond to risks residual in abstractness, strictness, universality, historicity, subjectivity; that is to be guided on the ground of ontological awareness, avoiding dispersal in ontic distractions. At the same time thinking so in-fluenced is to feel poetically, find fellow-feeling, be felt humanly. While poetic, it is to persist as thinking: not fabricating, not feigning, not fainting, not failing in fidelity to its founding sense. How are such perspectives: over-expected, easily suspected, or still respected? Pre-tending poetic powers: to make all of it would have to be presumptuous, and to say it all precarious, yet to see it at all pressing. Perhaps then not altogether out of reach, along the returning ray of Heidegger, will be carefully to pre-view, watchfully to pre-pare, hopefully to pre-sent poetic thinking conceived still in philosophy, but no more of philodoxy, flashy flirtation with meanings, rather with philanthropy, fluent fondness for humans.

This needs to be for our time poetic thinking, all of this, some of this, any of this. Thinking that discloses truths for the future rather than closes errors of the past. Thinking that lives with wonders, quests, dangers, rather than kills problems, puzzles, perplexities. Thinking that intimates delights, disappointments, decisions, rather than estimates costs, capacities, conclusions. Thinking that curves for curiosity, temptation, entrancement, rather than speeds for settlement, efficiency, success. Thinking that reflects frailties, doubts, improvisations, rather than deflects criticisms, reactions, dissents. Thinking that rings for destiny, for excitement, for joy, rather than knocks as fate, as rule, as law. Thinking that inclines when agreeable, when appealing, when adventuresome, rather than reclines when secure, stolid, sleepy. Thinking that projects to admire, to surprise, to extend, rather than objects to correct, to analyze, to disjoin. Not the thinking that implies, dictates, constrains, but the thinking that complies, unveils, frees. Not the thinking that tinkers, uses, services, manipulates, man-handles, but the thinking that inspires, widens, seduces, carries, graces. Not the thinking that insists on its self, its stature, its stability, but the thinking that exists toward the world, toward history, toward humanity. Not thought stopping eternally still but thought moving temporally onward.

Such thought is to turn; has turns; and turns. It is to turn with the rhythm of longing, of happening, of overcoming, of strain, of embrace, of ecstasy, of spread, of novelty, of life. It has turns in the dimensions of re-turns, re-quests, re-iterations, re-solutions; and blind-turns, burns, bewilderments, bypasses; and down-turns, delays, deceptions, defeats; and up-turns, attainments, accords, appropriations. It turns on me, when it bears on me, and bearing myself, fore-bears as mine; turns on you, when it feels for you, and feeling yourself, fore-feels as yours; it turns on us, when it plays with us all, and displaying all of ourselves, fore-plays as all ours. Such thought philosophizes for humans male or female, and poetizes for our time. It longs to be live-ly, love-ly, like-ly: it may be. Will it be?

As with the Will of Avon, the grave yields to the light.

Hear this: the thinker Unamuno – "For us to exist is to suffer"; the philosopher Nietzsche – "Man is the animal who suffers most and so discovers laughter"; the poet Beaumarchais – "Let us laugh; do we know if the world will be there another three weeks?"

A smile: as poetic as a sigh or a song. A smile to stay with and to think on. A smile at the end of a journey. I sight the end of a way, and look for the way to an end. This way I have gone at your expense has been long winding, perhaps not too long-winded. You may have felt it as a wild chase, perhaps not a wild goose chase. Our experience has turned out peripatetic, perhaps not pathetic, maybe even sympathetic. Taking an off chance on that, and by your leave, this is the end for my sight, where I intend to leave you, before leaving off.

Poetic thinking, thought poetically: which was, which is, which will be? It happens, for this time it may happen, it may be, it needs to be, it longs to be, it belongs to be. It happens this way on the ground of dis-maying needs of our time, our timely human longings, all belongings of the human being. These need poetizing as much as thinking, as ever before. The human being, ever, and in our time, is: more than manhood, more than womanliness, more than man-power, more than woman's liberation, more than man-kind, more than any human condition, more than an animal human species, more than a humane life, more than humanities and humanitarianism and humanism. The poet as the thinker may be aware and speak for: the human being. Needful or needless, happening or mishappening, longing or pro-longing, there is the human way to be, and was, and will be. The human being belongs to be. For a time, for our time, for any time, the human being intends a telos, as entelechy, to extend the way not to end but in being.

Being: a word or no word, a thing or nothing, a being or non-being. None of those in a way, and yet an end of a way, the human way. Being: the deepest void and the greatest vastness, mere emptiness and full evolution, sheer waste and all wealth. Being: in that beings inanimate may submerge to vanish, through that beings animate may merge to live, out of that beings human may emerge to exist, aware, ecstatic, clearing. Light extends human feeling alive to intend poetry, as clearing poetry intends thought, as thought intends being human in time. Not in-fluenced poetically, fluent thought may prosaically dry into dust; thought-less, the way may not long extend over human time; with no way, human being there may be nowhere. This is not to be. The human being belongs not to no end, but to an end in being.

Poetic thinking is to be: for the human being – on the way toward being.

# D. THE POET AND THE WORLD OF POETRY

L. M. FINDLAY

# FROM HELIKON TO AETNA: THE PRECINCT OF POETRY IN HESIOD, EMPEDOKLES, HÖLDERLIN, AND ARNOLD

## I

Heidegger tells us that "language is the precinct (*templum*), that is, the house of Being," and that "poetry that thinks is in truth / the topology of Being."[1] He arrives at such clearings in his own thinking in part through re-appropriation of the pre-Socratics and the poetry of Hölderlin. But how does language speak thus through Heidegger, and how does Heideggerian speaking differ from the speaking of language through the early Greek thinkers or nineteenth-century poets? What is the nature of the appropriative process at work here? Is it enigmatic, kerygmatic, both, or neither? Is it deferential, or ruthlessly self-interested, careful recreation or casual abduction, or what? is philosophical appropriation different from poetic appropriation? And what of the prominence of spatial terms in the title of this paper and the opening quotations from Heidegger? Are they extravagant or precise: a distorting supplement or a necessary tribute to Being as place and the places of Being?

Let us begin with spatial terms and what might be termed the topographical imperative. Job asks: "But where shall wisdom be found? and where *is* the place of understanding?" The answer comes sixteen verses later: "And unto man he said, Behold, the fear of the Lord, that *is* wisdom: and to depart from evil *is* understanding." This poetry of parallelism captures nicely our need and determination to know through placing, and also the realization that the seat of knowledge *is* the point of departure towards true knowing. Plato also is interested in the location of knowing, the *topos noetos*, yet he never goes beyond the formulation offered in the *Sophist*, in an exchange between Theaetetus and the Eleatic Stranger:

| | |
|---|---|
| *Stranger*: | But you surely, I suppose, will not grant the art of dialectic to any but the man who pursues philosophy in purity and righteousness. |
| *Theaetetus*: | How do they differ? |
| *Stranger*: | The sophist runs away into the darkness of not-being, feeling his way in it by practice, and is hard to discern on account of the darkness of the place. Don't you think so? |
| *Theaetetus*: | It seems likely. |
| *Stranger*: | But the philosopher, always devoting himself through reason to the idea of being, is also very difficult to see on account of the brilliant light |



*A-T. Tymieniecka (ed.), Analecta Husserliana, Vol. XVIII*, 119–140.

of the place; for the eyes of the soul of the multitude are not strong enough to endure the sight of the divine.

*Theaetetus*: This also seems no less true than what you said about the sophist.

*Stranger*: Now we will make more accurate investigations about the philosopher hereafter, if we still care to do so; but as to the sophist, it is clear that we must not relax our efforts until we have a satisfactory view of him.[2]

From this passage we learn that the place of sophistical not-being is dark, and the way of not-being is flight (*apodidraskon*) through empirical hazard and expediency. The philosopher, in contrast, is *in* the place of being, his only and perpetual activity a devoutly rational realization of the Idea of being. The distinction between mobile error and static truth is reinforced by the use of two different terms denoting place: the sophistical place of not-being is *choras*, whereas the place of being is *topos*: the former has strong pre-Socratic links with *kenos* and the void, the latter with partially occupied space and appropriate positioning – as of a soldier at his post.[3] Plato makes clear his preference through opposing a place that is no place at all (but rather a void and circumambient darkness where human figures ply their obfuscations in the palpable obscure), and a place which, as the locus for transactions with the realm of Ideas is partially occupied, though only by figures lost in an intensity of light sufficient to blind the vulgar. Wisdom is no more a sitting target for Plato than for Job. In the *Sophist*, as in the Old Testament account, what is given is simultaneously withheld in two clearly calculated expressions: *katanoesai chalepos* and *andamos eupetes ophthenai*. Hannah Arendt has offered two possible explanations for Plato's reticence on this question: either it is the result of his failure "to complete the trilogy of *Sophist, Statesman, Philosopher* or . . . he had come to believe that the answer was implicitly given in the *Sophist*."[4] The first point worth noting – and one that is highly relevant to a proper understanding of the Empedoklean poems of Hölderlin and Matthew Arnold – is how the challenge of apt ambiguity and essential incompleteness can be blunted by invoking a grand, though marred, expository project wherein (so the comfortable assumption goes) the places of thinking *could* have been fully disclosed. If, on the other hand, Plato gave us only what he could – trilogy or no trilogy – then we may conclude that there is no more than implication and difference; at worst a void, at best a partially occupied space, but certainly no plenary metaphysics of presence. The business of philosophy, described by Plato in terms of a determined adjacency to the Idea – *te tou ontos aei dia logismon proskeimenos idea* – may then be seen as installing secrecy, incompleteness and effort as permanent features of language.

II

So much, you might say, for a certain kind of philosophical language, exiled to a *terra incognita* of its own making. But what about "real" places? Places that existed for Pausanias and Strabo, and exist for tourists today. What about Helikon, and what about Aetna? Surely with them we know where we are, even if, like Hölderlin and Arnold, we have never seen these mountains for ourselves.

In *Theogony* Hesiod recounts the circumstances whereby his home on Mt. Helikon becomes a precinct of poetry and himself an instrument of divine inspiration. This is not the only and permanent home of the Muses – they reside also with Zeus on Mt. Olympos – but rather a great and holy mountain (*mega te zatheon te*) forever sacred to the Muses which frequent it. Human reverence may be a constant concomitant of the act of consacration (*kathiereusis*), but the presence of divinity is not.

When it comes to his acquiring of the gift of poetry, Hesiod resorts to language described by M. L. West as "curiously elliptical":

The Muses once taught Hesiod to sing
Sweet songs, while he was shepherding his lambs
On holy Helicon; the goddesses
Olympian, daughters of Zeus who holds
The aegis, first addressed these words to me:
'You rustic shepherds, shame: bellies you are
Not men! We know enough to make up lies
Which are convincing, but we also have
The skill, when we've a mind, to speak the truth.'

So spoke the fresh-voiced daughters of great Zeus
And plucked and gave a staff to me, a shoot
Of blooming laurel, wonderful to see,
And breathed a sacred voice into my mouth
With which to celebrate the things to come
And things which were before. They ordered me
To sing the race of blessed ones who live
Forever, and to hymn the Muses first
And at the end. No more delays: begin.[5]

This language is not "curiously elliptical" for Piero Pucci who has recently reread Hesiod in the light of Derrida.[6] The gift of poetry is conferred in terms true to the ambivalent genius of the place. There is no disputing the fugitive nature of wisdom, and the capacity of language for both revealing *aletheia* and for sophistical dissimulation. Knowledge is power realized in and through language, and man is not so much its master as mastered by it. Hesiod is singled out as a favourite of the Muses only after his fellow shepherds have been put firmly in their place – an anxious as well as gratifying moment for him. Hesiod's marks of distinction, the poet's staff and capacity to sing of the equal mysteries of past and future, are accompanied by a solemn ordonance to use his gift immediately and always to honour his benefactors. Hesiod readily complies with this instruction, reminding us in the process of human limitation. In particular, his present is created out of the past and future, for only the Muses within Olympos have the right to sing of what is past, *and* passing, and to come: *ta t'eonta, ta t'essomena, pro t'eonta*. Only the mighty mind of Zeus can cope with the full presencing of time, and even he seems to relish distraction more than prophecy, given the cares and tensions of his office. The Muses are his children by Mnemosyne, and, paradoxically, Memory thus contributes to the purchase of Oblivion:

> *Μνημοσύνη, γουνοῖσιν 'Ελευθῆρος μεδέουσα,*
> *λησμοσύνην τε κακῶν ἄμπαυμα τε μερμηράων.* (ll. 54–55)

This paradox is foregrounded by the placing of *Mnemosyne* and *lesmosyne* in the initial position in successive lines to emphasize the displacing powers of poetry. As we shall see more fully later, Matthew Arnold quotes this passage out of its cautionary context because he imagines Helikon to have been the ideal home of poetry, the precinct of poetic presence for human poets as well as gods. He follows the parallel between the Muses on Olympos and the poet in the world to imply – less circumspectly than Hesiod – that as divine poetry offers distraction to sorely pressed divinity, so human poetry ought to offer distraction to humanity. Now Hesiod says of the poet in the world:

> His voice flows sweetly from his mouth, and when
> A man has sorrow newly on his mind
> And grieves until his heart is parched within,
> If a bard, the servant of the Muses, sings
> The glorious deeds the men of old performed,
> And hymns the blessed ones, Olympian gods,
> At once that man forgets his heavy heart,
> And has no memory of any grief.
> So quick the Muses gift diverts his mind. (ll. 90–99)

The human poet does not engage with a world of present things, *ta t'eonta*, but with the pangs of recent sorrow, and, more generally, with the world as recollection or prophecy. In his case, as distinct from the case of Zeus, one memory supplants another in the play of anamnesis, in a process that knows neither presence, nor closure, nor permanent contentment. What Hesiod offers, and what Arnold will refuse to accept, is that the conditions of the poetic do not change in the movement, such as it is, from Helikon to Aetna; from the locus of poetic initiation to the locus of poetic immolation. The only thing that changes is the capacity of the poet to live with and through the paradoxes of his calling. He may find it convenient, of course; exhaustion and melancholy may make it expedient to imagine an originary time when things were otherwise, but that is to ignore the irony and tension in Hesiod's contention: "lucky is he whom the Muses love," "o d'olbios, ontina Mousai / philontai" (ll. 96–97).

If he could ignore the ironies of *olbios* in Hesiod, Arnold could not ignore the testimony of Empedokles: *olbios* and *deilos* are thus contrasted in fragment 132:

> *ὄλβιος, ὃς θείων πραπίδων εκυήσατο πλουπυ,*
> *δειλὸς δ', ὧι σκοτὸεσσα θεῶν πέρι δόξα μεμηλεν*
>
> Blessed is the man who has gained the riches of divine wisdom;
> wretched he who has a dim opinion of the gods in his heart.[7]

In particular, for a poet estranged from the Muses, no longer willing or able to meet the costs of song, the world becomes an alien and joyless place – *asunethea choran; aterpea choran* – and he himself, in so far as he puts his trust in *Neikos* (Strife), is an exile and wanderer from the gods: *fugas theothen kai aletes.*[8] What is left for such a man?

III

The story of Empedokles' suicide on Aetna has long been known to be apocryphal,[9] but it has obvious attractions for poets who feel almost overwhelmed by the burden of their gifts. During the romantic period Novalis, Friedrich Schlegel and others inaugurated a fresh career for the poetic fragment as the complete expression of the brave compelling synecdoche of Romantic art. New fragments were created and old fragments reassessed by a sympathetic restorative imagination that included in its purview not only the ruins of the Roman Campagna and the Elgin marbles, but also the fragments of Empedokles. Hölderlin may have been drawn to the Greek poet through his own Hellenism, or because of the intimations of facility, power and cosmological range evident in the ruins of much larger originals.[10] Whatever the specific motivation, what he found there, and in the accounts of Empedokles' life, was enough for him to embark upon his most ambitious poetic enterprise: *Der Tod von Empedokles: ein Traurspiel in fünf Akten*, none of the fragments of which appeared in print during his lifetime.[11]

There is not time here for detailed analysis of the whole of this great ruin, nor do I have much appetite for speculation about the poem's relation to Hölderlin's mental breakdown. But consider the following typical exchange from the second version of the tragedy:

Pau: Wie zahmen Hirschen,
Wenn ferne rauscht der Wald und sie
Der Heimat denken, schlug das Herz mir oft,
Wenn du vom Glück der alten Urwelt sprachst,
Der reinen Tage kundig und dir lag
Das ganze Schicksaal offen, zeichnetest
Du nicht der Zukunft grosse Linien
Mir vor das Auge, sichern Blicks, wie Künstler
Ein fehlend Glied zum ganzen Bilde reihn?
Und kennst du nicht die Kräfte der Natur,
Dass du vertraulich wie kein Sterblicher
Sie, wie du willst, in stiller Herrschaft lenkest?
Emp: Recht! alles weiss ich, alles kann ich meistern.
Wie meiner Hände Werk, erkenn ich es
Durchaus, und lenke, wie ich will
Ein Herr der Geister, das Lebendige.
Mein ist die Welt, und untertan und dienstbar
Sind alle Kräfte mir,
zur Magd ist mir
Die herrnbedürftige Natur geworden.
Und hat sie Ehre noch, so ists von mir.
Was wäre denn der Himmel und das Meer
Und Inseln und Gestirn, und was vor Augen
Den Menschen alles liegt, was wär es,
Dies tote Saitenspiel, gab' ich ihm Ton
Und Sprach und Seele nicht? was sind
Die Götter und ihr Geist, wenn ich sie nicht
Verkündige? nun! sage, wer bin ich? (p. 542)

For the naive but well-meaning Pausanias there is still both *Heimat* and the possibility of poetic *Heimkehr*. But his sense of displacement is aligned with that of stags together on the mountainside hearing the distant forest sounds. There is a consolingly social, natural quality to the graduation of stags from the one location to the other, and this, quite different in its pseudo-Hesiodic nostalgia from the poet's private reality, invites a bitter rejoinder from Empedokles. If the poet did indeed enjoy the omniscience and omnipotence ascribed to him ("Das ganze Schicksaal . . . zum ganzen Bilde"), he would cease to be human; and it is the resultant loss of identity which Empedokles portrays in his mockingly hubristic response. This version of self-transcendence stands in relation to the one he contemplates on Aetna as apotheosis does to suicide. Empedokles knows he was never able to complete any of his poetic scenes. The full presencing of time through art was not an option available to Hesiod or to him. The burden of prophecy and recall must now be shed in the plenary moment of self-immolation.

Hölderlin could not *complete* his tragedy. He might wish to do so, but the Empedoklean paradigm could not be denied. In successive verisons of his tragedy Hölderlin endeavoured to distance himself from Empedokles, but relinquished the task when he threatened to endanger the authenticity of his poem as an enactment and imminent termination of a particular poetic being-in-the-world:

*Emp*: Und viel vermag er und herrlich ist
Sein Wort, es wandelt die Welt
Und unter den Händen. (p. 543)

The lacuna is essential, a sardonic gloss on the creative *grasp* of "Wie meiner Hände Werk, erkenn ich es / Durchaus." Whenever the poet evokes the prospect of himself as *artifex mundi*, the fuller precinct of poetry manifests itself and the abyss opens before him.

IV

The remainder of this paper will be concerned in greater detail with Matthew Arnold's denial of his poetic being-in-the-world as unfolded in 'Empedocles on Aetna.'[12] Central to this work is the fear that poetry in the nineteenth century is *losing ground* to other literary forms, most notably the novel, and that the completion of a successful long poem is no longer possible. But had the precinct of poetry truly diminished to the meagre domain of the solipsistic lyric and the wreckage of grander projects, or was the sense of challenge and loss felt during this period based on a view of poets and of their work that never has been, nor ever can be true?

'Empedocles on Etna' is, at 953 lines, the second largest poem that Arnold wrote. But it is less than half the length of *Merope,* and is not much longer than *Balder Dead*, 'Sohrab and Rustum,' 'Tristram and Iseult.' Moreover, 'Empedocles' seems positively terse compared with the bulkiest works of the Brownings, Tennyson, Swinburne, Morris, or such misplaced tributes to poetic stamina as Bailey's *Festus* or Sir Henry Taylor's *Philip van Artevelde.* However, I will try to demonstrate that there is a good deal to be learned about 'Empedocles on Etna,' about Arnold's art, and about poetry more generally, by considering the poem from this vantage.

But then, it might be objected, 'Empedocles' is subtitled 'A Dramatic Poem.' In singling out its length as its most salient feature, can one safely ignore or play down the presumably important fact of its dramatic form? In subordinating poetic mode to verbal mass one may be seen to follow Arnold's own lead in the Preface to *Poems* (1853), where he collapses the distinction between epic and dramatic, and in the Preface of 1854, where he concedes to his critics that the earlier Preface had addressed itself to the problems of nonlyric poetry, whether epic or dramatic. Between 1852 and 1853, after all, he had substituted one long poem for another; the epic episode of 'Sohrab and Rustum' for the dramatic poem on Empedocles. For Arnold, especially at this time, drama and epic were readily interchangeable terms denoting past literary achievements of a high order, and a common project in the present – namely, the making of a long poem fit to stand beside its most eminent precursors. Epic and drama were of equal worth and shared a common destiny.

This brings us closer to the *problem* of the long poem. It needs to be emphasized in this connection that problems were posed for the Victorian literary imagination by the short poem, the novel, and by other literary forms, as well as by the long poem. It is always difficult to write well – whether in verse or in prose – the evidence of Victorian facility and copiousness notwithstanding. It is always difficult to live well – whether morally or intellectually – the evidence of Greek awareness and virtue nothwithstanding. This is the full burden of Arnold's quotation from Pittacus in the Preface of 1853: "this is no easy task – χαλεπὸν, as Pittacus said, χαλεπὸν ἐσθλὸν εμμεναι."[13] Arnold does not rest content with the coolness of a Wordsworthian litotes: "no easy task." There has never been a place for complacency in the quest for excellence, and the intensity with which he feels this truth, the urgency with which he expresses it, comes through in the repetition of the Greek term denoting obstruction, frustration and ire: χαλεπὸν . . . χαλεπὸν. In so closely identifying moral, intellectual and artistic excellence with the long poem, Arnold simultaneously defined his objective as a poet and, for reasons we will examine shortly, placed that objective beyond his reach. His formulation of an insoluble problem came *after* the publication of 'Empedocles on Etna,' *before* the appearance of *Balder* and *Merope*, and in the same volume as 'Sohrab and Rustum.' My intention in the argument and analysis that follows is to explain this sequence of events.

Thomas McFarland has given students of nineteenth-century literature much to ponder in his recent book on the romantic fragment.[14] Central to this study is the notion of "diasparactive forms" – that is, of literary forms that tear themselves asunder. This idea is obviously apposite to consideration of a poet who made so much of *architectonike*, and whose letters are full of sentiments such as these:

[To his sister Jane in the early 1850s] Fret not yourself to make my poems square in all their parts, but like what you can my darling. The true reason why parts suit you while others do not is that my poems are fragments – *i.e.* that I am fragments, while you are a whole. . . .

[or to Clough in October 1852 à propos of the Elizabethans] Keats and Shelley were on a false track when they set themselves to reproduce the exuberance of expression, the charm, the richness of images, and the felicity, of the Elizabethan poets. Yet critics cannot get to learn this, because the Elizabethan poets are our greatest, and our canons of poetry are founded on their works. They still think that the object of poetry is to produce exquisite bits and images.

[or to Jane again in September 1858] People do not understand what a temptation there is, if you cannot bear anything not *very good*, to transfer your operations to a region where form is everything. Perfection of a certain kind may be there attained, or at least approached, without knocking yourself to pieces, but to attain or approach perfection in the region of thought and feeling, and to unite this with perfection of form, demands not merely an effort and a labour, but an actual tearing of oneself to pieces, which one does not readily consent to (although one is sometimes forced to it) unless one can devote one's whole life to poetry.[15]

The misguided pursuit of local felicity for its own sake; or, the reluctant acceptance of imperfection of the life and of the work; or, total commitment of self to the creation of true poetic totalities: all three options can be explained in McFarland's terms. Indeed, when Arnold goes on to name Wordsworth as an example of the poet giving himself up completely to his art, we may feel a little surprised by the idea of the prolific Romantic tearing himself to pieces over and over again, but McFarland would not share our surprise. For him, even Wordsworth's longest poems are diasparactive, fragments of 'The Recluse,' while his shorter ones are flights from it. Consider these remarks about *The Prelude*:

I have elsewhere called 'The Prelude' the "only great poem that does not have a text, for its *substantia* hovers in *Aufhebung* somewhere between the facing page editions of the 1805 and 1850 versions." Actually, 'The Prelude' could as well be printed in, as it were, a Septuagint edition as one on facing pages, for it existed successively in two books, five books, thirteen books, and fourteen books, along with whatever tentative versions lay in between. The testimony of all this revision is simply and on its face that of an immense struggle with form and statement. In addition there are still further paradoxes that indicate an encumberment in the emergence of 'The Prelude.' The greatest poem of the nineteenth century, after being worked on approximately fifty years, was not even published by its author, nor was it even called 'The Prelude' by him. It was in fact not even designed as a prelude, but rather as a postlude or appendix – or in Coleridge's term, a "tail piece."

Furthermore, the poem is entangled with three other works, all of which are even more encumbered than it is. I shall not dwell on this here, other than to present the preliminary suggestion that 'The Prelude' relates exactly to 'The Recluse' as Coleridge's *Biographia Literaria* does to his *magnum opus*, that also 'The Prelude' is to the *Biographia* as 'The Recluse' is to the *magnum opus*, and that the proportion is strictly mathematical in its possible substitutions. [And there is] another kind of encumbrance . . . that imposed by [Wordsworth's] predecessor figure, or to use Harold Bloom's term, his "strong precursor," Milton. Indeed, I should like license to invoke another mathematical metaphor: that of a triangle in which the apex is 'Paradise Lost' and the two angles of the base are 'The Prelude' and Coleridge's 'Kubla Khan.' (pp. 237–38)

The effect of McFarland's remarks is for the most part salutary. It does us all good to feel the bibliographical ground give occasionally beneath our feet. But it is surely significant that McFarland's textual analogy with the Septuagint is offered less confidently than his mathematical metaphors. There is an apt wit behind the allusion to the earliest Greek translation of the Old Testament, originally thought to have been made simultaneously on the island of Pharos by seventy-two different hands, and now thought to be the work of Egyptian Jews working independently of each other at different times. There are suggestive affinities here with the movement from one Wordsworth and a unitary Prelude, to many Wordsworths and at least seventeen different versions of the poem. However, McFarland's "as it were" seems to admit that this analogy should not be too closely scrutinized or pushed too far. Behind this hint of nervousness may lie the recognition that, despite intermittent composition or translation, textual scholars and ordinary readers are perhaps in general agreement as to what is meant by the Old Testament and *The Prelude*. Perhaps Wordsworth's poem is less of a bibliographic curiosity than McFarland would have us believe, and less fragmentary than he suggests.

With the mention of "exactly" and "strictly" we are faced with a different order of affirmation. In rhetorical terms, as McFarland himself reminds us, we have moved from simile to metaphor; thematically we have moved from textual scholarship to mathematics in a show of pseudo-rigour designed to install the diasparactive principle at the heart of creation and criticism alike. Not only are individual texts "really" dismembered and dispersed, but such residual integrity as they may lay claim to is further eroded according to the Bloomian principle that there are no texts, but only relationships between texts. The finite mathematics of mensuration – how long is the completed work? – has given way before the interminable mathematics of proportion, permutation, relation. But McFarland cannot abandon mensuration altogether, for out-and-out deconstruction is not his purpose. Shortly after the long passage just quoted, McFarland glosses Elinor Shaffer's view of 'Kubla Khan' as a "collapsed epic": "Still, a collapsed or condensed epic is not exactly a true epic, for length is a necessary criterion of the epic form" (p. 238). Quite so, and this holds true for poets and critics of the nineteenth century no less than for those of any other century. McFarland believes in philosophical and moral foundations: he talks of Carlylean symbol as "the ultimate logic of the diasparact [and] also the ultimate logic of Romanticism as such" (p. 32), and of the " 'deep structure,' as it were" of the grounds for "Wordsworth's conservatism" (p. 195). But, as I see it, ultimate logics and deep structures are irrelevant and dangerous chimaerae which permit McFarland to deny unity and wholeness to 'The Prelude,' not on the basis of our actual experience of the work, but because of the history of its composition and the unfulfilled designs of its creator. In dealing with Arnold I will attempt to reverse this process while retaining the context of fragment and whole so vividly evoked by McFarland. In so doing, I will argue from my experience of 'Empedocles on Etna' as a successful, unified long poem profoundly threatening to its creator.

## V

The appearance of *Poems* (1853) under Longman's imprint and Arnold's full name, was preceded by the issue of two slim volumes under the imprint of his father's publisher, Fellowes, and the authorship of "A." These earlier volumes were *The Strayed Reveller and Other Poems* (1849) and *Empedocles on Etna and Other Poems* (1852). When Arnold publicly accepted responsibility for his poems, he also felt it necessary to suppress the title poem of his second volume and to issue – despite the advice of some of his friends and his own concerns about its adequacy and precision – a preface defending that action. 'Empedocles' remained suppressed until 1867, when it was reissued at the "request of a man of genius . . . Mr. Robert Browning." The 1853 Preface was retained in 1854, along with the much shorter preface of that year, and they were both reprinted in 1857, but then withdrawn until their appearance in *Irish Essays and Others* in 1882 at the instigation of "some of the readers of my poetry." Arnold added: "Exactly as they stand, I should not have written them now; but perhaps they are none the worse on that account."[16] To complicate matters further, Arnold's suppression of 'Empedocles' between 1852 and 1867 was not total. In fact, in 1853 he reprinted Callicles' song from act 1, scene 2 describing the metamorphosis of Cadmus and Harmonia; in the third edition of *Poems* (1857) he reprinted four more of Callicles' songs and 24 lines of Empedocles from act 2, actions that reconstitute the poem as predominantly serene and inherently fragmentary. Arnold offers us an editorial justification *ex post facto* of his act of self-censorship and the precepts of the 1853 Preface; an act of vandalism commends itself to us as an act of salvage. At the same time as he was cannibalizing "Empedocles," Arnold was also drawing on the preface for his inaugural lecture as professor of poetry at Oxford, 'On the Modern Element in Literature.' To cannibalize a polemical essay may very well secure for its leading ideas and expressions a richer, longer, more effective life; to cannibalize a successful poem is to destroy it.

The preface of 1853 is generally (and I think correctly) held to be an example of Arnold's prose at its finest. Composed over a period of five weeks, the preface reveals him at full stretch, making no concessions to his audience, and the result is rich and beautifully cadenced. Arnold is concerned throughout with authority: his own, his contemporaries', and the authority of the great poets and critics of the past. The connection between the terms "author" and "authority" is evident throughout as Arnold examines past and present varieties of poetic and critical discourse while finding his own critical style. He disowns the longest of his poems to date on the grounds that it fails to attain poetry's *eternal* objects; and at the same time he offers an analysis of his historical situation which effectively guarantees a leading role for his new critical prose. He confesses his lack of the grand style in terms clearly designed to establish his mastery of the grand manner of the great critic. The suppression of 'Empedocles' is thus characterized, not as proof of the acumen of its critics, but rather as evidence of Arnold's own critical abilities. Arnold repudiates in advance the suggestion of Tennysonian trimming to conciliate hostile reviewers. But his allegiance is not so much to his own poetic genius and artistic integrity as to the courage and incisiveness of his self-criticism.

It is in the very nature of polemical works to avail themselves of tactical distortion, and the preface of 1853 is no exception. Scholars have exposed Arnold's unfairness to some of his critics and reflected on his unfairness to his poem. No one seems to have paid sufficient attention to Arnold's distortion of his authorities, and the implications of such distortion. He remains fairly true to his principal theoretical prop, Aristotle, and, of course, any serious misrepresentation of *Poetics* would have been immediately evident to his audience and seriously weakened his claims to be custodian of the wisdom of the Greeks. In his third paragraph, however, in arguing that poetry should not only interest, but also "inspirit and rejoice the reader," he quotes Hesiod and Schiller in support of his views. Let us consider these citations more closely: "For the Muses, as Hesiod says, were born that they might be a forgetfulness of evil and a truce from cares.' " This is a romanticized version of the original, as had been his redaction of the *Theogony* in Callicles' last song. Arnold then quotes Schiller's Preface to *Die Braut von Messina* (1803): " 'All art,' says Schiller, 'is dedicated to Joy, and there is no higher and no more serious problem, than how to make men happy. The right art is that alone, which creates the highest enjoyment.' " By leaping across the centuries for his authorities, Arnold insinuates what Hesiod had permitted only to Zeus, access to immutable truths beneath the flow of existence. Schiller apparently reaffirms Hesiod's definition of poetry's true office, and in so doing authenticates those elemental passions and eternal objects essential to Arnold's later argument. Now it seems unlikely that Arnold skimmed Schiller's preface in search of an apt quotation. *Die Braut von Messina* does, after all, share a common location in Sicily with Arnold's poem, and a common interest in the symbolic significance of Etna. Cajetan sees the mountain as a false emblem of peace:

Auf der Lava, die der Berg geschieden,
Möcht ich nimmer meine Hütte bauen.
Denn zu tief schon hat der hass gefressen,
Und zu schwere Taten sind geschehn,
Die sich nie vergeben und vergessen
Noch hab ich das Ende nicht gesehn,
Und mich schrecken ahnungsvolle Traüme.

The Hermit, who wishes to purchase not oblivion but truth, feels that in that inhospitable precinct he is purged of grosser sense "Und von dem Berg der aufgewältzen Jahre / Hinabsieht in der aufgelöste Spiel / Des unverständlich krummgewundnen Lebens."[17] Moreover, as is suggested by the title of his preface ("Uber den Gebrauch des Chors in der Tragödie"), Schiller's play attempts to emulate both the conventions and the elevation of Greek drama. An embattled poet usually displays more than cursory interest in the work of an apparently kindred spirit. I therefore suggest that Arnold *deliberately* distorts Schiller's views by means of incomplete quotation. The final sentence of the paragraph from which Arnold translates runs thus: "Der höchste Genuss aber ist die Freiheit des Gemüts in dem lebendigen Spiel aller seiner Kräfte."[18] In other words, the highest pleasure of which art is capable is the result of the spirit's abandonment to the free play of all its faculties. Schiller alludes here to a line of aesthetic philosophy with its origins in Kant's formulation of art's purposive lack of purpose (Zweckmässigkeit ohne Zweck), its major development in Schiller's own *Letters on the Aesthetic Education of Man* (1793) and its culmination in Huizinga's *Homo Ludens*. It is this notion of free play that Arnold found inconvenient, if not threatening. It smacked of licence, uncertainty and that formalism he found so tempting each time he surveyed the mess of modern life. It also reminded him, perhaps, that 'Empedocles' was itself, despite appearances, produced "in dem lebendigen Spiel aller seiner Kräfte." In order to disown, it is often necessary to distort.

Arnold's use of authorities in the Preface is always clever, but in fairness one has to admit that his distortions, when they occur, are often minor or innocent. This seems to be the case, for instance, in the penultimate paragraph, where he uses Virgil to reassure himself and his readers that he has not been reacting to criticism *per se*:

> How often have I felt this when reading words of disparagement or of cavil: that it is the uncertainty as to what is really to be aimed at which makes our difficulty, not the dissatisfaction of the critic, who himself suffers from the same uncertainty. *Non me tua fervida terrent Dicta: Dime terrent, et Jupiter hostis.*

The words of Turnus in book 12 of the *Aeneid* serve to reinforce Arnold's distinction between taunting words and the hostile spiritual reality which occasions them. He seems to be playing innocently enough off the "fervida . . . Dicta" of an Aeneas eager for final victory, and his own earlier strictures on "a class of critical dicta everywhere current at the present day, having a philosophical form and air, but no real basis in fact." But there is more to this. The defiance of Turnus precipitates the action that successfully concludes the *Aeneid*. Seizing a huge boundary stone, he takes fatally uncertain aim at Aeneas, misses, and becomes an easy target. In aligning himself with Turnus, Arnold embraces not only defiance of mere *dicta*, but also certain failure. There is nothing Turnus can do to defeat the fury; courage is not enough: "Sic Turno, quacumque viam virtute petivit, / Successum dea dira negat." We may be forgiven for witnessing once again, though mutedly, what Arnold hauntingly called "the discouragement of Hamlet and of Faust."

It is only too easy to become lost in the laminations of Arnold's preface. However, it is necessary to consider two further instances of distortion: Arnold's response to what he considered David Masson's misrepresentation of Goethe, and Arnold's distortion of his own most deeply held view of poetry. First, Goethe: Arnold comment acerbically on Masson's essay ('Theories of Poetry and a New Poet') which appeared in the *North British Review* for August 1853:

> But the modern critic not only permits a false practice; he absolutely prescribes false aims. – "A true allegory of the state of one's own mind in a representative history,' the Poet is told, "is perhaps the highest thing one can attempt in the way of poetry." – And accordingly he attempts it. An allegory of the state of one's own mind, the highest problem of an art which imitates actions. No assuredly, it is not, it never can be so: no great poetical work has ever been produced with such an aim. *Faust* itself, in which something of the kind is attempted, wonderful passages as it contains, and in spite of the unsurpassed beauty of the scenes which relate to Margaret, Faust itself, judged as a whole, and judged strictly as a poetical work, is defective: its illustrious author, the greatest poet of modern times, the greatest critic of all times, would have been the first to acknowledge it; he only defended his work, indeed, by asserting it to be "something incommensurable."

This is an extraordinary lapse in what has hitherto been a coolly magisterial performance. The intensity of Arnold's aversion to allegory is based in part on his belief, expressed to Clough in 1847, that allegory is an outmoded rhetorical convention that "instantly involves you in the unnecessary."[19] However, this aversion is also fuelled by alarm at the recognition in Masson's comments on Alexander Smith and Goethe of an accurate, if unwitting, description of 'Empedocles on Etna.' Masson commends most highly what Arnold has chosen to repudiate. Arnold's tone becomes more strident and personal as he ascribes his own anxieties to the figure of the Poet, and then engages in a bout of self-persuasion. His mocking repetition of Masson's words indicates that we should relocate meaning in tone of voice. He then piles up the negatives before coming to terms with *Faust* and its creator. Arnold finds the figure of Goethe to be more tractable than his poem, just as he finds it easier to redefine his own personal objectives than to represent

'Empedocles' as a failure. Goethe, a poet-critic like himself, is esteemed more highly for his criticism than for his poetry, thus anticipating and justifying a similar shift of interest in Arnold. As Goethe is acclaimed as a greater critic than Aristotle, an enticing realm of possibility opens up for Arnold. But this occurs only after Arnold has implausibly attempted to distance himself from *Faust*. Goethe's masterpiece has somehow to be seen as less successful than the *Conversations with Eckermann* and *Dichtung und Wahrheit*, because if it is allowed in all its massive diversity to constitute a successful artistic whole, then the way is clear for acceptance of 'Empedocles' on its own terms with its relatively modest alternations between the lyrical and the cerebral. Arnold shared and understood Goethe's allegorizing tendencies – measure his praise of the Gretchen scenes[20] against his own Marguerite poems, and the stark absence of romantic love from 'Empedocles' – and in denying architectonike to *Faust* he felt he was the more emphatically damning his own poem. Having spoken through the generalized figure of the Poet, Arnold goes on to speak through and *for* Goethe, attributing to him the very powers of self-criticism he is himself in the process of exercising. *Faust*, part 2, was, of course, posthumously published, and Goethe never saw in print the whole of the masterpiece on which he worked so long and with intermittent intensity. But Goethe progressed from the piecemeal publication of episodes and larger sections to the final publication of the completed work, a process directly opposite to Arnold's diasparactive enterprise where the whole is in the process of being rent asunder. Arnold seizes on Goethe's description of his poem to Eckermann, when the conclusion of his work and life were both imminent, as "doch ganz etwas Inkommensurables." Arnold reduces this phrase to "something incommensurable," omitting the intimation of wholeness and consistency conveyed by the term *ganz*. In so doing, he fails to appreciate what can't be measured, and sounds like the insensitive chancellor in part 2, act 1 in the imperial residence. Goethe furnishes Mephistopheles with a fitting rejoinder:

> Daran erkenn ich den gelehrten Herrn!
> Was ihr nicht tastet, steht auch meilernfern,
> Was ihr nicht fasst, das fehlt euch ganz und gar,
> Was ihr nicht rechnet, glaubt ihr, sei nicht war,
> Was ihr nicht wägt, hat für euch kein Gewicht,
> Was ihr nicht münzt, das, meint ihr, gelte nicht.

"Was ihr nicht rechnet, glaubt ihr, sei nicht war." Goethe's "etwas Inkommensurables" is studiously ambiguous, designed to allow purists to betray themselves, while warning more flexible and adventurous readers against the moral, intellectual and aesthetic blight of blind allegiance to traditional notions of decorum. Arnold fails the test set by Goethe for posterity, and he fails more because of his need to deny the truth and efficacy of "Empedocles" than because of any innate inability to respond to the challenge of *Faust*.

## VI

Arnold's entire case in the 1853 Preface hangs on the notion of poetry as an art which imitates actions. He supports this view with apt quotations from the *Poetics*, by introducing a few protypical touchstones – Achilles, Prometheus, Clytemnestra, Dido – and by offering a characteristically tautological account of the grand elementary passions which move such men and women to poetic action, and their audience to experience of catharsis. Then in the fourteenth paragraph of the Preface he attempts to dramatize the mental processes of the spectator undergoing catharsis. The result is a strikingly *poetic* summation of the preceding argument, but surely also a disconcertingly static account of catharsis.[21]

> The terrible old mythic story on which the drama was founded stood, before he entered the theatre, traced in its bare outlines upon the spectator's mind; it stood in his memory, as a group of statuary, faintly seen, at the end of a long and dark vista: then came the Poet, embodying outlines, developing situations, not a word wasted, not a sentiment capriciously thrown in: stroke upon stroke, the drama proceeded: the light deepened upon the group: more and more it revealed itself to the rivetted gaze of the spectator: until at last, when the final words were spoken, it stood before him in broad sunlight, a model of immortal beauty.

Arnold was no doubt influenced in his choice of statuary as poetry's truest analogue by the ideas of Winckelmann filtered through Goethe and by his own temporary misreading of Lessing's *Laokoön* – "a little mares – nesty – but very searching." In Schiller's preface Arnold would have found a memorably sustained analogy between poetry and painting, but one that in its praise of colour and drapery, would perhaps have sounded rather too similar to Arnold's own descriptions of Elizabethan poetry.[22] Statuary, on the other hand, was clearly architectonic, synonymous with the genius of classical Greece and with a limited but attainable perfection. As Arnold says in the 'Epilogue to Lessing's Laokoön' (ca. 1864–65):

> Even in Greece,
> Where best the poet framed his piece,

Even in that Phoebus-guarded ground
Pausanias on his travels found
Good poems, if he looked, more rare
(Though many) than good statues were.

Later in the same poem we are presented with a composite image of the supreme challenge facing the poet, as opposed to the musician or painter:

But then comes his sorest spell
Of toil – he must life's *movement* tell!
The thread which binds it all in one,
And not its separate parts alone.
The *movement* he must tell of life,
Its pain and pleasure, rest and strife;
His eye must travel down, at full,
The long, unpausing spectacle;
With faithful unrelaxing force
Attend it from its primal source,
From change to change and year to year
Attend it of its mid career,
Attend it to the last repose
And solemn silence of its close.

Arnold italicizes both incidences of the word *movement* in this passage, thus clearly heeding Lessing's warning against confusion of the arts. Poetry exists in and reveals itself through time, and Arnold casts around for analogies which illuminate that situation. He draws first on weaving – on *texere* as process, not the static tapestries of 'Tristram and Iseult'; then on the alternation of emotional states, the flow of the stream ever seawards, the seasonally patterned passage of time, before drawing all these elements together in the rhyme of "repose" with "close." All this occurs in a poem which is not a self-styled "criticism of life" but a case of art criticizing art, offering a poetic supplement to Lessing's treatise. When his poetry becomes criticism in this specialized sense, Arnold's language effectively captures the *critical idea* of poetry as imitating actions dynamically enough to induce catharsis. However, this occurs only after he has abandoned any pretension to acting creatively upon this critical idea. Earlier in his career, when the challenge of Aristotelian mimesis was real for him, he too often abused the idea of action defined as mental receptivity, as in his comments in the Yale manuscript of 'Tristam and Iseult': "The misery of the present age is not in the intensity of men's suffering – but in their incapacity to suffer, enjoy, feel at all, wholly and profoundly; in their having their susceptibility eternally agacée by a continual dance of everchanging objects – the eternal tumult of the world mingling, breaking in upon, hurrying away all."[23] The facility of this disparagement is revealing. Even in his marginalia and random jottings Arnold shows a natural mastery of the idiom of Empedocles, yet even on private or informal occasions he chooses to spurn or stifle its expression.

Arnold's fear that action will precipitate chaos leads him to a version of catharsis in the preface of 1853 as progressively revealed statuary, a version that is finely imagined and executed but misleadingly static. Order and containment and repose are from the beginning of his account too much of a foregone conclusion. Far from evoking the experience of the *Oresteia*, he writes a recipe for the failure of his own *Merope*. The hopeless contradictions of his position become even more evident when one considers how groups of classical statuary such as the Laokoön seem to fit perfectly Arnold's description of 'Empedocles' and similar works as "those in which the suffering finds no vent in action; in which a continual state of mental stress is prolonged, unrelieved by incident, hope, or resistance; in which there is everything to be endured, nothing to be done." Arnold's prose at its most poetic evokes not the aesthetic deficiencies he ascribes to 'Empedocles' but rather the bases for despair which that poem so incisively explores. This is why his Lucretian tragedy beckoned so compellingly yet was never written. This is why he settled for a dramatic poem on Empedokles, only to discover – like Byron in that other dramatic poem, *Manfred* – that the message on the mountain was very different from the consolation Wordsworth chose to find on Helvellyn. When Arnold speaks in the preface of "the dialogue of the mind with itself," he is already at work supplanting the dialogue central to 'Empedocles' with a dialogue between the poet who wrote the poem and the critic who writes the preface. The result is poetry's loss and criticism's gain, for he cannot be in two places at once.

## VII

What I think of Arnold's 'Empedocles' has been available so far in this paper only through allegation and implication. The poem, I would argue, is tightly and effectively structured: moving from morning to night, from the base of Aetna to its summit, from youth through experience to death, focusing ever more exclusively on Empedocles while modulating astutely between lyrical and reflective verse. Each participant is given a distinctive voice. The action is predominantly in the realm of psychic events, but this bias sits well with the abilities and situation of the hero, asserting its own reality in a variety of haunting measures. The central figure is wonderfully apt, allowing Arnold to give us something more substantially ambiguous than *Manfred*, with a less mannered, less facile disdain for popular superstititions. Moreover, the vitality of 'Empedocles' is quite distinct from the polished glumness of Senancour's *Obermann*, and has much more in common with Hölderlin's rendering of the poet's limitations and Leopardi's brilliant reflections on Vesuvius in 'La Ginestra.' Arnold's poem may not induce a full Aristotelian catharsis or achieve the stasis of Laokoön in the course of its engagement with the infinite differentiations of Love and Strife. But this is not the point. Through his reappropriation of *Peri Phuseos* and *Katharmoi* Arnold discovers the power and probity of a poetry that scorns mere purgation for its own sake. Empedocles expresses this compellingly in the second scene of act 1:

> So, loth to suffer mute,
> We, peopling the void air,
> Make Gods to whom to impute
> The ills we ought to bear,
> With God and Fate to rail at, suffering easily.

Making a *topos* out of *choras* through the agency of language is a supremely human activity, but one to which, as a consequence, self-delusion is endemic. The true poet must act upon this knowledge, demonstrating as best he can that poetry need not be therapeutic to be good. Arnold shared in this knowledge and capacity, but could not live with such gifts beyond his reenactment of the death of Empedokles.

In creating an appropriate precinct for his own poetry, Arnold endeavoured to counter the threat of Aetna with reminiscences of an idyllic Helikon. To this end he gave Callicles the first and last words in the poem; but those words, this framing device, could not contain or subdue the menacing aspects of Empedokles. On realizing that the *Theogony* was not an antidote to 'Empedocles on Etna,' Arnold had to resort to further misprisions, and to the desperate remedy of brutal self-censorship. He could not match the acquiescence of Browning's Guido in *The Ring and the Book*, book 11: "Inscribe all human effort with one word, / Artistry's haunting curse, the Incomplete!" Nor could he match the mischief of Vischer's *Faust: Part Three* which professed to be "Treu im Geiste des zweiten Theils des Goetheschen Faust, gedichtet von Deutobold Symbolizetti Allegoriowitsch Mystikinsky." Instead,

Arnold offers us a version of poetic struggle which, like Hölderlin's, has much to teach those who would *be* poetically in the world of our own century.

*University of Saskatchewan*

NOTES

[1] Martin Heidegger, *Poetry, Language, Thought*, trans. Albert Hofstadter (New York: Harper and Row, 1971), pp. 132. 12.

[2] Cf. Job, 28: 12, 28; *Sophist*, 254. I quote the Loeb translation by Harold North Fowler (Cambridge: Harvard University Press, 1961).

[3] See, e.g., D. O'Brien, *Empedocles' Cosmic Cycle: A Reconstruction from the Fragments and Secondary Sources* (Cambridge: Cambridge University Press, 1969), pp. 104–27, 240–42, 246–47. This distinction is still clear in Philo Judaeus. Cf, e.g., *On the Confusion of Tongues*, xxvii; *On Flight and Finding*, xiv; *On Dreams*, xi.

[4] *The Life of the Mind*, 2 vols. (New York: Harcourt Brace, 1971), 1: 97.

[5] *Hesiod: Theogony*, ed. M. L. West (Oxford: Clarendon Press, 1966), p. 159. The translation given is by Dorothea Wender for Penguin Classics.

[6] *Hesiod and the Language of Poetry* (Baltimore: Johns Hopkins, 1977). I am especially indebted to Pucci's chapter on "The True and False Discourse in Hesiod."

[7] I follow the translation of John Burnet, *Early Greek Philosophy*, 4th ed. (1930; reprint, Cleveland: Meridian, 1964), p. 225.

[8] Ibid., frs. 118, 115.

[9] Cf. ibid., pp. 202–3.

[10] Cf. R. B. Harrison, *Hölderlin and Greek Literature* (Oxford: Clarendon Press, 1975), especially p. 121 ff.

[11] I follow the texts found in vol. 2 of *Hölderlin: Werke und Briefe*, ed. Friedrich Beißner and Jochen Schmidt (Frankfurt: Insel Verlag, 1969).

[12] Readers interested in the possible influence of the one poet on the other should consult Fred L. Burwick, 'Hölderlin and Arnold: Empedocles on Etna,' *Comparative Literature* 17 (1965): 24–62.

[13] *The Poems of Matthew Arnold*, ed. Kenneth Allott (London: Longmans, 1965), p. 604. All quotations from the poems and prefaces follow this edition.

[14] See *Romanticism and the Forms of Ruin: Wordsworth, Coleridge, and Modalities of Fragmentation* (Princeton: Princeton University Press, 1981).

[15] Quoted in *Victorian Poetry and Prose*, ed. Walter E. Houghton and G. Robert Stange (Boston: Houghton Mifflin, 1968), pp. 561–62, 564, 567–68.

[16] Quoted by R. H. Super in the first volume of his edition of *The Complete Prose Works of Matthew Arnold* (Ann Arbor: University of Michigan Press, 1960), p. 218.

[17] See *Friedrich Schiller: Werke in Zwei Bänden*, intro. Erwin Ackerknecht (Munich: Droemersche Verlaganstalt, 1954), 2: 28, 56.

[18] Ibid., 2: 506.

[19] *Letters of Matthew Arnold to Arthur Hugh Clough*, ed. H. F. Lowry (Oxford: Clarendon Press, 1932), p. 60.

[20] See 'A French Critic on Goethe' (1878); *Prose Works*, 8: 273.

[21] I am indebted to J. Hillis Miller's unpublished lecture on 'Empedocles' for this point, and for the inspiration behind most of my remarks on Arnold.

[22] See Schiller, 2: 509–10.

[23] Quoted by Allott in his edition of the *Poems*, p. 220.

CHRISTOPH EYKMAN

# WHAT CAN THE POEM DO TODAY? THE SELF-EVALUATION OF WESTERN POETS AFTER 1945

A survey of theoretical statements by contemporary poets on poetry reveals a variety of positions ranging from radical avant-garde to traditional. Such a bird's eye view, brief and selective as it must be, should, of course, be regarded only as a supplement to a full-range and comprehensive study of the poetry itself. What poets do in their poems and what they say that their poetry does might not always be the same thing. Yet the poets' reflections and definitions, though detached from the poetry proper, do shed some light on the state of the contemporary poem and thus deserve attention in their own right.

## I

According to the views of a number of ultra-avant-garde poets, the poem no longer expresses the thoughts or feelings of the individual author. Neither does it describe any aspect of the world. It is not "about" anything. It has no content, meaning, or message that could – by way of interpretation – be distilled from the text. The poem, instead, ignores the function of language as a medium and is based only on "the internal relations of the linguistic structure, on the network of mutual determinations created by phonological and morphological elements."[1] This is, basically, the poetic creed of a few (Swiss, German, Austrian, and South American) contemporary poets whom we associate today with the label "concrete" or "experimental" poetry. One of them is the Swiss Eugen Gomringer, whose work was influenced by the Swiss painter Max Bill who called his own nonrepresentational work "concrete painting." Bearing in mind the rather provocative poetological statements summarized above, one might expect concrete poetry to be but an empty and sterile play with words, sounds and structural elements of language. Even though the text is no longer a medium for something extra-textual, Gomringer claims – quite paradoxically, as it seems – that it still has a communicative function, even one that ultimately relates to its society and has an impact on its cultural development ("bewusst gesellschaftlich bezogen, von direktem einfluss auf neue kulturelle entwicklungen").[2]

Gomringer cites Mallarmé, Arno Holz, and Apollinaire as his poetic



*A-T. Tymieniecka (ed.), Analecta Husserliana, Vol. XVIII,* 141–156.

ancestors. Yet even more significant than this genealogy is his positive view of the age of technology and science which, in his view, needs a new more precise and more concentrated language, one that rivals the mathematical formula while being capable of establishing a harmonious bond between economic productivity and human relations under the guiding principle of functional efficiency.[3] The role of poetry is thus envisaged by this poet as one that is fully integrated into the technological age and its need for total organization and rationalization. Gomringer's theoretical writings, however, fail to demonstrate what the new role of poetry will be and why in an age of advanced science and technology the transmitter- and transporter-function of language, in other words: its role as a medium, should be abandoned while the poem is still supposed to communicate with the reader and through him with society.[4]

Like Gomringer, the German poets Franz Mon and Helmut Heissenbüttel call for the dissociation of language from subjective and objective reality, i.e., from the self and the world. The signs of language have been turned around as it were, now pointing back to language as a phenomenon per se. "The words were the houses of things," writes Mon, "now they are themselves a new kind of things."[5] Language in concrete poetry becomes auto-referential, but only to a certain degree since the word or phrase never totally shed their referential function. Mon's as well as Heissenbüttel's views spring, curiously enough, from premises contrary to those of Gomringer. On the one hand, the two authors state that language in our time shows signs of petrefaction and incrustation. It is no longer able to absorb and express our experiences for it has degenerated into the slogan or the cliché. Thus, one aspect of the act of writing poetry, according to both Mon and Heissenbüttel, must be to identify and expose the falsehood and emptiness of a language which has been abused for political propaganda, commercials etc., a language that has been turned into a mere tool for practical purposes, a language that has become merely "instrumental." The poem as a pillory showing the abuses of language – this is one of the missions the poem has according to the two authors. Another very significant mission is attempted in Mon's and Heissenbüttel's so-called text-collages. These contain word and phrase combinations which no longer follow the principle of mimesis but intentionally juxtapose heterogeneous elements in order to create out of a new constellation of linguistic elements a sense of a yet unmaterialized reality. Mon writes: "With the given materials, with the fragments of a reality we know all-too well, the collage generates by its method an "other" reality, which not only exposes the ugly side of our known world but designs at the same time patterns and rules of a new unused world which can be established only momentarily and maybe only in this artistic and artificial medium."[6] The technique of juxtaposing – in the true traditional spirit of the collage – words with heterogeneous semantic content, words that do not seem to "connect" while refusing to convey a "message," nevertheless indirectly yields a glimpse of an undefinable different reality. No blueprint for a utopia, yet shadows of an anti-world. Mon derives this technique from the structure of reality itself which is full of contradictions, ruptures, disharmonies.[7]

Heissenbüttel, on the other hand, proceeds in his reflections from a categorical denial of the existence of the individual self as a unified and centered being.[8] Since, instead of a self there is only a multiplicity of psychic phenomena, there can be no "expression" of the author as subject in the traditional romantic or even in the modern "expressionist" sense of the word. Not only does the poetic text *not* express the author's feelings or thoughts, it does not crystallize around a preconceived "idea" either.

Franz Mon does not go quite as far in this respect. According to his theory, the poet's self is not totally abandoned or dissolved: "There is a momentary totality for the individual self, not objectively but in playing the game of poetry, when it gives shape to the materials, structures, and phenomena it has seized in the creative act."[9] The disintegrated self reconstitutes itself in the creative process.

These rather radical views which we take as the point of departure of our survey, the discrediting of time-honored concepts such as expression, the lyrical persona, description, content, or the transmitter-function of language, closely parallel those held by the French *Tel Quel* group in the 1960s. French writers like Philippe Sollers, Jean-Louis Baudry, and Jean Ricardou, however, focus their attention primarily on the novel. Their ideas are – at least in part – rooted in Marxist theory which plays no significant role in the writings of Gomringer, Mon, or Heissenbüttel.

In the following paragraphs, we would like to demonstrate that, once we take a less close and more panoramic view of the way in which contemporary poets from various Western countries reflect upon their craft, it can be argued that their positions are – at least in theory – much more traditional than the ones of the "concrete" poets we just discussed. It might well be that such a survey that all but excludes the scholars' views, paying attention only to the poets' own assessment, yields a somewhat deceptive result insofar as it conceals the fact that ever since Rimbaud and Mallarmé, modern poetry has been characterized by an increased measure of hermetism, darkness, absence of logic, and dissociation from or derangement of reality as the object of "normal" sensory perception. Hugo Friedrich's book *The Structure of Modern Poetry*, which covers a field ranging from Baudelaire to the 1950s, cites many examples.[10] Nevertheless, what contemporary poets have to say *about* (and not: through) the poem, is equally valid and has to be part of a comprehensive analysis of the state of poetry in our time.

## II

The traditional view of poetry held by most contemporary poets reveals itself first and foremost in the belief that the poem expresses a personal "experience" of its author. According to James Dickey, poetry is "a transfer of human experience."[11] Frederick Cogswell defines poetry as "a response . . . to direct personal experience."[12] Alvin Aubert writes: "A poem is a verification . . . of experience in thought and feeling, but mostly the latter, for feeling is the means by which essential experience is transmitted."[13] The role of poetry is described as "an exploration of the possibilities of experience" (Raymond R. Patterson),[14] as "an incantation of exact experience that seizes the mind and the heart" (Lucien Stryk),[15] as "an attempt to catch the experience, preserve it" (Gwyn Williams),[16] as "a means of transmitting experience, feeling and emotion" (James T. Farrell).[17] According to the German poetess Hilde Domin, "one must focus upon the experience which one wants to express. . ."[18] Her compatriot Hans Bender considers it appropriate "that the poet ask himself how he could best communicate his personal experience, his subjective engagement."[19] Stanley Kunitz demands that the language of a poem "must do more than convey experience: it must embody it."[20] "I believe that a poem achieves its most effective articulation when it emerges from the intensity of a writer's lived experience," writes Shiv K. Kumar.[21] One of the motives for writing is – in the opinion of Donald Justice – "to recover and hold what would otherwise be lost totally – memory or experience."[22] Writing poetry is "a sort of description of a mental experience," writes the Norwegian poet Carl Frederik Prytz.[23] A poem can start in various different ways. One of these, tells us the American poet Galway Kinnell, "is through some experience from which I found words suddenly coming to my tongue."[24]

Numerous contemporary poets view the poem as a vehicle for the expression of emotions. The lyrical articulation of experience if often identical with the communication of the author's feelings. This is doubtlessly a countercurrent to the modernistic tradition represented by Valéry and his distrustful attitude towards "sensibilité" in the poem.[25] Already W. H. Auden stated, however, that the "final poetic order" was "the outcome of a dialectical struggle between the recollected occasions of feeling and the verbal system."[26] Christopher Hampton characterizes poems as "mediators, carriers, transmitters of the essences of feeling."[27] "The poet will aim to convey states of feeling," writes Louis Simpson,[28] and one of the entries in David Ignatow's notebooks reads: "in poetry I found my need of emotional expression gratified."[29] Charles Reznikoff states almost apodictically: "poetry is inherently an expression of feeling,"[30] and Allen Ginsberg defines what he calls "the poet's specialty" as "the rhythmic articulation of feeling."[31] Similar voices can be heard in Austria: "A poem is a reflection of a structure composed of thoughts and feelings"[32] (Erich Fried). Andreas Okopenko claims that "Emotionenmitteilung" (communication of emotions) prevails over "Faktenmiteilung" (communication of facts) in today's poem,[33] and Heidi Pataki calls poetry "eine Montage von Gefühlen" (a montage of feelings).[34]

III

Obviously, a writer who wishes to express his feelings or relate an experience to a reader by writing a poem cannot eliminate himself or the world in which that experience took place whatever transformations the self and the world might undergo in the creative process.

If the "world" is emphasized in its relation with the self, the mission of the poem is – as some contemporary poets see it – to reveal the true essence or identity of extratextual reality. Its function thus is an epistemological one, "Poetry is a search for essence, becomes essence, remains essential," states James Broughton.[35] "A poet," writes Jim Harrison, "deals in essences whether political, social or personal."[36] James Kirkup echoes this view: "In my poetry I have attempted always to express an essence both of myself and of experience."[37] The French poet Yves Bonnefoy explains in an interview (published in English) that the intention of the poem is "to tear away beings or things from the definitions or descriptions that fragment them, in order to give them back their proper identity."[38] Finally, the American Frank O'Hara makes a similar point when he declares: "It may be that poetry makes life's nebulous events tangible to me and restores their detail; or conversely, that poetry brings forth the intangible quality of incidents which are all too concrete and circumstantial."[39]

The poem's abilities to reveal identities and essences is not only directed towards the "world" but also towards the author's self. Here again many contemporary poets seem to turn away from the modernistic tradition established by Valéry and the early T. S. Eliot of 1919 (cf. his essay 'Tradition and the Individual Talent'). Far from surrendering that self as an entity that has been "lost" or dissolved – as the concrete poets would have it – many poets use the act of writing poetry for the purpose of exploring and defining the self. "Poetry invites us to the simplest and most difficult of all encounters, the encounter with ourselves," writes the German authoress Hilde Domin.[40] The Mexican author Octavio Paz compares the writing of poems to the research activities of the scientist. Both the poet and the scientist are searching for truth. Yet there is an important difference: "In the case of modern poetry, the subject of the experiment is the poet himself: he is both the observer and the phenomenon observed. His body and his psyche, his entire being, are the "field" in which all sorts of transformations take place. Modern poetry is an experimental process whereby the knowing subject is the object of knowledge. . . . "[41]

## IV

Poetic language is consequently defined by several poets as the most personal language. W. H. Auden states: "Poetry is personal speech in its purest form. It is concerned and only concerned with human beings as unique persons."[42] The American John Brinnin considers the poem to be "an autobiographical record."[43] His compatriot Richard Hugo likewise emphasizes the personal dimension in poetic language: "When you are writing a poem the words should be just a little bit more yours than anybody else's."[44] There is the language of communication as we all use it in our day-to-day dealings and then there is what the Australian poet David Malouf calls "our mysterious naming to ourselves what surrounds us."[45]

The contemporary poem, especially the one that is written in England and the United States, can do – so it seems – neither without the self of the author, nor without a "world." It cannot discard the notions of experience, expression (expression of emotion or expression in general), content or representation. This does not mean though that the contemporary poem uses language as a mere tool for the purpose of the communication of extra-textual content. While the poem still says "something," it says it – as mentioned above – in a highly personal poetic idiom. What is more, it becomes at the same time the most perfect and most highly developed form of artistic and creative language. Indeed, the poem is regarded by many contemporary poets as the crowning achievement of man's linguistic competence. "Poetry . . . is the perfection of speech . . . , " writes Octavio Paz.[46] Elsewhere he remarks: "Thanks to poetry, language reconquers its original state."[47] By offering his service to the words, the poet "returns them to the plenitude of their nature, makes them recover their being."[48] The German Hans Bender, in comparing poetic language with everyday language, finds that "poetic language must be more concentrated, more accurate, it must hit the mark–"[49] C. Day Lewis demands that the language of the poem be "a heightening of the common language."[50] It is the poet's task, says Lewis, "to purify the language of the tribe."[51] Similarly, the American poet Gary Snyder views the poem as "a special concentration of the capacities of the language."[52] According to Snyder, the poem has to achieve an "intensification and clarification of the implicit potentials of the language."[53] Writing a poem therefore means the "art of distilling language."[54] "Nobody has more precise scales for weighing words than the poet," states Hilde Domin.[55]

If poetry leads language to the heights of its potential, it also makes language encounter its limits. It is the poet who painfully experiences – just like (or possibly more than) the philosopher – that there are things for which words cannot be found. The contemporary poem frequently finds itself on the verge of silence. This is what the German poet Paul Celan means when he writes that the poem shows "eine starke Neigung zum Verstummen" (a strong tendency towards becoming silent).[56] The French poet Jean Claude Renard echoes this view: "Le language poétique se présente donc aussi comme la trace de ce qui ne peut pas être exprimé: la première présence possible (mais inqualifiable) de ce qui n'est pas présent dans les mots."[57] That which cannot be verbally expressed can be a sensory perception, or an inner image conjured up by creative intuition or possibly an emotion that is also beyond words. Yet even though the poem cannot adequately render such preverbal correlates, they can nevertheless leave a "trace," as Renard puts it, in the text. In the poem, writes Gary Snyder, you can hear an echo of those "non-linguistic, pre-linguistic, pre-verbally visualized or deeply felt areas."[58] Snyder's younger compatriot Mark Strand confirms this view: "I do think this: that very often poetry strives for experience that is preverbal or beyond words."[59] In the same context, Strand talks about the mystery that is always present "in the center of the known." Therefore, the poem contains "spaces of ignorance." If they become too apparent, they ruin the poem. "Poetry," defines Strand, "is a dance around a void, a debate with nothingness."[60]

Strand's position, however, is not exactly the same as Snyder's, since it seems that his "void" or "nothingness" denote the failure to know but not the failure to express something preverbally known. Whatever it might be, it is not present in the poem as a "trace," yet it does have – paradoxically speaking – a presence in the poetic text as that around or opposite which the poem speaks. This is what William Matthews means when he states: "I think of what poems don't say, the holes they leave, as being at least as important as the words. . . . "[61] The same poet also once says: "I use the words of a poem to organize silences. . . . "[62] Octavio Paz claims that language (taken in a broad sense) has two poles. One of them is represented by the utter precision and exactness of the mathematical formula. At the other pole we find poetry which is so to speak the spearhead of man's linguistic faculty intruding into the dark realm of silence. Between the two poles lies "the immense but finite realm of speech."[63]

The question of silence or of the inadequacy of language vis-à-vis preverbal experience is of course a very old one. Already Goethe said: "The most valuable part of our convictions cannot be expressed in words; there are things that language cannot render. Often we do not even know precisely what we see, intuit, think, remember, fantasize, or believe."[64] Later, Mallarmé, von Hofmannsthal (in his famous letter to Lord Chandos), and Rilke expressed similar views.

## V

The question of preverbal experience leads to the equally crucial question: how is the contemporary poem made? What do its authors have to say about the creative process?

Many contemporary writers insist that the poem originates from the unconscious. This, again, is a time-honored view, dating back to Plato or to Hume's and Kant's theories of creative imagination.[65] While the origin of the poem reaches down to the unconscious, there is, as many poets claim, a preverbal, nonrational prefiguration which is saturated with emotion and seems to have a preconscious rather than an unconscious status – a prefiguration of the poem itself before its verbalization. Some writers give it the (somewhat misleading) name: "idea." Although it must never be confused with rational conception, an intellectually devised outline, a "message" or intention, it is nevertheless the inner centripetal force or organizing principle of the poem. Some poets differentiate between such an "idea" and a "triggering" factor at the beginning of the creative process, a factor which might be quite peripheral in nature. It can be an experience, an image, a line or just a rhythmic cadence.[66] The German poet Karl Schwedhelm writes: "The layer into which the experience sinks is a subliminal consciousness. There, the experience is still present, yet not tangible. . . . At the same time, the experience is being integrated into an already existing 'stock' of remembered experiences ["in den vorhandenen Erlebnisbestand eingebaut . . . "] . . . Thus it becomes a theme. From the phase of mental presence [Vorstellung] this theme now has to be transformed into a verbalized entity, which is a painful process since the text that is taking shape steps out of the protective shield of its original preverbal state into the cold air of the rational world."[67] The "experience" here acts as the triggering device. Once fused with the poet's latent preverbal memory-traces of psychic experiences, it becomes a semi-conscious organizing "idea," here called "theme." Describing what he feels when he is about to start a new poem, Robert Duncan calls this idea or theme "pre-Gestalt": "It is like a musical feeling. Yes, you do have a primary feeling about fittingness . . . . It's fittingness is pre-Gestalt."[68] Here, too, the experience of a contemporary poet turns out to be but the continuation of a long tradition (if not the symptom of a timeless "law" of poetic creativity). In his letter of March 18, 1796, Friedrich Schiller writes to Goethe about the writing of poetry: "A certain musical mood precedes which in my case is then followed by the poetic idea."[69] Lew Welch observes a similar process involving a prefiguration – though this time not a musical one: "I'm trying to get out of the inside of my head some kind of wordless shape, and I take that shape and try to put it into words, and the result, if I'm successful, is a poem. Then I look with the same kind of discovery that any reader would, I say, 'Oh, that's what the shape is. . . . ' "[70]

Duncan's pre-Gestalt and Welch's wordless shape denote a preverbal generating eidos which is "a priori" to any differentiation into formal/structural aspects and semantic content or meaning. The manner in which various poets describe it may either emphasize its musical or rhythmic qualities, or its intuitive, image- or shapelike character. It might even approach the rational, conceptual dimension. Then it turns into a more or less preconceived intention or message. This third variety though is rarely acknowledged by contemporary poets who often state like Denise Levertov: "you can't start with the intelligence, if you start with the intelligence, you have nothing whatsoever."[71] The rational approach does not seem to further the creative process. "Intention endangers creation," says William Stafford.[72] Rational scrutiny in the form of "revision" only plays a part in the very late stages of the creative process in modern poetry – and even then it still is under the prerational tutelage of the pre-Gestalt. In Stephen Spender's terminology this pre-Gestalt is called "idea." It exists, so he writes, "clearly enough on some level of the mind where it yet eludes the attempt to state it. At this stage, a poem is like a face which one seems to be able to visualize clearly in the eye of memory, but when one examines it mentally or tries to think it out, feature by feature, it seems to fade."[73] What stems from the unconscious and presents itself in rhythmic form or as an image, is primarily of an emotional nature. It cannot be grasped rationally yet it can translate itself into a poetic text. Whether this text is close to the preverbal eidos, i.e., whether it fully transmits it, is another question. "The background of a poem is an unwritten poem", writes Walter Höllerer, "the written poem only gives you clues."[74] The poet's struggle to transscribe as closely and completely as possible the preverbal eidos is exactly what the French poet Guillevic describes when he notes: "Something is given, something which is a provocation, an opening. It is as if a curtain were lifted, or a wall broken down. I enter into contact with something. What is given to me is a coloration, a tonality, yes, a feeling for a thing. Then I begin . . . . In general I go without knowing where I'm going. I try to circumscribe a poetic thought, not the philosophic thought which hides behind this thing with which I have come into contact. . . . I search for what is concealed behind what I have found. . . . I remember what Braque once said: 'The painting is on the other side of the canvas.' Well, the poem is also on the other side of something, perhaps not behind the piece of paper, but it has to be located."[75]

As Guillevic states, the contemporary poet writes without knowing what he is going to say. This is not the automatic writing of the surrealists, yet the act of writing is controlled by forces which are prerational. Robert Creeley once said in an interview: "I have very little predetermination as to what literally is going to get said,"[76] and Maxine Kumin explains: "I have this recurrent phrase, or I have this insistent rhythm, or I have this concept, that I want to fiddle around with. And it isn't until I get the poem out that I find out what it was saying, what I wanted to say."[77] The same uncertainty as to what the poem is actually going to say, is expressed by Donald Hall: "In recent years, I have come to accept the beginning of a poem, or even a whole draft, without the slightest clue to the subject matter. Words come to me heavy with emotion, and I accept them even though I have no idea what they are trying to tell me."[78] In this case, not even the presence of a prerationally and preverbally organizing poetic "idea" is felt thought it might still be there after all as the mysterious inspirational voice that makes the words comes to the poet. Not only does the poet quite often not know what the lines of the poem will say. Once he has put words on paper, he does not understand their meaning. This understanding might come, however, at a later stage. As Allen Ginsberg states: "actually, in the moment of composition I don't necessarily *know* what it means, but it comes to mean something later, after a year or two, I realize that it meant something clear, unconsciously. Which takes on meaning in time, like a photograph developing slowly."[79]

## VI

The question "What can the poem do today?" can be understood as one asked by those who are looking for social and political change in their respective societies. In the light of the statements by contemporary poets on the creative process, which emphasize the role of the unconscious, the preverbal and the prerational, the fact that the poet often does not know what the text will say and what it means, it seems unlikely that we will find many proponents of a type of political poetry that embraces a cause or an ideology calling for action in a rational and unequivocal language. Indeed, most contemporary poets from Western countries deny that the poem can be a tool for political or social ends. The German poet Peter Rühmkorf writes: "If we ask the question whether a poem is capable of having an impact upon the realm of the extra-poetic, of society, to initiate social processes, to hatch revolutions, to make bombs fall later or more slowly, the answer has to be a regretful 'Never.'"[80] On the other hand, Rühmkorf insists that the poem can draw its inspiration from social and political reality. Rühmkorf's compatriot Hans Magnus Enzensberger shares Rühmkorf's views but adds an important point. The "political mission" of the poem is to "refuse any political mission, to speak for all where it does not speak for anybody, where it speaks of a tree, a rock, of something that does not exist. As long as it is present, it finds the state-of-the-nation address, the screaming of commercials, the manifesto, and the banner guilty of lying."[81] "Why write an ode?" asks the East German Günter Kunert, "if it only contains instructions and guidelines about how to practice solidarity; let's rather have the instructions themselves."[82] The American James Dickey emphatically proclaims: "The poets who have any kind of commitment to the high calling that they've chosen for themselves will not allow themselves to be stamped into official propaganda for any cause, no matter how admirable."[83] Such a statement is fully in accordance with Eugenio Montale's dictum: "ideological commitment is not a necessary or an adequate condition for a poetically vital world."[84]

Is contemporary poetry then totally a-political? Many authors would categorically deny this even though they wish to keep poetry out of the ideological rhethoric of political or social causes. Already Enzensberger's view that poetic language exposes the untruthfulness of the political jargon indirectly attributes a political function to the poem. The poet's "politics" then is to denounce the falsehood of political phraseology. This is what Philip Levine means when he writes: "I think the writing of a poem is a political act. ... Nothing is more obvious than what our politicians are doing to our language, so that if poets insist on the truth, or on an accurate rendition, or on a faithful use of language ... this is a political act."[85] The political poem is not one written by the party activist who transcribes political slogans into rhythm and rime, but by the poet who has absorbed political or social problems into his own personal world view with all its existential idiosyncracies. As Robert Bly, who wrote antiwar poetry himself, puts it: "The political activists in the literary world are wrong – they try to force political poetry out of poets by pushing them more deeply into events, making them feel guilt if they don't abandon privacy. But the truth is that the political poem comes out of the deepest privacy. ... The true political poem does not order us ... to take any specific acts: Like the personal poem, it moves to deepen awareness."[86] "A public issue is as legitimate a subject as any," writes Hilde Domin, "As long as a personal experience is presented as an exemplary, illustrative case ["insoweit eine Erfahrung exemplarisch formuliert wird"]."[87] The power of poetry, writes Howard Nemerov, is subversive, "in that it is unpredictable and beyond control or purposive application to the immediate situation: without meaning to, perhaps without especially wanting to, poetry changes the mind of the world."[88] The "awareness" about which Bly speaks, the changing of the world's mind, are an indirect and long-range effect of poetry which is achieved only if the political content is aesthetically mediated – and that means also: absorbed into the emotional and personal substance of the poetic "idea" which was discussed above. The poet Salvatore Quasimodo even goes as far as to claim that aesthetic perfection per se implies political effectiveness regardless of the poem's content: "Poetry is transformed into ethic, precisely because of its beauty: its responsibility is in direct proportion to its perfection. To write verses means to undergo judgement: and implicit in the aesthetic judgement are the social reactions to which a poem gives rise."[89] The German poet Walter Helmut Fritz expresses a similar view: "All literature is engaged", he writes, "even the most serene poem, because literature, by the fact that it is – among other things – devoid of practical utility and therefore opposes functionalism, schematization, rigidness, utilitarianism; because it creates freedom, permits involuntariness, eliminates tutelage, provides breathing room, and promotes doubt and the patience to be sceptical."[90] It would be easy to challenge the optimism expressed in such statements. But even though beauty, aesthetic lack of practical applicability may no longer be widely embraced artistic ideals, the question of the ethical impact of aesthetic qualities on the reader, listener, or viewer deserves to be raised again in our age.

Two implict premises attesting to the relatively traditional nature of contemporary poetology emerge from the poet's views on political poetry. First: the poem has a content, even if it is far from any political message, it expresses a "truth," which is aesthetically and subjectively mediated. And second: the poet as a self is an inalienable part of the creative process, especially if he is inspired by political or social motives.

What can the poem do today? Unless it belongs to the "experimental" variety, it will transmit the poet's experience, his emotions, it will strive to reveal – in true phenomenological fashion – the essence of the world and of the self. It is a heightening of the common language. At the same time, it wrestles with silence. Nevertheless, it achieves the transformation of a preverbal Gestalt or eidos into a text. It is political only insofar as a particular political or social issue is absorbed into personal experience. That a poem does present an experience, whether political or not, that it is the poetic verbalization of a preverbal eidos, and that it lays open the essence of world and self are by no means contradictory propositions. They are phases of the poetic process merging almost imperceptibly into each other and thereby constituting the contemporary poem.

*Boston College*

## NOTES

1 Eugen Gomringer, *worte sind schatten: die konstellationen 1951–1968* (Hamburg: Rowohlt, 1969), p. 291. All translations of quotes not originally published in English are mine.

2 Ibid., p. 287.

3 Ibid., p. 289.

4 Cf. Jürgen Theobaldy and Gustav Zürcher, *Veränderungen der Lyrik: Über westdeutsche Gedichte seit 1965* (Munich, 1976), p. 47, for a critical view of Gomringer's naive trust in technological progress. For a comprehensive view of "concrete" poetry see also Harald Hartung, *Experimentelle Literatur und konkrete Poesie* (Göttingen: Vandenhoeck & Ruprecht, 1975). Hartung correctly points out that Gomringer's own poetry is often less "concrete" than his theory insofar as it does refer to extra-textual realia rather than being autoreferential (p. 46).

5 Franz Mon, *Texte über Text* (Neuwied: Luchterhand, 1970), p. 135.

6 Ibid., p. 130.

[7] Ibid., p. 129.

[8] Helmut Heissenbüttel, *Über Literatur*, 2d ed. (Munich: Deutscher Taschenbuchverlag, 1972), p. 191.

[9] Mon, *Texte über Texte*, p. 39.

[10] Hugo Friedrich, *Die Struktur der modernen Lyrik*, 2d ed. (Hamburg: Rowohlt, 1968).

[11] Jame Dickey, *Sorties* (Garden city, N.Y.: Doubleday, 1971), p. 93.

[12] James Vinson, ed., *Contemporary Poets*, 2d ed. (London: St. James, Press, 1976), p. 266.

[13] Ibid., pp. 39 f.

[14] Ibid., p. 1173.

[15] Ibid., p. 1291.

[16] Ibid., p. 1687.

[17] James T. Farrell, *Literary Essays*, ed. Jack A. Robbins (Port Washington, N.Y., 1976), p. 134.

[18] Karl H. Van d'Elden, ed., *West German Poets on Society and Politics* (Detroit: Wayne State University Press, 1979), p. 84.

[19] Ibid., p. 120.

[20] Stanley Kunitz, 'Interview with Chris Busa,' *Paris Review* 83 (1982):221.

[21] James Vinson and D. L. Kirkpatrick, eds., *Contemporary Poets*, 3d ed. (New York: St. Martin' Press, 1980), p. 862.

[22] 'The Effacement of Self: An Interview with Donald Justice,' *Ohio Review* **16**, no. 3 (1975):50.

[23] Peter Orr ed., *The Poet Speaks* (New York: Barnes & Noble, 1966), p. 193.

[24] Robert Boyers, ed., *Contemporary Poetry in America: Essays and Interviews*, 2d ed. (New York: Schocken, 1975), p. 242.

[25] Cf. Jean Hytier, *La Poëtique de Valéry* (Paris: Armand Colin, 1953).

[26] W. H. Auden, *The Dyer's Hand and Other Essays* (New York: Random House, 1962), p. 68. Compare also quotes pp. 13 and 17.

[27] Vinson, *Contemporary Poets*, p. 631.

[28] William Heyen, ed., *American Poets in 1976* (Indianapolis: Bobbs-Merrill, 1976), p. 335.

[29] Ralph J. Mills Jr., ed., *The Notebooks of David Ignatow* (Chicago: Swallow Press, 1973), p. 90.

[30] L. S. Dembo and Cyrena N. Pondrom, *The Contemporary Writer: Interviews with Sixteen Novelists and Poets* (Madison: University of Wisconsin Press, 1972), p. 209.

[31] 'An Interview with Allen Ginsberg,' *Nimrod* 19, no. 1 (1971):24.

[32] 'Zustand 69: Eine Rundfrage und ihre Beantwortung,' *Literatur und Kritik* 32 (1969):74.

[33] Ibid., p. 84.

[34] Ibid., p. 85.

[35] Vinson, *Contemporary Poets*, p. 184.

[36] Ibid., p. 642.

[37] Ibid., p. 839.

[38] Serge Gavronsky, ed., *Poems & Texts: An Anthology of French Poems, Translations and Interviews* (New York: October House, 1969), p. 127.

[39] Frank O'Hara, *Standing Still and Walking in New York*, ed. Donald Allen (Bolinas, Calif. Grey Fox Press, 1975), p. 112.
[40] Hilde Domin, *Wozu Lyrik heute: Dichtung und Leser in der gesteuerten Gesellschaft* (Munich: Piper, 1968), p. 12.
[41] Octavio Paz, *Alternating Current*, trans. Helen R. Lane (New York: Viking Press, 1973), pp. 74 f.
[42] W. H. Auden, *Secondary Worlds: Essays* (New York: Random House, 1968), p. 130.
[43] Howard Nemerov, ed., *Poets on Poetry* (New York: Basic Books, 1966), p. 130.
[44] 'World Down to Size: A talk with Richard Hugo,' *December* 11, no. 2 (1969):197.
[45] Vinson and Kirkpatrick, *Contemporary Poets*, 3d ed., p. 979.
[46] Paz, *Alternating Current*, p. 172.
[47] Octavio Paz, *The Bow and the Lyre: The Poem. The Poetic Revelation. Poetry and History*, trans. Ruth L. C. Simms (Austin: University of Texas Press, 1973), p. 37.
[48] Ibid., p. 37.
[49] Van d'Elden, *West German Poets*, p. 121.
[50] C. Day Lewis, *The Lyric Impulse: The Charles Eliot Norton Lectures 1964–1965* (Cambridge: Harvard University Press, 1965), p. 23.
[51] Ibid., p. 23.
[52] William Scott McLean, ed., *Gary Snyder: The Real Work: Interviews and Talks 1964–1979* (New York: New Directions, 1980), pp. 44 f.
[53] Ibid., p. 70.
[54] Ibid., p. 165.
[55] Hilde Domin, 'Das politische Gedicht und die Öffentlichkeit,' in *Poesie und Politik: Zur Situation der Literatur in Deutschland*, ed. Wolfgang Kuttenkeuler (Stuttgart: Kohlhammer, 1973), p. 383.
[56] *Büchner-Preis-Reden 1951–1971* (Stuttgart: Reclam, 1972), p. 97. West German poetry after 1965, however, is more politically and socially oriented than Celan's and thus shows a naive trust in the communicative power of language. Cf. Theobaldy and Zürcher.
[57] Jean-Claude Renard, *Notes sur la Poèsie* (Paris: Editions du Seuil, 1970), p. 28.
[58] Ekbert Faas, *Towards a New American Poetics: Essays and Interviews* (Santa Barbara: Black Sparrow Press, 1978), p. 129.
[59] 'Mark Strand, Interviewed by David Brooks,' *New Poetry* 25, no. 3 (1977):35.
[60] Ibid., p. 35.
[61] Scott Chisholm, 'An Interview with William Matthews,' *Ironwood* 3 (1973):68.
[62] Vinson, *Contemporary Poets*, p. 998.
[63] Paz, Alternating Current, p. 69.
[64] *Briefwechsel zwischen Goethe und Kaspar Graf von Sternberg 1820–1832*, ed. F. Th. Bratranek (Vienna, 1966), p. 242.
[65] Cf. Mary Warnock, *Imagination* (London: Faber & Faber, 1976), Chaps. 1–2.
[66] Richard Hugo, *The Triggering Town: Lectures and Essays on Poetry and Writing* (New York: W. W. Norton, 1979), p. 4.
[67] *Auskűunfte: Werkstattgespräche mit DDR – Autoren* (Weimar: Aufbau Verlag, 1976), p. 80.
[68] Faas, *Towards a New American Poetics*, p. 76.
[69] Emil Staiger, ed., *Der Briefwechsel zwischen Schiller und Goethe* (Frankfurt: Insel, 1966), p. 198.

[70] *On Bread & Poetry: A Panel Discussion with Gary Snyder, Lew Welch & Philip Whalen*, ed. Donald Allen (Bolinas, Calif. Grey Fox Press, 1977), p. 23.
[71] David Ossman, ed., *The Sullen Art* (New York: Corinth Books, 1963), p. 74.
[72] Stephen Berg and Robert Mezey, eds., *The New Naked Poetry: Recent American Poetry in Open Forms* (Indianapolis: Bobbs-Merrill, 1976), p. 458.
[73] Stephen Spender, *The Making of a Poem* (London: Hamish Hamilton, 1955), p. 50.
[74] Walter Höllerer, 'Mauerschau,' *Akzente* 4 (1956):320.
[75] Gavronsky, *Poems and Texts*, pp. 84 f.
[76] Faas, *Towards a New American Poetics*, p. 173.
[77] Maxine Kumin, *To Make a Prairie: Essays on Poets, Poetry and Country Living* (Ann Arbor: University of Michigan Press, 1979), p. 24.
[78] Donald Hall, *Goatfoot, Milktongue, Twinbird: Interviews, Essays, and Notes 1970–1976* (Ann Arbor: University of Michigan Press, 1978), p. 21.
[79] *Writers at Work: The Paris Review Interviews.* 3d ser. (New York: Viking Press, 1967), p. 296.
[80] Hans Bender, ed., *Mein Gedicht ist mein Messer: Lyriker zu ihren Gedichten* (Munich: List, 1964), p. 153.
[81] Hans Magnus Enzensberger, *Einzelheiten II: Poesie und Politik* (Frankfurt: Suhrkamp, 1964), p. 136.
[82] *Auskünfte*, p. 467.
[83] James Dickey, *Self-Interviews*, ed. Barbara and James Reiss (Garden City, N.Y.: Doubleday, 1970), p. 72.
[84] Eugenio Montale, *Poet in Our Time*, trans. Alastair Hamilton (New York: Urizen Books, 1976), p. 82.
[85] Arthur E. Smith, 'Poetry and Politics: An Interview with Philip Levine,' *Partisan Review* 42, no. 1 (1975):69.
[86] Robert Bly, *Talking All Morning* (Ann Arbor: University of Michigan Press, 1980). pp. 98 f.
[87] Domin, 'Das politische Gedicht und die Öffentlichkeit,' p. 378.
[88] *National Poetry Festival*, held in the Library of Congress, October 22–24, 1962. (1964), p. 158.
[89] *The Selected Writings of Salvatore Quasimodo*, ed. and trans. Allen Mandelbaum (New York: Farrar, Straus & Cudahy, 1960), p. 14.
[90] Van d'Elden, *West German Poets*, p. 181.

TERESA GELLA

# POETRY AS ESSENTIAL GRAPHS

A subtitle might be "The poet's cry after his lost particles" or "The cry of the Thirdness after its replica in the Secondness." Moreover, the term "Essential" as I use it here is Peirce's definition of "Essence," namely: "*Intelligible element of the possibility of being.*" (I owe to Charles S. Peirce the idea of the possible correlation of "graphs" and "essential" as I put in the title – though I am not using his term "graphs" in the sense of "Existential Graphs" [the name of his system of logic diagrams]).[1]

Essence in Poetry became a visible existence. It constitutes traces and graphs of our identification. However, sometimes not only are they lost but also our need for their search.

Our contemporary civilization is not a good companion for man; he is man sometimes, and "in spite of" our civilization, just as the remedies of humanism are "in spite of" and not "thanks to" our civilization. Man should be on a constant alert so as not to be provoked into using the tools (no matter how nicely handed to him) of his own one-dimensional coexistential survival.

In the frame of that, the disappointments of poets and artists with contemporary civilization are caused by its inability to preserve the human species in an environment which is right for them: (1) in the exact physiological and mental sense and (2) in the sense of extrinsic structures in which man is now programmed to participate.

Beyond certain unfriendly philosophies by which we are surrounded, we are denied: first, the possibility to produce mechanisms; second, the possibility to produce those mechanisms which in turn would produce creative relations; third, the creative potentiality of those relations; and fourth, we are denied the Origin Matter.

*First*, by mechanisms I mean here sensitivity – of perception, apperception, or sensitive empathy – in other words, the mechanisms causing the cultivation of cultural, moral, esthetic sensitivity. Those mechanisms are shattered now at early childhood or even before birth, genetically, chemically, or environmentally. Our sensitivity is now much more dull – and here we proceed to the *second* problem, that of the mechanisms which are supposed to produce creative relations. Being already unprivileged, we meet the next oppressive element in our culture when the epistemological motiva-



*A-T. Tymieniecka (ed.), Analecta Husserliana, Vol. XVIII*, 157–159.

tions are now comprehended post-positivistically and are used with blessings only for science. However, in the humanities or in the social sciences or in the arts, the epistemological human drive is immediately processed by Behaviorism or Freudianism, whereas the target is nothing else but just this epistemological drive as well as its basic preconditions mentioned before.

*Third*, the possibility of potential creativity was taken away from us because everything is creativity now, which means that we must consume ourselves through the products we have not yet materialized through the creative act. It is so because one does not study for the beauty of knowledge; the first and the last motivation is the achievement of a better standard of vegetation. As for our basic abstract feelings such as happiness and joy, we translate them into, or correlate them with, the consumptive utilitarity of the economic market. This shows man with the "surplus" of emotions he has to spend. Consumption methodologically contradicts the process of comprehension, and therefore we have so far not misused the term. Now we are supposed to consume *Hamlet* or *King Lear* which therefore are treated as products to be prepared and consumed (which is perfectly shown in the school system and mass media). In this way, in our time, the "essential" stream of our existence-coexistence is possessed, or represented, by relativistically and behavioristically programmed particles or entities, which are, in turn, comprehended and absorbed by us the most important "essence" or "representaments" (vehicles). In this dimension, *King Lear* is a story about a father (the king), rich but stupid (because he didn't know how to handle his daughters). What is symptomatic is that the young person today is unable not only to create something like *King Lear* but also to watch those open, intuitive, ready-made units. He just rejects presented ideas or makes abbreviations of the drama according to the rule of linearity taken from the soap operas. In other words, the space between us and what we may comprehend is filled up with ready-made products of translations lacking in potentiality as well as our withdrawal from fighting it. This "unpotential matter" keeps us from the joy of being with nature, i.e., *potential nature*, as nature is. It happened that when we got rid of this nature's potentiality we rejected at that very moment our "essential" needs of identifying ourselves through the *epistimological generativity* of our various appearances. Hence, our recent contiguity with nature takes place not only at the moment of philosophical weightlessness, but also at the moment of the antibody vaccination of our deeper intrinsic structure.

*Fourth*, Origin Matter having been taken from us means the categories, such as "*quality*," "*substance*," "*relation*," – the basic principles of mental capability – are constantly misused by us.

In closing, all that is now denied to us, is what I call the "epistemological generativity" or possibility of human transgression (from the Latin "transgressio," and "transgredi") which is as inseparable from mankind as evolution is from nature.

Additionally, our civilization has produced another "existence-like" product, another "unpotential matter." This mass product is the stereotype or conditional response toward the natural or (for this reason) artifically produced stimuli. This "unpotential matter" within which we are programmed is invading our nature of thinking which means that it also has the capability of forming us because of its location within the structure of our "essential" particles.

## NOTE

1 *Collected Papers of Charles Sanders Peirce*, Ed. Charles Hartshorne, Paul Weis, and Arthur Burks (Cambridge, Mass.: Harvard University Press, 1931–1958).

# PART TWO

# THE EPIC COORDINATE
# IN HUMAN HISTORICITY

# A. HISTORICITY OF HUMAN EXISTENCE “CIPHERED” IN THE GREAT EPOS

L. M. FINDLAY

# THE SHIELD AND THE HORIZON: HOMERIC EKPHRASIS AND HISTORY

## I

It is now fifty years since Husserl formulated the "enormous task" facing transcendental phenomenology:

> the task of *carrying out all phenomenological investigations* within the unity of a systematic and all-embracing order by following, as our mobile clue, a system to be found out level by level, the system namely of all objects of possible consciousness, including the system of their formal and material categories – the task, I say, of carrying out such investigations as *corresponding constitutional investigations*, one based upon another, and all of them interconnected, in a strictly systematic fashion.[1]

I suspect that we are still floundering far short of such a comprehending system, and that such progress as has been made is under constant threat of deconstruction. However, particular areas continue to beckon compellingly from the horizon – and epic is surely one of these. Some thirty years ago Ernst Robert Curtius lamented the lack of a "comparative phenomenology of heroism, heroic poetry, and the heroic ideal."[2] The theme of this conference might well be seen as originating in Curtius' remark, and the discussions here cannot but encourage the completion of the great work required. However, my own objective is much more modest. I will examine the behaviour of one particular "mobile clue" among the temporalities of the *Iliad*, and suggest how and to what purposes it has been appropriated by later poets and translators in realizing their own historicity.

## II

I will begin with a consideration of Greek epic poetry and its relation to time and historicity. I approach this aspect of the *Iliad* via my own appropriation of Aristotle's *Poetics*, and not only because his concern for the uniqueness of literary genres sits well with the emphasis of this conference. Aristotle contrasts the action of tragedy, confined to (approximately) one revolution of the sun, with the unlimited temporal range of epic (ἀόριστος τῷ χρόνῳ).[3] He also notes that both tragedy and epic originally shared the same permissive



*A-T. Tymieniecka (ed.), Analecta Husserliana, Vol. XVIII*, 163–174.

time-frame, implying in characteristic fashion that they have evolved over time to their present appropriately differentiated forms. However, Aristotle does not mean to suggest that epic is entirely permissive. Thus, in his famous comparison of epic and historical narrative (1451b), chronicle is the slave of the actual, but epic must contend not only with the flexibility of the probable but also with the demands of the necessary (κατὰ τὸ εἰκὸς ἢ τὸ ἀναγκαῖον). The domain of poetry is the plausible, and, Aristotle believes, this means it is more scientific and serious than history (καὶ φιλοσοφώτερον καὶ σπουδαιότερον): heavy-weight words for Aristotle, as their analogues are for us. There is energy as well as earnestness contributing to σπουδαῖος, and to the special power and dignity of epic poetry. This potency also derives in part from epic's engagement, not with the particulars of history, but with general and universal terms and notions: τα καθόλου. For Aristotle the logical and linguistic force of this expression functions within the bounds of the probable and the necessary. We have to wait till Polybius for its application to the historian's craft in the idea of universal history: καθολικὴ καὶ κοινὴ ιστορία.[4]

Later in the *Poetics* (1459a) the sense of the uniqueness of epic is augmented by further comparison with tragedy and history. Epic shares with tragedy a structure such as is found in living organisms – each complete, unique, and so constituted as to produce its own appropriate pleasure (τὴν οἰκείαν ἡδονήν). Epic need not shoulder history's responsibility for a full record of the events of any period. The exhaustiveness of such accounts reveals neither probable nor necessary binding of particulars, nor the organic sequences of genesis, development and decadence that Polybius would discover in the materials of history,[5] but the causal interconnection of events (ὧν ἕκαστον ὡς ἔτυχεν ἔχει πρὸς ἄλληλα). As usual Homer knows better than other poets. The Trojan War would commend itself to the historian as a unitary topic, while the documentary imperative would produce a narrative stuffed with diverse incident. But epic is not casual serial held together by the mere facts of actual sequence. The epic action must be εὐσύνοπτος – easily taken in at a glance – and the *Iliad* thus selects one part of the Trojan story, incorporating other incidents only in so far as they are compatible with the aesthetic requirements of epic poetry. In making this point about episodes such as the Catalogue of Ships, Aristotle uses a tactful verb (διαλαμβάνω̃) that has apposite associations with the harmonious embrace of particulars.[6] Variety is not achieved at the expense of organic unity, nor by ignoring the integrity of incorporated incident.

The *Iliad* is ἁπλοῦν καὶ παθητικόν – simple and concerned with suffering

– and it increases its scope through simultaneity rather than sequentiality. Epic enjoys this advantage over tragedy that it can introduce simultaneous diversity, avoiding monotony while enlarging the narrative impressively. Aristotle again insists that this is compatible with the overall embraceability of the action in a single view (*συνορᾶσθαι τὴν ἀρχὴν καὶ τὸ τέλος*), before moving from matters of scope to questions of quality. Tragedy is once more the basis for comparison, and epic is seen as its superior in assimilating the mysterious, which is in Aristotle's view the main source of the marvellous. In seeing *τὸ θαυμαστόν* as dependent on *τὸ ἄλογον* (1460a) he introduces opportunely the connection of mystery with the limits of language as well as of reason. The visual emphasis of Aristotle's terms for epic unity is maintained in the contrast between what can be seen on stage in tragedy and what can be 'seen' in epic where the participants are invisible (*διὰ τὸ μὴ ὁρᾶν εἰς τὸν πράττοντα*). The foregrounding and coherent choreography of physical gesture in tragedy is positively characterized by Aristotle, but the mode of visualization in epic poetry is given only negatively and by implication. He provides further evidence of epic's management of the marvellous in the interests of pleasure, citing the widespread practice of embellishing stories in the belief that such enlargement of narrative always pleases. The length of the *Iliad* thus results from a refinement of a natural tendency of narrative to augment and elaborate. We may see this line of reasoning as consistent with the earlier definition of beauty as a function of magnitude and order (*τὸ γὰρ καλὸν ἐν μεγέθει καὶ τάξει*: 1450b 36–37), but it is difficult to accept the paradox of increasing *τὸ ἄλογον* by saying more, or accommodating the irrational or improbable in a narrative shaped *κατὰ λογον* (*κατὰ εἰκὸς*). Nor does Aristotle provide immediate clarification.

This apparently paradoxical situation may be better understood by considering further the mode of visualization in epic poetry. Aristotle cites the example of Achilles' pursuit of Hector, alleging that it would make no sense on stage to have the Greeks refraining from harming Hector because of some scarcely perceptible signal from Achilles, but that in book 22 of the *Iliad* the potential absurdity of the situation escapes our notice. How do we (and the text) bracket inappropriate dramaturgical (or empirical) possibilities? Aristotle commends Homer for briskly quitting his scenes so that we may experience events through the minds of the participants (1460a). Although this is often the case, it does not adequately account for what happens in Aristotle's example. After Hector takes flight in the face of avenging Achilles the narrative takes us three times in pursuit round Troy before a typical Olympian interpolation (22.166–78). Zeus is scolded by

Athena for pitying Hector. Athena then commences her descent from Olympos to set the imprimatur of Victory on Achilles.[7] At this point (188) begins the section containing Aristotle's example. Through whose eyes do we see what follows? We move swiftly from Athena to Achilles; his consciousness of certain victory, his houndlike visual fixing of his prey. However, the perspective shifts again to the chase as it might appear to an unaligned witness, whether human or divine. The description of the chase *has* been interrupted in the text but its simultaneity with the exchanges on Olympos is reaffirmed via a dream simile:

As in a dream a man is not able to follow one who runs from him, nor can the runner escape, nor the other pursue him, so he could not run him down in his speed, nor the other get clear.[8]

The employment of an oneiric type (*ὡς δ᾽ἐν ὀνείρῳ οὐ δύναται φεύγοντα διώκειν*: 199) evokes our experience of unresolved anxiety, while insinuating the private timelessness of this pursuit. (It could as well be a repeated motif on the side of a vase as in the continuum of a dream.) But it must end, as must the larger conflict of which it forms the core, for Apollo has interceded on Hector's behalf for the very last time (*πύμαστόν τε καὶ ὕτσατον*); and Achilles' men have refrained from letting fly at the Trojan only because Achilles signals with his head that this is *his* quarry. There seems to be no disproportion between Achilles' head held back in token of refusal, and the compliance of his men – partly because *we* contribute more actively to the visualizing process here than if we were responding to a dramatic performance based on this passage. We visualize when prompted by the text, especially in the dream simile's appeal to one of the mind's most fertile sources of imagery. We may also be inclined to accept the efficacy of Achilles' head movements precisely because they are visible (however minimally) to his men, unlike the similar gesture of refusal by Zeus in Book Sixteen to Achilles' prayer that Patroklos return from the fray unharmed.[9]

What Aristotle only implies, and what our experience of the *Iliad* continuously affirms, is not that the *sensus plenior* of epic poetry is attained by avoiding visual embarrassments by virtue of an intensified appeal to the ear. We *see* the poem as well as hear it. Its greater scope for the marvellous, whose chief source is the mysterious (*τὸ ἄλογον*), reveals our own mysterious capacity for creating meaning and coherence, as much as the text's intentional generic refusal to supply the *différence* between the potentialities fittingly realized in epic poetry and the actualization of a circumstantially similar dramatic text in performance. The dynamics of this situation are suggestively formulated in Hans-Georg Gadamer's *Truth and Method*:

> It is not being-in-itself that is increasingly revealed when Homer's *Iliad* and Alexander's *Indian Campaign* speaks to us in the new appropriation of tradition but, as in genuine conversation, something emerges that is contained in neither of the partners by himself.[10]

This is certainly closer to my own experience of the Homeric poems than the version given by Eric Auerbach in his seminal chapter on 'Odysseus' Scar' in *Mimesis*:

> To be sure, the aesthetic effect thus produced was soon noticed and thereafter consciously sought; but the more original cause must have been in the basic impulse of the Homeric style: to represent phenomena in a fully externalized form, visible and palpable in all their parts, and completely fixed in their spatial and temporal relations. Nor do psychological processes receive any other treatment; here too nothing must remain hidden and unexpressed . . . the Homeric style knows only a foreground, only a uniformly illuminated, uniformly objective present.[11]

Overzealous contrast with the style of the Book of Genesis leads Auerbach to ascribe an excessively cosy metaphysics of presence to Homer. Whether the phenomenology of the *Iliad* is essentialist or existentialist, it is certainly more complex than Auerbach seems to allow.

III

The *Iliad* is concerned with history, and its personae feel and struggle with the burden of the past. The poem depicts events already some five hundred years distant, while its historical digressions, in contrast to those in the *Odyssey*, are, in G. S. Kirk's words, "concentrated on the experiences of earlier generations."[12] The *Iliad* presents ten years of war that culminated in the fall of Troy, but it does not depend for its success on documentary accuracy and completeness. In fact its transparently partial appropriation of the past would seem inevitable and authentic to thinkers like Gadamer who view past and present as reciprocal creations.[13] The *Iliad* is unembarrassed by the anachronistic mix of detail in, for example, the description of armour and artifacts, for its subject is not so much archaeological accuracy or the war *per se*, but the wrath of Achilles. The conflict takes its course, as does the passion of the hero, and they eventually reintegrate more fully when he dons his new armour and avenges the death of Patroklos. Achilles is a constant unifying force, even when ostensibly absent from the action:

simultaneities proliferate, multiplying perspectives where appropriate, convincingly corrugating the surface of epic time, diversifying the action within the bounds of aesthetic order as the consciousness behind the poem establishes its massive but tactful dominion over human action and reflection.

One might hope to come to terms with time in the *Iliad* by means of some conveniently encapsulating myth, but one would be disappointed. As Jacqueline de Romilly notes in her fine study of *Time in Greek Tragedy*:

> Time was not a god for the Greeks. It only became one very late and in a very special form: in hellenistic times, it was worshipped under the name *Aiōn*; but *Aiōn* then meant a sort of sacred, eternal time, which is very differnt from the ordinary time, or *chronos*.
>
> As regards chronos, nothing similar occurs. That the word is never the subject of a verb in Homer means it is not even personified. And in ancient times *chronos* had nothing to do with Kronos the son of Uranus and the father of Zeus. Indeed, confusion between eternal and ordinary time does not seem to have ever been normal in Greek thought.[14]

The continuously graspable subject of the *Iliad*, considered in conjunction with its use of simultaneity and sequence, suggests a different course from that of the mythographer or allegorist: namely, to amplify the phenomenological implications of Aristotle's remarks on Homer so that we may recognize, *inter alia*, the force of Roman Ingarden's account of 'The Order of Sequence in a Literary Work.'[15] The *Iliad* does indeed exist simultaneously in all its parts; and there is also a "*one-sidedness of conditioning* in the constitution of its sequences." Patroklos surely must die and lose Achilles' armour to Hector *before* Hephaestus' forging of new and more glorious replacements for a hero once again disposed to use them. Nor is conversation, "genuine" or otherwise, reversible. Events in the poem are "founded" on those that precede them, even when those events intervene between the beginning and conclusion of a particular action or are paratactically adjacent to it, though each event also has elements peculiar to it alone, independent of the foundation it may enjoy. In other words, the *Iliad* may be apprehended *eusynoptically* as a structure outside temporal sequence; and it may also be realized incrementally along the linearity of its own internal sequences. That this poem is a self-assessing artifact, aware of the conditions of phenomenological existence and grounding its own aesthetic and model of consciousness thereon, is disclosed in an especially telling fashion in its use of one particular device – ἔκφρασις. It is here that Jacqueline de Romilly's comment about the Greeks never normally confusing *Aiōn* and *chronos* needs to be qualified by examination of the *Iliad*'s exploration of the boundary between the eternal and the transient.

## IV

*Ekphrasis* (literally, "speaking out," "uttering forth") denotes the use of language for description of, commentary or reflection on, or for making vocal an otherwise silent artifact. The description of the shield of Achilles in book 18 of the *Iliad* is usually given as the prototype of this device.[16] *Ekphrasis* is found in all literary genres in all periods, but the prominence of its role in epic is what concerns us here. Its later manifestations in epic are most often characterized, as George Kurman has noted, by emphasis on the features of "(1) historicity, (2) imminent quickening, and (3) timeless striving."[17] It complements rather than supplanting epic simile, dream and prophecy; reconciling the topoi of *Deus artifex, Natura artifex*, and *homo faber*. Such features and tendencies are already visible in the pseudo-Hesiodic *Shield of Herakles*, and shape and license many a rhetorical set-piece description in epic thereafter. However, the focus here will be on the Homeric prototype which seems to me "richer than all its tribe."

In Book Eighteen of the *Iliad* Thetis requests new armour for her grief-stricken son, and Hephaestus complies. An implement of war is commissioned by a loving mother, and becomes a thing of beauty produced by a god for the protection of a doomed mortal. The poem takes this opportunity to sing Creation, unfolding the details of the shield's structure and ornament in a way that places human action and divine intervention historically, aesthetically, cosmologically, while keeping constantly before us both the process and the product.[18] We apprehend making in the making: this artifact would not be possible without either its materials or strong-armed Hephaestus, while his response to this sudden request enlists physical skills, technical knowledge, and an *intuitive* architectonic sense. It is perhaps not only because he is a god that he does not stop to scratch his head or curse at any point. Genius is gratified in the exercising of it; the skills of the ἀοιδός are realized and perfected in performance. Hephaestus' values are, like the bard's, *integrated* in the work. The events depicted on their surface would be consumed irretrievably by *chronos*, the ideas of order-in-conflict informing their design would remain empty abstractions, were it not for the transcendent intentions realized in both shield and poem. As a maker, Hephaestus includes in his work sympathetic references to fellow makers – to the dancing-floor at Knossos made by Daedalos for Ariadne (591–92), the testing procedures of the potter at his wheel (600–601), about to begin his making at the moment when Hephaestus is close to concluding his commission. This commom participation of human and divine in artistic production is balanced by their share in the artistic product. Hephaestus thus depicts Ares and Pallas Athene as golden of feature and raiment, beautiful and conspicuously large in their armour because of their divinity (516–19). But the greater area of the shield is given over to humanity and nature, and a valuable gold is used to render ploughed earth, the vintage, herdsmen and their cattle, and the daggers of the Cretan dancers. The shield captures nature in artifice, juxtaposing the natural predation of two lions bringing down a bull (579–80), and the tragic bellicosity of man that gives such bitter point to the *personification* of Hatred, Confusion and Death (*ἐν δ' Ἔρις εν δὲ Κυδοιμὸς ὁμίλεον, ἐν δ' ὀλοὴ Κήρ*: 535). It balances the harmony of number expressed in music, dancing, marriage, the shared rhythms of agriculture and the seasons, against the discord of litigation and political strife. The immensely rich *ekphrasis* of the shield concludes as it began with the rim: the shining triple rim of the opening is now further characterized as the location of River Ocean, container of the infinite – a function so essential that Okeanos cannot leave his post to join the other gods on Olympos at the opening of book 22. This *ekphrasis* has explored the "phenomenology of roundness" in a way that would no doubt have gratified Gaston Bachelard.[19] Within the shield the circuit of dance and season and the sacred circle wherein the elders deliberate on seats of polished stone are effectively contrasted with the encirclement of military siege and the dismal cyclicality of strife. On the verge is Okeanos, both ultimate and without end, confinement and defence; and beyond him chaos is as certain as it is beyond the protective mass of the shield *qua* shield. Furthermore, in the context of the *Iliad* this *ekphrasis* fulfills an essential function as a deliberative episode, a respite of a kind from the rigours of war, but a respite that balances with superb tact the demands of life and the demands of art. The *Iliad* does not petrify imaginative process into utopian products of a craven or exhausted mind. *Ekphrasis* will build an invoy tower only for later and lesser makes.[20]

## V

In terms of Greek cosmology Okeanos is the horizon *in toto,*[21] but individual human beings exist and intend in face of only a segment of that totality alleged most compellingly by art. How one characterizes the relation between *Umwelt* and horizon is crucial,[22] of course, and it is no accident that Hephaestus, before fulfilling Thetis' commission, recalls his obligation to her in a way that prepares us for the *discordia concors* achieved in the shield:

She saved me when I suffered much at the time of my great fall through the will of my own brazen-faced mother, who wanted to hide me, for being lame. Then my soul would have taken much suffering had not Eurynome and Thetis caught me and held me, Eurynome, daughter of Ocean, whose stream bends back in a circle. With them I worked nine years as a smith, and wrought many intricate things; pins that bend back, curved clasps, cups, necklaces, working there in the hollow of the cave, and the stream of Ocean around us went on forever with its foam and its murmur. (18.395–403)

The making of art requires a haven, to be sure, but if the artist is too cosily closeted apart from the hurly-burly of existence it may well be he will produce little else but baubles – *δαιδαλα πολλά* whose contours naively imitate the bending back upon itself of River Ocean, instead of exploring the phenomenology of roundness; *κάλυκες* to adorn women, not armour for a hero like Achilles. Hephaestus is wiser and more practised now, but the description of his making of the shield touches such an important nerve in all of us that many of the *Iliad*'s translators have enforced their own sense of Being-in-the-World at the expense of the poem's. Comparison of Renaissance and Neo-Classical versions with the Lattimore translation may sufficiently attest to the processes at work in re-appropriations of the *Iliad:*[23]

*'Εν δ' ἐτίθει ποταμοῖο μέγα σθένος 'Ωκεανοῖο*
*ἄντυγα πὰρ πυμάτην σάκεος πύκα ποιητοῖο.*

All this he circl'd in the shield, with pouring round about
(In all his rage) the Ocean, that it might never out. (Chapman)

Thus the broad shield complete the artist crown'd
With his last hand, and pour'd the Ocean round:
In living silver seem'd the waves to roll,
And beat the buckler's verge, and bound the whole. (Pope)

He made on it the great strength of the Ocean River
which ran round the uttermost rim of the shield's strong structure. (Lattimore)

In the Greek there is a deliberate balance in the final position of the element signifying cosmological containment and coherence (*'Ωκεανοῖο*) and its aesthetic analogue (*ποιητοῖο*). There is an opportunity for the translator to preserve this balance, or to emphasize either order or its limits, what is coherently contained or that which lours beyond. George Chapman, whose "mighty line" impressed Keats so much, is true to Renaissance overreaching in his emphasis on the challenge facing the artist. For him Okeanos is more turbulent than tranquil, impatient of any attempt to tame and contain it. The parenthesis used – (In all his rage) – is insufficient assurance of Okeanos' good behaviour, and Chapman's version ends with a term ("out") that subverts the shaping impulse. Similarly, one can detect a Neo-Classical *Geist* ordering Pope's emphases. Not surprisingly, the stress is now on control and unity, with a clear cumulative suggestion that once elements are aesthetically arranged they are permanently redeemed from flux. However, this view involves a reduction of a different kind from Chapman's, namely an insistence on artificiality and semblance. One can be so confident about art's ability to control, only if one underestimates or profoundly distrusts the potent flow of *données* with which we interact. Lattimore's version is perhaps the most tactful of the three cited, but even his conservative approach incorporates a bias. In preferring a substantive to a verbal element he introduces a serviceable New Critical term, "structure," but in so doing reduces the dynamic quality that the Greek *ekphrasis* takes such pains to sustain.

In sum, the phenomenological aesthetic of the *Iliad* is more subtle and vigorous than these translators perceive or allow. *Ekphrasis* is one of the means whereby the poem reveals how art enacts interpretation of Being-in-the-World, and how individual works behave as objects of and contributors to apprehension and re-appropriation. Each process and product of ποιήσις establishes a locus of transcendence where *Aiōn* and *chronos* fight their duels; Logos strives against the great Lacuna, and, in the case of genius, whether individual or collective, *kairos* makes of opportunity an art.[24]

*University of Saskatchewan*

## NOTES

[1] *Cartesian Meditations: An Introduction to Phenomenology*, trans. Dorian Cairns (The Hague: Nijhoff, 1960), p. 54.

[2] *European Literature and the Latin Middle Ages*, trans. Willard Trask (Princeton: Princeton University Press, 1953), p. 170.

[3] *Aristoteles Poetik*, ed. Alfred Gudeman (Berlin: de Gruyter, 1934), 1449b 14; all other references to this edition are incorporated in the text.

[4] Cf. Aristotle, *De Interpretatione*, 7.1; *Metaphysics* 4.6, 2; *Rhetoric* 1.2, 15; Polybius, *Historia*, 8.2, 11; 1.4, 3.

[5] Cf. Lidia Storoni Mazzolani, *Empire Without End: Three Historians of Rome*, trans. Joan McConnell and Mario Pei (New York: Harcourt, Brace, 1976), p. 20. Polybius shares Homer's interest in motivation too, with interesting results: "La description psychologique prend l'allure d'une dialectique de l'action et l'historie elle-même devient à la limite la *phenomenologie* de la raison" (*Polybe: Histoires*, ed. and trans. Paul Pédech [Paris: Société d'Editions "Les Belles lettres," 1969], 1:xxvi).

[6] Cf. *De Mundo* 5.6.

7 For a fine study of this form of mediation between divinity and humanity, see Thomas Greene, *The Descent From Heaven: A Study in Epic Continuity* (New Haven: Yale University Press, 1963).

8 *The Iliad of Homer*, trans. Richmond Lattimore (Chicago: University of Chicago Press, 1951), 22.199–201. In accepting these lines as integral to the poem I follow E. R. Dodds, *The Greeks and the Irrational* (Berkeley: University of California Press, 1951), pp. 104–6 and notes.

9 Cf. 16.252, 254; *σόον δ'ανένευσε μαχῆς ἐξαπονέεσθαι*; and the appropriately greater physical emphasis of 22.205: *λαοῖσιν δ' ἀνένευε καρήατι δῖος 'Αχιλλεύς*.

10 *Truth and Method*, trans. William Glen-Doepel (New York: Sheed and Ward, 1975), p. 419.

11 *Mimesis: The Representation of Reality in Western Literature* trans. Willard Trask (Garden City, N.Y.: Doubleday, 1957), pp. 4–5.

12 *The Songs of Homer* (Cambridge: Cambridge University Press, 1962), p. 348.

13 For typical juggling with *erkennen, wiedererkennen* and *herauserkennen* see Gadamer's *Philosophical Hermeneutics*, trans. and ed. David E. Linge (Berkeley: University of California Press, 1976), p. 14.

14 Jacqueline de Romilly, *Time in Greek Tragedy* (Ithaca: Cornell University Press, 1968), p. 65.

15 See *The Literary Work of Art*, trans. George Grabowicz (Evanston: Northwestern University Press 1973), chap. 11. For a helpful introductory account of Ingarden's position, see Robert Magliola, *Phenomenology and Literature: An Introduction* (West Lafayette: Purdue University Press, 1977), p. 134.

16 For the dissemination of this device in antiquity, see, e.g., Armin Hohlweg, 'Ekphrasis,' in *Reallexikon zur Byzantinischen Kunst*, ed. Klaus Wessel (Stuttgart, Hiersemann, 1967), cols. 33–75.

17 'Ecphrasis in Epic Poetry,' *Comparative Literature*, Winter 1974, p. 3. This essay offers a sound introduction to the subject.

18 Werner Jaeger is typical of Greek scholars in his assessment of this *ekphrasis* as the "finest expression of the epic view of human life." See *Paideia: The Ideals of Greek Culture*, trans. Gilbert Highet (New York: Oxford University Press, 1945), 1:49.

19 See the concluding chapter of *La Poétique de l'Espace* (Paris: Presses Universitaires de France, 1958).

20 I have treated some features of the connection between *ekphrasis* and aestheticism in 'Aspects of Analogy: The Changing Role of the Sister Arts Tradition in Victorian Criticism,' *English Studies in Canada* 3 (1977):51–69.

21 See, e.g., Plato, *Phaedo*, 112 E; *The Shield of Herakles*, 314 ff.; and G. S. Kirk and J. E. Raven, *The Pre-Socratic Philosophers* (Cambridge: Cambridge University Press, 1957), pp. 11–15. For a sensitive account of spatial sense in the *Iliad*, see Thomas Greene, *The Descent from Heaven*, pp. 36–37.

22 Cf., e.g., Husserl, *The Phenomenology of Internal Time-Consciousness*, ed. Martin Heidegger and trans. James S. Churchill (Bloomington: Indiana University Press, 1964), pp. 106–9.

23 Chapman's *Iliad* was first published in 1611, Pope's between 1715 and 1720. I quote from *Chapman's Homer*, ed. Allardyce Nicoll (Princeton: Princeton University Press, 1956), vol. 1, and *The Iliad of Homer Translated by Alexander Pope*, ed. J. S. Watson (London: G. Bell, 1886).

24 *Kairos* has been popularized as a critical term for the enduring seasonability of art in Frank Kermode's *The Sense of An Ending: Studies in the Theory of Fiction* (New York: Oxford University Press, 1967), pp. 46–49, 139. For a much more probing and expert account of the term, see R. B. Onians, *The Origins of European Thought* (Cambridge: Cambridge University Press, 1951), pp. 343–48.

GILA RAMRAS-RAUCH

# THE MYTH OF MAN IN THE HEBRAIC EPIC

## I

The thematic in the epic genre is based on a perspective acquired in the past of a community, tribe or nation. The genre is often ahistorical, its content frequently beyond verification. Consciously or unconsciously, the epic will suspend all objective criteria and will dwell on its own self-contained materials. Thus its true correspondence is to the expectations of the community rather than to factual evidence. In its own creative fashion the epic genre will amplify, exaggerate, delete or reduce materials, subordinating them to the process of its own evolution. Here the epic genre approaches the timelessness of myth. In its mythopoesis, its aspiration to creative autonomy, the epic genre achieves its timeless quality. Mythical patterns are submerged in the epic materials themselves, and evolve around a historical memory – often a tradition rooted in pre-history.

The epic is therefore purposive, and it aims at establishing or maintaining ties with the past. In this, it also sets out to create certain models of human existence. Despite its generally conventional and stylized manner, and despite its non-psychological approach, the epic displays a pattern of human conduct and human existence.

All this should not obscure the fact that the epic is a work of art, not merely a document or the extension of a myth. It is a form which abides by its aims and by its aesthetic rules. Creativity and form are restricted by the memory and perspective embedded in its structure.

The Old Testament differs quintessentially from the typical epic and from the epic genre as such. First and foremost it is a testament, and its essential purpose is to reflect the basic covenant between God and the people of Israel. It is therefore anything but typical; rather, it is a unique creation which engulfs and incorporates past, present and future – all enclosed within a cosmic promise.

The process of secularization which is typical to the epic does not apply to the Old Testament. Its sanctity remains a viable element in the religious identification of Jews, Christians and Muslims. Yet it is at the same time capable of being considered as a literary document – as well as the source



*A-T. Tymieniecka (ed.), Analecta Husserliana, Vol. XVIII*, 175–183.

of archeological, sociological, economic and other data. We cannot clearly mark the line between the epic's use as ritual and myth, and its acceptance as proto-history. The Bible, in its comprehensive nature, includes all these elements. In the words of Erich Auerbach, "The Old Testament, in so far as it is concerned with human events, ranges through all three domains: legend, historical reporting, and interpretative historical theology."[1]

Our present concern is with the Hebraic epic as presenting a myth of man. An attempt will be made to delineate the unique nature of that epic, and to contrast the forces which shaped the Hebraic epic with those which shaped the Greek.

## II

The Bible aims at more than story-telling. One of its aims is the characterization of Man – and this is already in evidence in the episode of Eden. Here the Biblical man is launched on existence which is devoid of mythology, magic or the fantastic. He is shown as undergoing a transition from a state of unknowing to a state of knowing awareness, from being an unindividuated creature of nature to being a fully realized self. It is a transition from innocence to experience, and as such it adumbrates the idea of human existence as the actualization of individual potentialities. In essence, man is here at the stage at which he is able to perceive life as *Erlebnis*.

In Auerbach's distinction between the Greek epic and the Biblical, he points to the former as "fully externalized description . . . all events in the foreground, displaying unmistakable meanings. . . ." In the Biblical, on the other hand, only certain aspects are brought into relief, others left obscure, and there is a certain abruptness, with much left to interpretation. Yet we must bear in mind that despite the Bible's stylistic obscurity and terseness, what God expects of man is clear to Biblical man. The sons of Israel are told: "The secret things belong to the Lord our God; but the things that are revealed belong to us and to our children forever, that we may do all the words of this law" (Deuteronomy, 29:29).

There is an open dialogue which typifies the relation between God and man. Adam knows that his sin is in his disobedience. There is no obscurity here or multiplicity of meanings. God addresses man, and man responds. Dialogue characterizes the God/man relation from Adam to Job. No mythical chasm separates man from God. Rather, the Biblical hero is aware of his role, since he is presented with a divine plan in which he figures as central. "See, I have set before you this day life and good, death and evil. If you obey the commandments of the Lord your God . . . then you shall live and multiply, and the Lord your God shall bless you . . ." (Deuteronomy, 30:15ff.).

Eden and the subsequent banishment initiate man into selfhood and experience. The theme of banishment/exile extends throughout the book of Genesis and the Bible as a whole – and in this theme we see the growth of the Biblical hero into his preordained role.

The story of Eden is a model for the book of Genesis, and Genesis is the model for the Bible in its representation of Biblical man. In addition to exile and the experience emerging from the story of the Garden, another pattern for the model of man is established: trial. This is the situation of Abraham, Moses, David and Saul, Jeremiah, Jonah and Job. In the Bible as a whole, the prototype of the trial is the binding of Isaac (although it has its precursor in the story of Adam and Eve). The binding of Isaac (the unique ritual known as *Akedah*) is both a call to faith and a challenge to faith. In the Biblical epic, faith is the active element.

The operative quality of the Biblical individual is rooted in the fact of his being an individual within a situation, and in that situation there is no element of character to move him in one direction or another. In the Bible, there is no such person as an Odysseus "who is never at a loss." In this respect the narrative of the Bible presents an interesting interplay between "telling" and "showing." Thus human behavior is presented in its humanistic context, but without verbal characterization. Protagonists in the Bible are not labelled "cunning," "wise," "foolish" (as the Greek epic does). We know that Laban is not honest, but we infer this from the situation presented to us.

Moreover, Biblical characters are thrown into situations having no precedents. The situations are new and the protagonists must react to them as new. They may do right or wrong in the sight of the Lord, but in the ongoing process of human interrelationships they act as free agents. This also means that they are free of any expectations which *we* might have for them, i.e., expectations stemming from their earlier characterization. The Greek epic interposes the element of character between the individual and his actions, and we therefore know what to expect of him. In the Biblical epic, this gap is left open.

The two relations – man/God and man/man – are therefore problematic in the Bible, and they are made even more so by being contiguous. The tension between the two relations is inherent in the human, enterprising, active nature of Biblical man and in his preordained role in the wider context.

Another operative quality of the Biblical hero is rooted in the general facticity of man's fate. At a very primitive level, this is reflected in the hero's having to undergo exile or the life of the fugitive (Jacob, David), and in some this has a tragic dimension (Saul). Humiliation and fear play a role in the existence of some. In Biblical terms, they oscillate between the blessing and the curse, while experiencing existence at its fullest. Their personal commitment endows them with a dimension of credibility and humanity.

Yet perhaps the most unqiue presentation of human facticity is in the theme of chosenness. Here the Bible has no parallel in the Greek or any other epic. In his beginnings, the chosen one is found in a situation which is the diametrical opposite to his future state. His birth involves the overcoming of obstacles: many of the major figures are born to women who are initially barren, e.g., Sarah, Rebecca, Rachel, Hannah and the mother of Samson.

In addition, there is an element of dialectic in all this. There is, for example, a tension between the first born and the chosen one (in Hebrew, *Beḥora* versus *Beḥira*). The first born is not the chosen, as we see in the case of Ishmael and Isaac, Esau and Jacob, and others. As he grows into his role, the chosen one undergoes trial as he undergoes experience.

This is often expressed in the hero's name. Moreover, most of the central figures go through a change of name, or a new name is given to them as part of their experience and selection into the role of the chosen. Thus Jacob wrestles with the angel and is told what his new name is. "And he said unto him, What is thy name? and he said Jacob. And he said, Thy name shall be called no more Jacob, but Israel: for thou hast contended with God and with men, and hast prevailed" (Genesis 32:28–29). Thus the name is changed to one having a predictive significance: "Israel" means "he will strive with God", and this has the added significance relating to the struggle of every individual in the process of his becoming.

The distinction between the first born and the chosen leads to a reversal of the rule of primogeniture. This reversal, which runs throughout the book of Genesis, can be applied to the Hebraic epic in its entirety. The personal fate is emblematic of the fate of the nation. This is indicated by the bestowal of one and the same name on an individual and the nation: Israel. Yet even beyond this, we can see that parallel between a personal and a national epic in the fact that Adam, Abraham, Moses – and the nation itself – are in the state of being judged. Whether on the personal or the national level, it is the fate of the chosen to exist in a constant state of readiness for judgment. If there is anything which characterizes the Hebraic epic as a whole, it is this.

III

I have been arguing for the literary uniqueness of the Old Testament. Yet despite its uniqueness, the Old Testament is an epic. Here I would like to stress the dramatic character of its narrative, and the centrality of the element of "acting out" in it.

In emphasizing the dramatic aspect in the biblical narrative, we can juxtapose the state of the Hebraic protagonist to that of the Greek. In Greek tragedy, there is a movement from concealment to revealment. In this respect, Greek tragedy is ironic: a state of seeming will end in a state of being. Tragedy, along with this ironic transition, is absent in the Old Testament. As a conception, tragedy is foreign to the Hebraic world. The book of Job is often cited as an instance of tragic vision. Yet, as George Steiner points out,[2] the operative element here is justice, not tragedy. To the Biblical view, the ways of God are just and rational, not wanton or absurd. By contrast, tragedy is rooted in the idea of a blind necessity, of limitation, to which the Hebraic God could never be subject. This is why, as Steiner says, "To the Jew there is a marvelous continuity between knowledge and action; to the Greek an ironic abyss."

Thus Biblical man exists in a world of measure and divine design. This results in two opposed movements on the part of Biblical man: that of coming closer to God and to God's perception (Abraham), and that of running away from the presence of God (Jonah).

In juxtaposing the Biblical epic to Greek epic or drama, there is much to be learned in trying to apply a parallel terminology to the two areas. This has been suggested by Francis Fergusson,[3] who borrows from Kenneth Burke certain designations applicable to the tragic rhythm: *Poiema, Pathema, Mathema*. As he says, "They may also be called, for convenience, Purpose, Passion (or Suffering) and Perception."

I suggest that what unites these themes in the Hebraic epic is the notion of *trial*. In the Hebraic epic, Purpose is to be perceived as God's design for man and his world. That purpose is not superseded by motions of nature and natural law (as for the Greek). There is nothing of the idea of time as the moving image of eternity, as in Plato's *Timaeus*. Rather, the Biblical view of time is as the arena in which God's purpose is made manifest, and as far as man is concerned, that purpose is always perceived by him when in a situation of trial. Thus there is a direct line between Purpose and Perception: it is through trial that the perception is shared by man, and the purpose is revealed.

Passion is manifested in a series of trials. Actually, we may say that the trial is the meeting ground of two basic cycles present in the Old Testament: the divine cycle and the human/organic cycle. (The latter often suspends the socio-ethical cycle which is always there in the background.) Trial is a stage in individual existence wherein – through suffering, exile, doubt – the individual gains both knowledge and insight into the divine design and into the individual's role in the God-governed world. This is a vital part of the process whereby the Biblical protagonist is individuated into his role.

In essence, trial involves a movement: e.g., from innocence to experience (Adam and Eve), from a state of being accursed to a state of being blessed (David), from the state of being a fugitive to that of being a lord (Jacob). Yet there is no apparent change to be observed in the personal character of these individuals; they do not gain in saintly or semi-divine qualities. The Biblical individual is moored in his humanness, his mortality, facticity and vulnerability. This is part of the dialectical interplay between God's design and human purposiveness and self-assertion.

The Greek tragic hero is unique in being a captive of his own character, state or idea. This cannot be attributed to the Biblical individual. This is why the total subjection to one aim, and the suspension of all else, is foreign to him – however much the Biblical individual is typified by tenacity and zeal. It also explains why it is so difficult to speak of an epic in the Biblical frame of reference. That is, the Greek epic and Greek tragedy reflect the interplay between human character and blind necessity – neither of which are accepted as operative factors according to Biblical thinking. If anything, the Biblical epic can be said to reflect the interplay between (on one side) man in an indeterminate situation and (one the other side) God's will. These latter factors seem to parallel the Greek ones, but they do not, since they are altogether indeterminate. Thus, the Greek epic may begin with the words, "Sing, O muse, the wrath of Achilles" – and the outcome must be a foregone conclusion. Indeed, that predetermination is the substance of the epic grandeur. When the Biblical epic, on the other hand, begins with the words, "In the beginning, God created the heavens and the earth" – the outcome cannot at all be foretold.

The wide circumference of the Biblical narrative, as well as its structure, is epic in scope. Yet its unique character is embedded in the dialectical drama we have talked of, even if this is set amidst the epic and its dramatic manifestations. The freedom to choose and to move – this is a *donné* in the makeup of the Biblical individual. This freedom is, however, a human extension in a universe of divine design. Here, then, is the basis of the tension and its resolution which constitutes the story of the chosen one, the Hebraic epic. The Bible, then, is a narrative which tells the story of the growth of the individual in the sight of the Lord.

I have emphasized the uniqueness of the Biblical epic, *qua* epic. To balance this view, we ought also to point out the parallels between the Biblical epic and the epic genre as such. Both themes – the uniqueness and the similarity – have really to be considered as contiguous aspects.

The Biblical epic is of a dialectical nature: on one hand, it asserts the individuality of man, his capacity to choose and his spiritual autonomy; on the other hand, it expresses a divine design for which certain individuals are chosen for specific roles and to fulfill purposes which are not of their choosing. Thus the Hebrew Bible presents an interesting interplay between freedom and determinism – the determinism stemming from the twofold source of human limitation and divine foreknowledge. Yet the protagonists of the Biblical epic do not cease to exercise will, to freely satisfy appetite and indulge passion – even after their initiation into their designated roles in the history of their time. This is one reflection of the dialectical dimension in the Biblical epic, as well as the dialectical dimension in its protagonists – the paradigm example being found in the story of David and Bathsheba.

A further reflection of this dialectical tension is in the fact that the Biblical protagonist is constantly in a state of simultaneous revealment and concealment from the sight of the Lord – and the well-known scene in the Garden of Eden can serve as one example. Indeed, there is a special mode of the protagonist's self-assertion in response to God's question, "Where art thou?" – and here we can think not only of Adam but also of Cain and Abraham. Conversely, the self-assertion of the protagonist can also be manifested in relation to his own question as to who God is – and here we can think of Moses and others. In any case, the relation between revealment and concealment is decisive in giving shape to the character. Many of the characters "go down" – geographically, morally and physically – in the attempt to avoid the very sight of the Lord.

## IV

I have spoken of the idea of human models in connection with the epic in general. This holds true most characteristically for the classic epic. The concept of the classical is usually associated with the traditional or with a tradition, and tradition is associated with inherited models of human character, patterns which can be imitated and emulated. (The vast body of neo-classical literature can substantiate this assertion.) The question now is: Can we relate to the Biblical narrative in these terms? Further, can the form of the Hebraic tale itself be imitated? The Hebraic Bible does imitate itself – and here we have Deuteronomy and Chronicles as examples. That imitation and repetition within the body of the Bible is based on the quest for its interpretation – a quest that has continued for over 3,000 years. The objects of imitation one can point to are rhetorical and stylistic devices, the parataxis rather than hypotaxis, the art of parallelism, the verb-oriented rather than the noun-oriented emphasis, and others. One unique aspect in Biblical narrative is the terse, laconic description which is prevalent throughout. The contractive tendency in the Hebrew language gives the Biblical narrative a very tight and direct tone. There is no description for description's sake, and every element in the narrative serves the direct purposes of the narrative.

In Biblical poetry – the song of Lemech, the song of Miriam, the song of Deborah, the Psalms, the Song of Songs – we find ample use of the device of parallelism. There are three types of parallelism: synonymous, synthetic and antithetic parallelism.

Until the eighteenth century, Hebrew poetry was essentially sacred and liturgical. This originated in Byzantine Palestine in the sixth century, and its continuation in Iraq in the eighth to tenth centuries was based, thematically and structurally, on the Bible. The imitation was afflicted by an awkward didacticism, however, so that it lacked the elasticity and organic quality characteristic of the Biblical text. The Golden Period of Hebrew poetry in Spain (from the tenth to the fifteenth centuries) saw the development of both secular and devotional poetry. Despite the adoption of verse forms from the Arabic, along with metrics and other conventions, Hebrew poetry of the period was Biblical in its diction, allusions and associations.

Can we speak of a narrative tradition in the Bible? Were models of representation to be imitated? It seems to me that until the emergence of the Hebrew novel in the late nineteenth century, no deliberate attempt was made to capture the unique quality of the Biblical narrative. There is an *aggadic* literature which accompanied a thousand years of interpretation in the post-Biblical era, and this is essentially a non-legalistic Jewish literature. One can distinguish two general categories in it: first, there are folk stories and legends based on historical people and events – and here the Bible features centrally in supplying material; second, there is an interpretative literature, oral and written literature of an instructional nature – but in its didacticism, it reduced the dialectical quality of the Biblical narrative to fit the restricted models of human behavior. Here, the marvellous characters of the Bible sometimes seem to be characters in a morality play, serving the pragmatic needs connected to the survival of a people.

Yet no model of a heroic nature was established (formally or structurally) as a model to be emulated. Accordingly, it seems to me that in the Biblical epic the traditional quality of the protagonist is both his timelessness and his proximity to our experience as living individuals. The human quality – with the elements of excess, pity, fear and trembling – all enable the Biblical narrative to present a gallery of characters with whom we can identify, characters that come to life with every new reading. What we have, here, is a series of characters who are never remote enough to serve as models for imitation. Rather, they are close enough to the everyday consciousness and imagination to make imitation impossible. This, then, is the essential point of difference between the protagonist of the Biblical epic and the protagonist of the epic genre as such: the Biblical protagonist is never seen as *mythic* – even if the Bible gives us a myth of man.

*Ohio State University*

## NOTES

[1] E. Auerbach, *Mimesis* (Princeton: Princeton University Press, 1968), pp. 21 ff.

[2] G. Steiner, *The Death of Tragedy* (New York: Alfred A. Knopf, 1961), pp. 4–7.

[3] F. Fergusson, *The Idea of a Theater* (Princeton: Princeton University Press, 1972), p. 18.

JESSE GELLRICH

# ON MEDIEVAL INTERPRETATION AND MYTHOLOGY

> Like a thief in the disgrace of being caught,
> so will the House of Israel be:
> They, their kings, their princes,
> their priests and their prophets,
> who say to a piece of wood, "You are my father,"
> to a stone, "You have begotten me."
>
> Jeremiah 2:26–27

## I

Vituperation against apostasy, such as these words from Jeremiah, has been regarded as a repudiation of momentary lapses into fertility cults in Ancient Israel. Yet such criticism is also, Biblical scholars have shown, an expression of the distinguishing feature of Hebrew thinking: it is fundamentally opposed to mythological religions of immanentism. Because of the nature of this opposition, early Hebrew writing is not another mythology, but according to W. F. Albright, may be described as *archaic demythologizing.*[1] Following the thrust of this claim, H. N. Schneidau has argued the provocative thesis that later Biblical periods, too, carry out just as forcefully this opposition to myth and, furthermore, that Western tradition broadly may be distinguished by a characteristic demythologizing inherited from the Bible.[2] The impact is immediately obvious, for example, in the first century with St. Paul's admonition, "Have nothing to do with godless myths and old wives' tales," or again much later when scholars of the Reformation demythologized the past and their own culture in the letter and spirit of Jeremiah and St. Paul.

That Hebrew thought is opposed to myth implies, of course, that the vehicle by which it is communicated – the *style* of Biblical writing – is unmythological as well. Although the study of how stylistic features of Hebrew documents differ from mythology has not attracted much scholarly attention, one important contribution is Erich Auerbach's noted contrast of the styles "of the background" in the narrative of Genesis with those "of the foreground" in the myths of Homer.[3] Auerbach's argument is not



*A-T. Tymieniecka (ed.), Analecta Husserliana, Vol. XVIII,* 185–193.

without weaknesses, but it does suggest significant consequences that I would like to concentrate on: first of all, stylistic differences in Hebrew documents are not simply ornamental; instead, they are the preoccupations of a writing that is seeking to rid itself of mythology; secondly, it is by means of style that the activity of demythologizing makes its impact. If that is true, then we may assess the influence of certain Biblical documents not only by the familiar means of theological and philosophical history, but also in terms of a history of written styles begun by the Hebrews. A logical place to begin such a history is with the first full-range effort to respond to Biblical writing, namely, the allegorical reading of it composed during the Middle Ages. How is the style of Biblical writing received by medieval readers? Does the style of their allegorical interpretation carry out or reverse Biblical demythologizing?

## II

To consider these questions, I would suggest beginning with the document of the Davidic Succession (2 Sam. 7–20; 1 Kings 1), which was written at a time when Israel was quite close to her mythological neighbors. But unlike the influence recorded by Jeremiah, this early document about monarchy does not emulate the myths of the Egyptians and Babylonians whose kings were deities and whose courts replicated the kingship of the gods.[4] Instead the record begins with the fact that the queen is barren and leaves in the background any expression of divine will about a successor.

> It happened towards evening when David had arisen from his couch and was strolling on the palace roof, that he saw from the roof a woman bathing; the woman was very beautiful. David made inquiries about this woman and was told, "Why, that is Bathsheba, Eliam's daughter, the wife of Uriah the Hittite." Then David sent messengers and had her brought. She came to him, and he slept with her; now she had just been purified from her courses. She then went home again. The woman conceived and sent word to David, "I am with child." (2 Sam. 11:2–5)[5]

Whereas in Genesis we find at least some expression of Adam's fear and hear the divine judgment upon him, the absence of subordinating information in the account of David creates a sense of the matter-of-fact about an incident that is anything but a matter of simple fact. David's predicament has vast political and religious significance, but it remains unexpressed.[6] The one possible remedy – to send Uriah home from the battle-field to sleep with his wife – backfires when Uriah refuses out of his sense of loyalty to the king. "Are not the ark and the men of Israel, and Judah lodged in tents; and my master Joab and the bodyguard of my lord, are they not in the open fields? Am I to go to my house, then, and eat and drink and sleep with my wife? As Yahweh lives, and as you yourself live, I will do no such thing!" (2 Sam. 11:11). The impact of what is not being said in these lines creates a gaping void of the unexpected that eventually results in murder. The question of Uriah has the same rhetorical effect as the famous ironic question of Isaac. And as Isaac does not know that he himself is "the lamb for the burnt offering," Uriah's innocence of heart and his simple faith in his king prevent him from ever suspecting the adultery of his wife and the dishonesty disguised behind the king's favor. At the end of the account, all we are told is: "What David had done displeased Yahweh" (2 Sam. 11:27).

The resistance to justify David's behavior and express the divine judgment on him is evidence of the crucial difference between the Hebrew monarchy and Near Eastern mythologies of divine kingship. Studying the theology of this difference, Gerhard von Rad observes that the "Anointed" of Israel is clearly not a deity like Pharoah, nor is he especially pure, nor is his office intrinsically sacred. Unlike the monarchies of Egypt and Babylon, the Davidic dynasty did not "'come down at the bginning from heaven.' In this respect no mythic dignity of any kind attaches to it."[7] The assumption of the "presence" of the divine order in myth is contrasted sharply in the style of the Hebrew document. What Auerbach says of Genesis – "Everything remains unexpressed" – is paralleled nearly exactly by von Rad's remarks on the style of the Succession Document:

> The narrator only once draws the curtain back and allows the reader for a moment to perceive the divine power at the back of what appears in the foreground. . . . This historical work was the first word spoken by Israel about Yahweh and his anointed in Jerusalem; and it is an absolutely unmythological word. This realism with which the anointed is depicted, and the secularity out of which he emerges and in which he moves, are without parallel in the ancient East.[8]

Whereas Zeus, says Auerbach, is "comprehensible in his presence," and to touch Pharoah is to touch Ra, the Hebrew God made "Darkness . . . a veil to surround him" (Ps. 18:11).[9] The idea of this God is not rendered through mimetic representation, but in a style that demythologizes deities who exist in the foreground, are *immanent.*[10] The burning bush (Ex. 3:1–6) or the acts of David do not disclose the nature of divinity. The sign tells us nothing about the nature of the signified. The Hebrews speak of a God who is not known in idols, but who is always "at hand" in the background – *imminent*. The signs about him do not tell us *what* he is, but only *that* he is. All we hear in the episodes of Isaac and Uriah as the divine voice is an echo of a never-present totality, a sudden irruption that is only an accidental and arbitrary means by which Yahweh lets himself be known. Biblical writing confronts the idols of myth with the affirmation of the essential "otherness" of God, what a modern theorist would call the absolute "alterity" of transcendence.[11] The accommodation of uncertainty, the anticipation of the unexpected, the demand for change which this "book of alterity" exhorts, reveals by its opposition that the world of myth knows nothing alien, that it "exists for itself," as Auerbach argues, "contains nothing but itself; the Homeric poems conceal nothing, they contain no teaching and no secret second meaning. Homer can be analyzed as we have essayed to do here, but he cannot be interpreted."[12]

## III

However, Biblical documents demand interpretation. When we encounter medieval readings of passages like the Succession Document, the background of mystery in the text is only apparent; for behind the allegorical "veil" or through the "dark glass" the reader was taught to retrieve hidden meaning and "illuminate" it, like a painter of manuscripts. What may seem dark and obscure in the story of David and Bathsheba, for example, simply does not exist for the reader recorded in the *Glossa ordinaria*. In one word, *concepi*, (2 Kings 11:5), Bathsheba's confession that she is pregnant, the exegete sees a lengthy conversation.

> The pregnant woman goes to the king and says: "O king, I have been undone." And he says to her: "What is the matter?" "I am pregnant," she says. "The fruit of my sin grows, and inside I have the cause. In my womb I am making the betrayer. If my husband comes to me and sees, what will I say? What will I tell? What excuse will I put forward?"[13]

Hardly an enigmatic figure, Bathsheba is almost less important than the frustrations, fears and thoughts which are seen brightly illuminated and delineated by her. A context of motives behind David's questionable behavior is just as vividly in the foreground as the few details indicating that he was on his balcony "when kings go forth." For whatever reasons David did not go with his army to meet the enemy, exegetes saw *otiositas*, too much "idleness" or "leisure," as the explanation for his acts, and the old *sententia* from elsewhere in the Bible, "the devil finds work for idle hands," and from Seneca, "summa omnium vitiorum otiositas," are clearly intended by the Scriptural passage which the *Glossa* supplies as authoritative explanation between the lines of the quoted text in the "interlinear gloss," and on all four sides of it.[14] Not only does the *Glossa* add details to the text – namely, that David was strolling about his balcony, looking out of a window and gazing eagerly at Bathsheba bathing in the nude – but further, Gregory and Rabanus take us on a virtual tour of David's internal being, pointing out the image of Bathsheba imprinted on his mind as a primary cause of the "filth of thought" (*cogitationis immunditia*) inducing *oculorum illecebra*, an "enticement of the eyes" that ineluctably provoked David's soul to *concupiscentia* and his flesh to *luxuria.*[15] The ideas "behind" the events clearly have greater significance than the events themselves. That the baby conceived by Bathsheba dies suggests that God punished David for adultery by "taking" his child, but the comment in the *Glossa* elicits from this mere suggestion an elaborate conclusion informed by medieval morality. "Truly, those who were born after him [the dead child] God did not kill, because David married that woman."[16]

Although such allegorical readings have nothing to do with the figures and events that inhabit ancient mythology, neither does the allegory carry out the departure from it that characterizes the Succession narrative. For the medieval reader encounters the Biblical passage as a mimetic text, or more precisely, creates a text that is mimetic of the internal world of the moral and anagogical orders. Whereas the Biblical narrative demythologizes the mimesis of those orders that myth takes for granted, medieval allegory reverses that thrust. This reversal is clearest perhaps in the medieval response to irony. The biblical passage makes no attempt to explain away the ironic tone of events, but allegory, like myth, refuses to tolerate irony: all events are perceived as parts of a highly systematic and self-explanatory scheme. The overbearing irony of the entire narrative – that the king of Israel has to stoop to murder because of the devotion and chastity of one of his lowly soldiers – is no contradiction in the *Glossa*. For Bathsheba is a type of the "well of living water," *puteus aquae vivae* (because she is associated with water imagery in the narrative), David is a type of Christ because he is king of Israel, and Uriah signifies the Jewish people. As a type of the well, Bathsheba is identified with *Ecclesia*, which was carried as the ark by the Jews until that "marriage" was broken by Christ. David – "appearing as the Redeemer in the flesh" – took away "by strength of hand" (*manufortis*) that "wife" from her former spouse dedicating her to a new life of the spirit, no longer bound to the letter of the law but "conjoined" (*conjunxit*) to a new spiritual "husband."[17] The exegete is not really interested in righting

David's wrongs, but in the apparent conflict, for it reveals that even in an event reprehensible in itself God was giving "signs" to future generations about the superiority of Christianity to Judaism. The immorality is less important than the preoccupation to see common themes illustrated in new ways. In this case, the union of Christ and the Church, God and the soul of man by virtue of their unifying bond, *caritas*, are totalized in the text. While these meanings are not "literally" present, neither are they hidden in the distant background; from the medieval point of view, they were written "with Goddes fynger."[18]

The bright foreground of meaning in medieval allegory does not confirm the Biblical criticism of myth, but controverts it. What E. H. Gombrich has said about Western literature and art – that they "have preserved the peculiar habits of the ancient world to hypostasize abstract concepts" – gains unmistakable emphasis from medieval allegories of the Bible.[19] For as allegorization externalizes, it hypostasizes an imminent purpose; it makes that purpose present. Nowhere is this sense of the presence of the word more clearly in evidence than in Hugh of St. Victor's advice to his students to think of Divine Scripture "like a building":

> The foundation is in the earth, and it does not always have smoothly fitted stones. The superstructure rises above the earth, and it demands a smoothly apportioned construction. Even so the Divine Page, in its literal sense, contains many things which seem both to be opposed to each other and, sometimes, to impart something which smacks of the absurd or the impossible. But the spiritual meaning admits no opposition; in it, many things can be different from one another, but none can be opposed.[20]

The "House of Scripture" has no dark corners. Terms or episodes that seem absent of meaning merely postpone the act of interpreting from eventually catching up with the total sense that is *in* the text. Medieval interpreting is hypostatized as what is interpreted: the *interpretans* becomes the *interpretandum*, the book an object in its own right, a container of divine meanings. Although this development may appear to be distinctly medieval, it should be recognized for what it is – a structure of thinking that is grounded fairly and squarely in mythological thought.[21] There we see the hypostatized order of the divine – the kingship of the gods, what they look like, feel and think – and medieval allegory satisfies a similar mythologizing need.

But Biblical writing does not appeal to such needs. Instead it is devoted to challenging and testing them. It is not a container of meaning like the "House" of allegory, but a means for finding out spiritual meanings. In the Succession Document, Nathan's response to the story of David and Bathsheba is devoted exactly to this intention. His commentary on the intrigue and murder is not an allegory. He does not respond with an elaborate moral and theological network of answers specifying God's will for David, but with the parable of the ewe lamb and the rich man. The absence of an illuminated foreground of self-explaining significance in this fiction baffles its first listener so completely that Nathan's simple remark on the identity of the rich man points out that the intention of his fiction is precisely that bafflement; for it is the instruction – more powerful because of the absence of detail – that occasions profound moral searching in the heart of the king.

The fiction has an effect similar to the Succession Document itself, but opposite to that of allegory and myth. Whereas they are interested in classifying knowledge and making it present, Biblical writing provides the means for discovering it.[22] Nathan's fiction and the larger narrative history in which it is included are not self-explaining artifacts, as allegories about them sought to become the complete "text" of all received meanings. Rather the Biblical documents are a "pre-text" for inquiry, challenge, and change. In these interests they differ from myth and even oppose it, not only because it calls a tree "father" and a stone "mother." Biblical writing differs because its "meaning," like the kerygmatic claim of the New Testament, can have no words adequate to its value and sanctity, is not present in words at all, but comes into being only in an adequate response. Insofar as Nathan's parable provoked such a response, as do those of Jesus and the words of many Biblical documents, they have, in every sense, a grace beyond the reach of art.

*University of Michigan, Ann Arbor*

NOTES

1 *Yahweh and the Gods of Canaan: A Historical Analysis of Two Contrasting Faiths* (New York: Anchor, 1969), pp. 183–85. Albright refers to Hebrew writing between the thirteenth and the sixth centuries B.C.

2 *Sacred Discontent: The Bible and Western Tradition* (Berkeley: University of California Press, 1977); see especially chap. 1, 'In Praise of Alienation,' and chap. 5, 'The Bible and Literature.'

3 'Odysseus' Scar,' *Mimesis: The Representation of Reality in Western Literature*, trans. Willard Trask (New York: Anchor, 1957), pp. 1–20: "It would be difficult, then, to imagine styles more contrasted than those of these two equally ancient and equally epic texts. On the one hand, externalized, uniformly illuminated phenomena, at a definite time and in a definite place, connected together without lacunae in a perpetual foreground; thoughts and feelings completely expressed; events taking place in leisurely

fashion and with very little suspense. On the other hand, the externalization of only so much of the phenomena as is necessary for the purpose of the narrative, all else left in obscurity; the decisive points of the narrative alone are emphasized, what lies between is nonexistent; time and place are undefined and call for interpretation; thoughts and feeling remain unexpressed, are only suggested by the silence and the fragmentary speeches; the whole, permeated with the most unrelieved suspense and directed toward a single goal . . . remains mysterious and 'fraught with background' " (p. 9).

[4] See Henri Frankfort, *Kingship and the Gods: A Study of Ancient Near Eastern Religion as the Integration of Society and Nature* (Chicago: University of Chicago Press, 1948).

[5] *The Jerusalem Bible*, ed. Alexander Jones (Garden City, N.Y.: Doubleday, 1966), used throughout unless otherwise noted.

[6] Cp. Auerbach on Genesis; the journey of Abraham and Isaac proceeds "like a silent progress through the indeterminate and the contingent, a holding of the breath, a process which has no present, which is inserted, like a blank duration, between what has passed and what lies ahead" (p. 7).

[7] *Old Testament Theology*, vol. 1, *The Theology of Israel's Historical Traditions*, trans. D. M. G. Stalker (New York: Harper and Row, 1962), p. 308.

[8] Auerbach, p. 9; von Rad, p. 316.

[9] Auerbach, p. 9. Joseph Campbell, *The Masks of God: Occidental Mythology* (New York: Viking Press, 1964), pp. 108–9: "In any comprehensive view of the great and small mythological systems out of which the beliefs of mankind have been drawn, the Biblical idea of God must be clearly set apart, as representing a principle nowhere else exclusively affirmed; namely, of the absolute transcendence of divinity."

[10] Henri Frankfort, 'The Emancipation of Thought from Myth,' in *Before Philosophy: The Intellectual Adventure of Ancient Man*, by H. and H. A. Frankfort, John A. Wilson, and Thorkild Jacobsen (Baltimore: Penguin, 1973), pp. 241–44: "The God of the Hebrews is pure being, unqualified, ineffable. He is *holy* . . . all values are ultimately attributes of God alone. Hence all concrete phenomena are devalued. . . . Nowhere else do we meet this fanatical devaluation of the phenomena of nature and the achievements of man: art, virtue, social order – in view of the unique significance of the divine. . . . Only a God who transcends every phenomenon, who is not conditioned by any mode of manifestation – only an unqualified God can be the one and only ground of *all* existence. This conception of God represents so high a degree of abstraction that, in reaching it, the Hebrews seem to have left the realm of mythopoeic thought."

[11] E.g., see Jacques Derrida, *De la grammatologie* (Paris: Les editions de minuit, 1967).

[12] Auerbach, p. 11.

[13] *Biblia sacra cum glossa ordinaria et postilla Nicholai Lyrami*, 8 vols. (Paris, London, and Venice, 1590–1603), 2:575AB: "Praegnans facta mulier, vadit ad regem, et dicit: O Rex, perii. Et ille: Quid habes? Pregnans (inquit) sum. Peccati mei pullulat fructus, ac cusatorem intrinsecus habeo; in ventre profero proditorem. Si venerit et viderit vir meus, quid dicam? quid loquar? qua[m] excusationem praetendam?"

[14] Ibid., 2: 573–74.

[15] Ibid., 2: 575; *Patrologia Latina* (Paris, 1852), 113: 571; *Patrologia Latina* (Paris, 1852), 109: 99.

[16] Justin Martyr, ibid., 2: 581B: "Qui vero post eum nati sunt, non decauit, quoniam David illam duxit uxorem."

[17] Rabanus, *Patrologia Latina*, 109: 100: "Sed huic David uxorem abstulit sibique conjunxit, quia videlicet *manufortis*, quod David dicitur, in carne Redemptor apparens, dum se spiritualiter loqui legem innotuit, per hoc quod juxta litteram tenebatur, hanc Judaico populo extraneam demonstravit sibique conjunxit, quia per se illam praedicare declaravit"; see also: col. 99; *Patrologia Latina*, 113: 572; *Biblia sacra cum glossa ordinaria*, 2: 574–76.

[18] The reference is from Chaucer's *Summoner's Tale* (1.1890) in *The Works of Geoffrey Chaucer*, ed. F. N. Robinson, 2d ed. (Boston: Houghton Mifflin, 1957).

[19] '*Icones Symbolicae*, Philosophies of Symbolism and their Bearing on Art,' in *Symbolic Images: Studies in the Art of the Renaissance* (New York: Phaidon Press, 1972), p. 126. Cp. Harry Berger, 'The Ecology of the Medieval Imagination,' *Centennial Review* 12 (1968):306: the medieval sign is a "copy" which has "a magical ability to focus the reality and bring it into view."

[20] *The Didascalicon of Hugh of St. Victor*, trans. Jerome Taylor (New York: Columbia University Press, 1961), p. 140.

[21] Cp. Claude Lévi-Strauss, *The Savage Mind*, trans. George Weidenfield (Chicago: University of Chicago Press, 1966), p. 223: mythological thought "does not distinguish the moment of observation and that of interpretation any more than, on observing them, one first registers an interlocutor's signs and then tries to understand them: when he speaks, the signs expressed carry with them their meaning." See H. N. Schneidau, 'The Mythological Consciousness,' in *Sacred Discontent*, pp. 50–103.

[22] The opposition of Biblical writing to myth may very well be an important historical locus for recent discussion in literary theory of the difference between the contextualist notion of the work as an object that embodies significance and the post-structuralist claim that it is a system of signifiers that produces meaning; cf. Jonathan Culler, *Structuralist Poetics: Structuralism, Linguistics and the Study of Literature* (Ithaca: Cornell University Press, 1975). Similarly current arguments on the difference of fiction from myth ought to be considered in light of the Biblical opposition: Paul de Man, *Blindness and Insight: Essays in the Rhetoric of Contemporary Criticism* (New York: Oxford University Press, 1971), p. 18: "fiction is not myth, for it knows and names itself as fiction. It is not a demystification, it is demystified from the start"; Frank Kermode, *The Sense of an Ending: Studies in the Theory of Fiction* (New York: Oxford University Press, 1967), p. 39: "We have to distinguish between myths and fictions. Fictions can degenerate into myths whenever they are not consciously held to be fictive. . . . Myth operates within the diagrams of ritual, which presupposes total and adequate explanations of things as they are and were; it is a sequence of radically unchangeable gestures. Fictions are for finding things out, and they change as the needs of sense-making change. Myths are the agents of stability, fictions the agents of change." Much of this dynamic of difference and change is not a recent discovery; its unacknowledged debt is to Biblical writing.

VALDO H. VIGLIELMO

# THE EPIC ELEMENT IN JAPANESE LITERATURE

To begin with let us recall Sartre's emphasis upon the societal aspect of literature. No literature exhibits the impact of the social life more clearly than Japanese literature. My paper will treat largely the "epic element" in Japanese literature rather than the epic as such. Since the definition of what constitutes the epic derives wholly from Western literature, quite obviously I should have had to take great liberties in applying that definition to a literature which has had such a different development. A glance at any of the numerous handbooks or glossaries of literary terms will suffice to show that the common features of literary epics are in fact derived from the traditional epics of Homer. One of these handbooks[1] gives five major features of the epic: (1) the hero is a figure of great national or international importance; (2) the setting is ample in scale, sometimes world-wide, or even larger; (3) the action involves heroic deeds in battle, or a long and arduous journey intrepidly accomplished; (4) in these great actions the gods and other supernatural beings themselves take an interest and an active part; (5) an epic poem is a ceremonial performance and is deliberately given a ceremonial style proportionate to its great subject and architecture.

Only one work in Japanese literature comes even close to possessing these five features, namely, the thirteenth-century *Heike monogatari*, or *Tale of the Heike*, from Japan's early medieval period. For this reason at first I thought of focusing on this work alone, and examining especially those points wherein it resembles and differs from the Western epic. Upon considering the matter anew, however, I realized that this single work would not truly represent the epic element throughout the entire course of Japanese literature, for it contains a specific *Weltanschauung*, or world-view, not common to earlier or later periods of Japanese literature. (Of course I recognize that the same thing could perhaps be said of *Beowulf* or *Le Chanson de Roland* in relation to English and French literatures respectively.) In the case of Japanese literature the demarcations between periods are particularly sharp, and the nature of the literature produced in those periods comes to differ so radically that I decided to treat as well three other works from three other periods: the classical period, when the court aristocracy was in the ascendancy, immediately preceding the medieval age which produced *The*



*A-T. Tymieniecka (ed.), Analecta Husserliana, Vol. XVIII*, 195–208.

*Tale of the Heike*; the period of fossilized feudalism beginning in the seventeenth century, exhibiting some characteristics of the early modern period in the West with a rising merchant class; and finally the early twentieth century, when capitalism had triumphed, and the modern Japanese nation-state had attained virtual parity with the capitalist nations of Europe and with the United States. The specific works I have selected to represent these three periods are *The Tale of Genji*, or *Genji monogatari*, generally considered to have been written by Murasaki Shikibu (ca. 978–ca. 1031), or Lady Murasaki, in the early eleventh century, from the classical period, *The Life of an Amorous Man*, or *Kōshoku ichidai otoko*, by Ihara Saikaku (1642–1693) from the late seventeenth century, and from the early twentieth century a novel by the renowned author Natsume Sōseki (1867–1916) entitled *Kōjin*, or *The Wayfarer*. My main reason for selecting these three works, in addition to *The Tale of the Heike*, is that they all exhibit features which might be broadly termed epic, one of the most prominent being that they all contain a protagonist, or hero, who embodies many of the virtues – and occasionally some of the vices – of a major group or influential class of that particular period. Indeed, in their struggles, in their trials, in their vicissitudes we can often also see the strivings and aspirations of an entire segment of Japanese society at that time. Although none of these works could possibly be considered an epic in the narrow sense of the term, or even as much of an epic as *The Tale of the Heike,* they all can be said to contain epic elements in that the theme is a large one and the hero comes to grips with that basic theme. Moreover, these three works are considered to be among the most important of their respective periods. Certainly *The Tale of Genji* can be considered not only the greatest prose work of the Heian period, from the late eighth to the late twelfth century, but also the greatest prose work in all of Japanese literature. The work by Saikaku can similarly be viewed as the most important prose work emerging from the Tokugawa period, from the early seventeenth to the middle of the nineteenth century, and although *The Wayfarer* is not perhaps generally considered to be the *very* best of Sōseki's novels it would surely be considered *among* the best. Another basic difference between these three works and *The Tale of the Heike* is that whereas *The Tale of the Heike* is largely based on actual historical events – albeit with a considerable admixture of legend – the three other works are definitely fiction.

With this extremely broad definition of what constitutes epic elements, therefore, I shall proceed to an examination of the four works I have selected for treatment in this paper.

*The Tale of Genji* is a vast work, consisting of fifty-four lengthy chapters, or more than 1,100 pages in the renowned Arthur Waley translation[2] and slightly less than 1,100 pages in the more recent Edward Seidensticker translation.[3] Of the fifty-four chapters, only forty-one of them treat the life of the protagonist, Prince Genji himself. Genji is born to a favorite, though lower-ranking, concubine of the Emperor. Already at birth his beauty is extrolled, and he grows up to become a veritable paragon of masculine beauty. His elegance and grace as well are as outstanding as his beauty, and in the course of time he develops into a major practitioner of the traditional Japanese arts, such as calligraphy – especially in its relation to poetry – music, both instrumental and vocal, dance, and landscape gardening. He excels as well in his taste in dress and in his selection of the appropriate perfume. Indeed, his pre-eminence in these areas and his exquisite sensitivity to nature give him an almost godlike quality. Hence he is at once a real and an ideal figure. Significantly, his death is never actually described in the novel, and chapter 42 begins with the sentence: "Genji was dead, and there was no one to take his place" in the Waley version,[4] and "The shining Genji was dead, and there was no one quite like him" in Seidensticker's.[5] A phrase in Waley's translation, in the chapter entitled "A Wreath of Cloud," characterizes him extremely well: " . . . his life 'a tremulous causeway linking dream to dream.' "[6] (Seidensticker translates the passage somewhat differently, i.e., " 'A bridge that floats across dreams?' " but also does not identify the poetic reference.)[7]

The greater portion of the chapters dealing with his life relates his search for the ideal woman, one who will be beautiful, elegant, graceful, tender yet passionate, and accomplished in the arts – indeed, almost the female counterpart of himself, with the single notable exception that she will live only for him whereas he will continue his attachments to the many women with whom he has become involved. The novel gives a detailed account of this succession of lady loves, and it is almost as if all of the best of the moods of his loves remains somehow eternal with him. There is even the hint that contact with him brings ruin, for almost every woman suffers in some way through her encounter with him. One might say that it is almost as if a mortal cannot look upon the face of God and live.

Central to the novel and to the characterization of its protagonist is the aesthetic attitude of *mono no aware*, literally "the sadness of things." Ivan Morris, in his *The World of the Shining Prince,*[8] discusses this important term in conjunction with its major component, *aware:*

Though in Murasaki's time *aware* still retained its early catholic range, its most characteristic use in *The Tale of Genji* is to suggest the pathos inherent in the beauty of the outer world, a beauty that is inexorably fated to disappear together with the observer. Buddhist doctrines about the evanescence of all living things naturally influenced this particular content of the word, but the stress in *aware* was always on direct emotional experience rather than on religious understanding. . . .

Often the word appears in the phrase *mono no aware*, which roughly corresponds to *lacrimae rerum*, the pathos of things. It is when people perceive the connexion between the beauty and sadness of the world that they most poignantly sense *mono no aware*. The sensitive observer is moved to tears by the beauty of nature, or by its embodiment in art . . . not only because it is so impressive in itself, but because when confronted with such beauty he becomes more than ever conscious of the ephemeral nature of all that lives in this world of ours.

The aesthetic element is so prominent that it overshadows what some would consider to be the Buddhist theme of retribution for one's misdeeds. For example, Genji has relations with his father's consort, and a child is born from that relationship. Many years later, Genji's wife proceeds to be unfaithful to him and has relations with the son of one of his best friends, giving birth to a child who is presumed to be Genji's son. Hence we see Genji suffering the same fate which he inflicted upon his father. However, for the most part this Buddhist theme, at least in the portions of the novel relating to Genji, is secondary to the aforementioned aesthetic sensitivity of *mono no aware*.

The affinity of man and nature and man's sensitivity to nature are brilliantly expressed in many episodes. One that is particularly moving is the passage relating to a parting between Genji and one of his many lady loves, Lady Rokujo. I quote from the Waley translation:

At last the night ended in such a dawn as seemed to have been fashioned for their especial delight. "Sad is any parting at the red of dawn; but never since the world began, gleamed day so tragically in the autumn sky," and as he recited these verses, aghast to leave her, he stood hesitatingly and laid her hand tenderly in his.

A cold wind was blowing. The pine-crickets in neighbouring trees were whispering in harsh despairing tones, as though they knew well enough what was toward. Their dismal voices would have struck a chill to the heart of any casual passer-by, and it may well be imagined what cheer they gave to lovers already at the height of distraction and anguish. She recited the verse, "Sad enough already is this autumn parting; add not your dismal song, O pine-crickets of the moor." He knew that it was his neglect that forced this parting upon them. But now it was too late to make amends. Full of useless regrets, while the grey light of morning spread over the sky, he journeyed back disconsolately to the town, through meadows deep in dew. As she watched him go she could no longer restrain herself, and at the thought that she had lost him forever broke into a fit of reckless weeping.[9]

Here we can see how the emotions of the lovers at their parting are reflected in nature. The series of images from nature reinforce this sadness. The fact that the dawn seems to have been fashioned for their especial delight is of course ironic, and Genji expresses his sadness at parting at dawn, the dawn in an autumn sky adding a particular poignancy. A cold wind is blowing, and the pine-crickets are whispering in harsh tones. They seem to know what is going on – hence their empathy with the scene. Their dismal voices increase the anguish of the lovers. And finally through the gray light of morning Genji journeys back to the town, through meadows deep with dew. We must remember that the term "dew" was traditionally associated with tears, and in fact as Lady Rokujo watches him go she can no longer restrain herself and actually weeps. So here we have the interpenetration of the lovers' feelings and the natural setting, very much in keeping with the central theme of *mono no aware*.

The entire novel, then, can be seen as an expression of the aristocratic ideal of the courtier, and it is not surprising that this work held such fascination for the readers of Murasaki's generation and for the courtiers of later generations, those who looked back nostalgically to the flourishing period of the late tenth and early eleventh century when Murasaki lived. Indeed, despite the fact that almost a millennium separates modern Japan from the world of Murasaki and her Shining Prince traces of the intermingled beauty and sadness in her masterpiece can be found not only on the pages of the works of such novelists as Kawabata Yasunari – one of his novels actually has the title *Beauty and Sadness*, or *Utsukushisa to kanashimi to*[10] – but even in certain aspects of everyday life, thereby providing a welcome relief from Japan's ever-increasing ugliness and vulgarity.

About one hundred fifty years after Murasaki's death the dominance of the court aristocracy came to an end, in a series of wars that wracked the entire nation. It was out of that turmoil that the medieval world was born, and it was that same turmoil which provides the setting for the justly famed and much-admired *Tale of the Heike.*

*The Tale of the Heike* was preceded by two shorter war tales, *The Tale of Hōgen* and *The Tale of Heiji*, treating two interfactional disturbances in 1156 and 1159 respectively. From these conflicts, the Taira clan, or the Heike (if the Chinese-derived rather than indigenous Japanese reading is used), emerged victorious, under the leadership of the brilliant figure of Kiyomori. *The Tale of the Heike*, then, focuses on the Heike's swift rise, brief period of glory, and equally swift fall through devastating defeat at the hands of the rival military clan, the Minamoto, or the Genji. That this story is to serve as an object lesson on the evanescence of life and the vanity of human glory is shown by the oft-quoted prologue to the work. I quote from the A. L. Sadler translation:

> The sound of the bell of Jetavana [the monastery park of Buddha at Sāvatthi in India] echoes the impermanence of all things. The hue of the flowers of the teak-tree declares that they who flourish must be brought low. Yea, the proud ones are but for a moment, like an evening dream in springtime. The mighty are destroyed at the last, they are but as the dust before the wind.[11]

Despite the fact that the basic theme of the work is religious, it is certainly true, as Barbara Ruch indicates in a major review article,[12] that the work "is a compendium, as it were, of the religious, aesthetic, ethical, and psychological responses"[13] to the Heike-Genji, or Gempei, wars in the late twelfth century on the part of those who were so deeply affected by them. She goes on to say: "No other work before it spoke of or to so broad a spectrum of people."[14] Part of the reason for this is that before the work became available in written form it was chanted for many years by itinerant blind *biwa*, or lute, players known as *biwa-hōshi*. A fairly reliable account of the origins of the work shows the intent to have it chanted rather than read. Yoshida Kenkō's early fourteenth-century *Essays in Idleness*, or *Tsurezuregusa*, tells us that a minor official who became a priest, Yukinaga, with the support of his religious superior, Jichin, set down the *Heike*, but taught it to a blind monk, Shōbutsu, so that he might transmit it orally. Whether this is true or not is somewhat problematic, but there is complete agreement that the first text of the *Heike* that has come down to us was one which was set down more than a century later by an actual blind reciter of the *Heike*, a man named Kakuichi (d. 1371), who devoted the last years of his life to the task. It is the Kakuichi text which became the source of the modern editions of the work as well as that of the major translations. This tradition of chanting the *Heike* to the accompaniment of the *biwa* led to its widespread dissemination, and, moreover, as Ruch points out, "it created, for the first time in Japanese literature, a body of anecdotes and heroes who were to stand as touchstones for measuring and understanding all future Japanese encounters with the calamities of war, separation from loved ones, the rise and fall, the start and end of things, the loss of familiar ethical standards, and, of course, death."[15]

The language of the *Heike* presents an interesting mixture of Chinese-derived and indigenous words: in those passages dealing with warfare and the military aspects of the lives of the warriors we find more Chinese expressions, whereas in those relating to the personal and emotional lives of the characters we find that indigenous Japanese words predominate. Although this pattern of mingling Chinese and Japanese expressions did not begin with the *Heike*, the effectiveness of it in the *Heike* led to its adoption in most important prose literature therafter.

Obviously some may prefer the lyric to the martial passages, as the noted translator and scholar Edward Seidensticker does,[16] but for the purposes of this paper it is the passages relating to the warfare and political intrigue which are more important. But of course the warfare cannot be understood without an understanding of the personality of the man primarily responsible for it, namely, Kiyomori. In the early portions of the tale he is presented as a man of extraordinary arrogance and tyranny, which he manifests during the brief period of glory for his clan. Chapter 5 of book 1 (the entire work contains twelve books, and is about two-thirds the length of the *Genji*) is in fact entitled 'The Splendour of Kiyomori,' and contains the following descriptive passage:

> Now Nippon Akitsushima has but sixty-six provinces, and of these the domains of the Heike were thirty; almost half the land. Besides these the manors, rice-fields and gardens that they possessed were without number. In the multiplicity of their gorgeous costumes they were resplendent as the flowers of the field; the noble and illustrious crowded before their gates like a throng in the market-place: the gold of Yang Chow, the jewels of King Chow, the damask of Wu, the brocade of Shuh – of the seven rarities and the myriad treasures not one was lacking. For poetry and music, fishing and riding, perchance even the Mikado's Palaces were not more renowned.[17]

This period of splendor is followed inexorably by one of decline and then by the destruction of Kiyomori's descendants. It is rather ironic that Kiyomori himself dies some four years before the actual defeat of his clan, but some may feel that his premonition that the power of the Heike will not long outlast him is a major part of his punishment. Indeed, on his death-bed he declares that he does not desire any Buddhist rituals to be performed or temples or pagodas to be built upon his death. Instead, he wishes the head of his enemy Yoritomo to be hung before his tomb.[18] Significantly, this wish is not only not fulfilled but it is Yoritomo who vanquishes the Heike to become the next great military leader of all Japan.

If Kiyomori is the unquestioned villain of the work, just as surely Yoshitsune, Yoritomo's half-brother and the military genius of the contending Minamoto forces, is the hero par excellence. The latter portion of the tale is taken up with an account of his exploits and his prowess together with indications of his compassion for those less fortunate. For example, his forces intercept a messenger to the Heike, but Yoshitsune does not kill him, saying such would be a useless crime.[19] He emerges, therefore, as the true hero of early medieval Japan – indeed, of almost all of subsequent Japanese history – and it is not surprising that he should have been eulogized in numerous later works and that around him should have developed an entire network of legends.[20] Yoshitsune especially shaped what Helen Craig McCullough has termed "the image of the ideal Japanese warrior: a man of imposing appearance, magnificently appareled in costly armor and mounted on a stalwart war horse, who so valued family pride and personal honor that he stood ready to sacrifice his life for his lord, to kill himself to protect his good name, to risk death in order to be the first man to attack an enemy position, and to fight against overwhelming odds to take the head of an adversary of consequence."[21]

These two men, Kiyomori and Yoshitsune, then, are unquestionably the two giant epic figures of the tale, standing out from among a host of others, and it is unfortunate that because of the disparity in their ages, i.e., Yoshitsune is far younger, there is no episode in the work treating their direct confrontation.

The epilogue to the *Heike*, consisting of five brief chapters, focuses on the sorrows and distress of Kiyomori's daughter, the Heike empress dowager, Kenreimon'in, who survived the climactic disaster at Dannoura in 1185 when the major portion of the Heike forces was destroyed and her son, the young Emperor Antoku, drowned together with his grandmother, Nii-dono, Kiyomori's wife. She survives only to endure a tragic exile. She proceeds to take the tonsure, and goes from one retreat to another. Her fate, then, provides an elegiac note at the end of the tale, and the entire work closes with an account of her death. One passage especially captures the melancholy of her plight, and serves as a vivid commentary on the altered fortunes of this once-proud Heike lady and indeed of the entire Heike clan:

> Thus at the end of the ninth month she proceeded to the temple of Jakkō-in. As they went along she gazed at the beauty of the autumn tints, while the sun sank gradually behind the mountains. The dreary boom of the evening bell of the wayside temple, and the thick-lying dew on the grass as they passed drew tears from her eyes, while the fierce gale whirled the leaves from the trees in all directions. Suddenly the sky grew dark and the autumn drizzle began to fall; the cry of a deer sounded faintly and the shrilling of the insects was incessant. Nothing was wanting to add to the sum of her afflictions, which seemed indeed such as few had been made to suffer. Even when she was driven about from shore to shore and from island to island her melancholy was not to be compared to this.[22]

Thus at the end of the work we have come full circle from the tolling of the bell in the monastery park at Jetavana in India to the tolling of the bell at the Jakkō-in outside Kyoto where Kenreimon'in dies.

Although the theme of the evanescence of life and the transiency of all things is unquestionably the center of the work, it is interesting to consider the fact that for the rest of the medieval period – indeed, down even to the fateful year 1945 – this message was obviously not taken to heart by the rulers of Japan. In fact one might say that the conquering Minamoto, or Genji, clan attempted to make certain that they would not fall prey to the same kinds of disaster that the Heike had endured, and while they paid lip-service to the Buddhist theme of impermanence, they clearly made every effort to maintain their power even after the death of a particular military leader. Moreover, it is fascinating to note that the succeeding military dictatorships were all derived from the one established by the victor in the Gempei wars at the end of the twelfth century, namely Minamoto Yoritomo, the half-brother of the major hero of the work, Yoshitsune.

In sum, then, *The Tale of the Heike* comes closest to being a Japanese epic. Indeed, Ruch, taking Seidensticker to task for maintaining that the work is more lyrical than epic, invokes Robert Fitzgerald, the famed translator of the *Odyssey*, to support her thesis that the similarities between the *Heike* and the *Odyssey* are overwhelming. and hence that the *Heike* constitutes a genuine epic.[23] At the very least we can say that, genuine epic or not, *The Tale of the Heike* most certainly provides the foremost "epic element" in all of Japanese literature.

Five hundred years after the Gempei wars, in 1682, when Saikaku published his first work of fiction, *The Life of an Amorous Man*, Japan had clearly emerged from the somber medieval mood of *The Tale of the Heike*, although it had by no means shed its rigidly stratified feudal structure. The ruling Tokugawa shogunate, or military dictatorship, had brought a prolonged peace to the nation after more than a century of intermittent civil war. As a consequence trade flourished, and in the fast-developing cities merchants, artisans, shopkeepers, and moneylenders came to constitute a new class. They were placed at the bottom of the social herarchy. beneath the dominant – albeit now idle – samurai and even beneath the victimized peasant. And yet in contradistinction to their formal ranking in the society, the merchants enjoyed considerable prosperity, and developed a rich and diversified urban culture to their own liking.

It was this lively urban setting, with its theaters and pleasure quarters, which provided the backdrop to Saikaku's works of fiction. And in his first work he created a merchant-hero, a man appropriately named Yonosuke, or "The Worldly One." Combining techniques from *haikai* poetry, in which he was remarkably adept, with the realistic manner of earlier handbooks and travel guides, Saikaku created a new kind of fiction, perfectly suited to the tastes of the new merchant class of which he was a part.[24]

Already at the very beginning of the work Saikaku's emphasis on the rejection of the medieval temperament of lamenting the evanescence of life and his applauding of the pursuit of worldly pleasure are unmistakable: Yumesuke, who is to become Yonosuke's father, the owner of a silver mine, leaves the province of Tajima for the imperial capital of Kyoto, there to pursue "the charms of beautiful women and the pleasures of wine."[25]

Yumesuke in time falls in love with a beautiful courtesan named Kazuraki from the Shimabara pleasure quarter. He pays her ransom, marries her, and they become the parents of Yonosuke. Thereupon Saikaku proceeds to chronicle the life of their son, from a sexually precocious childhood to age sixty (the traditional end of a cycle in the Oriental way of counting years), when he sets off with his companions in profligacy to the fabled Island of Women beyond the horizon.

On the model of *The Tale of Genji*, the novel is divided into fifty-four chapters, although they are far shorter than those of the former work. Indeed, the entire novel can be viewed as a burlesque of the famous Heian work. The differences between Genji and Yonosuke are overwhelming, however, for whereas Genji seeks emotional and aesthetic satisfaction in his relationship with the many women with whom he becomes involved, Yonosuke seeks only sexual pleasure. Like his Western counterpart Don Juan, Yonosuke is interested primarily in the number of his conquests, and already at the end of the first chapter Saikaku tells us that by the time Yonosuke had reached the age of fifty-four "he had dallied with 3742 women and 725 young men."[26] Few readers would argue with Donald Keene that there is "nothing remotely endearing" about Yonosuke for "he seems to exist only in terms of his one obsession."[27] Nevertheless, as even Keene recognizes, Yonosuke *is* a hero, one who embodies – if not the ideals – at least the aspirations of the new urban-based society. He clearly represents the ebullience of the new age and the confidence of the rising merchant class. Such ebullience can be seen also in the fact that, as mentioned previously, Yonosuke, at the end of the novel, far from regretting his life of sensual pleasure, decides to set sail for the Island of Women with six of his cronies. The name of his ship, the Yoshiiro Maru, or as Keene rather humorously translates it, the S. S. Lust,[28] is proof of his undiminished sexual desire.

Saikaku, then, in *The Life of an Amorous Man*, created a Tokugawa period hero, whose prodigious – might one say epic? – sexual appetite can be said to provide a metaphor for the economic appeitite of the merchant class, which, less than two hundred years later, transformed into the modern capitalist class, firmly dominated Japanese society.

Turning now to the last of the four works selected for treatment, we find in Natsume Sōseki's *The Wayfarer*, the newspaper serialization of which was concluded in 1913, an intensely serious and somber novel documenting the profound spiritual anguish of a brilliant but tormented university professor, Nagano Ichirō, whose trials – although not overtly autobiographical – are surely akin to those of his creator.

The entire novel is related in the first person from the perspective of Ichirō's younger brother, Jirō, and in the first fourth of the work the story revolves around Jirō's activities in the Osaka region and his concern for his sick friend Misawa. It is only in the second major section of the work, entitled, significantly, 'Elder Brother,' or 'Ani,' that Ichirō makes his appearance, together with his wife, Nao, and his mother. Gradually, as Ichirō's difficult character is revealed, the focus of the novel shifts, until by the end of the work the reader is wholly engrossed in the dark twistings and turnings within Ichirō's soul. As if to underscore Sōseki's intent, he never lets us know the outcome of Jirō's search for a bride.

Ichirō's bad moods have affected the entire family, especially his relations with his wife. The two appear unable to communicate with each other effectively, and Ichirō begins to suspect Nao of being in love with Jirō.

During the family's visit to the Kansai region, the region around Osaka and Kyoto, unable to rid himself of his suspicions, Ichirō proposes a most unusual plan to his younger brother: Jirō is to sound out Nao's feelings toward her husband – indeed, he is to test her faithfulness toward Ichirō. Although Jirō naturally is quite shocked by the scheme, an occasion arises when, because of a storm, he and Nao are forced to spend a night at an inn alone. Here Sōseki may be accused of manipulating his material for dramatic – one might almost say melodramatic – effect, but this is merely another instance of the strange coexistence in his work of elements from third-rate popular fiction with those from the profoundest of psychological novels.[29] In any event, nothing overtly illicit happens between Nao and Jirō at the inn although Nao appears much freer with him than with Ichirō, and certain of her words and mannerisms suggest a desire for intimacy.

Jirō dutifully reports to Ichirō that there can be no doubt about Nao's integrity, but Ichirō, as one might expect from a man of his personality, is in no way relieved. On the contrary his gloom only increases, although he continues to have occasional intervals of good humor, and now he adds Jirō to the list of those whom he cannot trust.

After the return of the Nagano family to Tokyo, Ichirō's condition worsens until Jirō thinks that his leaving home to find a place of his own may ameliorate the situation. Before Jirō does so, however, he and his brother have a fierce exchange wherein Ichirō accuses him of being like their father in his shallowness and insincerity. He makes the following acid comment: "Present-day Japanese society, and perhaps the same might be said of western society, works in such a way that only superficial clever fellows can live."[30]

Gradually the number of such comments increases, until the reader becomes aware that Ichirō's particular plight constitutes an implicit as well as explicit critique of contemporary Japanese society. Precisely because the reader never really learns the exact cause of Ichirō's anguish – Ichirō's suspicion of Nao can be viewed as more symptom than cause – does he come to feel that it has been engendered in large part by the ills of modern society itself. Such a feeling on the part of the reader is only confirmed as he moves into the last section of the novel, entitled accurately 'Spiritual Anguish,' or 'Jinrō,' a Japanese word having Buddhist overtones.

Despite Jirō's taking up separate lodgings, not only does Ichirō's condition not improve, it deteriorates. Finally the family, with the help of Jirō's friend Misawa, prevails upon a Mr. H to have the latter persuade Ichirō to take a trip with him. They hope a complete change of scene in the company of a colleague of an utterly different, i.e., placid, disposition, will work a cure on the troubled man. At first Ichirō refuses but several months later he agrees and the two friends set out on their trip. Eleven days after their departure Jirō receives a bulky letter from H documenting Ichirō's state of mind which in fact constitutes the remainder of the novel. We never learn Jirō's or the rest of the family's reaction to the letter nor do we even have an epilogue telling us the outcome of Ichirō's case.

Almost all critics are agreed that this final section of *The Wayfarer* contains some of the most profound insights into the condition of modern man to be found in Japanese literature – or any literature, for that matter.

Ichirō's suspicion of his wife, his distrust of Jirō, and his more specific complaints against his father and others are now virtually forgotten as he confesses to H, in a series of remarkable conversations, his deep loneliness, his lack of faith, and his imprisonment within his overly intellectual personality. He voices as well his sense of profound insecurity in that whatever he does, becomes neither his end nor his means; and he ascribes this insecurity to science:

Man's insecurity stems from the advance of science. Never once has science, which never ceases to move forward, allowed us to pause. From walking to ricksha, from ricksha to carriage, from carriage to train, from train to automobile, from these on to the dirigible, further on to the airplane, and further on and on – no matter how far we may go, it won't let us take a breath. How far it will sweep us along nobody knws for sure. It is really frightening.[31]

Moreover, despite the fact that H attends him with the utmost concern and compassion, Ichirō feels his loneliness so keenly that at one point he rushes down a mountain path alone, exclaiming in German, "'*Einsamkeit du meine Heimat Einsamkeit*!' (Loneliness, loneliness, thou mine home)."[32]

Although Ichirō is treading a path to near-madness, very near the end of the letter – and hence the end of the novel – a scene is described by H wherein Ichirō is beginning to come out of himself and is observing intently some tiny crabs crawling in the grass. Whether such behavior will increase and ultimately heal the tortured Ichirō the reader never learns. It suffices that through Sōseki's magnificent characterization he has entered into the vivid experience of an alienated modern man, the suffering intellectual who embodies the spiritual crisis of the modern age.

It is not surprising, therefore, that Sōseki's later novels, of which *The Wayfarer* is such an excellent example, should be so widely read and should be viewed almost seven decades after their composition as having been so remarkably prophetic.

In conclusion, I must reiterate my thesis adumbrated at the outset: those artistic works which most clearly articulate the ideals, concerns, and worldview of a particular age are the candidates for the works most clearly containing an epic element. For the epic element does not transcend history, although of course it may point beyond history, but rather emerges from within it.

*University of Hawaii*

NOTES

[1] M. H. Abrams, *A Glossary of Literary Terms* (New York: Holt, Rinehart and Winston, 1957), p. 29.
[2] Lady Murasaki, *The Tale of Genji*, translated by Arthur Waley (London: Allen and Unwin, 1935).
[3] Lady Murasaki, *The Tale of Genji*, translated by Edward G. Seidensticker (New York: Knopf, 1976).
[4] Waley, p. 749.

[5] Seidensticker, p. 735.
[6] Waley, p. 370.
[7] Seidensticker, p. 337.
[8] Ivan Morris, *The World of the Shining Prince* (Oxford: Oxford University Press, 1964), p. 208.
[9] Waley, p. 194.
[10] *Utsukushisa to kanashimi to* (Tokyo: Chūōkōronsha 1973); translated as *Beauty and Sadness* by Howard Hibbett (New York: Knopf, 1975).
[11] *The Ten Foot Square Hut and Tales of the Heike*, translated by A. L. Sadler (Tokyo: Tuttle, 1972), p. 22.
[12] Barbara Ruch, 'Kakuichi's Complaint: Homer and the Heike Hazards' *The Japan Interpreter* (11, No. 2) Autumn 1976, pp. 229–236.
[13] Ruch, p. 229.
[14] Ruch, p. 229.
[15] Ruch, p. 230.
[16] *The Tale of the Heike*, translated by Hiroshi Kitagawa and Bruce T. Tsuchida, 'Foreword,' by Edward Seidensticker (Tokyo: University of Tokyo Press, 1975), p. xix.
[17] Sadler, pp. 26, 28.
[18] Kitagawa and Tsuchida, p. 369.
[19] Kitagawa and Tsuchida, p. 651.
[20] The most important work by far is the *Gikeiki*, a fifteenth-century work treating the life and exploits of Yoshitsune. It has been translated by Helen Craig McCullough as *Yoshitsune: A Fifteenth-Century Japanese Chronicle* (Stanford: Stanford University Press, 1966).
[21] McCullough, pp. 4–5.
[22] Sadler, p. 243.
[23] Ruch, p. 231.
[24] See Howard Hibbett. *The Floating World in Japanese Fiction* (Oxford: Oxford University Press, 1959), p. 42.
[25] Ihara Saikaku, *The Life of an Amorous Man*, translated by Kengi Hamada (Tokyo: Tuttle, 1964), p. 11.
[26] Donald Keene, *World within Walls* (New York: Holt, Rinehart and Winston, 1976), p. 179.
[27] Keene, p. 169.
[28] Keene, p. 172.
[29] See Howard Hibbett, 'Sōseki and the Psychological Novel,' *Tradition and Modernization in Japanese Culture* (Princeton: Princeton University Press, 1971).
[30] Natsume Sōseki, *Kōjin* (*The Wayfarer*), translated by Beongcheon Yu (Detroit: Wayne State University Press, 1967), p. 197.
[31] Yu, p. 285.
[32] Yu, p. 293.

BEVERLY ANN SCHLACK

# A LONG DAY'S JOURNEY INTO NIGHT: THE HISTORICITY OF HUMAN EXISTENCE UNFOLDING IN VIRGINIA WOOLF'S FICTION

Much of Virginia Woolf's reputation as one of the half dozen greatest prose writers of this century rests upon the depth with which she was able to render the struggle of personal destinies against time and death, within the context of society and history. Her persistent novelistic concerns were how to realize one's being in the midst of the massive structures of family, class, nation and era. She used the themes of love and marriage, premature death, war, madness and suicide, and exploited radical transformations of time, place, consciousness, even sexual gender, to serve this preoccupation. At the heart of Woolf's literary metaphysics is a three-fold reality composed of the interaction of the natural world, the social context, and individual experience, which interweaves the brute things of this world with men and women in the world. Her characters' ultimate aim is almost always the achievement of an ordered wholeness; Woolf projects that attempt at integration onto the inadequate or disintegrating reality (society, culture, history) around them. Unlike more conventional novelists, she does not place her characters within a societal soup tureen to observe them simmering. Rather, Woolf attempts to retrieve society and culture, time and history, *in* and *through* the human being. Thus, she is an unusual but profoundly historical writer. The characteristic symbolic patterns of historical fiction – patterns of destruction and rebirth – are patently present in her works,[1] as are an encyclopedic view of being and a multifaceted reality in which the past is always a relevant, indeed influential, given of the present.[2]

In this essay I should like to isolate and explore the developmental phases of Woolf's pursuit of her theme of the historicity of human beings, a phenomenon she often fights, even as she feels compelled to acknowledge its force.

The first of Woolf's novels deals only tangentially with larger issues of culture and society; its engagement with historicity is sketchy. In the second, *Night and Day*, Woolf expresses her characteristic inner/outer duality within a social setting which pits the self against the environment. As the title suggests, what is being explored are the conflicts between solitude and society as well as between conventional outer life and the inner life of the individual. There



*A-T. Tymieniecka (ed.), Analecta Husserliana, Vol. XVIII*, 209–224.

is full-fledged historicity here: Woolf has combined the social dimension of the individual's experience with dynamic analysis of his or her conscious life. She is looking for an ontological answer, or at least some workable compromise, that will allow the soul to grow and prosper within the framework of society, but much of this novel (as well as several of her subsequent works) hints that society is in direct opposition to personal fulfillment, that it answers activism and idealism with failure and sterility. Indeed, when Woolf's plot developments deprive the character Mary Datchet, a feminist working in a suffrage society, of normal happiness in love and marriage, substituting a life of lonely dedication to a cause, the "lesson" is that personal failure is the necessary price of participation in the social progress of political reform. Reality refuses to balance the claims of individual and society, Woolf implies. She was never able – here or in any of her other novels – to deny wholly or endorse wholly the notion that enlightened personal feeling is a more substantial and profound source of moral valuations than socially acquired beliefs, attitudes and goals. It is therefore not surprising that the end of *Night and Day* is equivocal. No clear-cut metaphysical victory belongs to Mary Datchet, who deliberately devotes her life to larger realities beyond personal satisfaction, who is alone yet not unhappy as she works for a higher cause within the stream of politics and history. But neither does any clear-cut superiority accrue to the choices of the nominal heroine, Katherine Hilbery, who opts for the subjective values of love and marriage and expresses her resentment at being involved in culture and history (1) through her love of impersonal, ahistorical, eternally true mathematics, and (2) with her complementary impatience with the culture-bound, crushed-by-tradition immersion in facticity and history which working on a biography of her famous grandfather entails.

A species of abortive transcendental subjectivity hovers over this novel, yet never blooms. *Night and Day* affirms the process of self-realization and offers the guarded claim that the human being may live fully and authentically within the social forms if he or she revitalizes them through strict adherence to the truths of personal existence (Katherine's way) or through explicit social reform when needed (Mary's way). The historicity of *Night and Day* is one of *being there*, of self-awareness, reflection and memory growing toward maturity, yet its embrace of facticity, the past, and societal norms is provisional, and its metaphysical allegiances blurred by Woolf's own Hamletizing hesitations.

*Jacob's Room* was Woolf's first technically innovative work and the first in which she projected a novelistic voice recognizably hers alone. If her first two novels suggested a dialectic between civilized society and personal fulfillment, in *Jacob's Room* society is a more villainous force, a direct and successful threat to actualization of the private self. The structure of *Jacob's Room* is a series of expeditions by its hero to discover the world and to actualize his unique potentialities. An epistemological tragedy, *Jacob's Room* is a novel in which friends seek but can never know him; the narrator seeks but never finds and communicates his essence; the reader finishes not "knowing" the protagonist. The structure of the novel parodies the alternating rhythms of Greek tragedy by contrasting episodes in which Jacob pursues his identity with episodes in which a chorus (the narrator or one of the characters) expresses frustration at not understanding him or his mission. The novel's *givens* offer the reader a series of literary snapshots of a vaguely interesting young man. The shapshots in the photo album that is *Jacob's Room* are isolated entities within finite frames, each very clear and in temporal sequence, yet important causal and psychological links are deliberately withheld by Woolf, so that the reader's experience approximates the apparent chance, enigma and disarray of life itself. Jacob is incompletely seen because he *is* incomplete, which the absurdity of his too-early death only accentuates. Woolf gives us a world in which individuality, action and communication are all sabotaged by modern social processes; the civilized routine is itself a process in which the individual is a mere replaceable adjunct. The cycles of life yield no clearer significance than do its random events. The flux is simply the inexorable process of biological growth and decay repeated over and over by each individual – what Woolf's narrator calls, referring to life, "a procession of shadows."[3]

Vague, unrealized Jacob Flanders conveys the novel's thesis that man's struggle for comprehension of self and world turn out to be elusive if not impossible. Jacob's room is both a literal room at Cambridge and the large, empty world-room in which he lives and dies. Government and politics, like society at large, run upon bureaucratic routines which reward conformity and discourage independent initiative and spontaneity. The society portrayed in this novel does not even support genuine authenticity, let lone honor its gadflies. It offers Jacob no ontological opportunities; it is instead a massive monolith of obstacles which merely engenders illusory social selves at the expense of authentic selfhood. Seen as an imposition upon the soul, organized society categorizes, restricts, even murders genuine Being. Most significantly, here, in Woolf's third novel, the darkest side of historicity – war, with its absurd waste and destruction – materializes.[4] Rendered by Woolf as a disjointed happening of singly unimportant events which nevertheless coalesce into an ominous reality, war is a jumble of Admirals and Ambassadors who make decisions and sign papers which decree "that the course of history should shape itself this way or that way" (p. 172). *Shape* is the key word: what politicians seek is "to impose some coherency .... to control the course of events" (p. 172). One specific result of the impersonal stream of political history is the death of Jacob Flanders (note the overtones of the surname; i.e., Flanders battlefield).

At no time does Woolf identify the historical event with the label of World War I; there are no sentences in which Jacob's death or the cause of it are explicitly stated – such is the nature of Woolf's impressionistic technique. In a one page final chapter, Woolf simply escorts us back into Jacob's room to find, amid the curtains blowing in the breeze, the flowers in a vase, and the wicker arm-chair, an *absence* among objects. In the street beyond the window "harsh and unhappy" voices cry out something "unintelligible." Jacob's mother, present in the empty room with his best friend, characterizes it as "'such confusion everywhere!'" (p. 176) as she holds a pair of Jacob's old shoes. In this vision of meaning and order dissolved, reality has chewed up its sentive, promising young men, leaving behind only a pair of boots in an empty room, harsh and unhappy voices, unintelligible cries in the street and confusion everywhere. The Great War to end all wars has murdered innocence and illusory ideals along with actual lives, giving rise to the emergence of the modernist sensibility. Woolf's hero wakes from the nightmare of history quite dead.

*Mrs. Dalloway* explores another facet of history as nightmare, society as fragmented. Its two mirror souls, Clarissa Dalloway and Septimus Smith, attempt to free themselves from the repressive influences of society, but can no more achieve their life task of inner unity than the modern world can achieve wholeness without. If society is a villain in this novel, so is science – science in the form of a psychoanlytic Establishment of tyrannical doctors whose reasonableness is mere camouflage for coercion, and whose vaunted discipline is an arbitrary power system. By taking two characters essentially alike in their emotional orientation, splitting them into male and female doubles, and bringing them to entirely different ends of sanity and madness, life and death, Woolf gives expression to her vision of modern consciousness as necessarily fractured.

Here as in the previous novel, human historicity enacts itself in the overwhelming presence of war. Political and scientific realms have conjoined to initiate and sustain war. War is the direct cause of Septimus' form of madness, shell-shock. Woolf would have us believe that Septimus' mad consciousness is not only effected but perhaps determined by his particular historical situation. Jacob Flanders, who might have been an exceptional man, did not survive society's war; Septimus Smith also disappears in the societally induced madness and destruction of someone else's war. Septimus' subjectivity was forced to constitute the world of which it was a part as a world of dark givens like destruction and death. Although he originally volunteered to fight in the war and "went to France to save an England which consisted almost entirely of Shakespeare's plays and [his teacher] Miss Isabel Pole,"[5] such poetic romanticisms were soon crushed by the awful facticity of the organized slaughter which maimed his psyche.

Woolf's fifth novel, *To the Lighthouse*, offers perhaps her widest phenomenolgy of character; human facticity is at its richest and most complex here. Woolf examines love, marriage, family, the roles of wife/mother, husband/father, daughter, son, sibling, friend, artist. The novel's emphasis on human relationships and interaction, on becoming a self in relation to others, does not mean that the historical context of outward events evaporates, however. If man *is* history, he also exists *in* history. The central section of the novel, entitled *Time Passes*, expresses this truth most clearly. The order and cohesion brought by Mrs. Ramsay to her family's daily life collapses into disunity with her sudden death. Time's impersonal corruptions, decays and disintegrations take over. Measured against nature's implacable processes, human importance dwindles into trivia and events are reported parenthetically, as mere interpolations. Woolf makes her very punctuation convey metaphysical truths: the death of Mrs. Ramsay occurs in a participial phrase within a bracketed sentence. The reader may infer the extent of human significance in nature's cosmic scheme.

The prying fingers of war – a metaphor from *Mrs. Dalloway* – also reach into the lives of this novel's protagonists: Andrew, one of the Ramsay sons, is "blown up in France."[6] This event, like Mrs. Ramsay's death and her daughter's death in childbirth, occurs in mere brackets which chastize human pride. "What power," Woolf's narrator inquiries, "could now prevent the insensibility of nature?" (p. 207). Woolf's answer is art. The fragile, mortal human manages to enact a significant self in spite of adversaries like death and nature at its most destructive: the poet Carmichael reading Virgil by candlelight (p. 189) is Woolf's emblem of abiding art, history's past pages surviving into the present. Moreover, a volume of Carmichael's poetry is published, and Lily Briscoe the painter arrives: "the house was full again" (p. 213). Human historicity, especially human creativity, is center stage again.

*Orlando*, the fantasy-novel-masque-allegory-parody which calls itself a biography in its mock subtitle, gives witty expression to Woolf's characteristically serious preoccupations with the relationships between inner and outer realities, tradition and the individual talent, life lived and biography written, and the coexistence of past and present. In her depiction of "the flow of historical consciousness,"[7] Woolf surveys English history, literature and society in terms of selfhood, time and sex. Her impressionistic treatment of history in *Orlando* takes on the totally radical dimension that one might expect from a fantasy which spans three and one half centuries from the late sixteenth century to midnight, October 11, 1928. Woolf insists on the intimate interaction of literature and history in this work, and to that end uses some 350 years of family history of the Sackville line, which had produced generations of poets. Defeating the determinisms of historical time with a vengeance, in one glorious rout of the reality principle, the character of Orlando begins as a young Elizabethan adult male, and ends as a thirty-eight year old twentieth century woman. This fantastic character has significance on the personal, national and cultural levels as the embodiment of nothing less than an English family, England itself, and English literature.

Since Orlando is a poet as both male and female, what he/she perceives and writes about on any given day reflects an always contemporary yet historical consciousness of the androgynous past that lives inside him/her. This in turn is being perceived and written about by a biographer-narrator with his own relations to time and historical truth. The complicated conceit upon which *Orlando* rests reveals the extent to which historicity has captured Woolf's full attention. In this work she synthesized several concepts relating to historical process which had hitherto merely strayed casually through previous novels. Until this work, transcendental consciousness has had next to no historical character; in *Orlando* history is fable and fable history. The conscious life of Woolf's protagonist takes place against the background of and is mediated by (1) a past of his/her own and (2) a social past. Woolf shows the historicity of conscious life by giving us an Orlando who is heir to a past created by the human beings who preceded him/her; moreover, Orlando moves in a sediment of tradition, and his/her being constitutes the unity of a history of its own.

The interpenetration of novel and biography, fantasy and realism that is the essence of *Orlando* required paradoxical, comic techniques to express the tensions between inner and outer. *Orlando*'s satiric stance toward conventional biography is the result of its assumption that biographers, those historians of individual beings in time, cannot adequately render the inner life – only the poet or novelist is capable of revealing that sort of subjective complexity. Hence the comic insufficiency of the narrator-biographer, who persists in trying to use his ridiculously limited genre to achieve the staggeringly difficult phenomenological task of describing Orlando, his/her literary compositions, and his/her journey through 350 years of a social milieu which is in itself also changing and needs to be charted along with the life existing in it. Against a hydra-headed phenomenology of Being – how, for example, to describe an adult male who falls asleep and wakes up a naked woman? – the biographer has only his pedantic trust in recorded history and objective facts, his penchant for fussy details and his retreats into pompous literary mannerisms. The metaphysical inadequcy of his weapons is painfully obvious, but he maintains scholarly exactitude – his slingshot against the phenomenological Goliath – to the very end, where readers are presented with what every good biography must contain, an Index. The inclusions and exclusions of this Index, as irrelevant as it is incomplete, are a monument to arbitrary selectivity masquerading as scholarly thoroughness. The "philosophic" humor of the Index contains no little malice toward the small-minded academic mentality. Woolf's biographer is historicity personified as a bit of a fool. He has his quest for historicial truth and he tries bravely to consult the evidence of existence (although coping with the facticity of Orlando's sex change certainly disconcerts him), and to do justice to the framework of history through the centuries. The character is a successful comic illustration of history as a happening discovered by man's critical (often bewildered) reflection on the past.

Like *To the Lighthouse, Orlando* offers an aesthetic answer to the problem of historicity – art as creative solution. Yet lives are lived in the flux of nature – as, for instance, in the unceasing activity of the sea. If *Orlando* attempts to explore its protagonist's place in cultural history, *The Waves* explores the human being's place in the cosmic scheme of things. It addresses a historicity which by this, Woolf's seventh novel, is transcendental and archetypal. One important theme in all Woolf's fiction is what Mircea Eliade has called the "revolt against concrete, historical time;"[8] it surfaces in Woolf's use of myth and archetypes, and in her drift toward a species of archaic ontology which attempts to locate the most significant human existence in a nonhistorical, philosophic "primitivism" where meaning is not bound to time, chronology, or specific cultures, but is recurring, cosmic, and paradigmatic. In *Jacob's Room* time was a continuum and Woolf reflected upon Greco-Roman culture. In *Mrs. Dalloway* the present is briefly contrasted with a future time when London will exist only as archeological ruins, and a ragged old woman streetsinger symbolizes the primordial beginnings of human history. In *The Waves* Louis' sense of history emphasizes "the long view of historical perspective and continuity,"[9] but perhaps even more importantly the meditations of Bernard, the novel's principal character, best illustrate Woolf reaching for the essential, the exemplary human gesture beyond the confines of chronology. The transcendental consciousness pervading *The Waves* has little if any ordinary historical dimension.[10] The soliloquies spoken by the characters add to the sense of timelessness: in the conventional soliloquy, event precedes speech; here the characters speak not in the progressive present but the pure present tense, conveying a sense of constant possibility of action beyond imprisonment in specific time realms.

*The Waves* takes philosophy as the model for the highest truths obtainable from historical forms of knowledge; its pursuit of universal truths is grounded in abstract contemplation of patterns of existence, not particularities of living in a specific culture or society. While the suggested milieu is recognizably British upper-middle class (nominally as limited a social world as Proust's), the characters' inner world is so resolutely universalized as to counteract any feeling of specific contemporaneousness. The enigmatic character whose life and death effects the six protagonists is named Percival – an allusive identity summoning up past centuries of literary culture and the wholeness of an assumed moral tradition largely lost to modern man. The implications of the Percival figure are comparative and ironic, much as Jacob Flanders' invocation of classical culture served only to diminish modern man's accomplishments by highlighting how far from the glorious past of Greek heroes and seekers after the Grail contemporary society had come. Decidedly, for Virginia Woolf, historical *process* (the brute facticity of *this happened*) is not the same as historical *progress*.

The history of the self in history becomes an increasingly tragic spectacle for Woolf after *The Waves*. In her final two novels, the self's own contradictions and difficulties are further compounded by the difficulties of the self as part of family, nation, history and mankind. The complexities with which Woolf grapples take on an ominous quality as human historicity darkens into a sense of impending apocalypse. From the cosmic symbolism of the life-as-wave trope Woolf progressed (if that is the word) to *The Years*, a novel in which the sea of life has narrowed into a small cluttered stream. The chapter titles, named years ranging from '1880' to '1918' and the 'Present Day' (i.e., 1933–1936), seem to represent more than the structured presence of time. Rather, they mark the human being's entrapment in history, in the tight facticity of chronology. These jagged-edged, broken pieces of time, fragments from a whole lived life, are what now remains of personal and national history. In *The Years*, lyric passages describing weather changes and the cyclic pattern of the seasons but dimly recall *The Waves*. Here they have no metaphysical logic; the natural rhythm has been disrupted, coherence and sequence have vanished. Instead of returning to the sense of flux that pervaded *The Waves*, each section of *The Years* offers not renewal and continuity but recurring death. At the conclusion of the first seven chapters, which cover the years 1880–1913, no less than seven persons have died, including various family members, the old dog Rover, the King and Parnell – private and public worlds are both steeped in death. Where Woolf's best works accomplish a synthesis of her two orders of reality, *fact* and *vision*, in *The Years* facts swamp vision, and individual sensitivity proves no match for war, failure, society's corruptions, familial inadequacy. Despite several beautiful episodes, *The Years* reveals the evil, squalor and cruelty of organized society as never before in Woolf's work. The novel concretizes a world in which family ties are in dissolution, even a state of decadence: in its opening pages, Colonel Pargiter visits his sleazy mistress, who lives in a rundown house with an eczema-marred dog and "dirty clothes hanging on a line in the back garden,"[11] even as his wife lies dying in their bedroom at home, where his sons and daughters live frustrated, isolated and unhappy lives. The decline of the Victorian upper middle class is to be inferred through this depiction of its representatives, the Pargiter clan. Analogously, the nation has become an arena of discord, ugliness and aggressive hatreds. The individual in such circumstances suffers from unhealable alienations. By the novel's end, the various family groups have been all but disbanded, many of the children remain unmarried or unfulfilled; the evils inherent in the system and entrenched tradition have done their grim work. Woolf discovers the *solidity* of the Philistine world, its entrenched injustices, with agony and despair: her novel is in many ways an attempted debuking of this hateful society, a society caricatured by one of the characters as an ugly beast combining the shapes of the Queen Mother, a bird, a tiger, and an elephant. The sketch evokes a social body of vulture-like greed and egotism, tigerish materialism and possessiveness, and elephantine animality, with the whole wearing the royal approval of Victoria's suffocating domesticity. This grotesque portrait is a terrible mockery of the characters' yearning for the Heavenly City and their repeated toasts to "a new world" (p. 329).

While the first half of *The Years* takes place in the historical period leading up to World War I, the novel's remaining portion leads to the 1930's world threatened by dictatiorship and global warfare. The "real" years during which

Woolf worked on *The Years* are scarcely less grim than the world vision her fiction communicates. In 1936, with Hilter's army on the Rhine, she notes in her Diary the presence of war "in our private life again. . . . even though I go on, like a doomed mouse, nibbling at my daily page." [12] The Spanish Civil War had claimed her nephew Julian Bell, killed in action one year before *The Years* was published, and in 1937, the year of publication, the Exhibition of Degenerate Art took place in Hilter's Munich. In a literary work (and a world) so cluttered with failure, decay and death, Woolf nevertheless attempted to sound the theme of renewal and rebirth in a particularly philosophical, perhaps even Hegelian manner: from the dissonance and contrapuntal conflicts a harmonious resolution of opposites results, just as dawn emerges from darkness. In the rising sun of this dark novel's concluding pages, there is a crucial, new factor. The present is not now an end but a means. *The Years* offers not the phenomenology of Consciousness Triumphant, which Woolf often outlined in her greatest works, but only a consciousness of *possible* triumph, postponed into a problematic future. As the significant characters in this novel, whatever their ages, strive for this sense of an open future, the formal ending of the novel assays a complementary vision of a better world through its invocation of a benign natural world. The sky is described as one of "extraordinary beauty, simplicity and peace" (p. 435). These are the novel's last worlds – "beauty, simplicity and peace" – but they speak a harmony which is only rhetorical – a formal, lyric possibility, not an achieved, existential experience. Woolf's ruthlessly honest, hence disappointed vision has been forced into contingency (upon what the future *might* bring). She has struggled to make the investment of faith in human progress and historical development that shines faintly in the closing paragraphs of *The Years*.

Woolf is also struggling to exorcise "the terror of history" [13] that was Europe in the 1930's. In 1938, when she began work on her last novel, *Between the Acts*, she and her husband were issued gas masks.[14] In the following year, she lived through Franco's victory in Spain, Mussolini's occupation of Albania, Hilter's takeover of Prague, his September invasion of Poland, and the British declaration of war on Germany.[15] Her letters are full of references to "aeroplanes always around us; and air raid shelters." [16] As she persisted in the writing and revising of *Between the Acts* war became part of her daily life. The air-raid shelter described in *The Years* as "a large cellar with its crypt-like ceiling and stone walls" (p. 289) was literal as well as literary. From September to the close of 1940, fighter bombers flew over the Woolf home at Rodmell. German planes were often intercepted over

Sussex:[17] without exaggeration, Woolf watched a world at war overhead. Both of her residences in London, at Mecklenburgh Square and Tavistock Square, were bombed.[18] In the midst of constant air raids and bombings, a pump for extinguishing incendiary bombs had to be installed in Woolf's bedrom; she was taught first aid and practised fire-drill procedures.[19] Yet, "listening to bombs dropping on other poeple,"[20] Woolf continued to work on *Between the Acts*, fending off the historical horror and pursuing continuity as intensely as she had in *To the Lighthouse* or *The Waves*. In *Between the Acts*, she preserved the same pattern of generational continuity, familial and marital representations, she had used in *The Years*. The time structure of the novel is threefold: the *present* day reality of the characters' lives is joined to the *past* of historical Pointz Hall and human ancestors and the pageant, which presents the Elizabethan through the modern eras, then both are enriched by a sense of *prehistoric* time, rendered through imagery and allusions to be discussed below. The title itself suggests at least three simultaneous dramas: a personal one, played out principally by the married couple, Isa and Giles; a historical drama enacted in the pageant play; and a societal drama involving the impending war. The significant spaces to which the title's preposition alludes are not only the intermissions of the play itself, but the interstices of history – between the wars – and Woolf's own life. The tone of this novel cannot help reflecting Woolf's increasing despair over the destruction and death that surrounded her. The historical facts are, once again, in deadly opposition to what one character as far back as Woolf's second novel called "the faith in an illusion" necessary for survival.[21]

*Between the Acts* suspends civilization between catastrophic moments of history. Woolf's masterstroke is the organization of the novel around an inept pageant play performed on the lawn of Pointz Hall, the country house. Perhaps Woolf is saying that the lies of history and biography are analogous to the artful lies of drama, and are subject to the particular virtues and vices, the changing perhaps contradictory interpretations, of each new generation. The arbitrary accidents and random developments, as well as the planned aspects of the pageant of history are not much different from those of the amateur pageant play production.

Sadly, the play is more parody than celebration of England's past; its reduction of that past to literary caricature suggests that history is stale and finished, absurd when seen from a disinterested perspective, and little more than the repeated mistakes of man's existence proved. Throughout the play, the voice of history frequently fails, slipping away into vague mumblings. Nature drowns out art (literally, with a rainstorm), or with distractions such as birds, dogs, or cows;[22] the stage voices are not audible; the phonograph does not work properly. More ominously, the needle gets stuck in the record groove, to repeat endless incoherences. Any significant unity art does manage to achieve is shattered by and in the intervals – that is to say, between the acts, when the temporary unity and community of the audience (read also society) falls apart into separate beings. The procession of English history, Woolf seems to be saying, is playing its last foolish act, for the reconstruction of civilization cannot be accomplished by mere constructs of the mind, whether they be pageant or novel or theory. Woolf's terrible suspicion is that history and art are both illusions which get stuck in the same grooves and have no intrinsic power to bring what they describe or desire into existence.

This novel recapitulates the inadequacies of all the other agents of unity proposed in Woolf's earlier works: the family, love and marriage, societal institutions, the church, politics, art itself, are weighed and found wanting. War is no longer impressionistically oblique, as in *Jacob's Room* and *Mrs. Dalloway*, nor is it rendered parenthetically, as in *To the Lighthouse*. Literal bombs fall in this novel as they did in *The Years*. Because war, which fragments man's psyche (Septimus Smith in *Mrs. Dalloway*) when it does not shatter his body, so shrivels humanity's soul, the abortive personal relations in *The Years* and *Between the Acts* both *mimic* war – sexuality as a battlefield of love and hate – and *indict* it, as the external reality which chops human lives into the "scraps, orts and fragments"[23] to which the novel repeatedly refers.

*Between the Acts* is so explicit about the war's desolation of the promise and achievement of England's earlier days, and so painstakingly charts the degeneration of the sense of community, that it seems to be seeking to demonstrate the historical inevitability of fragmentation. It does this, in part, by framing English history within the even larger frame of prehistory, thus expressing a new version of the dialectic of history with which Woolf had previously flirted.

By this stage of her writing history, Woolf has moved through three major stages or aspects of historicity: (1) a localized history of particular individuals and/or places (Jacob Flanders, for example, is much taken with travelling abroad and observing the relics of the past), (2) a larger, national history, such as pervaded *Orlando*, and (3) what one might call an archaeological dimension of history. The latter, by far the most complex of Woolf's explorations of historicity, is not an altogether new feature of her fiction.[24] Yet in *Between the Acts* her urgency to escape history's nightmare manifests itself in an obsessive fascination with primordial time. Seeking to understand the very beginnings of history, *Between the Acts* traces the historically determined world back as far as the original life-world upon which it is built. Hence, Woolf's frequent recourse to primitive plant and animal imagery and the allusions to a book entitled *Outline of History*, in which Lucy Swithin reads of "mammoths, mastodons, [and] prehistoric birds" (p. 217) and an England that "was then a swamp" (p. 218).

Even as the shadows darken outside her own cave/house, Lucy reads of "prehistoric man . . . half-human, half-ape . . . [in] his semi-crouching position" (p. 218). It is as if night has cancelled evolutionary progress and the history of animals has replaced human history. The jungle is reconstituted, symbolizing the continuation of primeval life and recapitulating the chaos at the beginning. The agitation of life must be endured each day, to die each night and be born again tomorrow in the immense redundancy of historical and cosmic processes: reality is replicated nightmare. Lucy's older brother, rising from his armchair to go to sleep, repeats the movements of prehistoric man about which his sister has just finished reading. How far indeed has man evolved from the apes in Lucy's historical outline?

Very different indeed from the affirmation of the cyclic cosmic processes which took place in the growing dawn at the end of *The Waves*, the final paragraphs of *Between the Acts* present the failure of culture as a conscious force, and the triumph (if only by default) of the instinctual and the unconscious. The animal nature of human beings is what remains in the final paragraphs. Woolf has worked her way toward a rather bleak affirmation of the sheer biological continuity of the human race, represented by the marriage of Giles and Isa, a none too happy couple joined by *both* "enmity" and "love" (p. 219). "Before they slept, they must fight," Woolf observes (p. 219). She adds that "from that embrace another life might be born" (p. 219), but it is more the animal embrace of reproduction than the human embrace of sexually expressed love. Woolf insists on its animal analogies, comparing Giles and Isa to "dog fox" and "vixen," stressing that their union occurs after fighting, "in the heart of darkness, in the fields of night" (p. 219), a night so primordial that it is likened to "the night that dwellers in caves had watched" (p. 219). Human history is anonymous biological reproduction; life is a cycle of replication which continues despite our ignorance of its ultimate meaning or value.

As the reproductive cycle goes on, the female author of the pageant play, Miss La Trobe, is gearing up to write her next play. To this extent, some species of humanity beyond the merely biological also endures. The conclusion of *Between the Acts* does not generate the unifying epiphanies of *To the Lighthouse* or *The Waves*, nor does it make any larger gesture of affirmation than to use Art's creative births as its weapon for fighting darkness to a stalemate. Beneath the divisions which a decayed and trivial social order has bred, there persists both an ANIMAL energy which overcomes division and a CREATIVE energy that keeps trying, however exhausted or discouraged. The full cycle must repeat, and that cycle includes history as *written word* (the copy of the *Outline of History*) and *evolutionary deed*.

The last line of the novel, "Then the curtain rose. They spoke" (p. 219), implies both human survival and artistic continuity. Isa and Giles are themselves figures in the artist's book of life. If the curtain is going up on another World War, it is also rising on Woolf's primary value, creativity (in both its biological and aesthetic senses).

Virginia Woolf took her own life in March of 1941 at the age of fifty-nine. It can be argued that her suicide by drowning in no way mitigates her art's greatness, nor its affirmations. She met the intellectual and artistic demands of self-fulfillment as few others have; her fiction and nonfiction works are characterized by unstinting intellectual integrity. In his moving eulogy, E. M. Forster asserted, "It is wiser, it is safer to regard her career as a triumphant one." He further claimed for her perhaps the greatest of successes for any writer, that "she pushed the light of the English language a little further against darkness."[25] The legend she constructed in her literary works of art was both her own personal myth and a myth of the self in history. Woolf is a remarkable illustration of Eliade's contention that "modern man can be creative only in so far as he is historical; . . . everything is denied him except the freedom to make history by making himself."[26]

As another critic justly puts it, Woolf "takes her place among other heroic figures of culture for it was . . . on the larger more resonant stage of history that she pursued her exemplary role."[27] Her epic of selfhood and her quest for literary excellence show how genuinely *heroic* is the mapping and revealing of the human condition through art.

## NOTES

[1] Others who have called attention to Woolf's constant awareness of the historical dimension include Werner J. Deiman, 'History, Pattern and Continuity in Virginia Woolf,' *Contemporary Literature* 15, No. 1 (Winter 1974): 49–66.

[2] Even Woolf's characteristic allusiveness – a technique I have documented and analyzed

in my *Continuing Presences: Virginia Woolf's Use of Literary Allusion* (University Park: Pennsylvania University Press, 1979) – is a function of her historical mind. I took the title phrase from *A Room of One's Own*; it is a revealing bit of historical ontology à la Woolf.

3 *Jacob's Room and The Waves: Two Complete Novels* (New York: Harcourt, Brace & World, A Harvest Book), p. 72. Pagination henceforth will appear in the text.

4 *Night and Day*, published in 1919, makes no reference to global warfare. Prior to *Jacob's Room*, Woolf's experience of the First World War is documented only in her diary and letters. "One has come to notice the war everywhere," she observes in January of 1918 (*The Diary of Virginia Woolf: Volume One, 1915–1919*, ed. Anne Olivier Bell (New York: Harcourt Brace Jovanovich, 1977), p. 100). Her 1917 entries repeatedly describe her subjection to air raids and bombings. (See *Diary*, pp. 84–85, 93, 116, 124). In her 1916 through 1918 letters, she referred to raids in which German planes flew over her house, causing her and her husband to take refuge "in the cellar" (*The Letters of Virginia Woolf: Volume One, 1888–1912*, ed. Nigel Nicolson and Joanne Trautmann (New York: Harcourt Brace Novanovich, 1975), p. 185; see also pp. 112 and 202). By 1918 she was "without meat and butter" (p. 215); it was "impossible to get any milk at all" (p. 225); the constant air raids – "three raids on three nights running," p. 222 – sometimes require sleeping on the kitchen floor until the all-clear signal (see pp. 214, 217).

5 *Mrs. Dalloway* (New York: Harcourt Brace and World, 1925), p. 130.

6 *To the Lighthouse* (New York: Harcourt Brace and World, 1927), p. 201. Pagination henceforth will be included in the text.

7 Jack F. Stewart, 'Historical Impressionism in *Orlando*,' *Studies in the Novel* 5, No. 1 (Spring 1973): 72.

8 Mircea Eliade, *Cosmos and History: The Myth of the Eternal Return*, trans. Willard R. Trask (New York: Harper Torchbooks, Bollingen Library, 1959), p. xi.

9 Deiman, p. 55.

10 For a discussion of the philosophic dimensions of *The Waves*, see my essay 'Creative Consciousness and the Natural World in Virginia Woolf's The Waves,' in *The Phenomenology of Man and of the Human Condition*, Volume XIV, *Analecta Husserliana.*

11 *The Years* (New York: Harcourt Brace & World 1937), p. 7. Pagination henceforth appears in the text.

12 March 13, 1936 entry, *A Writer's Diary*, ed. Leonard Woolf (New York: Signet Library, 1953), p. 248.

13 Eliade's title of the final chapter of *Cosmos and History* (pp. 141–162). The problem of how history relates to issues of freedom and creativity, which Eliade discusses, are precisely those confronted by Viriginia Woolf in her final decade.

14 *The Letters of Virginia Woolf: Volume Six, 1936–1941*, ed. Nigel Nicolson and Joanne Trautmann (New York: Harcourt Brace Jovanovich, 1980), p. 261.

15 Ibid., p. 308.

16 Ibid., p. 347.

17 Ibid., p. 428. She also describes seeing "a plane shot down on the hill besides Lewes" (p. 429) and the "direct hit on top" (p. 429) of the house across the street. From Sussex she wrote of raiders who "machine gunned the next village" (p. 422).

18 She writes of seeing "London all blasted, that too ranked my heart" (*Letters*, VI 431) and took expeditions to London with Leonard in which they wandered sadly through demolished streets.

[19] *Letters*, VI, pp. 408, 396.
[20] Ibid., p. 413.
[21] *Night and Day* (London: The Hogarth Press, 1966); p. 271 Cf. Mircea Eliade's argument, in the 'Despair or Faith' section (pp. 159–62) of the 'Terror or history' chapter of *Cosmos and History* that history's horrors cannot be successfully transcended "unless we accept a philosophy of freedom that does not exclude God" (p. 160). Religion is a prominent theme in *Between The Acts.*
[22] Compare Eliade on the opposition of accepting the nightmare of history *or* reidentifying, as did archaic man, "with modes of nature" and "nature's eternal repetition" (p. 155) in hopes of retrieving spontaneity and creative freedom.
[23] *Between the Acts* (New York: Harcourt Brace & World, 1941), pp. 188, 189, et passim. Pagination hereinafter included in the text.
[24] For example, even Woolf's initial novel, *The Voyage Out*, postulated the imaginary destination of a primitive South American setting teeming with plant and animal life; this background allowed Woolf to ponder the relationship of civilized consciousness to the prehistorical past.
[25] E. M. Forster, the 1941 Rede Lecture, rpt. in *Recollections of Virginia Woolf*, ed. Joan Russell Noble (New York: William Morrow, 1972), p. 198.
[26] Eliade, p. 156.
[27] Sonya Rudikoff, 'How Many Lovers Had Virginia Woolf?' in *The Hudson Review* 32, No. 4 (Winter 1979–80): 549.

# B. THE EXISTENTIAL SIGNIFICANCE OF RHETORIC

ANGEL MEDINA

# THE EXISTENTIAL SOURCES OF RHETORIC: A COMPARISON BETWEEN TRADITIONAL EPIC AND MODERN NARRATIVE

To the memory of my friend James Dagenais

As an introduction to the following reflections, a distinction between higher and lower figures of speech must be established here. The study of rhetoric as tropology has traditionally consisted in a definition and enumeration of images in a text. Images, in turn, have been seen, following Aristotle, as extensions that contradict the semantic correctness dictated by the control that ontological categories exercise over the division and coordination of the parts of the speech. Thus, metonymies, metaphors, synecdoches, and even personifications are seen as substitutions of words for words within the sentnece. Figures like irony and hyperbole, which rely on tone and emotion, appear as equivocations that require a context larger than the sentence, but can be constructed by linking several phrases.

Because of this philosophical prejudice, the contemporary process of reviving and systematizing of rhetoric has been based on the proposition that metonymy and metaphor are not just extensions of grammar but the tools of crystalization of grammatical structure; further, the rhetorical codes that help in the organization of narrative texts are seen as a continuation of the main parts of the speech (roughly, characters or persons are a continuation of names or nouns, actions an extension of verbs, and settings a weaving of adjectives or qualities). In a previous work, *Reflection, Time and the Novel,*[1] the author of these notes placed literature strictly within a communicative context; the thesis of that work was that if human interaction is reflective, that is, characterized by a process of interchangeability of consciousnesses, then communication does not presuppose language but rather constitutes it. The interaction that founds reflection is, within this theory, a mode of reason that is essentially narrative, and the interchangeable selves and courses of life that are born through narrative reason must be considered to be the ultimate source of the understanding of language. In this context, grammar can appear only as a formalization and division of the rhetorical codes (persons, actions, and settings) that rise reflectively from narrative reason and not, as contemporary views assume, as the origin of those codes.

With the above notions in mind, it is possible to assert that the codes



*A-T. Tymieniecka (ed.), Analecta Husserliana, Vol. XVIII*, 227–240.

themselves, not in their formal function as tools for the organization of texts but as the very content of the narrative, are the higher figures of speech. Persons, actions and settings should not be seen in this regard in their utmost generality as concepts but as individually constituted within a particular narrative style and text. It may appear strange to find, defined in rhetorical terms, realities that were traditionally conceived as the principal metaphysical and epistemological elements of man's view of the universe but the fact remains that these realities are often today altogether denied, or else defined as epistemic syntheses resulting from basic faculties or capacities of association first manifested in the production of metonymies or metaphors. These basic syntheses, the lower figures, which in their basic associative functions are simple divisions of the formal codes, should therefore not be seen as the sources of atomic grammatical meaning (parts of the speech), or of its subversion (literary images). They are, in fact, never given in isolation; between the codes and the lower figures one can encounter the true text, the intermediary complexes or configurations of each code out of which, by a process of analysis and formalization (recursiveness), it is possible to extricate context-free syntheses of meaning such as simple metonymies or metaphors.

Before examining those peculiar configurations of the codes that constitute the guidelines for a true and natural analysis of texts, it is necessary to characterize the most common practices of literary criticism on the basis of their treatment of the higher and lower figures of speech. Following up on the criteria established above, it can be stated without fear of sweeping generalization that literary criticism, especially in present times, is concerned primarily with five tasks. (1) *Explication de textes*, an unsystematic account of lower figures of speech and their conglomerates; this account is carried out by chance discovery. To this is normally added an account of topical or conventional themes, that is, of themes that can be exemplified in scores of previous texts. (2) Compositional or "stylistic" analysis, an exhaustive inventory of basic images and their sources in the literary tradition, what used to be called *florilegium*. The abundance of some images over others (of metaphors over metonymies, or of similes over metaphors) is given as the principal means of determination of the style of a work. Themes are usually not discussed except when they coincide with moods (serene, desperate, tragicomic, for example) contributing to the description of situations or settings. Sometimes compositional analyses concern themselves with detailed discussion of a type of image (the famous study of *figura* by Auerbach is an example) as distinctive of a literary period. (3) Thematic discussions. Themes are usually established by reference to genres or, in a moral way, by the use of cultural or philosophical motifs as received by men of letters. Unfortunately, most classifications of themes have recently become formalistic dissections of the codes with no consideration of the existential matrix inherent in the intersection of these, the larger of the formal categories. The relation of persons to their actions, for example, is more important for the understanding of a period and style than any possible structures within the code of action that would be valid for all times and cultures. (4) Theoretical generalizations about the use of images in literary discourse, or about general categories, such as plot, genre, voice, and others, abstracted by a formalization of discourse or by appeal to literary history. This practice also includes aesthetic analysis of major genres, such as satire, tragedy, novel, or essay, usually by the same means as above. (5) Consideration of the rhetorical dimensions of texts that arise out of their communicative aspects. The most advanced of these methods in present criticism presuppose, however, that certain fixed courses of action and forms of interaction, as determined by the faculties of the human species, precede communicative, narrative interaction (such is the case, for example, in various modalities of reader response, and in Bremond's "specular criticism"). However, since what is specifically human in human conduct is the pursuit of unique goals related to, and freely established in, interaction, the current formalization of those goals and attendant motives destroys the uniqueness of the codes in their particular configurations corresponding to a style or work.

It is important, then, to understand correctly the "typicality" of the human goals that are encountered in literature as well as in life: *these goals constitute modes of totalizing existence in interaction.* Fulfilment is here communicative (it occurs in the giving and taking of courses of action) rather than biological or passional. Thus, literary motifs such as love, death, or suffering are "meaning events" that coincide only partially with (indeed, most of the time drastically alter) the corresponding, "real," biological or psychological events. From a communicative point of view, the unique human goals that determine the courses of human existence are also the primary temporal events, in the sense that out of the interchangeability of consciousnesses emerges a distended awareness of present, past, and future as ecstases, phases that transcend each other yet coexist with each other. Communicative human goals are, lastly, reflective. In this respect also, the communicative standpoint imposes some changes in the conceptualization of an area of experience that was previously envisioned in subjective, psychological terms. Communicatively, reflection is the presence of an entire course of existence, as totalized by its ends, in each one of its moments. Since totalization is carried out for the purpose of interaction, of fitting one course of life into another, the reflective expanse covers not just one biological individual endowed with consciousness but all those involved in interaction with whatever limitations are required by the means and circumstances of this involvement. These considerations will be relevant in the upcoming discussion of the modes of selfness or personalization that constitute the peculiar configurations of the code of persons.

How do contemporary schools of criticism confront the five above mentioned tasks? It must be first indicated that they are never performed together because critics can specialize in one of them and take it as far as they wish in the name of the division of labor, or because they are, by training or temperament, not prepared to do more than one thing, whether it be positivistic induction, or formalistic generalization, or moralistic preaching. The fifth task is, of course, rarely performed by anyone and, when attempted, inadequately carried out. Formalist, structuralist and deconstructionist critics do mostly 1 and 2; they reduce 3 and 4 to the first two tasks (even a "philosophical" critic like Derrida does this) under the principle that "rhetoric reduces to grammar." On the other hand, literary historians do only 3 and 4, either in the name of comparativism or by studying the diachronic variations of synchronic literary elements. However, since many of them come originally from philology, they also subscribe to the reduction of rhetoric to grammar. Finally, hermeneutists and aestheticians do mainly 4, though occasionally also 3. These critics understand themes in tolerably adequate existential terms, but because they do not fully realize the implications of existential primacy (Ricoeur, in particular, has abandoned existential hermeneutics in favor of a detour through the ontological implications of language), they are incapable of articulating, either thematically or formally, the relations of the higher to the lower figures of speech, and ultimately to grammar. Sartre in his biographies/critical studies of Baudelaire, Genet and Flaubert took a long, if incomplete, step in this direction, a step that has not been fully understood in its true significance.

In a recent study, *Joseph Conrad and the Figures of Speech: Rhetorical and Existential Symbolism in the Novel*,[2] I arrived at a method to define the functions of existence and rhetoric, content and form, whole and part and, at the same time, clarify their relations within literary texts. The insights reached there constitute a, hopefully open ended, systematization of literary stylistics. They can be summarized as in Figure 1 which will be examined in detail in the remainder of the paper.

The earlier discussion of the distinctive goals of human existence that can be discovered through an ontology of communication opened the door for

| | Code of Persons | Image Clusters | Code of Settings | Image Clusters | Code of Actions |
|---|---|---|---|---|---|
| Interactive Boundaries: Love, Birth, Sacrifice, Redemption. | Splitting | Inverted Similes, Metonymic Chains, Extended Images, Families of Images, Metonymic Objects, Isolated Objects, Fixations | | Peculiar Activities and Social Relations, Peculiar Space/ Time Configu- rations: prolepsis, metalepsis. | |
| Repressive Boundaries: Struggle, Death, Suffering, Guilt. | Doubling | | | | |
| Higher Figures | Irony<br>Hyperbole | | | | |
| Lower Figures | | | Synecdoche<br>Metonymy<br>Metaphor | | Synecdoche<br>Metonymy<br>Metaphor |

Fig. 1. Relations between Codes, Boundaries, and Figures of Speech.

further consideration of the code of persons in terms of a more precise view of those goals. The category that best fits the proper description of human ends is the old existential notion of boundary situation. As is well known, Heidegger, in *Being and Time*, pointed clearly to the boundary of death, and less clearly to that of guilt, as constitutive of the being of man (a being-unto-death revealed by the call of conscience). Sartre, in *Being and Nothingness*, focused on birth (the formation of our self-identity in fateful choices); he also discussed in various contexts the role of guilt which constitutes existence in bad faith. But it was Karl Jaspers that initiated the systematization of the boundaries of existence by proposing and elaborating a theory of four boundaries: death, love-struggle, suffering and guilt. His systematization was, however, incomplete. Jaspers wrote[3] about boundaries only as unavoidable ends of existence (an unclear indication of their reflectiveness, of their presence in all moments of a concerned *Existenz*); and he gave a certain pre-eminence to the boundary of love, which he paired with struggle in order to indicate that, paradoxical as this may seem, interaction can fuse human lives either through cooperation or through conflict. Yet Jaspers, in spite of his intellectual debt to Kierkegaard's theory of "indirect communication," did not fully perceive the originary character of *love* in relation to the rest of the boundaries. He did not fully understand either, in spite of his technical training in psychopathology, the repressive role of *struggle* in the formation of some boundaries that dissolve the reflectiveness of loving interaction.

When all of these pertinent considerations are taken into account, it becomes necessary to develop, as Figure 1 reflects, a theory of human goals that begins formally with eight boundaries. Such a formal acceptance of a limited number of boundaries should not lead anyone into forgetting that, in peculiar combinations potentially unlimited in number, these boundaries have had a real, non-formal role in shaping Western life styles in various historical periods. It must be also born in mind that non-formally all the boundaries are direct or indirect forms of love: *love*, in the previous definition is indeed the will to a total, and hence narrative, sense of existence through the interchangeability of courses of action. The interactive particularizations of love here proposed are: *birth*, the completion of an incomplete course of life by the establishment of an origin that explains human yearning for love; *sacrifice*, the curtailing of a course of existence that is hence to be completed not in itself but in another (sacrifice brings into a reflective foreground the spiritual meaning of violence that begins with a hopeful personal, or sometimes communal, self-limitation and can hence, by transfer, be inflicted on others. In a truly spiritual sense this transfer and the consequent violence are, of course, always symbolic);[4] *redemption*, the fulfillment of *sacrifice*, in an active sense, by the realization of the goals of a person by another, or by a community, acting as restorers of betrayed expectations, and, in a passive sense, by the abnegation of certain goals and the acceptance of others by persons who consider themselves redeemed in giving a new sense to their existence. Time is what is redeemed in both cases; in the active redemption, through the fulfillment of long delayed expectations while, for those who are redeemed, time loses the sense of inertness that comes from an unacknowledged loss of existential and moral bearings.

The boundaries related to *struggle* stand for the possibility of the breakdown of interactive existence. Since this existence always involves at least two, but often more, biological individuals of the human species, the failure of communicative relations affects some of these individuals more intensely than others depending on their maturity at the moment of crisis and on the depth of the emotional, perceptual and behavioral formations through which the crisis crystallizes. The breakdown, however, can never be fully acknowledged, either because the individual affected has never known anything but fragmentary or distorted human relations (as happens to children or culturally dependent individuals), or because he does not want to give up the exercise of truly interactive relations (as happens to oppressors and instrumentalizers of interaction). For these reasons, the boundaries that particularize *struggle* must be called repressive boundaries and, although psychoanalysis claims it, they cannot represent a true regression to a pre-existent state of consciousness which might have been, at one time, standard for any human group or for the species as a whole. The essential result of repression must be either the denial of one of the interactive boundaries (by avoidance or by resentful overemphasis), or the displacement of one boundary by another with the consequent modification of the total set or profile of boundary relations shared by individuals of a certain culture or age group. *Struggle* proper hinges on the loss of the reciprocity that must, by love and design, accompany the domination or subordination of one course of life to another. With the disappearance of reciprocity all the beneficial reversals of activity and passivity in interaction are eliminated; domination becomes sadism, subordination masochism. *Death*, as a repressive phenomenon, is simply the confrontation of the fact of not having been born to narrative existence. It is accompanied by a generalized loss of esteem for the living functions in oneself and in others. Direct acknowledgment of human mortality is not in itself repressive, thus the existential feeling of care is not related to the fear of death but to concern for the preciousness of an existence that can be lived only once either meaningfully or worthlessly. A mortality that can be absorbed by *sacrifice* or *redemption* is never accompanied by deep-seated fear, on the contrary, it is the source of the resolution that can overcome all fears, while a repressive confrontation of death is the source of all fears, and hence is always indirect. Repressive *death* is psychoanalytically defined as mourning behavior and, in more common experience, as the constant destruction of good relationships in favor of impossible ones. *Suffering* is the active expression of the unacceptability of *sacrifice* when one refuses to undergo, as one often must, it for others, or when one is unjustly sacrificed by others. As an awareness, yet an unacknowledgment, of impotence, true human suffering manifests itself in a dialectic of struggle without hope against others or against the circumstances. One cannot cease to fight, yet there is no longer a belief in the possibility of achieving the goals for which one fights. *Guilt* is the struggle against meaning and completeness in the course of one's existence. It is motivated by a fascination with either the past or the future. Because of this fascination, the past, its purposes and its deeds, can never be accepted through repentance or reorientation, while the future, in the form of an abysmal freedom, must always rob us of whatever harmony or merits we may have created in loving interaction. As the total loss of the will-to-meaning concerning the course of existence, *guilt* has been well diagnosed by Jean-Paul Sartre when he shows as its direct manifestation the abdication of responsibility, and also the rationalization of irresponsibility, concerning the actions that make up the total course of one's existence.

From the reflective definition of the course of human existence in terms of its boundaries, and from the proper understanding of the boundaries themselves in an existential rather than a metaphysical or psychological context, it is possible to develop an ontological definition of the human person as a consciousness that reflectively limits itself by receiving from its community, and individually modifying, its boundary profiles. Since the course of a person's existence is constructed with and for others, its ontological/narrative definition will overlap with a rhetorical definition: the modes of narrative combination of boundary profiles, their creation and modification *in narrative*, will coincide with both general and singular rhetorical figures for the presentation of persons. Personification must be hence elevated from its narrower rhetorical forms (comparisons of humans and animals, humanizing of natural forces) and become the correlate for a character's existence as based on its boundaries. In the study of Conrad's rhetoric mentioned above, two boundary profiles were found, one centered on the boundary of *birth* and the other on that of *guilt*. Some of the early short stories and novels leading to, and culminating in the masterful sea tale, *The Nigger of the Narcissus*, and some of the later autobiographical stories such as 'The Secret Sharer' and *The Shadow-Line*, reflect the preocupation of the novelist with his spiritual origin and his growth to full moral and vocational stature. Their thematic connection of adventure tale and *Bildungsroman* links these narratives to a clear line of generic content that precedes and follows Conrad in the history of the novel. This linking suggests that literary classifications of novels and, by extension, of other genres, should be pursued to their foundation in particular boundary profiles that mold the configuration of the narrative development of the theme in question. With regard to the novel, one should not see its development as an exercise of the imagination but as a form of existential elucidation on the part of an author, even when the plot is not explicitly about his own existence. A second dominant boundary found in Conrad's work can be seen already in some of the early masterpieces, such as 'Heart of Darkness' and *Lord Jim*, both of which still share in the boundaries seen as typical of the novel of adventure/education, which are dominated by *birth*. But this second boundary, *guilt*, is most fully and devastatingly present in the three grand masterpieces of maturity: *Nostromo, The Secret Agent*, and *Under Western Eyes*. The boundary profile dominated by *guilt* is clearly repressive, as is demonstrated by the profound moral and mental disintegration of the characters of these works. The predominance and the modes of presentation of *guilt* in this second set of narratives can help us establish a better definition of this boundary in rhetorical terms[5] and, at the same time, give us some indications as to the real thematic content of the psychological novel, which is truly concerned with *guilt* in the above sense of a preoccupation with the worth of existence.

The above examples of existential boundary profiles in the work of Conrad have been discussed here to ullustrate the reflective/rhetorical definition of persons introduced in the previously mentioned works and expanded in these notes. The rhetorical side of that definition can be clarified, as Figure 1 shows, by the differentiation of the code of persons in the modes of personalization called there *splitting* and *doubling*. The person as a splitting consciousness appears in narrative through the division of the centers of interest and responsibility for action into various psychological and social poles and the consequent dialogue or struggle amongst these poles. It must be remembered that in the communicative/reflective definition proposed here, all personal poles must be a part of the interacting of consciousnesses. They become differentiated *only in and through* interaction. However, in certain narrative genres whose import is autobiographical and confessional, the conflict and dialogue of selves is transferred to personified psychological powers within a single consciousness. Such was the particular form of the code of persons in traditional allegories, which originated with the interpretation of myths by classical authors as narrative expositions of the conflict between personified passions. *Splitting* continued to be the prevailing mode of personalization in the early types of the novel, the adventure tale and the social narrative. The second mode of personalization proposed above, *doubling*, is marginally present in those early types, but it is as present and operative as *splitting* in the symbolist novel, and finally becomes the prevailing mode in postmodern narrative. *Doubling* can be defined as the coexistence of conscious poles in interaction, but without dialogue or final reconciliation: conscious personae struggle with each other and alternate as centers of responsibility for interaction, but this coexistence is directed toward mutual displacement without interchange of points of view or courses of action. It may be surmised from the dramatic struggle of *doublings* just described that this form of personalization marks the prevalence of repressive boundaries in a narrative, as the earlier use of the word dialogue indicated the prevalence of interactive ones. In psychological terminology, *splittings* are neurotic personalities while *doublings* are psychotic ones.

The above defined existential psychological forms of personalization become rhetorical when the corresponding poles or centers of self-consciousness and action acquire a *voice*; persons are then fully constituted as personifications or characters. There are two principal types of rhetorical voice: the first, *irony*, manifests both the duplicity and the hope for reconciliation typical of *splittings*; the second, *hyperbole*, manifests the exaggeration of self-importance and the tensions typical of struggling *doublings* in all situations. As with the mixture of boundaries and personalities, there is also a mixture of voices; *irony* and *hyperbole* coexist in allegorical literature (whether mythological, epic or tragic), in the early and late forms of the novel, and in postmodern narrative,[6] but the ironic mood prevails in all the genres up to the symbolist novel, from Flaubert to D. H. Lawrence, while the hyperbolic is the main determinant of the postmodern canon.

Since this code constitutes the ontological foundation for the other two, it was important to specify, with a certain amount of detail in this discussion, the relations among boundaries, personalization and rhetorical figures in the code of persons. The code of action, both as plot and as a repertory of conventional or unorthodox engagements of the characters, is constituted by fragments, in various series and orders, of the courses of life established in the first code. On the other hand, the code of settings can be considered, as it originally was, either an external extension of the characters and their attributes or, as is more common later on, as the most adequate environment for certain actions. Through their changes in various genres and periods, settings have been associated with qualities that persons perceive, and with their predisposition to perceive them.

These important connections among the three codes indicate their nature as higher rhetorical figures. What is then the practical effect of the codes upon the composition of the text? The relations (never to be confused with the workings of the imaginative skills of the author or with any other form of psychological causality) that control the development of literary units in a text are reflective in three ways. (1) In a communicative context, texts are parts of human lives to be fitted with other parts in making up narrative wholes. (2) In that reflective fitting, the interacting consciousnesses appear to each other in the manner in which the codes require it, that is, first in their totally projected courses of existence as circumscribed by the boundaries (persons); then in their temporal display through interactive routines (action); and thirdly in the external distribution of their actions and perceptions (settings). (3) Finally, reflection dictates the narrower syntheses of elements from the three codes, syntheses that constitute the real units of textual content. Non-formally, then, the particular lexies of a text do not emerge within anyone of the codes but are the result of peculiar relations among the codes; these relations determine the genre and style of each text.

Thus we reach the important conclusion that the style of a text, the nature and structure of the formations that give it its identity as a genre and as an individual work, is always the result of modifications in the internal disposition of the codes as they mingle with each other. In the end, these modifications are demanded by changes in the boundaries of persons as they are found both in narratives and in ordinary life. Such changes in the boundaries are the basis of all new cultural attitudes, hence they constitute at the same time the narrativity of human existence and its historicity. On the other hand, there are formal units of the texts that emerge from simple idealizations of structural and temporal connections (e.g., substitutivity for metaphor, association for metonymy) within the span of the codes. By isolating these connections and then applying them to ever smaller contexts, it is possible to reach the simple rhetorical units that have come to be known as the lower figures of speech. Hence, while the intermediary complex lexies that constitute the content of a text always belong in two codes or in all three of them, the lower rhetorical figures are analytically established within single codes. By themselves, the lower figures can never be taken as part of the content but only as decorative complications of basic grammatical structure, and this alone is, as was initially seen, the truly acceptable "content" in formal/analytic theories of meaning. An important part of a communicative/reflective theory of meaning will be the conclusion that the true figures of speech, the higher ones, are not aesthetic or presuasive alterations of logical meaning but the very substance of relevant human messages before they are frozen into informative grammatical formulas.

After this brief excursus concerning reflection, content and style, it is necessary to return to the reflective formation of the complex lexies of the text. If the previous research on Conrad that has been developed in this essay showed that *irony* and *hyperbole* were the rhetorical particularizations of the code of persons, what are the specific rhetorical formations that can be found in the conjunction of the various codes? A meticulous account of the rhetorical composition of Conrad's works shows the presence of frequent inverted similes, metonymic chains that link external and internal perceptions, families of images that channel his intense awareness of competing interactive and repressive boundaries such as *birth* and *guilt*, extended images that miniaturize the larger thematic allegories, and a great abundance of diverse object symbolism of an obsessive nature. All of these lexies emerge, of course, in the description of settings but show the penetration of settings by persons, and a corresponding thematic interpenetration of the related codes. Between persons and actions we encounter peculiar activities and social relations, the acts of adventurers, anarchists and cowards, and the magnificent clash of different cultural types (a clash that reaches inimitable heights in *Nostromo*). Between actions and settings, the most common lexies found in the work of Conrad exhibit peculiar configurations of space, e.g., narrow or huge expanses that oppress the soul, acts of falling, sliding, struggling with enormous weights, or dizzy spells, often revealing parallel states of the mind and its surroundings. Frequently, space sensations are conveyed in families of images connected with boundary situations or in object symbolism (mountains, clouds, darkness, silver, ivory preside like a silent destiny over the course of the action). Space configurations mirror each other at all levels, from the simplest images to the large scale geography of certain novels; they even enter into the disposition of chapters and sections of the narrative, e.g., the famous untold parts of the action, a vacuum in the middle of *The Secret Agent* that reflects the empty lives and concealed depths of the characters. Time has also a central role in Conrad's works; its patterns and dislocations are always the result of communicative interaction or of its interruption. Through various displacements of the narrator's role, the characters and even the readers are involved in the communicative game; thus, very extreme forms of prolepsis and metalepsis are, as in the thaumaturgical genres of old, not just elements of the plot but a part of the effect that the story is to have on all who share it. The early tales of Conrad display mythical configurations of time, such as decay or cyclical patterns. The adventure stories show a double time orientation, one back toward the unknown origin of the hero, the other toward the outcome (redemption, failure) that confirms his identity as a character. As an indication of the weight of destiny, the two patterns turn out to be symmetrical. Finally, in the *guilt* stories, time is disjointed and full of holes, thus it reflects the personality structure of their characters as *doublings*.

In comparison with these lexies that are typical of the conjunction of the codes in the symbolist novel, an analysis of the epic would show a supremacy of the interactive boundaries evidencing attitudes of unanimity or reciprocity in authors, characters and audience. Reflectively, these boundaries would control synecdochal images (*pars pro toto, totum pro partibus*) within and among the codes. The most commonly found lexies would thus be: epithets that abbreviate the narrative presentation of a character or define his communal role, elucidation of individual feelings by comparison with species (animal) feelings; these images would comprise the code of persons. In the code of action, the most common lexies would be: stories within the story, that is, stenographic capsules of the plot. In the code of settings, the principal components would be iconic objects functioning as tokens of the communal will (idols, sacraments), and part/whole correspondences suggesting the integrity of the cosmic order. This comparative analysis cannot be further pursued here; let the above examples serve as summary proof of the effectiveness of the communicative/reflective method as it applies to rhetoric. An important corollary of this proof would be the extension of the method beyond literature to other arts, particularly painting and music which, in the opinion of this writer, have also a rhetorical base, and would respond well to a non-formal treatment of their higher rhetorical components.[7]

There is no space in these notes for a further exemplification or defense of the proposed system of existential and rhetorical analysis. As a method, it attempts to correlate ontological, psychological and rhetorical factors emulating the comprehensive hermeneutical spirit of Northrop Frye or Harold Bloom. In comparison with this spirit, even the hermeneutics of Ricoeur and Gadamer appear too one-sidedly committed to the ontological implications of textual grammar. In Bloom and Frye, however, the correlations between ontology, psychology and rhetoric remain problematic. Indeed, Frye, with the ontology of Aristotle implicitly behind him, reduces psychology to rhetoric. Bloom, on the other hand, with a brilliant but scrambled ontology always ahead of him, reduces rhetoric to psychology. In a communicative/reflective system, there is never a question of reduction to a grounding level of meaning; there is, hence, never a question of whether this level would be objective and textual or subjective and existential. Psychology (existence) and rhetoric (text) are strictly correlated in the interaction (repression) that constitutes each actual narrative whole and its formal and partial elements.

*Georgia State University*

## NOTES

[1] A. Medina, *Reflection, Time and the Novel* (London: Routledge and Kegan Paul, 1979).

[2] A. Medina, 'Joseph Conrad and the Figures of Speech: Rhetorical and Existential Symbolism in the Novel,' Forthcoming.

[3] K. Jaspers, *Philosophy*, trans. E. B. Ashton (Chicago: University of Chicago Press, 1971), vol. 2, pt. 3.

[4] For an outstanding treatment of violence as a human phenomenon resulting from the institutionalization of sacrifice, cf. René Girard, *Violence and the Sacred*, trans. Patrick Gregory (Baltimore: Johns Hopkins University Press, 1977).

[5] Paul Ricoeur has attempted to separate *guilt* from an existential definition of man, thus neglecting the interactive/rhetorical dimensions of this phenomenon and placing it in a genetic context that coincides with, yet transcends, the development of reason. By using the ontological categories of possibility (faculty) and actuality (choice), on the one hand, and those of responsibility (internality) and causality (externality), on the other, Ricoeur develops an objective explanation of evil. Since all of these categories translate existential reflection into metaphysical analysis, the result has been, in the opinion of this writer, a disastrous retreat from phenomenological ontology to a linguistic ontology based on grammar rather than rhetoric and, in the final analysis, speculative.

[6] The distinction between modern and postmodern narrative can be rigorously explained in terms of the concepts introduced in this essay. The formation of the postmodern canon would be due to the overwhelming presence of *doublings*, first only as characters but increasingly also as authors. With a *doubling* in the role of author, the meaning of writing and reading changes radically. The purpose of writing ceases to be the interactive exploration of crises with a triumphal or sacrificial resolution in view. The writer now enters into a game of display and concealment, and the plot can have no possible end in sight. Thus, reading becomes an investigation of the enigma of the writer/character in his passage from one role to another, and of the emotional keys to the juxtaposition of the various roles. The exhaustion of the *will to believe* in meaningful ends, interpersonal reconciliation and complete narrative wholes, represents the end of narrative as *fiction*. The rhetoric of postmodern texts includes a great variety of metonymic objects, of fetishes invested with suppressed or unattached emotions, of rituals that give a semblance of order to the action, and of metaphoric substitutions among things, qualities and actions whose only similarity lies in the common emotional response that they provoke.

[7] Naturally, the advocates of a genetic method of explanation of objects and acts of meaning will ask for proof of the precedence of the higher over the lower figures of speech; literary texts are after all meaningful contexts. It must be clearly understood that the precedence upheld here is primarily ontological, and only in a qualified sense genetic and temporal. It is a necessary condition of the *reflective* formation of conscious wholes (of the acts and objects of *self-consciousness*) that the conscious agent or agents involved be in possession of their own horizons. But the power of conscious association, whether elementary or complex, is always subjected to the influence of outside aims or of conscious responses to them. Human goals become internal to the process of interaction, and this process becomes essentially communicative when, regardless of how it is supported in the environment, its formal aim becomes communicative interaction, that is, the interchangeability of the courses of action. In this interchangeability, all relations between parts and wholes envision the existential boundaries.

PATRICIA M. LAWLOR

# METAPHOR AND THE FLUX OF HUMAN EXPERIENCE

Human experience *is* flux, process, change. It is in the nature of human beings to try to understand this experience (is this not man's eternal quest for meaning?), but how can we understand this experience, or even talk about it, if this experience is in continual flux? This problem is rendered even more complex with the realization that any attempt to understand human experience becomes, itself, part of the experience. We seem caught, then, much like Virginia Woolf's record player needle, in a vicious circle that is endlessly retracing itself, a movement which underlines the hopelessness of our ever grasping its significance. Even if some way were found to stop this movement in order to explore it, that too would be futile since the experience would no longer be the same one (flux, process) that we had sought to understand, but some *other*.

It would seem that this experience could be grasped – not arrested, but understood enough to be discussed – only by acknowledging and taking into account this continual movement. Since this movement is one of ebb and flow (already a metaphor), a moving from the "present" toward the future while trying to make sense of (interpret, give meaning to, incorporate) the past, one would have to be wary of simply following this movement, of being swept away by it and thereby losing the detachment (distance) necessary to discuss it. In addition, when we talk about human experience we are often bogged down by language that is inadequate. It is precisely because the human experience transcends language that we have such difficulty putting it into words (it is too vast to be "put" anywhere at all). How do we discuss it then? Arthur Koestler tells us to "invoke a metaphor – what else can one do when faced with the unutterable?"[1] But why a metaphor? This, of course, brings me to the main focus of this paper. Through a discussion of metaphor – what it is, how it functions – I hope to show that metaphor is intimately related to human experience as well as to commentary on human experience, that metaphoric language is language transcending itself and cannot be translated or paraphrased. With a clearer notion of metaphor, its relationship to the flux of human experience (which is "beyond words") will hopefully be made evident.

Let us remember, first of all, the etymology of the word itself, from the



*A-T. Tymieniecka (ed.), Analecta Husserliana, Vol. XVIII*, 241–248.

Greek *meta-* (that which goes beyond, expressing succession, change, and participation) and the verb *pherein* (to carry, transport). The notion of movement and change is in the word itself, and we shall see how this notion has evolved from its originally limited idea of transfer of meaning to a broader and more complex idea of creation of meaning. Metaphor has, in fact, been the subject of heated debate and discussion for centuries, and the question is far from being resolved. Metaphor has become a topic of interest and concern in almost every domain that seeks to understand man and his universe (one can wonder, parenthetically, if the renewed interest in metaphor especially within the last thirty years does not have some relation to a growing sense of urgency to understand and make sense of our existence). With this heightened interest in metaphor, a term often used but seldom defined, much confusion has understandably arisen as to what is meant when one says metaphor.

The difficulty, in fact the impossibility as Derrida acknowledges, of defining metaphor becomes evident as soon as one makes a serious attempt to do so; this impossibility seems inherent to the nature of metaphor. Ricoeur expresses this paradox quite neatly: "metaphor is a figure and the word figure is a metaphoric word."[2] One often feels impeded by language when trying to explain this very important aspect of language. Consider Monroe Beardsley, for example, and note his two deviations from "literal" language:

> If we turn from the objects referred to in the metaphor, and consider the significations of the words themselves, we must look for the *metaphoricalness* of the metaphor, *so to speak*, in some sort of conflict that is absent from literal expressions.[3]

We always seem to end up speaking metaphorically about metaphor, a paradox which complicates an already complex notion and which, I think, resembles the paradox of our attempt to understand and discuss the human experience becoming in turn part of that experience.

Traditional theories of metaphor are based on the earliest one that has come down to us, namely Aristotle's, and have for the most part changed very little since then. According to Aristotle, "metaphor is the transfer of the name of one thing to another thing, the transfer of genus to species or of species to genus or of species to species or by means of analogie."[4] In her chronological study of the principal theories of metaphor since Aristotle (*Etude sur la mètaphore*), Hedwig Konrad shows that these theories are merely variations on the theory set forth by Aristotle. Gertrude Buck arrives at the same conclusion:

> all existing definitions of metaphor are practically identical. . . . Aristotle's conception of this figure has simply repeated itself in each succeeding generation of rhetorical

treatises. To him metaphor was essentially a transference of meaning from one word to another over the bridge of analogy. [Metaphor within her definition of metaphor!] Later definitions only phrase somewhat differently the same idea. Metaphor is the use of a word denoting one object to indicate another object resembling the first in some particular; or it is the "comparison" of two objects; or it is the affirmation of the identity of two objects. In spite of superficial diversities, it is evident that all these definitions come in the end to the same thing. Each is based upon a recognition of the composite character of the metaphor. Each resolves the figure into at least two parts. There is the thing meant and the thing said. . . .[5]

The basis of the Aristotelian theory, and of those which follow from it, is therefore the idea of transfer of meaning, of substitution of one word for another. This transfer is sometimes necessary, sometimes desired, for reasons which Scott has formulated this way:

Transfer of meaning takes place either because it is necessary or because it is pleasurable. Philologists and philosophers have, for the most part, emphasized the hypothesis of the poverty of language, rhetoricians that of the desire for ornamenting speech.[6]

Even though such definitions and explanations have never been actually satisfactory, each new attempt to clarify the concept of metaphor has ended by coming back to this theory handed down by tradition.

In more recent times, however, much of the debate concerning metaphor has centered around the notion of substitution as the fundamental criterion of metaphor, and it is this notion that has gradually been refuted. Traditional theories had been based on the idea that there existed a proper (literal, real) term in relation to which metaphor would be a deviation. Metaphor would then be in opposition to the norm (the real or literal). However, just as poetry is not merely deviant or fancy prose, metaphor is not merely ordinary language that finds itself decorated, but another language, as proper and real as ordinary language. As Max Black states, "Metaphorical statement is not a substitute for a formal comparison or any other kind of literal statement, but has its own distinctive capacities and achievements."[7] It no longer suffices to "resolve the figure into at least two parts:" instead of dividing the figure into its constituting parts – the thing *meant* and the thing *said* – in order to *name* what comprises it, one can rather concentrate on the new entity created by a certain combination. The operation of dissection can be abandoned in favor of the study and contemplation of the new *proper* that springs into being from a particular *mise-en-contact*. It will therefore be necessary to renounce efforts aimed at fixing metaphor in one neat and concise definition, for discussing metaphor engages us in a project not of *reduction* but of *expansion*.

Rather than a substitution or a transfer of meaning, metaphor is the manifestation of a relationship that springs forth from the interaction of two or even several terms (words, sentences, texts, etc.). This relationship is founded on the juxtaposition (*rapprochement*) of terms that have no obvious rapport but that bring forth, by their *rapprochement* and interaction, new *correspondances*. It is not sufficient, of course, to group words together haphazardly, for neither a simple lack of logic nor absurdity constitute a metaphor. Rather, the creation and use of metaphor are the act of the artist (in this paper the poet, in particular) who sculpts his language to express what is inexpressible in non-metaphoric terms; according to Ricoeur, metaphor has the capacity to provide untranslatable information.[8] The poet puts in contact terms without any logical rapport, and from this contact springs an unexpected rapport – "a third entity which can be expressed in no other way"[9] – which sustains itself by an internal tension.

Metaphoric rapports exist on the level of ideas and on a linguistic level. Verbs and adverbs lend themselves to metaphoric use as easily as nouns and adjectives do, which enhances the virtuosity of metaphor: "metaphor makes a more complete use of linguistic structures than other kinds of figurative language."[10] Since all categories of words lend themselves to a metaphoric use, the possibility of creating new rapports (between texts, etc., not only between words) are tremendous. "In the land of metaphor,"[11] as Wallace Stevens would say, rapprochement of terms is not limited by grammatically correct organizations, but is instead free to create new organizations – adverbs modify nouns, prepositions modify adjectives, nouns function as verbs, etc. (e.g., Dylan Thomas' "a grief ago"). Christine Brooke-Rose tells us: "Metaphor is not merely the perception of similarity in dissimilarity, it is the changing of words by one another, and syntax is rich in methods of doing this, each with different effects."[12] Even when the terms of a metaphor are identifiable and named, they are transformed both semantically and according to their function insofar as they are part of the metaphoric rapport. Even though the terms exist outside this rapport, they have a particular (but not necessarily only one) meaning and function within this rapport. The process which is metaphor transforms the terms as well as the relationships of the terms among themselves; the "new reality" thus brought about is of course inexpressible in any other way. The particular combination and reaction set in motion by the poet gives life to that which otherwise would have been undetected, or at least unexpressed.

There are of course degrees of metaphor (simile-cliché). The most interesting are those where the rapport between the terms is least evident. As Terence

Hawkes notes, "indeed the tenuousness of the connection contributes greatly to its effectiveness."[13] The greater the distance between the terms of a metaphor, the greater the tension required to sustain the rapport between the terms, and it is in the space (*écart*) of this tension that springs forth the metaphor; indeed, as Brian Wicker tells us, "metaphor has an inbuilt *tension*."[14] In many cases the distance between the terms is so vast that no connection can be found, hence the futility of any attempt to "explain" a metaphor by reducing it to its component parts. As Max Black states: "It would be more illuminating in some of these cases to say that the metaphor creates the similarity than to say that it formulates some similarity antecedently existing."[15] Going even further, Paul Ricoeur tells us: "To say that a new metaphor comes out of nowhere is to recognize it for what it is, namely a momentary creation of language, a *semantic* innovation. . . ."[16] Metaphor, then, is a function which transforms language, and the greater the gap between the terms of a metaphor, the greater the metaphor's motivating potential, for the tension required to sustain the rapport motivates the language to fulfill its function as transformer: "The test of essential metaphor is not any rule of grammatical form, but rather the quality of semantic transformation that is brought about."[17] Insofar as it effects semantic transformations, metaphor reveals itself to be a "semantic operator,"[18] a motivating force, both that which motivates change of meaning and the new meaning which results, both discovery and creation.

As process and function, metaphor is in continual movement and undergoes a continual change as it destroys ordinary rules and meanings and creates new ones. (When we talk of metaphor as a process of destroying old meanings and creating new ones, it is understood that the old meaning is not completely annihilated but that some shred remains – otherwise the new could not be perceived as new – and the tension and distance between the old and new reflect and sustain this co-existence.) The movement of metaphor, engaged in this double function of destruction and creation, is itself double as it is directed inward toward transforming its own terms and rapports and outward toward creating new ones. It is a movement both centrifugal and centripetal, as Marcus Hester notes: "[Metaphor tends] to implicate wider contexts and to focus inwardly. . . . [C]entrifugal designates the plurisignative nature of metaphor; centripetal designates the conjunctive use of several metaphors to focus on the same phenomena."[19] This ambiguity, and the tension which it provokes, motivate the language.

When metaphor is perceived in this way, its rapport with human experience is quite clear: as human experience is constant change, a continual movement, so is metaphor the process of change within language. It is in the gap between ordinary and metaphoric language that man can seek to grasp his experience, not by reducing it to what he knows but by discovering in it a new awareness. To do this we must, of course, as Ricoeur puts it, "entertain two different points of view at the same time."[20] The tension required here is analogous to that inherent in metaphoric language, which never completely loses its quality of reference but which is stretched taut, to its limits, and wherein new "meanings" are as strong as the old (ordinary, literal).

The flux of human experience, then, can be discussed most clearly and most completely in language that does not diminish or interfere with its quality of flux, and does not merely reflect or imitate this quality, but in language that is its counterpart – in language that is itself flux, change, and process. And according to Peter Handke, language itself is revitalized by the experience of poetic creation. Any discussion of or attempt to understand human experience is, it seems to me, not an attempt to limit or contain the experience but rather to see it in a specular fashion in order to expand our vision and our horizon. This attempt to grasp the significance of human experience of course engenders new experience and becomes a part of the *quest*ion rather than its solution. This is part of its intrigue and appeal, and recalls the necessity of defining and discussing metaphor in metaphorical terms. This paradox, however, is not a dead end but an example of the double movement of metaphor inward upon itself and outward toward new rapports and meanings. This extensive (ex-tensive) quality of metaphor, this property of tending *outward*, of implying greater and vaster contexts, the plurality which it implies by definition, is a reflection of the human experience of moving from past to future and back, of our reflecting on the past in order to give it meaning in the hope of giving meaning to our future. This return to the past and projection (whether yearning or dread) toward the future is both what constitutes our present and what prevents us from grasping it. As Marshall McLuhan notes: "Each extension creates a new environment that inflicts change and new motivation upon the old one. The old and new environments are 'twins' that perpetually impel us onward in a non-stop process of transformation."[21] This change is also an *exchange*, derived from the tension inherent in the movement. In a literary work of art, this exchange between the writer and the world or, as Handke might put it, this "poetic permeance into the world," is accomplished by metaphor, thanks to which, Ricoeur tells us, "individual life and universal life grow together."[22]

Our own focusing inward upon ourselves in our effort to achieve greater self-interpretation is accompanied by a movement outward in every *sens* (meaning *and* direction) in order to gain greater understanding of our own historicity. These contradictory movements (counter-currents) are of course tension-producing, and it is within the space of this tension that new insights are to be found. These contradictory movements are mutually nourishing and mutually destructive:

> The cells which build each other mutually . . . also destroy each other mutually. . . . [T]he organism grows in consuming and dissolving itself. "We live our death" says Heraclitus, in order to express the vanishing of the discontinuous in the continuity of the process.[23]

Far from being limited to the notion of comparison, or of transfer of meaning, metaphor is a process of tension and energy, a dynamic force that allows us to grasp, express, and share experience. Dylan Thomas tells us that poetry is the poet's "attempt at an expression of the summit of man's experience."[24] According to Handke, language itself is revitalized by the experience of poetic creation: indeed, he sees poetry as the putting of shapeless forms into new shapes. The experience of the metaphorical language in a poem may be one of frenzied propulsion or of supreme peace and silence; either way, it is a capital experience. Indeed, as Richard Shiff notes, "metaphor is not a means to other ends but one of the main ends of life . . . sharing metaphors becomes one of the experiences we live for."[25] Trying to grasp and interpret and share the flux in which we are involved, we come to realize, with Brain Wicker, that "metaphor is not just a way of describing things but is a way of *experiencing* them."[26]

*Tufts University*

## NOTES

[1] Arthur Koestler, 'Cosmic Consciousness,' *Psychology Today* (April 1977): 104.

[2] Paul Ricoeur, *La mètaphore vive* (Paris: Seuil, 1975). Translation mine.

[3] Monroe Beardsley, 'The Metaphorical Twist,' *Philosophy and Phenomenological Research* 22, no. 3 (March 1962): 298; my italics point out the coined word "metaphoricalness," and the formula "so to speak," which always introduces a linguistic deviation.

[4] Cited by Jacques Derrida, *Marges de la philosophie* (Paris: Editions de Minuit, 1972). Translation mine.

[5] Gertrude Buck, 'The Metaphor: A Study in the Psychology of Rhetoric,' in *Contributions to Rhetorical Theory*, vol. 5, ed. Fred Newton Scott (Ann Arbor: Inland Press, 1899), p. 1.

[6] Fred Newton Scott, ed., *Contributions to Rhetorical Theory*, pp. 2–3.

[7] Max Black, *Models and Metaphors: Studies in Language and Philosophy* (Ithaca: Cornell University Press, 1962), p. 37.

[8] Ricoeur, 'The Metaphorical Process as Cognition, Imagination, and Feeling,' *On Metaphor* (1979), p. 141. My use of metaphorical language is intended to underline the necessity and untranslatable quality of "non-literal" language.

[9] David Miller, *The Net of Hephaestus: A Study of Modern Criticism and Metaphysical Metaphor* (The Hague: Mouton, 1971), p. 57.

[10] Owen Thomas, *Metaphor and Related Subjects* (New York: Random House, 1969), p. 60.

[11] Terence Hawkes, *Metaphor* (London: Methuen, 1972), p. 6.

[12] Christine Brooke-Rose, *A Grammar of Metaphor* (London: Secker and Warburg, 1958), p. 93.

[13] Hawkes, p. 19.

[14] Brian Wicker, *The Story-Shaped World, Fiction and Metaphysics: Some Variations on a Theme* (Notre Dame: University of Notre Dame Press, 1975), p. 108.

[15] Black, p. 37.

[16] Ricoeur, *La métaphore vive*, p. 126.

[17] Philip Wheelwright, *Metaphor and Reality* (Bloomington: Indiana University Press, 1962), p. 71.

[18] Jean Cohen, *Structure du langage poétique* (Paris: Flammarion, 1966), p. 15. Translation mine.

[19] Marcus Hester, *The Meaning of Poetic Metaphor* (The Hague: Mouton, 1967), p. 17.

[20] Ricoeur, *On Metaphor*, p. 152.

[21] Marshall McLuhan and Wilfred Watson, *From Cliché to Archetype* (New York: Viking Press, 1970), p. 142.

[22] Ricoeur, *On Metaphor*, p. 152.

[23] Martin Foss, *Symbol and Metaphor in Human Experience* (Princeton: Princeton University Press, 1949), p. 39.

[24] Dylan Thomas, *Quite Early One Morning* (New York: New Directions, 1954), p. 193.

[25] Richard Shiff, 'Art and Life: A Metaphoric Relationship,' *On Metaphor* (1979), p. 114.

[26] Wicker, p. 11.

CHRISTOPH EYKMAN

# THE LITERARY DIARY AS A WITNESS OF MAN'S HISTORICITY: HEINRICH BÖLL, KARL KROLOW, GÜNTER GRASS, AND PETER HANDKE

The diary as a literary form has, in recent times, enjoyed an increased measure of popularity among writers. In this paper, we shall not deal, however, with works of fiction which have taken on the form of a diary but with genuine diaries, i.e. the periodically recorded testimony of autobiographical experiences, diaries thus, which are (in spite of certain fictious inserts) not "invented" but can be regarded as documents of real life. Yet, at the same time, these diaries are written as *literary* works, they are the product of artistic creation, they have been shaped into an esthetic "Gestalt" without thereby losing their biographical authenticity. That means of course that they were all written for publication.

We have selected four fairly recent literary diaries by three West German authors and one Austrian. They were all published between 1957 and 1977. All four cover a relatively short period of time, ranging between six months and three years. In spite of the brevity of the time-periods covered, these diaries all have their own way of recording the steady flow of time and thereby revealing man's historicity. Since they all differ greatly from each other, they demonstrate, that, with regard to both form and content, the literary diary is a genre of great diversity. Its phenomenology has yet to be written.

Heinrich Böll's *Irish Diary*, published in 1957, contains a composite report about several journeys to Ireland and presents a multifaceted picture of the country and its people. Descriptions alternate with anecdotes or short stories that highlight the life style of the Irish who, seen through the eyes of a German, appear as a slightly exotic species of humans. The diary hardly ever focusses on the author's ego.[1] The text directs the reader's attention almost exclusively to its central topic: that foreign country Ireland.

Inevitably, the encounter with Ireland becomes an encounter with history. Not only churches and other monuments speak of history, traces of the past can also be found in the slums of Dublin: "In the dark backyards, which Swift's eyes saw, decades and centuries have deposited this dirt, the depressing sediment of time."[2] Time leaves its sediment, yet time also erodes what man has built, for example that village of empty houses, this "skeleton of a village,"[3] as Böll calls it, which men left over a hundred years ago. Rain, sun, and wind have eaten away everything except the building stones. History



*A-T. Tymieniecka (ed.), Analecta Husserliana, Vol. XVIII*, 249–260.

is present and visible everywhere, one just has to be able to recognize it: "He who does not have the sharp eye of an archeologist, will not be able to distinguish the walls built in the twentieth century from those built in the fifth, they are all covered with green leaves sparkling in the golden sunlight."[4] Yet the author's mind does not just function as the receptacle of realities outside his own self. At times, he feels personally touched by the mementos that the flow of history left behind to be beheld by posterity. On a bench in St. Patrick's Cathedral, Böll looks at some enamel signs which exhort the visitor to pray for the soul of a deceased person.

Pray for the soul of Michael O'Neill, who died at the age of sixty on January 17, 1933. Pray for the soul of Mary Keegan, who died on May 9, 1945, aged eighteen; what a pious and clever extortion: the dead become alive, the date of their deaths awakens in the reader of these signs associations with his own memories of that particular day, that month or that year. Grimacing nervously, Hitler was waiting for power when here, at the age of sixty, Michael O'Neill died; when Germany capitulated, Mary Keegan died at age eighteen. "Pray," so I read, for Kevin Cassidy, who died on December 20, 1930, at the age of thirteen, and it hit me like an electric shock, for in December of 1930, I myself was thirteen years old: in a large dark apartment in South Cologne – in 1908 one would have called that an upper middle class apartment house – I was sitting holding my Christmas report card in my hands, vacation had just started, and through a hole in the cinnamon-colored curtain I was looking down upon the street.[5]

Grand political history and the seemingly less significant history of the unknown individual are evoked simultaneously by means of memory, and yet the passage contains the implicit question how much impact one might have had on the other.

Böll has repeatedly emphasized the stark discrepancy between the "grand" history of the history books and the untold private sufferings of the individual. In a speech geven by him on the occasion of the West German equivalent of Memorial Day in the same year in which *Irish Diary* was published, he focusses his audience's attention upon those countless mothers, sons, daughters, and wives of those killed in action, i.e., upon the survivors about whom the history books are silent:

Perhaps they don't dare ask the monuments to return the dead to them, those monuments in the shape of pillars carved of rock or in the shape of pompous statues of bronze which never cease to claim death as a public property. Those monuments which transmit a lie to posterity when they say that those soldiers died for future generations and which demonstrate to the survivors that death is owned by history as documented in marble or brass.[6]

Böll's diary ends with the description of a dream in which grand history, in compressed form, floats past the observer, one giant procession of the relics of past centuries:

'The entire holdings of the National Museum . . . took on a magic distinctness and rigidity like the props of an exhibition of waxworks in this no-man's-land between dream and memory; like on a ghost train which children ride at a fair, we were riding downhill: St. Brigid's shoe was shining in the dark, big black crosses were consoling and threatening us, freedom fighters in pretty green uniforms, wearing puttees and red caps showed us their shotgun wounds, their passports, were reading to us in a child's voice farewell letters: "My dear Mary, Ireland's freedom . . . "; a saucepan from the thirteenth century swam past us, so did a canoe from prehistoric times . . . monuments were pushed past us or we past them, likewise men of bronze, serious, holding swords, pens, parchments, bridles or a pair of compasses; . . . the waters of the Liffey turned black carrying history's flotsam out into the ocean: documents the seals of which were hanging down like a sounding lead, contracts with initials full of flourishes, documents heavy with sealing wax, wooden swords, cannons made of cardboard, lyres and chairs, beds and closets, inkpots and mummies whose bandages had come off and who were floating through the water like dark palm branches. . . .'[7]

Man's capacity to absorb history and to live with it and its remnants is no match for the abundance of relics and monuments compiled in a museum. Böll's dream signals in its own humorous way the defensive reaction of the modern mind to the burden of history which has become too heavy: the entire historical heritage is flushed out into the ocean, i.e., into oblivion.

The title of Karl Krolow's *Poetic Diary* (written 1964–65, published in 1966) indicates that this work is different from the traditional diary which includes – at least potentially – the full range of a writer's experiences in the real world. The title does not only mean that a diary has been elevated – by means of form and style – to the status of a literary work. It means in Krolow's case that its content has been restricted almost exclusively to the realm of *esthetic* experience. Indeed, Krolow's diary interweaves in the most artful fashion three strands: the writer's thoughts on writing, memories of literary works read or paintings seen, and thirdly: a highly sensitive observation of nature as it reflects the sequence of the four seasons. One might almost call this book the "Diary of an Esthete", since it so strictly excludes any non-esthetic experience or thought. We learn nothing about the private life of the author or about the socioeconomic and political problems of his time.

And yet, the dimension of historicity is present throughout the work. Impressions of the outer world as perceived through the senses immediately find an echo in the writer's mind which has stored a wealth of experiences from literature, music, and the fine arts. Thus a sensory perception coupled with a certain emotional response is instantaneously linked with the memory of a line from a poem, a passage from a prose text, the shapes and colors of a painting or the structure of a musical composition. In Krolow's mind a continuous synthesis of remembered esthetic experience and present sensory perception of nature takes place. Both correspond to each other and confirm each other. The memories of literature, music, and art constitute a historical dimension, since many of these memories refer to works, truths, feelings, and attitudes from past centuries. The validity of these esthetic experiences, however, as they are integrated into the perception of the surrounding phenomenal world at a moment of real time, is a timeless one, there is hardly a sense of distance or alienation from the documented experience of past writers or painters.

Time and the notion of change are also present in that third strand we mentioned above, namely, in the perception and contemplation of nature, in the meticulously and sensitively observed and recorded transitions from summer to fall to winter to spring and back again to summer. Let us take a look at one entry in Krolow's diary which is representative of the entire work:

> The sky of June, in the morning, on a clear day is a blue cubic house. Roses and dust are living under its protection. The mornings in early summer! In 'Le Grand Ecart' by Jean Cocteau one reads: "It is the animal and the plant in us who think for us at the moment of awakening." This plantlike or animallike (or maybe also minerallike) moment of coming to (so to speak) will only later take on a certain significance as does the unconscious body in its warmth, in the Euclidian clarity of a forenoon with its tension between the light and the color of the eyeball. The blue has the nuance of a painter's breakfast in Barbizon. Blue like the "Rome with its shades of blue between the pinetrees as painted by Poussin," which is from a line of Rafael Alberti's poem 'A la Pintura,' which I remember and thus rescue from oblivion. It is the geometry of an ancient blue, which spreads its light on houses and faces, while the minutes go by, eyes filled with golden light, and while on the tables shaded by leaves, vegetables and fruit symbolize the image of life. The first grass, ready to be cut, is like breathed air. Late summer's heat waiting with a red wick in a jungle of leaves, is, fortunately, still far away. Yet it is lurking, a match between its teeth, on the other side of solstice.[8]

The gradual progression of time (from early to late summer) is alluded to. More importantly, the passage shows the symbiosis of perceived reality and the associations it awakens from the memories of literature and art. Past artistic and literary experience – preserved in the writer's mind – fuse with the present moment of real experience to create a fuller and richer impression in which past and present are one.

Günter Grass' *From the Diary of a Snail* (1972) is a peculiar blend of genuine diary, fiction, and historiography. It was written while the author campaigned for the candidates of West Germany's Social Democratic Party in 1969, and it documents this half-year period in the author's life. The work constitutes therefore a piece of both private and public historiography with all its numerous references to names and events of this time, to what the German language calls "Zeitgeschichte". Thus we hear about Willy Brandt, chancellor Kurt Georg Kiesinger, about the astronauts' first landing on the moon or about the death of the philosopher Th. W. Adorno.

But this is only one layer of the work. In another, Grass reaches back into a darker chapter of German history and, as a historian, reports about the persecution and expulsion of the Jews from the city of Danzig (Grass' hometown) during the period of National Socialism. The diary turns into the chronicle of events not witnessed directly by the author. Here historical facts are fused with fiction. The constant confrontation of the present with the past gives the present, i.e., Grass' political campaign, a historical perspective.[9] Much of the author's political philosophy, his plea for reforms, his social concerns, his humanitarian attitude, are derived from the dark times of which he gives an account. He wants to remind his fellow Germans of their past, which they are all-too-willing to forget in the midst of their affluence. At this point, historiography turns, at times, into fictional narrative as Grass invents the character of a Danzig high school teacher who helps the Jews and hides out in the cellar of a bicycle repair shop until the war is over. In addition to the levels of the past and the present, the diary-form allows its author to jot down any thoughts or ideas as they come to mind, regardless of any time levels. Thus it is not surprising that Grass reflects upon history and, in doing so, reveals a good deal of scepticism about the subject.[10] For example, he parades before the reader a number of empty phrases containing the word "history," phrases strung together without comment: "When history demands it. When we feel the breath of history. When history surges over our heads. When history provides great tasks for us. When history, in which, as Hegel says, only those peoples matter which have formed a state, will measure us in terms of our sacrifices."[11] The terror of recent history has given all those pompous phrases a rather hollow ring.

In a more concrete context, Grass criticizes the smug self-confidence of affluent West German who looks back upon his country's achievements after 1945, conveniently blocking out any reference to the Third Reich. Once again, this attitude expresses itself in fragmentary cliché phrases, the broken molds of a self-congratulatory phraseology: "and so we have after more than twenty-five years. Have out of rubble and ashes. We have from nothing. And today we are again. Without being immodest. Which all over the world. Nobody expected it. Can show itself with pride."[12]

And yet, Grass is far from denying the importance of history and its impact on us. His political campaign is, after all, in its own modest way, an attempt at shaping history, and the diary which records the daily vicissitudes which constitute this attempt represents an act of artistic faith in human historicity. The symbol which Grass found for this faith in history (i.e., the history of the past and the future history yet unborn) is the snail. The snail is a political symbol in Grass' book. It stands for slow and partial progress as opposed to the revolutionary radicalism or reactionary conservativism of some of Grass' contemporaries. It is the Social Democratic Party that represents, in the author's view, exactly this commitment to progress at a snail's pace. Thus he categorically states in one of his campaign speeches: "Progress is a snail."[13] He even compares himself to a snail: "I am the civil snail turned human. Given my desire to move ahead, to withdraw to the interior of my shell, my tendency to dwell, to hesitate, to adhere to a surface, my restlessness and my rash emotional reactions, I am like a snail."[14]

By means of the symbol of the snail, Grass is able to link all the various levels of narration with each other. For example, he presents his Danzig high school teacher as an avid collector of snails and snail shells. No doubt that the snail is a time symbol, and that means: a symbol of man's historicity. It represents the cautiously utopian aspect and the will to help shape future history even if progress is ever so slow. But it also represents history as time that has passed and is now open to retrospective scrutiny:

> Across everything that stands and costs money, across debit and credit, across everything that is idling or perpetuating mere habit, the snail is gliding, leaving behind its trace: a layer of slime which dries up with a crackling sound. it remains transparent and makes us still and visible, respectable as we are, i.e., as we have again become.[15]

History lets us see men and events as they really were; it is a preservative and a translucent shrine at the same time. It is itself a snail, as Grass once humorously remarks in his diary: "The truth is that Hegel did not have in mind somebody on horseback but a rider on snailback when he talked about the 'Weltgeist' (the world spirit)."[16]

Peter Handke's diary *The Weight of the World* was written during the time period between November 1975 and March 1977 and published in 1977. Of the four diaries discussed in this paper, Handke's is the most unusual. He spends sixteen months in the city of Paris for the sole purpose of writing his diary. It does not record events or experiences in the traditional sense of the word. Neither does it follow the pattern: "On such and such a day I did this and that (or: this and that happened to me)." There is no linear sequence of any "story," there is no string of events in Handke's notes. He just receives and records – at random – impressions which, at first sight, seem hardly noteworthy. They are unspectacular and tell us little about any coherent purposive actions which might constitute the meaning of the author's life. What is being recorded in the diary is almost "subliminal" in the sense that it normally escapes conscious perception even though perception takes place to a minimal extent. Handke's poetic achievement lies in overcoming that subliminal status by means of intensifying and deepening the visual experience and making it communicable. Here are a few examples:

Somebody who has finally identified himself with his sadness, goes willingly along with every move of the bus, of the road, of the other passengers while holding on to the pole in the bus.[17]

The old man in the restaurant: he holds the empty wine bottle over his glass until the last drop detaches itself from the bottle (I felt as though I was looking at a self-portrait).[18]

A man stoops down to a child as if he were to try mouth to mouth resuscitation.[19]

The old woman runs towards the waiting train: while she is unable to run faster, her face is hurrying showing an expression of panic, especially her mouth.[20]

Birds are invisibly flapping their wings in the bushes in the evening without flying away.[21]

The dead cars in front of the window at night.[22]

People who are walking on the street early in the morning as though they were still wearing slippers.[23]

A new season: those thin wooden sticks which once were attached to ice-cream bars are again lying in the sand on the paths in the park.[24]

What is the purpose of collecting such observations, such inconspicuous microscenes of the neglected moments of everyday life, scenes which never integrate themselves into a macro-image or story? The puzzled reader might assume that an author has ventured to write a parody of the diary and at the same time a book of anti-history describing the fleeting moments of the unglamorous anonymous daily life when it is least note- and newsworthy, including his personnal idiosyncracies, thoughts, and feelings. And yet, the seemingly negligible object or event observed fills itself with poetic significance once the author's mind and the described realia form a synthesis by means of the written word. This is what Handke calls in his diary "the poetic permeation of the world."[25]

It is worth nothing that Handke emphasizes the role of feeling in that synthesizing process. "It is possible to describe the world when feeling finally becomes associated with an object (the epic genre)."[26] Feeling turns a merely perceived object into what Handke calls "Erscheinung" (appearance) and, once lifted to that higher level, observed reality allows for a successful transformation into language, a language that is more than the act of naming or categorizing by means of abstract concepts. The following sentence which describes the *failure* of the creative process, sheds some light on Handke's esthetic theory: "A day on which the appearances remained just things and the things just names; the names are ambiguous. Everything has turned bad, even the sunset."[27]

In his introductory remarks to his diary, Handke describes his literary efforts:

> I practiced reacting through language to everything that happened to me and I noticed how at the moment of the experience language revitalized itself for a while and became communicable. A moment later it might have been again the language we hear every day, meaningless in its familiarity with reality, the helpless you-know-what-I-mean language of our age of communication. For a moment, the vocabulary running through my mind day and night became real. Whatever I experience appeared in this 'moment of the language' freed from all privacy and thus of universal validity ["allgemein"].[28]

In the light of this statement it makes sense that Handke never mentions names in his diary but only anonymous expressions like "a woman," "someone," "a person," or abbreviations like F. and even for his own daughter A.

Thus the diary which contains a haphazardly assembled collection of seemingly negligeable everyday impressions takes on a validity as a human document which rivals any diary filled with "noteworthy" and "important" events. The author focusses on observations quickly discarded and forgotten by the average mind. Handke's diary indeed constitutes a very subtle form of historiography recounting dispersed fragments of the micro-history of normally neglected inconspicuous events of our daily lives.

But Handke has a very low opinion of history. He refuses to attribute any meaning to the history of the individual person: "I could not say who I am, I do not know anything about myself. I am someone without any origin, without history, without a country – and on that I insist."[29] "It is nice," he writes, "not to feel one's history any more, not to have the present happiness contested by what one was at an earlier time, as a child, as an adult, etc."[30] He asks himself: "Why always this urge to find in one's personal history a cause, an explanation, even a justification of all sensitivities, expectations, desires – as though they were in need of an explanation or justification."[31] Once Handke even equates memory with physical pain which grows into a death-threat: "I must get rid of my memories. Against Proust and Benjamin and the well-sheltered bourgeois consciousness with its tendency to seek pleasure in memory (my struggle against memory which has limited my existence ever since my childhood. . . . "[32]

Not only personal history but also the history which transcends the individual is viewed negatively by Handke: "That people always want to drum into our minds the 'lessons of history,' so that we may become human beings – as though I were unable without history (and indeed for me the only way is: without history) to understand what should be done and what not."[33] History evokes in Handke a sense of alienation which he feels for example when looking at a photograph which represents for him a fragment of a reality totally estranged from the viewer.[34] "Experiencing history," Handke writes, "means to me: to free myself from it or: to already have disentangled myself from it."[35] And yet, in his own special way, Handke is after all a historiographer, even though he is not tracing lines between the past and the present. Instead, he searches for what he once calls "the common denominator of our hidden life."[36] His search focusses on isolated and fleeting moments in the everyday life of the anyonymous human being, far below the stirrings of political history. "While looking at clouds, fields, and people in a landscape," he writes, "all of a sudden I see with breathtaking clarity the history which happened behind the pompous official history throughout the centuries. It is the history of the suffering of these people, of their deaths, of their being worn down, the actual history, my history, my work."[37] This "history" manifests itself at any street corner, in the subways, the parks, the apartment houses. Only the sensitive look of the writer can penetrate into the full essence of what he sees. Therefore, we must not overlook the *subjective* aspect of this special type of "historiography." Handke claims in his introductory remarks that the purpose of his diary is a continuous effort towards the self-penetration of his mind. It is thus an ever progressing act of self-discovery through the discovery of the outside world, it is indeed the discovery of "my own history about which I know so little."[38] In spite of his remarks to the contrary, Handke does not deny the historicity of his own self: "Triumphant though," he once states in his diary, "I have a history, and I shall continue to have one."[39]

The creative process of writing appears therefore as one which arrives at an image of both personal and transpersonal historicity in the most basic sense.[40] An a priori historicity dislocated from the official history of the historians, a history never defined since its manifestations remain disintegrated, fragmented and open-ended. Even Handke's punctuation shows this openendedness. None of his entries in the diary end with a period. This is how Handke himself describes the act of writing:

> The feeling as though almost everything I have heard or seen in the past loses in my mind immediately its original shape. It is neither directly describable by means of words nor can it be represented by images, but it changes right away into something entirely shapeless. I feel that the purpose of my writing is to transform those numerous shapeless experiences into new shapes. The latter would still be linked by my feeling to the original experiences. They would be the mythological images of these authentic, factual but insignificant things, images produced by my consciousness and my existence.

It seems that all four literary diaries we discussed reveal an ambivalent attitude towards history. Böll is fascinated by the remnants of Irish history, yet the historical heritage overwhelms the mind which – even though only in a dream – attempts to free itself from it. The very fact that Krolow's diary narrows the focus of memory and perception to the esthetic dimension, might be interpreted as an expression of indifference vis-à-vis full scale history. Even if one views Krolow's preoccupation with both art and nature as an escape from history, both his experience of the arts and the phenomenal world contain the element of time and change. They are intimately linked with the author's emotional and intellectual existence which, in spite of the dominance of the esthetic dimension and in spite of its position above "Zeitgeschichte," is nevertheless essentially historical. Günter Grass launches biting attacks against certain pompous clichés about man and history thereby exposing the contempt of human dignity inherent in such clichés. Man becomes the mere object, the victim of a history whose greatness and splendor turns out to be a meaningless succession of wars and destruction. Yet the very image of the snail as a symbol of slow progress reveals a cautious, "pedestrian" optimism with regard to man's ability to fashion history in a meaningful way. Otherwise, the author of the *Diary of a Snail* would not have made those political campaign speeches he so frequently mentions in his book. Finally, Handke's diary is of two minds as regards man's history. His numerous statements denying the importance of both personal and transpersonal history are contradicted by the literary accomplishment of the diary itself which can be interpreted as a micro-history of the inconspicuous daily existence of the author and his anonymous fellow human beings, a fragmented, disparate and discontinuous history. A history that is incoherent and atomistic, not suited for the history books, yet one that lets us experience "the weight of the world" as the title of Handke's book reads. "Weight" means both: a heavy burden and at the same time something weighty, important, significant.

The four diaries are, within the framework of German literature, in the tradition of a pessimistic attitude towards history, an attitude typical of many contemporary writers. It has its own history, since it can be traced back via Dadaism to Friedrich Nietzsche.[41] But the four diaries also reaffirm the notion of man's historicity, even though it has been scaled down and deflated (since the breath of Hegel's "Weltgeist" is gone). The pompous clichés of past historiography have been discredited. Nevertheless, man still has a history.

## NOTES

1 Cf. Cecile Cazort Zorach, 'Two Faces of Erin: The Dual Journey in Heinrich Böll's *Irisches Tagebuch*', *GR* 53 (1978):128. Zorach notes at this point the absence of first-person expressions. Her paper does not deal with the theme of historicity in this work.

2 Heinrich Böll, *Irisches Tagebuch*, in *Werke*, vol. 3, *Romane und Erzählungen*, ed. Bernd Balzer (Cologne: Kiepenheuer & Witsch, 1977), p. 20.

3 Ibid., p. 31.

4 Ibid., p. 77.

5 Ibid., p. 23.

6 Heinrich Böll, *Werke*, vol. 1 *Essayistische Schriften und Reden 6 (1952–1963)*, ed. Bernd Balzer (Cologne: Kiepenheur & Witsch, 1980), p. 222.

7 *Irisches Tagebuch*, pp. 87 f.

8 Karl Krolow, *Poetisches Tagebuch* (Frankfurt: Suhrkamp, 1966), pp. 122 f.

9 Cf. Jürgen Rothenberg, 'Grosses "Nein" und Kleines "Ja": *Aus dem Tagebuch einer Schnecke*,' in *Günter Grass Materialienbuch*, ed. Rolf Geissler (Darmstadt: Luchterhand, 1976), p. 148. Rothenberg calls this "Einschwärzung der Gegenwart im Sinne von 'Einbräunung,' ein Verdrängen des Aktuellen auf das Gestrige hin." The West German present is revealed as an extension of its recent past in the sense of "sich erneuernder Nazizeit" (ibid.). The Nazi spirit, though transfigured, is presented as a factor in post-war West German society.

10 Cf. Fritz Raddatz, 'Der Weltgeist als berittene Schnecke: Günter Grass' kleine Hoffnung – aus grosser Melancholie,' in: *Grass: Kritik – Thesen – Analysen*, ed. Manfred Jurgensen (Bern; Francke, 1973), p. 196. Raddatz writes about Grass' "Skepsis gegen die Geschichte, genauer gesagt: gegen ihren peristaltisch-dialektischen Ablauf." A rather self-contradictory observation since peristalsis (the alternation of constriction and dilation in muscles) is obviously irreconcilable with the notion of dialectics.

11 Günter Grass, *Tagebuch einer Schnecke* (Neuwied: Luchterhand, 1972), p. 57.

12 Ibid., p. 134.

13 Ibid., p. 49.

[14] Ibid., p. 76.
[15] Ibid., p. 135.
[16] Ibid., p. 76.
[17] Peter Handke, *Das Gewicht der Welt: Ein Journal*, November 1975–March 1977, p. 43.
[18] Ibid., p. 48.
[19] Ibid.
[20] Ibid., p. 276.
[21] Ibid., p. 56.
[22] Ibid., p. 57.
[23] Ibid., p. 257.
[24] Ibid., p. 101.
[25] Ibid., p. 93.
[26] Ibid., p. 98.
[27] Ibid., p. 206.
[28] Ibid., p. 6.
[29] Ibid., p. 151.
[30] Ibid., pp. 14 f.
[31] Ibid., p. 180.
[32] Ibid., pp. 85 f.
[33] Ibid., p. 118.
[34] Ibid., p. 189.
[35] Ibid., p. 138.
[36] Ibid., p. 154.
[37] Ibid., p. 218.
[38] Ibid., p. 271.
[39] Ibid., p. 57.
[40] Cf. Manfred Jurgensen, 'Peter Handkes "Journal" *Das Gewicht der Welt* (1975–1977),' in *Handke: Ansätze- Analysen – Anmerkungen* (Bern: Francke, 1979), p. 182. Jurgensen emphasizes the subjective factor in Handke's attempt to derive "neue Mythen . . . mit denen ich mich neu anfangen kann" from his observations (p. 185). Nevertheless, the objective aspect of this literary venture, the painstakingly accurate rendering of observed reality, should not be overlooked. Neither should the aspect of (ever so slight) mental insanity "eine Typo- und Psychologie des Kranken") be overemphasized. It is true that Handke presents us with a "Aesthetik des Schmerzes" (an esthetics of emotional pain), which does not in itself constitute insanity.
[41] Cf. Christoph Eykman, *Geschichtspessimismus in der deutschen Literatur des zwanzigsten Jahrhunderts* (Bern: Francke, 1970).

VICTOR CARRABINO

# THE FRENCH *NOUVEAU ROMAN*: THE ULTIMATE EXPRESSION OF IMPRESSIONISM

In his *Social History of Art* Arnold Hauser claims that "Impressionism is the last valid European Style." [1] A reappraisal of such a statement is in order, for it is essential to understand the method and goal of Impressionism before one can accept Hauser's statement.

Impressionism, as an artistic movement, stresses the importance of man's sensory response to nature. It is a subjective study of nature as it appears to the individual.

The Impressionists attempted to introduce into painting independent appearances, free of cultural, intellectual and mental ties. Their will to submit themselves to physiological vision was for them a search for an uncompromising realism, which ironically was to lead beyond realistic painting. From pure look to pure painting was but a step which was once taken by nonrepresentation art. The nineteenth century academic painters conceived a composition as a theatrical scene. The subject was depicted from a spatially removed viewpoint. The Impressionists, however, negated this corporeal separation of the element.[2]

Academicians scrupulously respected the old methods, chiefly balanced composition and perspective. Objects were centered, the painting was a close entity within itself, each detail taking part on the arrangement of the work. Compact, solid, dense objects were separated one from another. There was no confusion of material.

Light illuminated the object with no attempt either to dissolve it nor to blend it with the others present. It was this static scheme, the reflection of a society which wished to be eternal, that the Impressionist painters were going to explore, sweeping away conceptions both of space and of matter totally unadapted to the scientific and technical revolution of the 19th century.

This dismantling of the classic framework was hinged around six principle axes:

(1) Beginning with Degas, point of view shifts.

(2) Objects and environment interpenetrate, de-materializing them. The disparate components of reality mingle in light.

(3) Color acquires its independence.



*A-T. Tymieniecka (ed.), Analecta Husserliana, Vol. XVIII*, 261–270.

(4) The artist no longer seeks to *compose* his work. Instead of being an ensemble of forms balanced within the canvas, the painting becomes a section of nature which extends to a limited setting.

(5) The painter minimizes perspective and brings it back to its two dimensions. Depth becomes indistinct and hazy. Landscape verticalizes itself. Painting relates itself to the picture plane.

(6) The artist solicits the spectator's optic nerve.[3]

As one can see, a revolutionary art, a new way of "seeing" is born to tear down the old pillars of Renaissance aesthetics where man was *Deus in terris*. With the artistic revolution man was no longer the imperious arranger, the organizer.

Contrary to realism or naturalism which viewed reality from a distance, by offering an objective, impersonal reproduction of the world of objects, impressionist reproductions of reality are subjective, relying mostly on the senses. As Cézanne explains: "For an Impressionist to paint from nature is not to paint the subject, but to realize sensations."[4] The impressionist breaks up then from the immutable order of classical nature and responds to the pleasures of immediate sensations.

By rejecting the mechanistic, deterministic, pragmatic world expressed in artistic works of the second part of the nineteenth century, Impressionism finds its purest expression in Bergson's theory of duration. Time is of primordial importance for the Impressionist. It is his eagerness to capture reality in its fluidity, changeability, and flexibility. The world of flux and its ephemeral reality is then captured in the instantaneity of the moment – a basic expression of Heracleitus' theory of flux. Impressionism is fascinated with change. Philosophically, reality is not in a state of *being*, but of *becoming*. Therefore, the world as viewed through the eyes of the Impressionist is fragmentary, and eternally immersed in a constant metamorphosis. The Impressionist is then interested in capturing the truth of the moment. The primacy of the moment is the only temporal dimension in which the Impressionist finds his artistic expression.

As we have mentioned, as an artistic movement, Impressionism has done away with the classical Renaissance line and contour. Instead, it places its major emphasis on colors, movement, and rhythm. Dissolution of forms and content has given way to a pulsating, vibrant, ephemeral reality. However, this dissolution has not been solely an extrinsic one, for the psychological world of man has also been shattered. Man's fragmentation of his interior world echoes man's aesthetic fragmentation of his exterior world. However, the most important element in this artistic expression is the subjectivity of the artist which colors his work. A new perspective is imposing itself in the artistic world, a monocular perspective. The painter, then, realizes not reality but his experience of reality. It is not a reproduction of what the eye sees but the manner of seeing which is important. The painting becomes a document on the fragmentary state of the vision of a certain moment.

A geometrical, formal conception gave way to a subjective, intuitive one. Reality has finally found the autonomy from man's visual and mental habits. Each creator in his own plane, Cézanne, Mallarmé, Débussy, Ravel, later even Freud, offered a demonstration of the same truth. The artist breaks up from the immutable order of classical nature and responds to the pleasures of immediate sensation. The rebirth of landscape found expression in the capture of the instantaneous. The artist exalted the chemistry of pigments and shortened his perspective by revealing a new subjective space.

This emphasis on the subejctive vision and description of reality shares with phenomenology the same penchant in viewing reality not as it should be, but as it is, as it appears to the perceiver's eyes, once it has been bracketed from its spatial and temporal surroundings.

The purpose of this study is in fact to develop this theory of convergence of arts, and perhaps conclude that Impressionism did not perish with artistic masters such as Monet, Manet, Pissarro, or composers such as Débussy or Ravel, or even literary figures such as Verlaine, Rimbaud, Mallarmé, and Proust. Impressionism, as a subjective way of expressing reality, seems to have found its ultimate expression in works by most representatives of the French Nouveau Roman such as Michel Butor, Nathalie Sarraute, and Alain Robbe-Grillet, just to mention a few.

Robbe-Grillet's novels are the prototype of the phenomenological novel. This internalization of the object, before final expression, in any artistic medium thus shares with phenomenology the same aesthetic preoccupation. The Impressionist's way to transcribe reality is then a phenomenological one, for it stresses the subjective expression of reality.

According to phenomenology the world exists. Man has no direct or indirect control over it nor has he any innate knowledge of the world around him. The world simply *is*. Man should not try to impose a new meaning on the world nor should he begin his perceptive effort with any prejudices. Man should look at the world and try to describe the formal structure of the phenomenon without any interpretation or evaluation. Since it is a man who sees, the egological presentation of the phenomenon is significant. The phenomenologist underlines the fact that man and the world are not totally divorced. The world is seen as a compliment to consciousness. Husserl states that "tout état de conscience, en général, est, en lui même, conscience de quelque chose."[5] Husserl stresses the "intentional" experience on the part of the subject. He proposes then a theory which accentuates the importance of the factual world in relation to man's consciousness. As Merleau-Ponty states, the world can be known through the externalization of man in the world. Man needs the world to understand reality, as the world needs man to have any meaning. Innate ideas are banned from phenomenology.

In his major work *Phenomenology of Perception*, Merleau-Ponty emphasizes perception as a primordial tool for discovering the meaning of exterior reality. He wishes to reconcile the subjective structure (noesis) with the objective structure (noema) through the corps-sujet (the body) as an intermediary function. As a true existentialist Merleau-Ponty states that the world *is* before any dependence on the subject's consciousness. The body – immersed in the world of change and facticity – is, however, the source of any knowledge.

The French Nouveau Roman, also known as *l'école du regard*, stresses the paramount importance of sight (perception). Robbe-Grillet, for example, claims that "le regard apparaît aussitôt dans cette perspective comme le sens privilegié."[6] Merleau-Ponty emphasizes that "le réel est à décrire, et non à constituer."[7] In 1893, Ferdinand Brunetière stated in his *Le Roman naturaliste* that with Impressionism "il s'agit comme un perpetuel courant d'impressions entre le monde extérieur qui agit, l'homme physique qui *est agi* et l'homme moral qui reagit. . . . du dehors vers le dedans."[8] Charles Bally reiterates that in impressionist art "todo se ve a través de un temperamento."[9] Robbe-Grillet voices these statements with his constant assertion that his novels present a narrative consciousness and reality as entirely subjective. Jules Lemaître, reports Lionello Venturi, claims that "it is proper and necessary to begin criticism with the sympathetic reading of work without any preconceived idea, if possible."[10] This statement echoes the phenomenological experience of reality.

In Impressionism, the object is viewed and then expressed in personal and subjective forms, just as the phenomenologist deals with the contingent world. This proclivity to present reality, phenomenologically, is the realm of both Impressionism and the Nouveau Roman.

The new novelist as much as the phenomenologist and the impressionist is very belligerent in his criticism of classical analytical methods. Such a rejection of past techniques reflects the phenomenologist's preoccupation with the elimination of a priori notions and innate ideas of the world.

The author's role is to describe, to observe the exterior world – the source of any perceptive knowledge. This is the reason why the novelist limits his narration to precise nouns and avoids, for the most part, colorful metaphors, just as the Impressionist did away with the classical Renaissance concept of line and contour. By stressing the importance of a subjective reality, the new novelist, as much as the phenomenologist, stresses the importance of the perceiver's consciousness and the description of the phenomenon. In fact we have in novels by Michel Butor, Nathalie Sarraute and Alain Robbe-Grillet a reporting consciousness. The classical character manipulated by the strings of the omniscient author, such as Balzac and Mauriac, is not subject to a causal determination of events. His perception is limited only to his unique way of seeing. The report he offers the reader is tinted with his emotional, sensorial and subjective colors.

This destitution of the old myths of depth, as proclaimed by Roland Barthes and later echoed in Robbe-Grillet's theoretical essays, has in fact given to perceived reality a certain elasticity and fluidity, whose equivalent can only be found in Impressionist aesthetics. The emphasis on perception, vision, optical representation is what defines twentieth century modernism.

If we follow the method of the new novelist, the manner in which he creates and builds his character, not to mention the structure of his work, the novel itself, the similarity in methods and approaches between the Impressionist and the Nouveau Roman is striking. The author of the Nouveau Roman no longer penetrates into the psyche of the character, just as the Impressionist discredited the classical notion of the preorganized world. The New Novelist presents the character's interior reality through a phenomenological bracketing technique, which consists in describing and reporting the manner in which the phenomenon appears to the narrator's consciousness. Once perceived the phenomenon is filtered through the perceiver's consciousness and the report of it is a product of subjective exegesis. Any description is then limited to the needs and the normal or distorted personality of the narrator who is the only reporter in the novel. As Merleau-Ponty states: "Tout ce que je sais du monde, même par science, je le sais à partir d'une vue mienne et d'une experience du monde. . . . je suis la source absolue."[11] Man is the point of departure of any investigation.

Likewise, just as Monet painted the Rouen Cathedral at different points of the day, to capture the ephemerality of time in space, thus capturing the sensuality of colors, the new novelist's preoccupation does not claim absolute reality because the narrator in his novels is constantly molded and modified by the surroundings. The report, however, is a personal one, just like Monet's painting the Rouen Cathedral is only Monet's. Consequently, both the spectator in viewing a painting, and the reader of the novel, are walled up in the narrator's consciousness. In the case of the novel, the vision is limited to the narrator's report, while the viewer of a painting is limited to the reality as molded and described by the painter who sees reality in his own way. Both are caught in the ephemerality of the instant. Both are caught in an amorphous reality. The product is a reality sifted through the subject's emotions, sensations, and perceptions.

Contradictions, multiplicity of characters, slow crystallization and nuances, appearances and disappearance of characters, a story just about to be made yet never finished, always in a state of becoming – these are the same elements which parallel Impressionist techniques and goals.

To enumerate those points and to show their application and relevance in the New Novel, would be an enumeration of parallels. We should determine to what extent the Nouveau Roman has gone beyond the precepts of Impressionism.

As Proust discovered, and later explained by Georges Poulet, the Nouveau Roman discovers that the problem of Art and of Reality is a problem of vision. The real depends less on a objective reality rather than a system of focal variables with which we pretend to see or register it. The universe as seen by both Impressionist and New Novelist is a floating one. Likewise, images are constantly floating in an atmosphere of change and ephemerality.

In *La Modification* of Michel Butor, the reader is faced with a shifting narrative consciousness. We experience in fact a psychological odyssey of Léon Dalmont who is torn between passions, emotions, and guilt. The canvas on which Butor records his story displays an atmosphere of distortion (psychological disorder), unfinished sentences, pivotal objects. These objects in turn trigger in the character's consciousness other objects as one color on Monet's canvas blends and triggers another color. Léon's consciousness is triggered in a myriad of psychic associations resulting in an apparent reconciliation with his behavior. His orphic quest, tinted with numerous mythological references, his mental assocciations, leave us with a painting of unfinished reality, for we don't known whether Léon Dalmont will, the next day, the next week, the next year, leave his wife and join his mistress Cecile in Rome. His indecision, his inability to objectify and concretize his feelings create a sense of haziness, a mental fog. One element which is extremely important is the temporal dimension, the durée in which Butor has bathed his works. Not only have we a chronological time, the excursion in Paris-Rome, but we have the mental time with all its changeability, flashbacks, flash-forwards, variants of Léon's psyche – a perfect example of his mental disorder.

If we pass briefly to Nathalie Sarraute, we have another interesting example of impressionist literature, this time a pointillistic presentation of the human unconscious. The narrative voice is embedded in a network of tableaux, the character's personal "tranche" of psychological life. With Sarraute, we are in Neo-Impressionism which according to Felix Fénéon is a "screen, a sowing of small spots."[12] Seurat comes first to mind in the pointillistic technique. Just as the reader of *Tropismes*, by Nathalie Sarraute, identifies with certain scenes, gestures, movements, the viewer of Sarraute's tableaux, chooses his distance, chooses the version of the painting he wishes to see, all in spots, of a synthetic version whose volume forms anew and with distance acquires their representational significance. The Neo-Impressionist work is understood by an *action*, an experience, and no longer merely by passive contemplation. The pulverization of figure and the surface, the atomization of reality, reflect the instability of matter and the relativity of all perception.

Seurat's *croquetons* (small but rapid oil sketches) parallel Sarraute's *Tropismes*:

> Those movements, of which we are hardly cognizant, slip through us on the frontiers of consciousness in the form of undefinable, extremely rapid sensations. They hide behind our gestures, beneath the words we speak, the feelings we manifest. They seemed, and still seem to me, to constitute the secret source of our existence, in what might be called its nascent state.[13]

"Undefinable" and "extremely rapid" as these sensations are, it is Sarraute's task to define them and to slow them down. As Sarraute states: "it was also necessary to make them break up and spread out in the consciousness of the reader the way a slow-motion film does. Time was no longer the time of real life, but of a hugely amplified present."[14] These undefinable and rapid sensations define Sarraute's pointillistic technique, her prismatic view of reality, in addition to her eagerness to capture the richness of those ephemeral thought currents as they pass through the narrator's consciousness. Sarraute's world is a world reduced to the state of amebic, microscopic reality. She deals with fugitive movements, gestures, reactions, nuances as expressed by a movement, a gesture, a blink of the eye. Those inner dramas she portrays, reminiscent of Dostoevski's *univers romanesque*, point out to Sarraute's emphasis on language, elaborated in her essay "Conversation, sous-conversations." It is in fact the interior language, the language of the unconscious which Sarraute scrutinizes with a microscopic precision and which in turn is kaleidoscopic.

As was the case with Dostoevski, Sarraute, as Vivian Mercier claims, "is searching in the murky depths for those authentic yet unidentifiable impulses toward truth." [15]

Sarraute's novel parallels Seurat's pointillistic techniques. Both achieve the same result, a fragmentation of reality, seen through spots and points in Seurat, while Sarraute embroiders her tapestry with ephemeral sensations, fleeting movements, marine-like vegetation which undergo the uncanny ocean of the unconscious. In *Entre la vie et la mort*, the sensations, works, fragments, conversations, fleeting thoughts, are reported through the indecisive consciousness of an author, not described, not even named.

The internal realism which typifies this new artistic expression reaches its most explosive point in Robbe-Grillet. Robbe-Grillet's work displays a pure point of view narrative. As with the other new novelists, this new point of view narrative suggests a subjective narration. Robbe-Grillet is also preoccupied with an objective vision of reality not omnisciently maneuvered but presented in its multiple aspects without the author's interference. Reality is reported in a matter-of-fact way. Robbe-Grillet's phenomenological technique describes reality with the impartiality of a cinematographic camera – by presenting exactly what is seen without any interpretation or analysis. It is a descritpion of pure temporal and spatial exteriorization of human behavior. While it is true that a camera is objective if no one points it, the camera, once directed by someone, assumes an element of subjectivity. Robbe-Grillet does not deny this subjective element in any cinematographic view of the world. Actually, it is Robbe-Grillet's intention to create a character in his film as one creates a character in a novel. Likewise he wishes to report the narrator's consciousness in a novel with the objective impartially reminiscent of cinematography. Robbe-Grillet's camera eye, as Astruc labeled it, leads to an impressionist vision, this time, however, reconciling the distance, the space, between the subject and the object arriving thus at a subjective objectivity. Robbe-Grillet's method consists in fact of a subjective approach to objectivity. It is with this reconciliation between subject and object that Robbe-Grillet has surpassed the limitations of the Impressionist school. His presentation of reality displays a certain neutrality, coldness of description, even impersonal, without any intervention on the part of the narrator nor the character in the novel. Yet, Robbe-Grillet's novels are an excellent display of phenomenological narrative. Behind all the impartiality of the reportage, the neutrality of the persona, there is a human eye that perceives, that absorbs, that molds reality, that spies, that is jealous, that prepares his own alibi – in other words – a man who molds and changes reality according to his own evanescent moods. In both films and novels by Robbe-Grillet, reality is in a constant shift. Reality is amorphous, just as reality appeared to the eye of Monet in his "Impression, Sunrise." This ultra-impressionist uncertainty with which Robbe-Grillet presents his reality, translates a world always in the process of becoming. Space has finally been temporalized while time has been spatialized in the perpetual fixation of the eternal present – the moment – the fraction of time which is in itself rich with life, rhythm, movement, and harmony. Reality is then seen in the instant. It is only through this phenomenological bracketing of both space and time that Robbe-Grillet has been able to arrive at this subjective objectivity. In this world of shifting reality, we are in a world of ultra-impressionism.

In *Last Year in Marienbad*, for example, the reader/spectator will never be able to ascertain the lover's reality which he imposes with cold crudity on the heroine. All takes place in the male character's consciousness; all sensations, impressions, emotions imposed on her, are nothing but a direct reflexion of his consciousness through a simmering process which has taken place, thus bracketing any information which would be detrimental to his pleas. In *La Jalousie*, the narrator's consciousness, clothed with suspicion, jealousy, reaches a climax of pure subjectivity. The whole narration is centered on the object – the centipede – and the reactions of this object in his consciousness. Modern narrative, having reached the "zero degree of writing" in the words of Barthes, unfolds a new language which in Robbe-Grillet's work becomes a visual dialogue. His presentation of objects, as the case of the slice of tomato in *The Erasers*, for example, fills the language with pictorial images, actually invents the color of the object on the page itself.

If the novel is no longer seen and told by an author, there are only two elements left: the thing in itself, the object, and consciousness. This extreme subjectivity and extreme objectivity is what Robbe-Grillet has been able to reconcile. Robbe-Grillet has successfully blended the two.

A similarity in perception between phenomenology and Impressionism exists. To cite the many examples which would show this parallel would be superfluous in this case. Let us stress, once again, that the legacy that runs through Impressionism and the Nouveau Roman, has found its logical ultimate expression. This legacy is still felt in modern novels which are labeled Nouveau Roman novels by Simon, Pinget, Wittig. However, this legacy rests mainly on the importance of man's consciousness through which all reality is reported, thus stressing once again the complexity of truth, of man, and most of all, the nebulosity and labyrinthian structure of man's consciousness.

*Florida State University*

## NOTES

[1] Arnold Hauser, *The Social History of Art*, vol. 4 (New York: Vintage Books), p. 175.
[2] René Huyghe, *Impressionism* (Secausus, N.J.: Chartwell Books, 1973), p. 6.
[3] Ibid., pp. 7–8.
[4] Cézanne in ibid., p. 28.
[5] Edmund Husserl, *Méditations cartésiennes*, trans. Gabrielle Peiffer and Emmanuel Levinas (Paris: Vrin, 1966), p. 28.
[6] Alain Robbe-Grillet, 'Nature, humanisme, tragédie,' in *Pour un Nouveau Roman* (Paris: Gallimard, 1963), p. 26.
[7] Maurice Merleau-Ponty, *Phénoménologie de la perception* (Paris: Gallimard, 1945), p. iv.
[8] Ferdinand Brunetière, *Le Roman naturaliste* (Paris: Lévy, 1893), p. 88.
[9] Charles Bally, *El Impresionismo en el lenguaje*, vol. 2 (Buenos Aires: Colleción de Estudios Estilísticos, 1936), p. 73.
[10] In Lionello Venturi, 'The Aesthetic Idea of Impressionism,' *Journal of Aesthetics and Art Criticism* 1 (Spring 1961):44.
[11] Merleau-Ponty, *Phénoménologie*, pp. ii–iii.
[12] Félix Fénéon, in *Impressionism*, p. 275.
[13] Nathalie Sarraute, *Tropisms*, trans. Maria Jolas (New York: George Braziller, 1967), p. vi.
[14] Ibid., p. vii.
[15] Vivian Mercier, *the New Novel: From Queneau to Pinget* (New York: Farrar, 1971),

# PART THREE

# TRAGEDY – THE TRAGIC FEELING AND THE DRAMA IN THEIR EXISTENTIAL SIGNIFICANCE

# A. THE EVERLASTING TRAGIC ELEMENT OF HUMAN EXISTENCE AND OF DRAMATIC ART

MARLIES KRONEGGER

# THE BIRTH OF TRAGEDY OUT OF THE SPIRIT OF MUSIC: CLAUDEL, MILHAUD AND THE *ORESTEIA*.

> L'ordre est le plaisir de la raison, le désordre le délice de l'imagination
> L'ìambe en musique, sanglot, l'homme blessé au pied
> (Journal, 1907)

Claudel thought of himself not simply as a European, but as a citizen of the world. He was an artist far more than a social or political thinker. For creative renewal he turned, like many playwrights, to the myths and legends of Greek drama, and to the tragic conflicts in the house of Atreus, in particular. Whether Orestes or his sister Electra is made a central figure, whether the action is presented by Aeschylus, Sophocles, Euripides, Racine, Goethe, Alfieri, Hofmannsthal, T. S. Eliot, O'Neill, Jack Richardson, Jean-Paul Sartre, Giraudoux, Cocteau, or in Richard Strauss's operatic masterpiece, it has a terrifying simplicity. Claudel is mostly moved by the lyrical, musical flow of the Greek language, not present in most translations of Aeschylus. Claudel explains:

> I began to read Aeschylus in the translation by Leconte de Lisle, which I thought horrible . . .
>
> I said to myself, it just isn't possible. Aeschylus can't be like this. So I decided I must read him in the original . . . Aeschylus was giving me the education in prosody which I needed.[1]

The world of Aeschylus is to Claudel this whole where each "part", when one takes it for itself, suddenly opens unlimited dimensions – becomes a total part. Paul Claudel is among the few poets of the twentieth century who had not lost his sense of being at one with nature or with the divine. Both myth and poetry rose from the depths of his psyche, stripping away conventional appearances which appeal only to the rational faculties.

The art of Aeschylus was revealed to Claudel in Boston (1894) and in China (1894–1904), Milhaud's[2] musical transpositions of the *Oresteia* mostly in Brazil (1914–21) and the classical theaters of Nō, Kabuki, Bugaku, Bunraku in both China and Japan (1921–25). These two poles of theatrical experience made Claudel aware of the fact that the distance between French and the Aeschylian Greek or Oriental theater is the distance between "conceptual"



*A-T. Tymieniecka (ed.), Analecta Husserliana, Vol. XVIII*, 273–293.

and "pre-conceptual" thought. He realized that the semantic units of these plays are so predetermined that the denotation-connotation distinction virtually disappears: all meanings are primary and more or less explicit. Like a musical score, they are normally not written only to be read but to be performed.

As a theatrical text can only produce an effect when it is performed, it is virtually impossible to describe this effect without taking into account the audience. Claudel, in translating the *Oresteia*, realizes that the poles of threatrical text and audience, together with the musical interaction which occurs between them, form the ground-plan on which he has to build a theory of theatrical effect and response. The subject matter of his inquiry is into what happens while the theatrical text is performed with the musical transposition acting as intermediary between play and audience. The theatrical text becomes, in Claudel's view, a macro-sign, as its meaning is constituted by its total effect. The approach has the advantage of emphasizing the subordination of all contributory elements to a unified textual whole and giving full weight to the audience as the ultimate maker of meanings. Thus, the theatrical performance in its entirety consists of the dramatic text as "thing" or ensemble of material elements, and whose significance is the aesthetic object residing in the collective consciousness of the public, well acquainted with the tragic myth of the *Oresteia*.

To Claudel, *nomos*, moral, social and political law, is the same word as *nomos*, a musical mode. Claudel sees in the *Oresteia* the *nomos* of divine law, patterning the *nomoi* of musical modes, and metrical speech, into an ordered segment of enacted time. Thus, *la parole* is the act of speaking or performance itself, the historical moment at which all the rules of *la langue* come into play in the real world as unrepeatable, temporal events.[3] There is *harmonia*, the fitting together of parts, and *rythmos*, a musical recurrence and proportion. Rhythm, transformation, and wholeness are the basic mechanisms of myth.[4] The fact of recurrence implies retrospection, the containment of the past in the present, persisting in memory.

Temporarily, the tragic conflict of the *Oresteia* comprehends three levels: the epoch of myth, the epoch of Aeschylus, the epoch of the spectator. The tragic conflict is a conflict between the spheres of thought and feeling, between the spheres of song and speech. Conflict in the *Oresteia* is at once cosmic and personal. Greek tragic poetry is the expression of the divine mystery of the cosmos. In the music of the spheres there is harmony in discord. The making concordant of the discordant forms a major structure of counterpoint of the *Oresteia* as we hope to illustrate in this paper.

Claudel offers us a phenomenological description of the stage production which brings to light the elementary operations which the theatrical text of Aeschylus, transposed into music, activates within the audience. Aesthetic effect and response will therefore be analyzed in terms of a dialectic trinity: theatrical text-audience and the interaction between the two. Claudel calls it aesthetic effect because it brings into play the imaginative and perceptive faculties of the audience in order to make them adjust their own focus:

> a great poet, by creating a fictitious plot, gives the heart the chance to act and to express itself (the two most important things) in full.
>
> Music speaks only to the senses. Poetry calls into being all forms of intelligence, character, feeling, and speech. Because speech also is capable of giving joy. . . .
>
> This delight arises not only from the appeal to our feelings, . . . but from an enchanting synthesis of these different feelings, not only static, as in a piece of sculpture, but dynamic, moving logically – with a logic that is at least all its own – from beginning to end, not counting the complications and subplots it brings with it.[5]

Thus, theater is taken to refer to a complex of phenomena associated with the performer-audience transaction: that is, with the production and communication of meaning in the performance itself and with the systems underlying it. The epithet theatrical is limited to what takes place between and among performers and spectators. The need for the audience to share in the action of the play is essential to Claudel. Perhaps it is even the basis of his work on the art of the theater, as he says: "Everything happens within the audience, which feels both involved and remote at the same time."[6] Claudel feels that the French audience of classical drama prefers those mental codes and expected norms of cultural models which it assumes to be real in a system of intelligibility. Claudel is very disappointed with this approach:

> The conception of our own great classical tragedy is quite different. The Frenchman lives in valleys and loves horizons. Line, continuity, dignity are all-important to him. Rather than to feel, he likes to know and understand. He likes to see before him a complete and comprehensive whole, well constructed and composed, on which both eye and mind can take and keep hold. He likes things that last, things firmly fixed, that come to life as time flows past them, as when we follow the Rhône Valley and watch the double line of hills and mountains moving with us like a melody constantly broken and constantly renewed. In the same way tragedies like *Bérénice, Le Cid, Britannicus*, and *Polyeucte* move like a sort of insistent counterpoint. The individual passions of the characters mark a painful and more or less successful effort toward the solution or at least the clarification of a universal crisis or conflict of ideas. The play has done its work when it has found a compromise between the opposing forces which stem from necessity or duty; and its characters are enlisted for the solution of a problem far greater than themselves. . . .[7]

The traditional French audience only asks for what Claudel considers dramatic texts, designed for stage representation and constructed according to particular dramatic conventions. The epithet "dramatic" indicates the network of factors relating to the represented fiction. The performance, at least traditionally, is devoted to the representation of the dramatic fiction. Language is thought of as self-effacing in "un langage-tableau," explaining, analyzing, for instance, the theme of justice. Rather than having reference to conceptual definitions of Justice,[8] the Greek language covers for Claudel a field of meaning based on a sensuous feeling of experience which suffuses whatever judgments the mind makes, with the result that the judgment and experience become intertwined, each invoking and expressing the other. Claudel does not ask himself, who among these characters are just personas in contrast to the unjust,[9] yet rather is aware that *dikē* covers a field of meaning which he proposes to exploit, especially in his adaptation of the *Oresteia*.

Music and drama in the process of birth is an expression often used by Claudel. Like Aeschylus, Claudel intended the mobility of myth in time, space and structure of the text to be matched to music, so that sound and word would oscillate from choral-orchestral numbers to straight speech with mounting emotion, and would sift out from the drama an enactment of myth which goes beyond the personal experiences to the moral and historical meaning of human existence. He says: "I want music that appears to emanate from the words rather than accompany them, passing imperceptibly from the realm of feeling to that of sound."[10] The tragedy itself, by its own intrinsic logic, creates the world around it, whether this world borrows elements from reality or is purely imaginary. The poet gradually builds up the vast background, with the chorus, so that when human action emerges, we see it transformed into its superhuman equivalent: the depths of mythical narrative unfold in drama and music. For this reason, Claudel admires Wagner, especially his *Tannhäuser:*

> This conception of drama helps us to understand the part played by music. Richard Wagner's claim to glory is that he understood that every sound, from speech to song, is linked by subtle ties, and that music is inherent in everything which takes place in time; it may do no more than impose a rhythm on it, or it may gradually color it with different timbres until it opens out at last into the full glory of orchestra and chorus.[11]

Similarly, he discovers the birth of drama out of the spirit of music in the Chinese and Japanese Nō theater, and the link of language and gesture at Hellerau, as I hope to illustrate in the following pages.

Music is anterior to spoken language, and it is in the musical tapestry of images and sayings that an "exclamation, the cry, the reverberation of the character's emotion" can appear "until it returns to him again."[12] Words are necessary to construct a transcendental consciousness which does not correspond to something already existing, but merges from the ambient creative silence. There is not an actor speaking, but speech acting:

> The first of my notes concerns the use of the Chorus. The Japanese dramatists realized that the great human issues, the typical confrontation of thought and feeling, do not develop only between two people linked vaguely by events or words. . . . The Word is creative. Every word man's lips utter, every truth he expresses, causes a series of echoes, replies, memories, appeals, and contradictions in vast concentric circles, opening around it like a flower. A circumstance is modified by an idea, a sound can shake a soul, and a sentence uttered at one moment in time has its repercussions at great distances. The Chorus in Nō drama (and I mean the chorus of instruments, or orchestra, as well as that of voices) has a double role. First of all it is the listener. It takes a word from the lips of the actor and weaves it into a tapestry of images and sayings. It adapts thought to speech. But both above and below poetry and music, Nō drama adds a third element, which is the exclamation, the cry, the reverberation of the character's emotion until it returns to him again. That is what is called in Japan the *Ah*! Laughter, sobs, grumbling, protestation, invective and the lash of the whip – all the reactions of a sensibility directly assaulted.[13]

With the discovery of both Aeschylus and the Nō theater, Claudel defines tragedy as a direct description of experience in which man's essence is resituated in the context of his existence. In Aeschylus, the rediscovery of a naive contact with the world, which precedes the reflective epistemological act, constitutes a primary expression which is always located within lived space and time. The return to this primordial expression operates a phenomenological reduction in which the self is apprehended not as a Cartesian ideal form but by the lived (*vécu*) tragic experience.

For Claudel, tragic wisdom has emotional and imaginative powers such as purely discursive statement in conventional legend or history lacks. Drama, says Claudel, has the same license as legend, and does the same work at less length. Reality is only a sketch which the artist has the right to complete. The signposts of history are few and uncertain, and it is the dramatist who shows us the road.[14] For Claudel, in agreement with Aristotle, tragic theater is more true than history: it can give at a glance an essential insight into some aspect of the human experience which, we ourselves, will never experience, never could experience, except in imagination. A performance brings about a coherent change in a musical structure in the sense of the original text each time it occurs. It is the text which appears, it is the performance which is real. The language of Greek poetry, the language of the imagination unveils the truth or being of things. Man is inside Being, endowed with being. Body and soul are conceived as two aspects of one single, visible, sensible world which is connected with Being which is the source of all things. What is called the soul or essence is not implanted in the body, but it is an inherent part of the projection of Being, as speech is to man the saying of essence. There is merely the emergence of a hidden presence which is Being. Claudel would agree with Heidegger that tragic language or poetry performs a rite. The purpose of tragic poetry is to establish respect for the meaning of human existence. The poetic language of Aeschylus with its subtle feeling for timbre and harmony, its images and its movements, reach our soul.

Language as intentional act is for both Claudel and Merleau-Ponty a gestural action: it is an embodiment, an incarnation of meaning. Embodiment is densest and richest in poetic language because in it conceptual and non-conceptual elements (*sens emotionnel*) play roles approximating their importance in the total life of the human being. For language that expresses concretely instead of just conceptually, Claudel reserves the term *la parole*.

In translating Aeschylus, Claudel sees that there is no dichotomy between perception and language. Writing is a bodily expression which is part of the ongoing primordial process of perception. The act of naming arranges a series of perceptual data, inaugurating rather than exhausting order. With the Greek poet, Claudel listens to the music in himself which determines the nature, the rhythm, the organization of the words to come, created out of silence, as he confirms in two letters to Milhaud:

15 August 1938

> Once more we shall have to discuss this thorny question of music *in the process of birth* and music springing from poetry as poetry from prose, and prose from silence and the formless mutterings of the mind. It is very difficult, and we shall have to reach perfect understanding on the subject.

5 September 1938

> I should like to see inspiration rising from the murmur to the vowel, then the consonant, the word, the note, then finally to song and from thence descending again.[15]

Claudel's translation creates the impression of an emerging order, of an object in the act of appearing, organizing itself before our eyes. Inspiration is emerging from the color of a word, it is an act of seeing.

The art of Aeschylus, lyrical in essence, reveals to Claudel an art of nuances and half-tones, absorbing the *rhythmos* of inner life within the structure of myth. The Greek language moves Claudel like music. It is music that takes him to the text itself. He sees that it awaits the as yet non-existent speech which makes it possible. Since music is its condition, it is inevitable to Claudel that certain complex structures can appear in the spoken language only when the musical language has already, to a certain extent, rendered them current. Music is in the process of birth: Claudel realizes that the musical sounds are not the signs of musical meaning, but the very substance of Aeschylian song descending within him. Claudel feels the thought of a passage in Greek, such as the Watchman's prologue being "spread out," insinuating itself from sentence to sentence.

Claudel recreates the Aeschylian Iambic verse, because it gave his theatrical text the atmosphere, the environment, the dignity and distance, which speech in its arid nakedness could not provide. Realizing that the recitation on the Attic stage was of musical character, sustained, intonated with a slow and measured stateliness, the Iambic metre of Greek verse seemed to Claudel best suited for the French language.

The dramatic or lyric verse form *par excellence* is the iambic. All great dramatic poets have written iambic verse – the Greek tragic dramatists, Shakespeare, and the great lyric poets, Pindar, etc. All their verse is based on the iambus, that is to say a short syllable followed by a long one: tick-tock, tick-tock, or else tick-tick-tock, tick-tick-tock.[16]

Claudel wishes the division into lines which he has adopted to be based on the rhythm of breathing. It breaks "the phrase up into units which are not logical but emotional; it should . . . make study easier for the actor. When we listen to someone speaking, we hear the voice rise at a variable point about the middle of the sentence, and fall at the end. These two climaxes and the inflections which come between, form the lines"[17] of the verse of Aeschylus, as Claudel explains in a letter of 27 May 1913:

My dear Milhaud,

Thank you for your very discerning letter, which contains all the objections I have made myself. And yet . . . I *feel* (as an artist feels, quite apart from any reasoning) that the dialogue between Clytemnestra and the choir cannot be simply declaimed without actually becoming music. It was not for nothing that Aeschylus completely altered the rhythm at this point, and made Clytemnestra speak in lyrical verse. At this moment Clytemnestra is filled with a savage, almost devilish joy. She is possessed by a spirit that is almost cannibalistic (in the Greek manner, without ever going beyond the bounds of moderation). Her words must not *sing*, but *dance*. The rhythm must be hammered out with a brutality which ordinary declamation could never achieve. We need a kind of "music" reduced purely to its rhythmic element, for example the beating of drums and other percussion instruments or staccato blasts on the trombone. Oh, it's so difficult to explain! And yet I *feel* that at this moment speech alone is not enough, although, like you, I find it quite enough for the other parts of the *Agamemnon*. I would like you to read the *Oresteia* all through (in Leconte de Lisle's translation for example, bad as it may be), particularly the *Eumenides* which I imagine ending with a full-throated chorus in the style of Händel.[18]

On the subject of the "spoken choruses" of the *Choephori*, Claudel declared:

Rhythm, the succession of long and short vowels and the distribution of the stress accents are at least as important to poetry as the sense of the words. . . . That is what I tried to make clear in my translation of Aeschylus. I have attempted to adapt Greek meters to the French tongue.[19]

In two scenes, 'Présages' and 'Exhortation,' Milhaud employed a large battery of pitchless percussion as background music to a woman narrator who declaims the text in strict synchronization with the music, and the chorus is required to whistle, groan and shriek. The lyrical element of these scenes is not musical, but rather savage. Milhaud makes use of measured speech, divided into bars, and conducted as if they were sung: "the 'Présages' and 'Exhortations,' two scenes so violent in character that they created a problem that I solved by having the words spoken in time with the music by one woman narrator, while the choruses uttered words or disjointed phrases, the rhythm but not the pitch that was indicated. To support all these various elements I used percussion instruments having no definite pitch – quite ordinary instruments. . . ."[20]

Claudel, for his performance of the *Oresteia*, was very inspired by the stage productions at Hellerau (Germany). He would agree with Merleau-Ponty that perception is also the basis of the performer's reflections, and the body, anchored in the world, is the center of perceptions. It is perceiving and being perceived, a creative structure developing in time:

Perception opens the world to me as the surgeon opens a body, catching sight, through the window he has contrived, of the organs in full functioning, taken *in their activity*, seen sideways. It is thus that the sensible initiates me to the world, as language to the other: by encroachment. *Ueberschreiten.*[21]

For this reason, Claudel was fascinated to see the stage production of Hellerau: Music there is "alive and visible in the human body. Between music and the body there is a common element, movement, a common measure, time, and a common expression, rhythm . . ."[22] Therefore, Claudel wishes the actor to listen to music with the whole being, the whole body. The aim is to recapture the genesis of existence, and that can only be done through the body which is the basis of perception. For Claudel, the body is a structure open to the world, correlative with the world, it is in the world. There is, therefore, a continuous interpenetration between the body and the world, both bathing in the same substratum of *rhythmos* which unites them.

How is the actor capable of recreating the tragic situation? The use of masks of the Attic stage is not possible on a European stage. Claudel suggests that there is nothing "more beautiful and more tragic than complete immobility, " always preferring" attitudes to gestures. . . . In dramatic moments a gesture which unfolds slowly and tragically is better than an outburst."[23] Thus, in his production of the *Choephori* in Brussels, he tried to fit a whole lyrical speech with the movement of the arms slowly lifting an amphora, then crashing it down which "made a considerable impression."[24] Antiquity, so it seems, acted as twilight does: it reduced the historic person to that sublime state of gloom which suited the views of the poet who wanted only the situation. The mask or a certain immobility, then, are in harmony with the scene, for the passion of each actor is essentially fixed throughout, and unfolding in the context of the performance.

While Claudel insisted that the tragic chorus maintain a stubborn, almost liturgical immobility and should not dance as the Greek chorus did, he realized that the twentieth-century spectator needs equivalences, as Roland Barthes explains when rejecting Barrault's interpretation of Claudel's *Oresteia* in Paris:

> By restoring to the chorus, by liturgical analogy, its function as a literal commentator, by expressing the massive nature of its interventions, by giving it quite explicitly the modern attributes of wisdom (seat, lectern), and by rediscovering its profoundly epic character as a narrator, Claudel's solution seems to be the only one which might account for the situation of the tragic chorus. Why has it never been tried?[25]

To both Claudel and R. Barthes, the chorus is the hard chore of tragedy, because it focuses our attention on being rather than things: on how to understand the meaning of existence.

When the overall transposition of the *Oresteia* into the arena of a performance is seen globally,[26] myth and music are no longer perceived in their own right but contribute to the impression of an emerging order out of silence: with the chorus, the word and the music, there are gestures, emotions in the act of appearing, organizing themselves before our eyes. In a letter to Milhaud, Claudel explains:

> Je crois, dit-il, qu'il y a une soudure à chercher entre la parole et le chant. De même que j'ai montré que tout est poésie et que des choses les plus basses et les plus grossières aux paroles les plus sublimes il y a suite et continuité, de même il faudrait ajouter à ce

domaine de l'expression parlée celui de la musique, et que tout parte du même fonds et naisse l'un de l'autre, sentiments, bruits, paroles, chant, cri et musique tantôt se cédant, tantôt s'enlevant la place.[27]

The Chorus, being sung, formed a considerable portion of the drama, and took more time in the performance than spoken dialogues. Claudel always insisted that the impression imparted to the audience in a Greek tragedy was a musical one, because he discovered Aeschylus, the musician, the choral lyricist who creates works prompted by a mood of profound inner excitement, the mood of a composer, one which precedes articulate poetic ideas.

The first thing to do for the chorus, then, is to create the tragic atmosphere. Claudel does not break up the action of the *Oresteia* externally into Aristotelian formal divisions (*prologos, parodos, epeisodia, stasima, exodos*), but has these parts of the action and the formal division of the tragedy shadow one another: the *rhythmos* of their proportional relation doubles all the more powerfully, because the correspondences between them are only accidentally exact. Claudel writes to Milhaud:

Musically, I think the play is made for you, because of all its wildly dissimilar and contrasted elements which you have the art to fuse together and mold into something harmonious.

There are three elements:

1. The crowd, filling with constant movement the sort of bowl made for it, accompanying the whole play with its uproar. (I think we could do something wonderful with the vote scene.)

2. The Olympian element which gradually reveals itself and ends by submerging everything in a great flood.

3. The shrill pathetic chatter of the human race, spoken, but accompanied rhythmically. The phrasing and movement must not be solemn and ponderous but violent and rapid. From this point of view the lines of my verse might lead you astray. It must all be poured forth almost in one breath.[28]

The time of tragedy, according to Claudel, objectifies its submission to the universal principle of *nomos* by being ordered into a *rhythmos*, a proportion among its segments of time. Claudel's translation of the *Oresteia* tries to recaputre the melodic patterns which make us aware of eternal time, the indivisible presence outside ourselves. In Aeschylus, as Jacqueline de Romilly remarks, the word *chronos*, time, occurs four hundred times or so in extant tragedy.[29]

Time accounts for the structure of the *Oresteia*; time also brings out justice. The correspondence which Claudel believed to exist in Aeschylus between sounds, tone combinations and cosmic phenomena, seasons of the year and parts of the day, sun and moon cycles, growth and weather, man an woman, birth and death, etc. is brought into relationship with human temperament.[30] Man, like nature, is torn between two extremes: one striving for clarity, temperance, and moderation, the other driving him toward the fantastic and orgiastic, the cult of Dionysus. Claudel himself was torn between these two extremes, having two souls in his breast. Greek tragedy is the expression of his inner self: he wishes to make visible how mythical reality touches him; he waits for his creation to come to life for other people.

Tragedy interposes the plastic world of myth between the universality of its music and the Dionysiac disposition of the spectator, and in so doing creates the illusion that music is but a supreme instrument for bringing to life the plastic world of myth. The delight created by tragic myth has the same origin as the delight created by dissonance in music. The tragic myth is reborn from the matrix of dissonant music. Myth structures pictorial space. The structure of myth is preserved or enriched by the interplay of transformation laws which speak for the very dynamics of myth. Piaget's definition of structures as having the qualities of "totalité, auto-réglage et transformation"[31] is particularly basic to an understanding of Claudel's perception of myth and tragic justice. It demarcates different levels of a unified cultural phenomenon and it compares to a painting in the sense that it has the greater part of its life in front of it.[32]

The character of structured wholes depends on their laws of composition. These laws are structuring: the aim of tragedy is to give the audience an experience of the ongoing and formative foundations of what makes them a people. Perception, for Claudel, is not first the perception of concepts, but a perception of elements (such as water, light, air) which raise questions within our life, within our history. Therefore, the text of the *Oresteia* had a completeness, an inevitability, a wholeness which permits and justifies its paradigmatic character and its poetic superiority to real history, while our own existence lacks these qualities altogether.

Claudel's text moved toward the theatrical musical text where the text itself is affirmed as its own reality, that is, a text concerned with representing the protagonist's consciousness through which tragic experience is filtered and by which mythical reality is presented in theatrical form. Mythos, the kind of reality we recognize in Greek tragedy, treats of the most deep-seated concerns and involvements which we have as human beings. This leads Claudel to the insight that not "something" but "someone" happens in both the *Oresteia* and Nō theater:

In drama something happens, but in Nō drama someone happens. It is rather like the door which appeared in the early days of Greek theatre as a means of communication through the wall with the invisible world behind, and in which the characters of the *Oresteia* appeared one by one.[33]

Consciousness is the point of departure of any perception. When Paul Claudel scrutinizes the indestructible tie between hours and places, at the outset of his *Art Poétique*, he wishes to locate himself in time, as do Orestes and Electra, as we all do:

Now and then, a man raises his brow, sniffs, listens, considers, finds his position: he thinks, sighs, and, taking his watch out of his vest pocket, reads the time. *Where am I? What time is it?* these are the inexhausible questions we ceaselessly ask the world. *Where am I? What have I achieved?*[34]

At the same time, the protagonist/actor, be it Orestes or Electra, is surrounded by an "ambiance" of witnesses: the chorus/audience:

Every voice, every word, every action, every event brings its own echo and reply, calling forth and spreading that anonymous murmur like the sea, which is the collective voice of the generations that watch and listen. This is what I call the *Chorus.*[35]

Without a spectator, without a witness, without being "watched," consciousness would lose its being, its essence, its past, and would be reduced to its momentary fleeting spontaneity – to what just now is, namely, its nothingness and ignorance.

The chorus expresses visually the emotion that strikes the keynote of the whole trilogy – the waiting in darkness for the light of deliverance. Both music and myth universalize in a timeless situation, the particular suffering of Orestes' limited existence in an illusory world of compromise, throbbing, and surging around them, like a homogeneous ocean, as Claudel observes in both Racine and Shakespeare.

The greatness of both Aeschylus and Racine, is the intervention of superhuman elements for Claudel:

But what gives the drama its poignancy – because it is not only Phaedra's, but Racine's too – is the question it asks of the conscience of anyone who is inspired, who is both victim and accomplice of an unknown, ambivalent, and questionable power. At the very beginning of tragedy, on the very threshold of that terrible door from which so many masterpieces were to emerge, old Aeschylus set up the imposing figure of the Trojan prophetess, accusing her seducer Apollo:

*Apollo! Apollo! God of the Gate, my Death-dealer! For thou hast destroyed me utterly, this second time! . . . Why then do I wear these mockeries of myself, this scepter, and these prophetic chaplets around my neck? I will destroy you before I meet my fate! Go, accursed that you are ! . . . Go and enrich with doom another instead of me.*[36]

For Claudel, the poet like the protagonist/performer of Greek tragedy is in a "situation." That is to say, in ultimate unity with his surroundings. All elements have an effect on each other, as do colors in a painting, notes in music. And in the field of tone, the same note may become the keynote in which music is written, and as such expressing a movement toward the universal. Similarly, Claudel's synesthetic apperception in explaining *Hamlet*, for instance, bears witness to the idea of world harmony as all senses converge into one harmonious emotion. While Hamlet disposes cognitively of a core of perceptual awareness in relation to society and nature, affectively, his emotional tensions, moods, attitudes, concerns or desires suggest modes of being in the world and become expressions of his global existence, as much as the color yellow within its particularity as yellow and through it can become a universe or an element. Claudel explains:

there is a pleasure rather like that of painting. Painting gives pleasure through the juxtaposition of colors, as music does by that of timbres . . . at least that is one of the elements of music. Similarly, in the theatre we derive pleasure from the juxtaposition of the characters and the different reactions they provoke in each other. It is interesting, for example, to see the effect of the different characters of *Hamlet* on each other. Even apart from absolutely strict mechanical logic, there is a harmony to be found in the relationships of the characters as such, reacting one upon the other as yellow can react on blue and red on green, but with infinitely more detail, because it does not appeal to only one of our senses, sight. It speaks also to our understanding and to our heart.[37]

Orestes has learned to identify with the foundations of his being, its pain and contradiction, yet he becomes one with the forces of nature, suffering in their destructive power and rejoicing in their creative activity. The growth of moral awareness of Orestes, the people of Argos, the Olympian gods, is based on conflict.[38] Crisis (from *krinein*, meaning to judge or decide) is the continuation of what has happened in the mythical past. The *agon* of these protagonists typifies certain common situations. The *agon* of the *Oresteia* serves the edification of the audience not only to judge, but to learn how to understand and how to control their lives. Orestes emerges in front of the audience/chorus, witnesses in heaven, the city, the underworld: his consciousness objectifies itself; it can reflect on its own acts, and taking on an essence in the performer, playing the role of Orestes, there arises "an amazing paradox": the sentiment is not within the actor, but the actor within the sentiment. He acts out his own thought before us, and bears the witness to his own expression.[39] The substance of the emotional power of music is with the performer at all times. The performer is no longer the actor, he is Orestes, the result of a perfect fusion of subject and object.

Claudel is convinced that Greek tragic poetry is the expression of the divine mystery of the cosmos. In the *Oresteia*, Claudel discovers that the Greek mind has been able to see harmony in discord: the making concordant of the discordant confronts us on various levels. Conflict in the *Oresteia* is at once cosmic and personal, because the central figure, Orestes, is the human counterpart of triumphant Zeus:

> The play is only the means by which the event, the moment of change, the essential and central conflict on which all human life is based, is isolated, outlined, completed, illustrated, and given its place in the domain of both the general and the particular. It takes a certain potential for contradiction among the forces present, and transforms it into action in order to reach a conclusion.[40]

Without good, and without the desire for good, there can be no evil. Without good sentiments, there can be no bad ones. And without struggle, there can be no progress. . . .[41]

In Aeschylus, Claudel discovers the good-evil dichotomy of the human being which he attempts to explain both mythically and in Christian terms. According to this notion, the soul is immortal and divine but imprisoned in a mortal, Titanic body. Claudel's tragic picture of the universe, then, as discovered in Aeschylus, Racine and Shakespeare, postulates a limited free will: with Orestes, Phèdre and Hamlet, all wrongdoing is involuntary. They are powerless to interrupt a chain of disasters. They cannot determine the pattern of events, but they are responsible for the initiation of an evil chain latent in their situation. Only suffering and their indomitable passion make them what they really are. While Orestes finds justice in relation to the state and to those around him, Phèdre's and Hamlet's longing for self-transcedence can be rewarded in death, and only there, where they affirm themselves in perfect self-equality. They integrate their existence into the structure of the cosmos. Since the just individual – Orestes, Hamlet, or Phèdre – must aspire to attain Apollonian self-control and self-knowledge, his suffering is truly the universal condition of the individual who painfully arrives at wisdom.

In the *Oresteia*, in both *Hamlet* and *Phèdre*, the Greek *nemesis* happens impersonally, yet brings harmony to both the cosmos and the characters themselves. Harmony consists of opposing tensions, and Milhaud's polytonalities of the trilogy set to music present this insoluble conflict as an existential mimicry, an emotional reality to be lived by the audience.

The Orestes of the Greek and Claudelian stage asserts the freedom and ultimate greatness and dignity of man and life. Suffering and struggle are productive forces that are not to be abolished, but overcome. Conflicts are means of self-conquest, self-creation, and self-definition. With the aid of Apollo and Athena all Dionysian discord and dithyrambic excess are given harmony, clarity, form. With Claudel's version of Aeschylus, the aim of tragedy is not defeat, resignation, despair, the evocation of pity and terror, but rather the victory of the human spirit over the inexorable laws of the furies. The discordant Dionysian manifoldness of revenge acts is transformed into forces of harmonious unification: this explains how Dionysus and Apollo imply and complete each other: the triumph of the one is the fulfillment of the other. It is through the workings of Apollonian musical triumph that even the realm of myth takes plastic shape before us.

With Claudel and Milhaud we have come to interpret Greek tragedy as a Dionysiac chorus which discharges itself in Apollonian images: the ultimate reality of Aeschylian tragedy is synthesis, reconciliation, unity, and in Claudel-Milhaud this unity becomes a triumph. Claudel's notion of justice, then, refers to notions of equilibrium, balance, equality. Harmony consists of opposing tensions. For Claudel, deeply inspired by both Greek and Christian ideas of world harmony, all reality is seen in an antithesis: Good is the mirror opposite of Evil, and darkness is the mirror of light. The equality of Good (Orestes', Electra's, Hamlet's, or Phèdre's striving for purity and perfection) and Evil (they are both victims and accomplices of an unknown, questionable power in search of both purity and perfection in a corrupt world of compromise) constitutes justice: it is identical with harmony within the protagonist's soul, within society, and within the cosmos.

In the Claudel-Milhaud version of the *Oresteia*, music has become the real actor, a composite character with many different voices blended into one, playing the part of the chorus. As such, it can express the feelings of the audience, and reply to the action: music listens. "It intensifies the rhythm, opens the heart, in the same way as the chord spreads to every passion, impregnates the spheres of feeling and thought, and both speaks and listens at the same time."[42] Referring to his collaboration with Milhaud, Claudel explains:

> Milhaud and I have tried to show how the soul gradually arrives at music, how the phrase springs from rhythm, the flame from fire, melody from speech, and poetry from the coarsest reality, and how all kinds of expression in sound, from speeches and dialogue supported by percussion to the full glorious outburst of voices.[43]

There is a move from a state of semi-obscurity and unknowingness to light and knowledge. Silence is not the opposite of speech, but the surrounding and therefore the setting-off element or climate of speech, which surges from, and returns to silence. Language and silence are as closely related as shadow and light.

The enactment of the myth of the house of Atreus recreates and animates the relationship between the individual and his culture. Animated by the life and drama of the music, by the lyric cry and vital escstacy of the chorus, the myth of the *Oresteia* undergoes a collective transformation: it is Milhaud's musical transposition which permits speech to be understood: the verbal polyphony presupposes musical polyphony. Myth and music behave as structures – they act as wholes, regulate themselves and transform themselves in various colors: in music, there are psychological colors, function colors, modal colors, tonal colors, geographical colors, historical colors and others: so too, and more splendidly in the Claudel-Milhaud version of the *Oresteia*.

In 1913 Milhaud had begun his settings of Agamemnon. Milhaud effected a novel type of transition from speech to song. The choral meters, the Expectation, the Arrival, the Context, the Murder, and the Justification – all draw the formal divisions into themselves. Of these sequenced events,

> the Expectation takes up proportionally almost half of the play (1–809), heavy as it is with the past. The Contest between Cassandra and Clytemnestra and its aftermath (1035–1342), heavy as it is with the future, lasts longer than the Arrival and the Murder taken together. The imbalance of these proportions sets the central event, the Murder, onto a plane where a vastness of time dwarfs the short action in itself, while magnifying it in its isonomic relation to the larger law of action that the play is ordering.[44]

Milhaud subtitled *Les Choëphores* (1913) 'Harmonic Variations', intensifying the expressive power of harmony with polychords: a polychord being to his ears "more subtly sweet and more violently potent"[45] than diatonic harmony. Here is the use of two keys simultaneously. Such bitonality or polytonality, as it is called when several keys are involved, became one of Milhaud's favorite modes of expression. Various polytonal planes are distinguished by different instrumental timbres, shocking and stimulating our feelings by variations of pitch and rhythm, more brutal and direct than speech, as Milhaud explains:

> "Libation," . . . first attempt to write a chorus in two simultaneous keys, with the lines of chords in the male voices set over against the women's voices, and with both forming a background for a soprano solo; an "Incatation" sung by Electra (soprano), Orestes (baritone), and chorus before Agamemnon's tomb.[46]

Once again, Arrival is given far more space proportionally (1–837) than both Murders and their Aftermath (838–1076). The major portion, then, is the dialectical confrontation taking place in the present.

Extension of the polytonal principle produces the complex dissonances of the closing scene of the opera *Les Euménides* (1917–22). Milhaud builds up to six simultaneous different keys, reduces them gradually to two, and finally resolves on the single key of C major. The play looks to the future: dialogue and chorus stand in equalized, proportional alternate relation to one another. We see the action from farther away in preparation for a solution. A large portion of events to which the audience listens is invisible, yet felt. It takes place only in memories and reflections on prophecies. The action is divided into three equal parts: 'The Pursuit at Delphi' (1–234); 'Trial at Athens' (235–751); and 'Acquittal with its Aftermath' (725–1047). The direction of the musical movement is defined by rhythms and melodic lines, while the arrival at the final goal is made climactic by the resolution of dissonance into consonance.

Justice is the turning of darkness into light. Light and darkness are recurring alternatives, but the action of the play is an establishing of light, permanent, using the work of men to make darkness bright. Like music, light brings to life the being it envelops, and works together with him. For the Christian audience, Claudel/Milhaud regenerate equivalent sources of religious emotion, philosophical insights, and aesthetic enjoyment to those conveyed by the Greek version. The poet, on behalf of the people, purges the masses of the unformed aspirations that struggle confusedly within them. This seems to have been the idea of Greek classical theatre with its single protagonist emerging from the anonymity of the chorus and going back to it again. The final hymn to Justice, is a hymn to light, to truth, to harmony within the city, to world harmony. As soon as harmony, for Claudel, an action of grace, is obtained, tragedy vanishes, yet Christianity has not been able to shake the validity of Greek myth. Without knowing it, the spectators through their participation in this act of communal ritual, absorb the prophecy and vision of Greek dramatic art at its highest reach thanks to both Claudel and Milhaud.

Further investigation of Claudel's translation is needed. It is obvious that we find in his translation a Christian vocabulary, such as "action de grâce,"[47] and Christian images of "lumière," "joie," "espérance"[48] which have a pagan meaning in the Greek text. The use of modern instruments, such as the trumpet,[49] transforms the religious songs of the Greek chorus into a Christian hymn of praise of world harmony. Claudel repeatedly has asked Milhaud to create something like the note on which the Epistles and the Gospels are chanted.[50] Claudel could have found in the hymns of Ambrose a performance, an incarnation of that world harmony about which the Greeks had speculated. Claudel realizes that the Church, which was represented in these hymns as echoing the music of the universe, served, actually as the theater for the performance of these hymns, as it was to serve later as the original stage of medieval drama.

The mysterium is by its very nature not unlike Greek tragedy. The mystery play grew out of a religious cult and out of music, that is, from the sources as Greek drama. Claudel explains:

> it is that same Chorus after the triumph of Christianity, when it was admitted into the church to become the intermediary between priest and people. Between the silent crowd and the drama unfolding on the stage – or, if I may say so, at the altar – an officially appointed spokesman was needed.[51]

In Milhaud's version, then, music is the real actor, a composite character with many different voices blended into one. Gradually, and with mounting emotion, it sifts out from the drama itself the material of the great final hymn, suggesting the triumph of *nomos*, of musical harmony which carries us away with it. The time of Aeschylian tragedy objectifies its submission to the universal musical principle of *nomos*. These harmonies and polyharmonies are what make a literary work of art, and they require from the audience an "aesthetic attitude." Such a mind is aware that *nomos* covers a field of meaning, and its design is to exploit the field, not tidy it up.

We have tried to show that Claudel and Milhaud saw in the Greek mind harmony in discord, and finally the triumph of "symphony" over the discordant voices, the triumph of cosmos over chaos, a tension and triumph viewed by Nietzsche as "an Apollonian embodiment of Dionysiac insights and powers." With Claudel-Milhaud, Greek tragedy has become in essence the musical embodiment of the spirit and contents of Greek mythos. The artistic construct of the Claudel-Milhaud version of the *Oresteia* expresses the continual rebirth of existence.

*Michigan State University*

## NOTES

1 Paul Claudel, *Paul Claudel on the Theatre*, ed. Jacques Petit and Jean-Pierre Kempf, trans Christine Trollope (Coral Gables: University of Miami Press, 1972), p. 178; hereafter cited as Claudel.

2 Their collaboration has lasted for forty years. Claudel's works set to music by Darius

Milhaud are: 1912–13: *Sept Poèmes de la Connaissance de l'Est*; 1913–19: *Protée*; 1915: *Les Choéphores*; 1917–17: *Quatre poèmes pour Bryton*: 1917–22: *Les Euménides*; 1918: *L'Homme et son désir*; 1918: *Psaume 136*; 1928: *Christophe Colomb*; 1932: *L'Annonce faite à Marie*; 1935; *Le Signe*; 1936: *Cantique du Rhône*; 1937: *Fête de la Musique: Celestial Ballet*; 1937: *Cantate de la Paix*; 1940: *Cantate de la Guerre*; 1942: *L'Annonce faite à Marie* (new music for play); 1950: *Le repos du septième jour*; 1952: *Christophe Colomb* (a new version for play).

[3] Albert Cook, *Enactment: Greek Tragedy* (Chicago: The Swallow Press, 1971), pp. 99–117.

[4] J. Piaget, Le *Structuralisme* (Paris: Que Sais-je? 1968) p. 78. Paul Claudel, *Mes Idées sur le Théâtre* (Paris: Gallimard, 1966).

[5] Claudel, pp. 174–75.

[6] Claudel, Nō,' p. 52.

[7] Ibid., p. 36.

[8] J. Walter Jones, *The Law and Legal Theory of the Greeks* (Oxford: Clarendon Press, 1956).

[9] Josée Van de Ghinste, *La recherche de la justice dans le théâtre de Paul Claudel* (Paris: Nizet, 1980), and Eric A. Havelok, *The Greek Concept of Justice* (Cambridge: Harvard University Press, 1978), p. 279. Much of Greek tragedy turns on this theme, but it is a particular preoccupation of the *Oresteia*, whose diction is saturated with the symbol *dikē* and its correlatives as these are applied in a great variety of contexts to explain what has happened, is happening, or will happen. These form a kind of symbolic system. In the *Agamemnon* the noun (*dikē*) occurs seventeen times, accompanied by the adjective "just" (*dikaios*) thrice; "justly" (*dikainos*) thrice; "nonjust" and "nonjustly" (*adikos adikos*), one each; two verbs of "adjudication" (*dikazo, dikaio*), one each; a noun "adjudicator" (*dikastes*), once; and a variety of compound derivatives (*anti-dikos, pro-dikos, sun-dikos, en-dikos, euthu-dikes*, one each; and *dike-phoros* twice); for an overall total for the play of thirty-six. This leaves outof account fourteen instances of *diken* used as an adverbial preposition governing the genitive, meaning "just like."

[10] Claudel, p. 107.

[11] Ibid., p. 79.

[12] Ibid., 'Japanese Nō Drama,' p. 103.

[13] Ibid.

[14] Paul Claudel, ' "L'Orestie" d'Eschyle,' *Oeuvres en prose* (Paris: Gallimard-Pléïade, 1965), 417–521.

[15] Claudel, pp. 116–117.

[16] Ibid., p. 178.

[17] Ibid., p. 14.

[18] Ibid., p. 23.

[19] Ibid., p. 122.

[20] Darius Milhaud, *Notes Without Music* (New York: Knopf, 1953), p. 66; originally published in France as *Notes sans musique* (Paris: Julliard, 1949).

[21] Maurice Merleau-Ponty, *The Visible and the Invisible*, trans. A. Lingis (Evanston: Northwestern University Press, 1968), p. 218; originally published as *Le Visible et l'invisible* (Paris: Gallimard, 1964).

[22] Claudel, 'The Spell of Hellerau', pp. 17–18:

We go to Hellerau to see two things:

1. Music, alive and visible in the human body.

Between music and the body there is a common element, movement, a common measure, time, and a common expression, rhythm. The aim of M. Jaques-Dalcroze is not to teach his pupils dancing or gymnastics. He teaches them to listen to music, not passively, but participating with their whole being, their whole body. For instance, while their steps follow the melody as it quickens or diminishes its speed, the movement of their arms marks the essential rhythm. The musical phrase animates and is itself animated; it follows, governs, inspires, unfolds, and develops the gestures, attitudes, and steps which come naturally to our bodies, but which in daily life have to chance to find more than timid, random and stilted expression. I have seen girls conduct a choir or an orchestra, not with a baton like the usual black-coated gentleman, but with their whole bodies, every movement following and leading at the same time, listening to every part and sweeping the whole thing forward. When the body has completely submitted itself to music, it is sometimes as though it had been created by it.

Such training probably corresponds to what the ancients meant by *music*, and its educational advantages are many and great. It teaches concentration, proportion, self-possession, the control of our impulses, and harmony with all the other movements around us. The body grows pliant and obeys the soul, which in its turn respects the instrument which has to give it visible life.

[23] Ibid., p. 15.

[24] Ibid., p. 122

[25] Roland Barthes, *Critical Essays*, "Putting on the Greeks" (Evanston: Northwestern University, 1972), p. 64.

[26] Claudel imagined a performance of the *Eumenides* in which the audience, seated on the steps of the Roman theatre at Oragne, represented the people of Athens summoned to judge Orestes.

[27] Paul Collaer, 'Paul Claudel et les musiciens" *Bulletin de la Société Paul Claudel* 63, no. 3 (1976).

[28] Claudel p. 26.

[29] Jacqueline de Romilly, *Time in Greek Tragedy* (Ithaca: Cornell University Press, n.d.), p. 59.

[30] Leo Spitzer, *Classical and Christian Ideas of World Harmony* (Baltimore: Johns Hopkins Press, 1963), pp. 80–107. Spitzer associates *temperare* not only in the Greek context with the expressions for order and musical harmony, with both "the right time" and nontemporal meanings, but also with health, harmony, balance, but also in the Christian context with classical equipoise, "to harmonize oneself," to imitate God in bringing about harmony in our soul which is the image of the divine soul.

[31] *Piaget*, chap. 1. See also: Eva Kushner, 'Greek Myths in Modern Drama: Paths of Transformation,' in *Literary Criticism and Myth*, ed. Joseph P. Strelka (University Park: 1980), pp. 198–209.

[32] Maurice Merleau-Ponty, *L'Oeil et l'esprit* (Paris: Gallimard), pp. 92–93, would agree with Claudel that the first paintings went straight to the future.

[33] Claudel 'The Japanese Theater,' p. 52.

[34] Paul Claudel, *Poetic Art* (New York: Philosophical Library, 1948), p. 4, and for an explanation see M. Poulet, 'Claudel,' *Entre moi et moi* (Paris: Corti, 1977).

[35] Claudel 'Drama and Music,' p. 88.

[36] Ibid., 'Last Words on the Theater,' p. 187.
[37] Ibid., p. 174.
[38] Richard Kuhns, *The House, the City, and the Judge* (Indianapolis: Bobbs-Merrill, 1962).
[39] Op. cit., 'The Japanese Theatre,' p. 55.
[40] Op. cit., p. 137.
[41] Op. cit., 'Le "Soulier de satin," p. 137.
[42] Op. cit., p. 113.
[43] Op. cit., 'Drama and Music,' p. 87.
[44] *Enactment: Greek Tragedy*, pp. 100–101.
[45] Darius Milhaud, *Notes without Music* (New York: Knopf, 1953), p. 66.
[46] Ibid., pp. 66–67, 121, 187.
[47] Paul Claudel, *Théâtre I* (Paris: Pleiade, 1956), pp. 880, 928, 978–80.
[48] Ibid., pp. 871, 872, 909.
[49] Ibid., p. 967.
[50] Claudel p. 137 and *Oeuvres en prose*, p. 1453. Claudel writes to Milhaud: "la finale à ma grande surprise a une tournure étonnamment chrétienne. Il n'y a rien à y changer pour en faire quelque chose qu'on pourrait chanter littéralement dans nos processions."
[51] Claudel, p. 187.

ANNA-TERESA TYMIENIECKA

# TRAGEDY AND THE COMPLETION OF FREEDOM

## INTRODUCTION

(a) There is no more compelling, stimulating, and intimately evocative word in the language of human existence than that of "freedom." Its use in every facet of human life is evidence that from his birth till his death man continuously claims his right to freedom. We could say that man's life is a chain of strivings to accomplish that freedom in innumerable ways. I submit that tragedy as an artistic form, in which the deepest concerns of man are brought to life, consists in manifesting precisely his – apparently – "fatal" failure to accomplish it.

What, then, is the significance of freedom in the pursuits of human life? What does tragedy as a literary genre reveal to us about it?

With reference to my conception of human beingness as consisting in *man's self-interpretation in existence*, I will attempt to treat this question by differentiating between the "life-significance" of freedom – when we consider the human individual striving for "freedom" within the real life-world – and the properly "existential" significance of freedom, as tragedy reveals it.

Life-significance manifests the role of "freedom" in the actual strivings of man to unfold his self-projected life-course; whereas the existential significance illuminates this striving from the perspective opened up by the question: "What does freedom mean for the nature of human existence as such?"[1]

(b) Alberto Casella's preface to the comedy, *Stormy Weather* (which in the Italian original has the more suggestive title, *Le Ombre del cuore*) brings the essence of our argument to light:

> Nella vita quotidiana, i fatti hanno un'importanza mediocre. Si crede che ne abbiano molta, ma non è vero. Essi agiscono sui nostri sensorie e li impressionano vivamente fino al punto di sopire, per un instante, l'indagine del pensiero. Resta, invece, attivissima, l'indagine del pensiero, sul quale agiranno le proiezioni dei fatti, nel tempo e nella profondità, con un sviluppo di associazioni, di meditazioni, di scrutinii: sicchè avviene che spesso il fatto di cronaca accaduto avanti a noi, dopo averci violentemente colpiti nei nostri sensorii, sparisce quasi della memoria, e ci accordiamo che il nostro pensiero evade dagli scarsi limiti del l'accaduto per dilagare in zone assai più vaste, investendo tutti i campi d'azione della nostra esistenza psichica e spirituale.



*A-T. Tymieniecka (ed.), Analecta Husserliana, Vol. XVIII*, 295–306.

Perchè la vera tragedia dell'uomo è nel suo cuore segreto e nella sua mente taciturna. Le più grande disperazioni si compiacciono di una splendida solitudine.

Ma il dramma più penoso è quello che l'uomo ignora, e che improviso gli scaturisce dall'animo, mentre l'illusione gli aveva potuto far credere che la libertà più sconfinata lo proteggesse dalle insidiose passioni.[2]

Indeed, the main point which our argument will corroborate is that while in real life, the interpretative *life-significance* falls upon the facts – in contrast, it is shown within the literary work of art, that the "external" facts are merely a concrete groundwork, a pretext for unfolding the significance of the "interior facts." While there is a strict correlation between the two, yet there is also a significant divergence: exterior facts – events, acts, actions, intrigues, etc. – seem to have a matrix with sharp contours; a clear-cut profile (they happen, unfold various processes, and find an end in one of them). On the contrary, the matrix of the "interior facts" stretches as far as the spread of the conflicting action as in the depth of sentiments which endow them with meaningfulness. Yet inward conflicts, tendencies, moral sensitivity, and moral exigencies which enter into the process of its meaning-bestowing do not lead to sharp, concrete conclusions. They establish a more or less rationally graspable pattern as an axis of interior life which remains, however much the eye of a searching heart is in a flux. Where does the contour of the interior drama lie? Casella quotes Sarcey, who at the end of a comedy raised the issue: "e poi?" Indeed, the poignant doubt arises: Is the death of the protagonist the conclusion of a tragedy? Is it the solution of the tragic conflict? Or is it simply "a way of finishing the drama?" In the perspective of my conception of man's *self-interpretative* life-course, with each of his acts (as well as his thoughts) playing the role of a mesh within this advancing weave, the tragic conflict once knotted continues to reverberate within man as long as life itself.

This clear distinction between the matrix and life of the "external facts" and that of the "internal life" of their meaning-bestowing by the protagonists, indicates an interpretative process that does not find its completion in the factual "dénoûement" of the intrigue or plot, that is, which does not find its solution in real life. If the real tragedy – the one experienced by the protagonists – finds its full significance in the tragic conflict and tragic feeling in the work of art, to what specific real-life situation confronting man in his self-interpretation in existence does it correspond? It is obviously a situation in which the innermost striving of human existence finds itself in insoluble conflicts; man has to choose and yet no alternative offers a solution. That is, real-life tragic sentiment – "tragic" with respect to the life-significance of the human strivings, struggles and aspects – is re-created in tragedy as a work of art to reveal its full-fledged significance *within the Human Condition*.

The tragic conflict expresses the existential situation of *human freedom*. Indeed, it is by understanding the significance of freedom in the Human Condition that we illuminate its situation with respect to the external facts of life.

We will then focus our attention, first, upon the tragic feeling; the existential significance of human freedom will then come to light.

## *PART I*

## *THE LITERARY PATTERN OF TRAGEDY*

### *As the Context Setting its Sense*

Tragedy stands out from all the other literary forms by the immediacy with which it manifests the essence of concrete life. When we discuss tragedy as a literary genre, or when we analyze and interpret a specific tragedy, we may emphasize as its essential moment either the "tragic knot" of the dramatic plot, or the "tragic feeling," which focuses more on the interior state of the protagonists than dramatic action itself; lastly, we may, with Aristotle, see the essence of tragedy in the unique type of response which it releases in the spectator, that is, the "catharsis."

These three essentials of tragedy are inseparable, although distinct. They represent, in fact, the three functions of the human consciousness, forming the tragedy as a work of art. As a work of art, tragedy is understood by the reader, spectator, and critic as being spontaneously interpreted with reference to the three basic, meaning-giving functions of the human being. One of these is the intellect, which structurizes in the intentional acts, and processes with their concatenations the cognitive (sensitive, emotive and rational) data into an objective system of a *life-simile*. It is due to the function of the intellect that the drama is plotted and that we discover, in interpreting it, its tragic knot of facts: of events, actions, and psychic conflicts among the protagonists. The tragic knot ties the strings of the forces of life into a conflictual pattern with such artistic devices that their role in the fatal failure in which the deepest human strivings terminate is manifested. Why is this failure "fatal"?

However, no configuration of facts could be "tragic" unless it released a "tragic conflict" of *feelings*; it is conversely in the propensities for the tragic feeling to be experienced by the protagonists that the tragic conflict resides.

No matter what various and nuanced forms it may assume, tragic feeling – as poignantly manifested in the classic drama of Racine and Corneille – is always co-significant with the moral conflict; that is, it draws upon the mode in which the protagonists interpret the turn of events adverse to their profound tendencies, desires and intentions. Whether the tragedy consists in their being bound to choose between two contrary but equally noble causes – as is the case in the classic drama – or whether they are shown as having no choice between the alternatives and either have to submit to the force of circumstance or to withdraw from the context renouncing it, since all issues out of the situation appear to be blocked by doom, life, or their own nature – as it is in the contemporary drama (especially in the theater of the absurd) – the tragic feeling expresses the *moral conflict and the impossibility of solving it*.

That is, the tragic feeling – as the very life of the tragedy – is brought in by the second fundamental function of the human being, who uses it to interpret his life transactions: the "moral sense."

Now, let us turn to the structural aspect of tragedy as a work of art. Roman Ingarden's analysis of the literary work as a polyphonic symphony of aesthetic values shows that it is an intellectual skeleton of a skillfully constructed schemá of a life-simile, devised from elements of feelings, facts, and nature, that is, the work of the intellect, the sense of which it incarnates. But beyond that it shows the work of art as a *symphony of aesthetic experiences*, which the consumer draws from them. I see in this aesthetic essence of the drama the workings of a special sense: *aesthetic enjoyment*. As the fourth meaning-giving function of the human being, this aesthetic enjoyment endows rational and moral sense with the final significance, by turning the rationally/morally devised plot of the life-simile into a work of art. The catharsis is the fruit of the aesthetic enjoyment function.

Bringing this altogether too brief outline of what the tragedy as a literary work of art consists of, and reaching a conclusion, it may be ascertained that in its intrinsic composition it brings together the raw material of facts, feelings, ideas, and nature put into intellectually devised sets of interconnections and thus endowed with a new intellectual sense as a substructure of the new order with respect to real life. Furthermore, tragedy releases moral significance through these artistically interpreted human transactions, as lifted from its merely mater-of-fact meaningfulness of events and experiences into an aesthetically enjoyable orchestration; doing so it creates a new significant context. Distilled through its significant filters of the structural interpretation of facts, the moral interpretation of experiences and aesthetic significance, tragedy brings about a novel *existence-significant moment*. By this succinct account we have outlined the interconnections among the main points which we intend to discuss in the present study.

Indeed, from the above it appears that the interpretative approach to the literary work indicates retroactively that its root-elements stem from the real life-world. It means that originally they represent dispersed fragments of the four basic meaning-giving functions of man: the *vital sense*, the *intellectual sense*, the *moral sentiment*, and *aesthetic enjoyment*.

I also propose that the literary work brings about a new significant complex, with respect to the vital significance of life. What is the *new* significance of the work of art? What does the significance of the tragedy consist of?

These are the two questions which we will aim to answer. Anticipating their treatment, I submit that:

1. The significance of the work of art is a new type of life-significance: *existential significance*.

2. That it is the specific significant (form) complex which the moral sense assumes in the *tragic feeling* that constitutes the essence of the tragedy.

3. That the moral conflict underlying the tragic feeling corresponds directly to the moral experience as operative within the factual life; it plays the role of a two-directional transmission factor between the innermost strivings of the human being to enact his self-interpretative life-course by his own device in the real life, on the one hand, and his interpretative exploration of its existential constraints and unattainable aims in the dramatic form of literature, on the other.

4. Indeed, to conclude, I will show that the rationally insoluble dilemma of freedom within the real life-world experience finds its existential significance in the tragic feeling: the *completion of freedom*.

## *PART II*

## *TRAGEDY AS EXFOLIATING INTERPRETATIVELY THE EXISTENTIAL SIGNIFICANCE OF FREEDOM*

It is for a good reason that I chose Casella's *Stormy Weather* to open my investigation. As intended by the author, the low-key dramatic setting of the external facts in which the dramatic plot is set makes them remain inconspicuous. The light falls upon the "drama of the heart." And even there, the choice between two loves – two loyalties equally honorable and worthy – does not present strong contrasts of life-alternatives. In the significance of the facts, the heroine has the freedom to choose, and neither of the alternatives is menacing with grievous consequences, nor does either offer great rewards. Yet, in the "language of the heart" it is not so. The possibility of choice, the so-called "freedom to choose" as her husband calls it, is not of her own doing; she is thrown into it against her will. To choose between two clear-cut alternatives means to falsify her inner state; means to impose an unduly determined form upon strong but delicately balanced feelings; means to break down by an outside decision the natural direction of her inner moral commitment; means to distort a most subtle balance between love and trust, on the one hand, and love and friendship, on the other.

To choose, means to accept the challenge which itself expresses lack of trust, as well as to destroy the equipoise between love and friendship by submitting it to an unworthy test that should decide which one will win over the other.

The freedom to choose – in the "language of facts" – turns, in the interpretation of the *inner sense*, into a tragic conflict of a unique type: accept the challenge or do not accept; choose or do not choose. The seeming freedom to act is insignificant, futile; the conflict which its pattern creates is tragic: neither alternative is acceptable. Both defeat the inward truth of the person.

In this conflict, there is no place to affirm freely this truth. It has to remain retained within, concealed. The striving for the expression of this truth has to be self-conditioned, retained in the experiencing person. To retain it wilfully and in silence within, while it cannot be exercised without: the tragic feeling is thus completion of the freedom within.

This refusal to make any choice or enter any contest reveals the tragic significance of human freedom and offers the basic elements for its exfoliation. This expression of the tragic feeling pervades classic and romantic drama. It is even more accentuated in contemporary tragedy, in which precisely the inner life of the protagonists comes into a particular focus.

Let us turn now to romantic tragedy as represented by Friedrich Schiller's *Kabale und Liebe*. This builds its tragic conflict within the same pattern. In fact, the tragic conflict which is coated in the social form of an intrigue devised by the powerful personage of a father in order to discourage his promising son from contracting a socially unacceptable marriage of love, centers in the tragic feelings of the two lovers/protagonists involved; both Ferdinand and Luise are thrown into the conflictual situation created by the intrigue which violates their natural dispositions toward each other. The love/disappointment, which falsifies the facts as much as it puts Ferdinand's feelings to a test, makes him prefer her death over her life in her assumed unworthiness, unbearable to him. If his murderous act meant an exercise of freedom aimed at the implementation of justice – he fails in its aim.

In contrast, Luise, who is caught between a loving duty to protect her father's life and a duty to the lover – whose lack of faith in her means his betrayal – finds herself in a tragic conflict: if she convinces him of her innocence, she will break her promise on behalf of her father. She is apparently put to a test, revealing which of the two loyalties she would betray. Yet the choice does not represent freedom. On the contrary, it only accentuates its tragic feeling. Her tragic feeling is expressed by her words: "Ein entsetzliches Schicksal hat die Sprache unserer Herzen verwirt." She will not choose to talk; nothing will set the situation straight. Her freedom would not have been accomplished had she cleared up the suspicious doubt her lover has about her; would she regain thereby her absolute faith in *his* confidence and faith *in her*, which she deserved? Not acting but nurturing the inward truth of this tragic *sense* of the conflict – not having found an unshakable faith in the heart of the beloved, in the assumption of which the sense of the entire plot originally lay – she preserves the freedom of her inward self intact.

We find the same silent tragedy in Othello. Whatever dramatic effect created by the misunderstandings between the furiously jealous Othello and Desdemona, the tragic sentiment which lifts the drama from a pedestrian level to its measure of art is to be found in the heart of Desdemona. Although no inward conflict is apparent yet it is the confusion of the significance of facts and passions that present a tragic conflict beyond repair. (Discrepancy between the suspicion-stricken Othello and her innocence, in which she ignores even the possibility of such a possible suspicion, creates the tragic tension.) Othello's passions impair his freedom of judgment; while he exercises the choice, he destroys what he longs most to retain. His act does not make him free – on the contrary, it robs him once and for ever of knowing the truth of the facts; of setting things right.

Indeed, in such classic dramas as Corneille's *Le Cid* or Racine's *Bérénice*, the protagonist makes a choice between two morally significant alternatives and thus violates his freedom: no matter whom he chooses, he renounces by the same stroke that which is equally precious to him; his alternatives are not of his own doing; he is thrown into them. Their very opposition violates his inner state.

To the numerous interpretations of Hamlet's tragic feeling expressed by his utterance, "To be or not to be that is the question," which summarizes the conflict, we may add another. Since there is no way to set things aright, "to be or not to be" loses its meaning. Neither life nor death solves the conflict.

It may be argued that it is the special type of drama (Greek, classic, romantic, etc.) which consists in taking the life-enactment with its existential axis and bearings as its fulcrum; the Human Condition, which reveals itself by these variants, would appear in their specific perspective. However, this might be neither its only nor its "essential" mode of revealing. In fact, when we take into consideration the contemporary drama of Ionesco, Beckett, Brecht, Mrozek, etc., we see that, first, the dramatic schema is not set upon the previously discussed model of "the tragic knot – the tragic feeling – catharsis." At the extreme opposite, it is the *tragic factor of the Human Condition* that is brought forth directly as the pivot around which are set the entanglements and disentanglements of human life in their bare essentials. It is not through the unwinding of the threads of life that in reflective interpretation we establish for ourselves the perspective in which the revelation of the Human Condition should appear, but we face directly the Human Condition in terms of every move, breath, and wish by which, as human beings, we are irremediably bound. The enactment of the "life-simile" as the means to stimulate and guide reflection is dispensed with and is replaced by an appeal to meditate upon the Human Condition directly (e.g., Ionesco's *Le roi se meurt* or Beckett's *Waiting for Godot*).

Second, the drama as a specifically representational art no longer consists in the "enactment" or "performance" of the life-simile. This latter seems, in a way, of secondary significance – as much as is also secondary the life-simile setting: decor, costumes, stylization, etc. On the contrary, the entire enactment complex is played down to the strict minimum in order that the meditative revelation may be released directly.

The question thus emerges whether in such a reduced form we still have to do with a "drama" or with its fragment. If the meditative reflection/presentation dispenses with the stage production altogether, will we still have to do with the dramatic art?

More significantly, we must in view of this deviation from the emphasis upon the enactment of a life-simile to the meditative manifestation ask, "Can we gain a complete and adequate access to the Human Condition through a perspective which is already an interpretation – however penetrating and profound – in terms of a selection of some universal threads of human existence?" Such an interpretative appeal, which solicitates a specific response, is the fruit of the human mind in its most intricate form. Does not the Human Condition reveal itself more adequately and completely without an interpretative screen but through the pulling of the chords of man's ever-recurring crucial functioning within the life-world? Whether man's destiny is ultimately bound to a master/slave model, or to that of an unfulfilled expectation or to a quest for an "assignment" which never is transmitted – is it any of these existential views which makes him undertake tasks, struggle for life, conquer or fall, suffer or feel joy? Indeed, it is not in our meditation *about* existence but in its *enactment itself* – that is, in man's life-functioning and not in thought – that there lies the key to cognition itself.

Sartre appears to have seen it well. Even Orestes (*The Flies*) cannot find in his mind any reason to undertake anything, to choose, etc., yet he is prompted with urgency to act. Since his mind fails to provide him with any direction within the life-world he kills himself. This act of taking his own life is still an affirmation of the force of life itself.

We may set aside here the question of the most appropriate form for a dramatic presentation of tragedy – in which the genre of tragedy would be maintained (a drama remain a drama). Yet we may observe that the life-enactment has an irreplaceable role within the drama: it is precisely through the life-enactment that the Human Condition in its irreducible and unmasked nature *reveals* itself. To be – and to become – human is to undertake, to pursue, to win or to lose, to fall and, like Sisyphus, to take up ever and again the same never-to-be-accomplished task.

## *PART III*

## *THE LIFE-SIGNIFICANCE VERSUS THE EXISTENTIAL SIGNIFICANCE OF FREEDOM*

At this point the question arises: "What is the nature of this 'freedom' which appears as the crux of real life and reappears as a crucial factor in tragedy?" "What significance does it acquire in the patterns of human transactions within the social world, on the one hand, and how does its exfoliation in the tragic form manifest its significance in human existence, on the other hand?"

To begin with, we must state that in general we understand by the notion of 'freedom' the unhindered possibility of spontaneous life-enactment, whether through actions, undertakings, or self-manifestation. In the context of my conception of human life as man's *self-interpretation in existence*, we may distinguish two main phases of this self-interpretation which relate directly to the question of freedom.[3]

In the organic/sentient phase of the individual's self-interpretation the significance of freedom is "vital"; it is the capacity of the living being to satisfy the requirements of his vital course without undue external hindrances or coercion which might lead him to deviate from his intrinsically prescribed course or to an outright violation of his vital strivings. Whereas in the conscious/moral phase of the *specifically human self-interpretation in existence* freedom is morally significant. It is understood as man's unhindered possibility of unfolding his existential course in a self-invented (creatively devised) action, undertaking expression, etc., within the social world.[4]

It is this second significance of freedom within the real world which is our present concern; yet, as is obvious from our foregoing analysis, the first significance of freedom tacitly underlies the conditions of its fulfillment.

The dilemma of the Human Condition, at the crossroads of which lies the question of freedom, is thus apparent. On the one hand, the living being is – to use an expression of Heidegger, with whom, however this analysis shares no affinities – "thrown into existence." Man chose neither his initial endowment nor the situation into which he is projected. He is thrown into the play of anonymous forces, neutral energies and adverse currents. To individualize himself from them into his own course of beingness means to wring out of these neutral or adverse factors the indispensable satisfaction of his vital needs, while maintaining his own course. To do this, he has to struggle with innumerable obstacles: to neutralize their impact, to adapt, to succumb, to transform them into his allies, to win, to fail.

On the other hand, the human being who strives to enact his inventively devised life-course in transactions with other beings – transactions, which draw their very sense from the *moral sentiment* – has to assume a set of basic moral commitments (principles, rules, postulates, etc.). These commitments interfere directly with the promptings of his innermost spontaneities. Furthermore, these spontaneities themselves pose a question. In fact, man is not the master of his dark impulses, unforeseeable emotions, instincts, drives, and violent passions, which trouble the clarity of his thought and impair his judgement; but even his judgement, which is of paramount importance in the self-devising of his course, as well as in the direction of his creative powers which give form to the innermost intuitions, ideas and nostalgias which he desires to convert into the meaning-giving factors of his life, is by no means reliable. Lastly, and no less importantly, he is constrained by moral commitments.

The moral commitment of the human individual as giving sense to human transactions is basically threefold: (1) Man faces God, whom he is free to accept or to reject; (2) He stays "under law" of the social group, community, state, to which he belongs as a social being; (3) He is bound by a concrete network of interrelation with other human beings according to the more or less explicit principles which he is expected by them to respect.

In fact, it is only in regard to God that the human being within the social world can be and should be – but not always is – unconditionally *free*: God does not impose concrete sanctions within the life-world, the way human beings do! But the two other types of "commitment" impose a network of restrictions upon man's spontaneities; restrictions working already within himself.

Indeed, when we raise the question of "being free" to accomplish a self-invented task, aim, role within the social world, this question contains the postulate that the individual – while he will try to counter or outwit the adverse forces to make himself a way among the infinitely intricate situations as well as to assemble all the elements indispensable to accomplish his aim – has to discharge himself from the above commitments. His proposition then appears futile. He is in no position to master the forces of the social world; but is he master of himself? If he actually does fulfill the limited task of daily survival, or accomplish in a very limited way his creative projects, dreams and nostalgias, it is, first, never in full; second, it is not in virtue of *his power* but in virtue of *favorable circumstances*.

Now when we turn to the tragic feeling with which art pinpoints and illuminates the game of life, we see that it exfoliates precisely this situation of human strivings and dreaming, while it endows life precariousness with a special sense.

Our tendencies and undertakings are, indeed, dwarfed before they materialize; our passional groundwork, which under the promptings of circumstances, on the one side, and of our moral commitments, on the other side, releases adverse impulses, feelings and imaginings, obscures the real significance of the inner truth; no human act can set it straight again. And yet, the craving to solve the "insoluble" prompts us to act.

Thus the significance of freedom does not lie in the accomplishment of our aims. Yet the human being never ceases to seek beyond that which is directly available and within his reach. In principle he cannot but fail – but others, in other circumstances, for whom it will be no "aim" anymore, may succeed.

His ever-repeating strivings are the only way he has to express his "freedom." The life significance of freedom resides in this exercise.

The existential significance of freedom reveals to us in the tragic feeling this inner truth of man. It belongs to the self-inventive, creative mode of *man's self-interpretation in existence* to seek ever-new aims, tasks and ideals for his innermost nostalgias by the realization of which he believes he gives a new, nobler and higher meaning to his existence.

His very *beingness* and his grandeur consist in inventing and striving to accomplish them.

*World Institute of Advanced*
*Phenomenological Research and Learning*

NOTES

[1] For the "existential significance of art", cf. A-T. Tymieniecka, *Poetica Nova at the Creative Crucibles*, a monograph in *Analecta Husserliana*, Vol. XII, 1982.

[2] Alberto Casella, *Le ombre del cuore*, in *Teatro Italiano Contemporaneo*, No. 10, Rome, Casini ed., 1955.

[3] Cf. *Poetica Nova*, op. cit.

[4] Cf. A-T. Tymieniecka, 'Nieboska Komedja – struktura a wizja poetycka' ('The Undivine Comedy – Structure versus Poetic Vision'), in *Szkice Filozoficzne; Romanowi Ingardenowi w Darze*, Krakow, PWN, 1964, pp. 444–470.

SHERLYN ABDOO

# HARDY'S JUDE: THE PURSUIT OF THE IDEAL AS TRAGEDY

All tragedy is grotesque. (Thomas Hardy, *Life*, August 13, 1898)

## INTRODUCTION

Virginia Woolf's tribute to Thomas Hardy was written shortly after his death on January 11, 1928. In it she said: "if we are to place Hardy among his fellows, we must call him the greatest tragic writer among English novelists." She goes on to assert that although it is "the most painful" and "pessimistic" of his novels, *Jude the Obscure* "is not tragic."[1] Hardy, himself, in the 1895 Preface to the First Edition of the novel referred to *Jude* as "simply an endeavor to give shape and coherence to a series of seemings, or personal impressions . . . not of the first moment."[2] Superseding, however, is his later statement, which under the stimulus of the early critical attacks on *Jude*, identifies the novel's central interest for him:

> the greater part of the story – that which presented the shattered ideals of the two chief characters, and had been more especially, and indeed most exclusively, the part of interest to myself – practically ignored by the adverse press.[3]

The purpose of this paper is exegesis, a critical analysis of Thomas Hardy's last and greatest novel. My aim is a comprehensive interpretation of the novel, a look into its heart – to the origins – of Hardy's view of the human situation: the inevitable defeat of the human spirit by a powerful force which allows no redemption, whether in this life or elsewhere.

In *Jude*, Hardy recapitulates and interweaves the major themes of his prior novels: idealization of the beloved; man's alienation from himself and isolation from others; God as unknowable; that man's fate – puppet-like – is controlled by a capricious hand; that life, particularly marriage, is a trap from which death is the only escape; the self-proclaimed believers are among the most impious members of the human community, wherein modern religion is itself an hypocrisy; the injustice of a social order favoring a privileged class – while, others, the working classes most especially, must strictly adhere to conventional propriety in order to retain even an aura of respectability; the



*A-T Tymieniecka (ed.), Analecta Husserliana, Vol. XVIII,* 307–318.

juxtaposition of the ordinary with the grotesque; Pride, the dominant human character trait of Hardy's protagonists, as major contributor to their tragic ends; and finally, the immense living presence of Nature as She provides both a backdrop to and a reflection of, the unredemptive human tragedy, which is simply, being alive.

The primary issues this paper addresses and tries to resolve are the following: (1) whether *Jude the Obscure* is, in fact, a tragic novel; (2) the ways that the novel's tragedy is supported by Hardy's antecedent prose writings and extant biographical materials, including Florence Emily Hardy's *Life of Thomas Hardy* (which we now know he wrote himself), and the recently published *Personal Notebooks* and *Letters*; (3) how Hardy's preoccupation with the heroic man and the idealistic feminine became objects of his intellectual pursuit; (4) the degree to which Thomas Hardy, as poet/novelist, pursued the elusive, ideal woman of his dreams (much to the distress of his two wives) in his novels and poetry – and, in so doing, he became his own hero: (5) the alliance in Hardy's novels between Evil and the grotesque; (6) the metaphor for Life itself in Hardy's novels is the Greek "web," the inexorable pattern that binds his characters to their doom; and finally (7) "tragedoia," the "goat-song" – as the primary factor in the origin of tragic genre – is, at its most primitive, raw, excruciating, half-mad sense, the mimetic impulse that Hardy's dramas were created to express.

This paper's position is based on the premise that with *Jude* Hardy had ultimately fulfilled his purpose as a novelist; he had given full expression to his understanding of the tragic in the human condition; and that his abandonment of novel-writing for poetry, therefore, was an artistic necessity. As A. Alvarez comments: "After *Jude the Obscure* there was no other direction in which he could go."[4]

Many have tried since Aristotle's *Poetics* to explain what the tragic experience is. At best, we can expect only to receive a second- or third-hand description. However, what we can hope to gain in understanding through art, is a moment's glimpse – an approximation – of individual suffering which may help us bear our own loneliness and fear.

To begin at the beginning necessitates a return to the Greek origins of tragedy, in this case primarily to the understanding of Hardy's sources as he was an avid, life-long student of Greek tragedy. But first let's consider some of his own thoughts on tragedy and identify those aspects of his earlier novels which are particularly relevant to Jude's tragedy. His early note (November, 1885) that: "tragedy exhibits a state of things in the life of an individual which unavoidably causes some natural aim or desire of his to end in a catastrophe when carried out" (*Life*, p. 176), became, just three years before *Jude* (October, 1892) more specific: "The best tragedy – highest tragedy in short – is that of the WORTHY encompassed by the INEVITABLE" (*Life*, p. 251).

Jude's tragedy is bound to his two failures: (1) failure to achieve admittance to Balliol College, Christminster, and the education he desires, and (2) failure to achieve complete possession of Sue Bridehead, the woman he successively idealizes, loves, pursues, loses, regains, lives with, has children by and finally loses again, forever. Throughout the course of the novel Jude is psychially sustained by his concurrent struggles to achieve education and to win Sue's love. His efforts with these involvements provide both the impetus and the reasons for his continuing to live. Jude's ineligibility for a university education, when he is finally rejected, is due not to his lack of scholarly ability, but to his low social status; he is, after-all, a laborer, a restorer of churches. And while he finally loses Sue altogether because of her overwhelming guilt after the hangings of their children, Jude *never* derived full satisfaction from their sexual life together. Sue was incapable of completely giving herself to him. To do so would have made her totally dependent on Jude for her identity, a state of being that her already too fragile self-image could not tolerate. In short, Sue Bridehead is terrified of losing her self-identity in Jude – of becoming a non-existent personality.

Jude's frustration remains unabated while he has hope of realizing his goals and he doggedly struggles to achieve them. But when his idealizations are denied him Jude's will to live evaporates and he dies. Obtaining a university education and possessing Sue Bridehead are identified here as idealizations in the Platonic sense. For Jude education represents knowledge, the ultimate achievement that will open the door to worldly success; possessing Sue is the possession of something especially fine, a quality above his rough experience. To claim Sue for his own is for Jude equivalent to touching the ethereal or other-worldly – similar in fact to experiencing a transcendent religious experience. In Sue's case, her lack of sexual desire and her inability to love Jude completely creates a barrier beween them. Jude is forced to live out his Platonic idealization of the feminine, while Sue's denial keeps "his passion as hot at the end as at the beginning, and helps to break his heart" (*Life*, p. 272). On the other hand, Jude's struggles to educate himself in preparation for the university are, in the face of his desperate situation, only keeping alive his hope for a better time. Had it been more than a dream, would he have waited so long to find out the truth? Was it not then the struggle against adversity, the struggle towards education, towards possession of the most beautiful, wonderful woman he knew that keeps Jude alive?

Hardy's original idea for Jude's story was as follows: "A short story of a young man – 'who could not go to Oxford' – His struggles and ultimate failure. Suicide" (*Life*, pp. 207–8). And in a letter to an unidentified friend after its publication, Hardy comments on the novel's reviews and tries, retrospectively, to clarify his original thesis:

> It is curious that some of the papers should look upon the novel as a manifesto on "the marriage question" (although, of course, it involves it), seeing that it is concerned first with the labours of a poor student to get a University degree, and secondly with the tragic issues of two bad marriages, owing in the main to a doom or curse of hereditary temperament peculiar to the family of the parties. (*Life*, p. 271)

Now that we are closer to Hardy's purpose, let's consider the actual sequence of events. The boy Jude lives with his maiden aunt and is inspired by his teacher, Phillotson, to advance himself by studying so that he will eventually be accepted to the university at Christminster. Suffering the rough indifference of his aunt, Jude grows up to become a stone mason – all the while teaching himself Latin and Greek and looking towards a brighter future, a time when he would be in Christminster – the "heavenly Jerusalem" (*JO*, p. 18). Along the way he becomes temporarily distracted by his awakened sexuality and marries Arabella Donn. But she soon tires of Jude and leaves him. When he finally does reach Christminster it is to meet his cousin, Sue Bridehead, the woman he fell in love with when he saw her photo at his aunt's house. But he also again meets Phillotson, his old teacher, and introduces him to Sue. Sue becomes attracted to Phillotson when he offers her a job as pupil teacher if she will attend a teacher's college to become qualified. Instead Sue runs away from the college and shortly thereafter Phillotson marries her, nonetheless. The marriage is an altogether excruciating event for Jude who is forced, by virtue of the fact that he is Sue's only living relative besides Aunt Drusilla who is against marriage, to give her away at the ceremony. Sue goes off with Phillotson, but their marriage is never consummated. Meanwhile Aunt Drusilla dies, Jude meets Arabella again and in his despair has a brief affair with her, becomes inspired by religious music and begins ecclesiastical studies. Sue, having met Jude at Aunt Drusilla's funeral and knowing she cannot stay with Phillotson, who is getting impatient with her excuses, declares her love for Jude and runs away to live with him. At this point Jude abandons his religious studies. Eventually Jude and Arabella divorce as do Sue and Phillotson; Phillotson's teaching career is, meanwhile, ruined by the results of his disastrous liaison with Sue. But it is Sue's fear that Arabella will win Jude back that compells her to agree to marry Jude for convention's sake. Circumstances arise, however, which cause postponement until they abandon the idea. At this time Jude learns of the existence of his child by Arabella – Little Jude, nicknamed "Little Father Time ... because [he] looked so aged" (*JO*, p. 221). The child arrives to live with Sue and Jude and three years pass. Sue by this time has succumbed to Jude's physical desire, has borne him two children and is pregnant with a third. When Little Father Time learns of Sue's pregnancy he is horrified with what seems to him her irresponsible fecundity. He chides her remorselessly and in a fit of despair hangs the two infants and himself. Sue's remorse and guilt drives her away from Jude and back to Phillotson who she has convinced herself is her only true husband. Arabella now a widow, seduces the ill, distraught Jude and tricks him into marrying her for the second time. Jude's last unsuccessful attempt to get Sue back is preceded and followed by a suicidal walk in the freezing rain. He dies reciting Job while Sue lives in self-punishment, submitting to Phillotson's lust.

The connective thread running between Jude, Sue and his educational pursuits is Phillotson. Jude's initial desire for university education is stimulated by Phillotson's early interest in him as his boyhood teacher and friend. Ironically, it is the same Phillotson – the only male figure who takes an interest in him – that frustrates the boy Jude by promising but not immediately sending the Latin text-books, who marries the woman he loves and then after Jude has won Sue back and they have produced several children, takes her away from him again. Thus, Phillotson is the menacing, powerful father figure who must prevent the threatening son from replacing him. It is Phillotson's inner weakness, though, that keeps him, like Sue and Jude, isolated and dependent on others for his own sense of personal identity. Phillotson's role (whether or not consciously known to Hardy) is that of the castrating father in competition with the son. Similarly, Sue's relationship with Phillotson is an Oedipal one; for her he also at first becomes a father figure by virtue of his age (he is forty-five when they marry, "old enough to be the girl's father" [*JO*, p. 86]). Her instinctive revulsion for his physical person is a self-protecting act against incest.

The novel's structure is largely constructed around the tensions of the couplings and separations that take place among the participants of two sets of love triangles: Jude-Sue-Phillotson, and Arabella-Jude-Sue. But this pattern is an already familiar one to Hardy; he used it in virtually *all* his prior novels.[5] In fact, we can state that in Hardy's case the love triangle is a compulsive, repetitive literary device. But this device is more than a convenient structure. It is the portrayal of Hardy's own inner neurotic state, whether or not he ever realized it. The similarities between Hardy and Jude, despite his disclaimer that "no book he had ever written contained less of his own life" (*Life*, p. 274), must be considered before the central problem can be revealed and resolved.

To move on to this paper's central premise that *Jude the Obscure* is a tragic novel, we must recognize that Jude is also St. Jude, martyr and patron saint of hopeless causes; his goal is Christminster, "the home of lost causes" (*JO*, p. 66). Jude is the sacrificial scapegoat of both the pagan and Christian ethos. The novel's progression travels historically from the pagan sacrifice to the Christian sacrifice, but in neither case is redemption achieved. The first sacrifice comes with the discontent and break-up of Jude's first marriage to Arabella. It is the death of erotic love that is so vividly portrayed in the pig-killing scene. Arabella, after-all, threw the pig's pizzle that first captured Jude's attention. So it is the agonized death of the dumb creature, so clearly representing their earthy, natural coupling, that appropriately joins Jude's fate to the pig's: "the white snow, stained with the blood of his fellow-mortal" (*JO*, p. 55). And later, when he and Sue discover their two children hanging on garment hooks and "little Jude . . . in a similar manner" (*JO*, p. 265), Jude cries out in recognition: "O my comrade, our perfect union – our two-in-oneness – is now stained with blood" (*JO*, p. 267)! The children's deaths are virtual sacrifices; the three little bodies a crude emulation of the crucifixion and the result is Sue's final desertion of Jude for Phillotson in an attempt she says, to "mortify the flesh – the terrible flesh – the curse of Adam" (*JO*, p. 272)! The pagan sacrifice of the pig marks the death of Jude's marital relation to Arabella and the physical expression of his nature; the Christian sacrifice of the children marks the death of Jude's relation to Sue and the spiritual/redemptive hope that he lives for. He dies finally neither pagan nor Christian, but identifying totally with the suffering Job:

> Let the day perish wherein I was born, the night in which it was said, there is a man child conceived. . . . Why died I not from the womb? Why did I not give up the ghost when I came out of the belly? . . . For now should I have lain still and been quiet. I should have slept: then had I been at rest! . . . Wherefore is light given to him that is in misery, and life unto the bitter in soul? (*JO*, p. 320)

Hardy's juxtaposition of the pagan, Christian and Old Testament sources as only briefly noted here, reveals his ambivalence about religion. The issue at hand is basically to identify those elements of Greek tragedy beginning with Aristotle's definition in the *Poetics* that *Jude* incorporates, and to show that despite the later consequences of Christianity vis-à-vis Darwin's *Origin of Species, Jude the Obscure* is nonetheless a tragedy following in the classical tradition.

From Plato who identified the tragedian as "an imitator, whose product is at least three removes from nature . . . and the truth,"[6] we can consider first Aristotle's initial statement towards a definition of tragedy:

> Tragedy, then, is an imitation of an action that is serious, complete, and of a certain magnitude; in language embellished with each kind of artistic ornament, the several kinds being found in separate parts of the play; in the form of action not of narratives; through pity and fear effecting the proper purgation of these emotions.[7]

Jude's tragedy, then, arising from Hardy's imagination, is an imitation twice removed from the experiences of his personal life and three times removed from the experiences of friends, relatives and strangers as related to him second- or third-hand, and common hearsay embellished and altered to fit the situation. The dramatic agents of the tragedy are principally Jude Fawley and Sue Bridehead. Their actions reveal the arrangement of incidents – in short, the plot. The novel's action is certainly "serious, complete and of a certain magnitude," if by "serious" we mean that it deals with issues or events that can alter the course of a life, result in poverty, physical disability, or death, "complete" because it relates a single story or statement with a beginning, middle and end, and "of a certain magnitude" because of the powerful emotional effects it has on the reader – due to reversal (or turn of the plot) and recognition by the hero/protagonist of his true situation and the consequences thereof. In *Jude*, the elements of surprise triggering the plot's reversal are several. First of all Arabella tricks Jude into marrying her with a false claim of pregnancy; second, Sue Bridehead's sudden decision to marry Phillotson when she discovers that Jude is already married; third, the appearance of Little Father Time; and fourth, the grotesque hangings of Jude's and Sue's children. Jude's recognitions dealing with his erotic nature are two; in each case he sees the human failings of the two women in his life and realizes that they do not measure up to his original idea of them. In Arabella's case: "He knew well, too well, in the secret center of his brain, that Arabella was not worth a great deal as a specimen of womankind." But, in an effort to repress the true state of his feelings he "kept up a fictitious belief in her. His idea of her was the thing of most consequence, not Arabella herself, he sometimes said laconically" (*JO*, p. 48). And later, Jude accuses Sue in a moment of revelation: "Sue sometimes . . . I think you are incapable of real love" (*JO*, p. 192). He recants almost immediately though, accepting in place of the passion and warmth that he craves, the Sue he imagines her to be: "you spirit, you disembodied creature, you dear, sweet, tantalizing phantom – hardly flesh at all; so that when I put my arms round you I almost expect them to pass through you as through air!" (*JO*, p. 195).

Just as important is Jude's recognition of the value of his work as a stonecutter in light of his ambition for obtaining a university education:

For a moment there fell on Jude a true illumination; that here in the stone yard was a centre of effort as worthy as that dignified by the name of scholarly study within the noblest of the colleges. But he lost it under stress of his old idea. (*JO*, p. 69)

To continue, Hardy identified the tragic hero as one who was the "WORTHY encompassed by the INEVITABLE" (*Life*, p. 251) – one whose misfortunes and suffering inspired in the reader not only feelings of fear and pity, but by association to one's own situation, results in a therapeutic purgation of the emotions. We may question why our feelings of fear and pity are aroused? We feel fear because of the closeness of the protagonists' situation to our own and pity because of the undeserved misfortunes that plague a decent man. Purgation is a temporary cleansing of our emotions through weeping and lamentation. It is precisely because of this temporary relief to our emotional tensions that the entire process of purgation must be repeated again and again. Hence, the significance of repetitious or cyclical ritual sacrifice. It is the spectacle of watching someone else suffer, the sadistic satisfaction (sometimes amounting to lust) we feel in knowing that it is not us doing the suffering, and the exhilaration of having survived yet another test of time that we enjoy. Again, a public display of grief allows the spectator to empathize with the slain or suffering hero and expiates the guilt he feels. The realization that death is so close to the living experience intimately joins us to the pathos of the victim's plight. It is the survival of the fittest that counts, or perhaps more accurately, the struggle between the will-to-live undermined by the subconscious death-wish. Here again, the idea of a sinless hero is an important distinction; for, while a despicable, worthless hero would inspire neither feelings of fear or pity, nor can we share our most profound experiences with a perfect man/god who has the power to control his own destiny. Rather, he must be like us, a human man with feelings and flaws.

Hardy's concept of nature and man's place in the universe obviously deviated from the traditional acceptance of a Christian God who offered redemption from sin in an afterlife beginning only with death. Darwin's *Origin of Species* appeared when Hardy was a young man; its impact when he read it was immediate and so affected his beliefs that when Darwin died Hardy attended his funeral. Though his family was long a church-going one, Hardy nevertheless found it difficult to continue believing in a Christian God who could allow good people so much earthly suffering. Of God's existence he says: "I have been looking for God 50 years, and I think that if he had existed I should have discovered him" (*Life*, p. 224).

In addition, his early and continued study of Greek literature introduced him to an alternate view of the universe and man's relation to its nature. His search for a cosmological order led him, an avowed agnostic, to devise a plausible answer to the question of what force, if not God, determined the natural order of things?

His search carried him through the writing of fifteen novels, two volumes of poetry and one volume of short stories (*Wessex Tales*) before he could offer the explanation he finally gives in *The Dynasts*. In *The Dynasts* Hardy creates the Immanent Will or the Prime Cause as the force in nature which determines the course of events. Humanity for Hardy was an immense creature, "a monster whose body had four million eyes and eight million heads" (*Life*, p. 136). It is a "collective personality" (*Life*, p. 416), "one great network or tissue which quivers in every part when one point is shaken, like a spider's web if touched" (*Life*, p. 177). And man, compelled by his own nature, marches his path blindly until he is woven so tightly into his fate as the threads of a carpet, that he becomes trapped like the fly in a spider's web. No amount of struggling will undo or dislodge him from the pattern that the Immanent Will has created by pulling the strings of the men/puppets to weave the tapestry of its universe.

The structure, then, of Hardy's universe, this giant tapestry of humanity and nature, allows no deviation. Once Jude has chosen his path, he is compelled to follow it to the end of his life, which he does. It is no accident that Hardy had a cobweb phobia.[8] He would search the corners of his house every night before retiring so that he wouldn't sleep in the vicinity of them. The importance of the web image in Greek tragedy is the symbol of entanglement; it is what binds Agamemnon to his fate – it is the carpet that Clytemnestra uses to capture and disable him with. In trying to understand the workings of a universe in which the Christian God's existence was for him unknowable, Hardy borrowed from Aeschylus a metaphor, the web or net, which he then translated into a whole fabric of nature into which each man weaves his own destiny.

To return at last to the original mimetic impulse that Hardy was trying to express in his tragedies, we must consider "tragedoia," the Greek word for tragedy, meaning "goat-song." As we know the goat was used as a sacrificial animal in primitive ritual to represent the slain god Dionysus and helped promote the theory that the cult of Dionysus lamenting over its ritual slaying of its dead god was the root of tragic drama. Here I am concerned with the 'song' aspect of "goat-song," the threnody, the lamentation, the agonized inhuman scream of Oedipus when he puts his eyes out. In *Jude the Obscure* there are at least two instances when Hardy connects Jude's suffering to this primitive expression of tragic song, the vocal expression of pain and grief.

The first example is found in the well-known pig-killing scene early in the novel. As Arabella ties the pig down in preparation, its "repeated cries of rage . . . changed. . . . It was . . . now . . . the cry of despair; longdrawn, slow and hopeless." After Jude stabs the pit in its neck, "The dying animal's cry assumed its third and final tone, the shriek of agony" (*JO*, p. 54). And secondly, after Jude's marriage to Arabella failed and he is then discouraged from applying to the university, his mental suffering caused by these two failures is compared to the agony of Virgil's Laocoön when he is squeezed to death by serpents in punishment for violating the sacred wood:

> And twice about his gasping throat they fold.
> The priest thus doubly chok'd, their crests divide,
> And tow'ring o'er his head in triumph ride.
> With both his hands he labors at the knots;
> His holy fillets the blue venom blots;
> His roaring fills the flitting air around.[9]

The parallel passage in *Jude* is:

> If he had been a woman he must have screamed under the nervous tension which he was now undergoing. But that relief being denied to his virility, he clenched his teeth in misery, burying lines about his mouth like those in the Laocoön, and corrugations between his brows. (*JO*, p. 152)

*Jude the Obscure* is a tragedy not because Hardy created it for the purpose of giving "shape and coherence to a series of seemings, or personal impressions . . . not of the first moment." It is a tragedy because he has offered a post-Darwinian cosmological structure in which humanity is a "collective conscious"; he offers an unChristian, unredemptive conclusion to his novel, in fact an ending totally without hope; he borrows the Greek/pagan metaphor of the web or net of entrapment to express his idea of the human situation; and finally, he gives us, through first the pig-scream and second the suffering Laocoön, an imitation only of Jude's agony. For these reasons I must disagree with Virginia Woolf's statement that *Jude the Obscure* "is not tragic," and declare that it is most definitely a tragedy, particular to and coincident with Hardy's own life experience.

*Columbia University*

## NOTES

1 Virginia Woolf, *The Second Common Reader* (New York: Harcourt, Brace & World, 1960), pp. 230–31.

2 Thomas Hardy, 'Preface to the First Edition,' in *Jude the Obscure* (New York: W. W. Norton, 1978), p. 5. The following abbreviations of Hardy's novels and biographical materials are used intertextually, followed by the appropriate pagination where quotations or related passages may be found. References are keyed to the Wessex Edition of the novels: *A Pair of Blue Eyes* (*PBE*); *Far From the Madding Crowd* (*FFMC*); *The Return of the Native* (*RN*); *Two on a Tower* (*TT*); *The Mayor of Casterbridge* (*MC*); *The Woodlanders* (*TW*); *Tess of the d'Urbervilles* (*TD*); *Jude the Obscure* (*JO*); Florence Emily Hardy, *The life of Thomas Hardy*, 1 vol. ed, 1962 (*Life*).

3 Thomas Hardy, 'Postscript,' *Jude the Obscure* (New York: W. W. Norton 1978), p. 6.

4 A. Alvarez, 'The Poetic Power of *Jude the Obscure*,' in *Jude the Obscure*, ed. Norman Page (New York: W. W. Norton, 1978), p. 423.

5 The love triangles in Hardy's novels are chronologically the following: *Desperate Remedies*: Cytherea, Springrove, Manston; *Under the Greenwood Tree*: Fancy, Dick, Maybold; *A Pair-of Blue Eyes*: Knight, Elfride, Smith; *Far From the Madding Crowd*: Oak, Bathsheba, Troy; Troy, Bathsheba, Boldwood; Fanny, Troy, Bathsheba; *The Hand of Ethelberta*: Ladywell, Ethelberta, Christopher; *The Return of the Native*: Wildeve, Eustacia, Clym; Eustacia, Clym, Mrs. Yeobright; Wildeve, Thomasin, Venn; *The Trumpet-Major*: John Loveday, Anne, Bob Loveday; *A Laodicean*: Somerset, Paula Power, Dare; *Two on a Tower*: Swithin, Viviette, Constantine; *The Mayor of Casterbridge*: Henchard, Lucetta, Farfrae; Henchard, Susan, Farfrae; *The Woodlanders*: Giles, Grace, Fitzpiers; Mrs. Charmond, Fitzpiers, Grace; Marty, Giles, Grace; *Tess of the d'Urbervilles*: Angel, Tess, Alex; *The Well-Beloved*: Avice, Pierston, Marcia.

6 Plato, *The Republic*, trans. G. M. A. Grube (Indianapolis: Hackett Publishing Company, 1974), 597, e2–5.

7 Aristotle, *Poetics*, trans. S. H. Butcher, introd. Francis Fergusson (New York: Hill & Wang, 1961), VI, 2.

8 *Concerning Thomas Hardy: A Composite Portrait from Memory*, ed. D. F. Barber (London: Charles Skilton), p. 119.

9 John Dryden, *Virgil's Aeneid*, p. 43.

## BIBLIOGRAPHY

Aristotle. *Poetics*. Translated by S. H. Butcher; introduction by Francis Ferguson. New York: Hill & Wang, 1961.

Else, Gerald F. *The Origin and Early Form of Greek Tragedy*. New York: W. W. Norton, 1965.

Gittings, Robert. *Young Thomas Hardy*. Boston: Little, Brown, 1975.

——. *Thomas Hardy's Later Years*. Boston: Little, Brown, 1978.

Hardy, Florence Emily. *The Life of Thomas Hardy 1840–1928*. Hamden, Conn.: Archon Books, 1970.

Hardy, Thomas. *The Dynasts*. New York: St. Martin's Press, 1965.

——. *Far From the Madding Crowd*. New York: Signet, 1960.

——. *Jude the Obscure*. New York: W. W. Norton, 1978.

——. *The Mayor of Casterbridge*. New York: W. W. Norton, 1977.

——. *A Pair of Blue Eyes*. New York: St. Martin's 1975.

——. *The Return of the Native*. New York: W. W. Norton, 1969.

——. *Tess of the d'Urbervilles*. New York: W. W. Norton, 1979.

——. *Two on a Tower*. New York: St. Martin's 1975.

——. *The Woodlanders*. New York: St. Martin's 1974.

Houghton, Walter E. *The Victorian Frame of Mind 1830–1870*. New Haven: Yale University Press, 1957.

Jaeger, Werner. *Paideia: The Ideals of Greek Culture*. Translated by Gilbert Highet. New York: Oxford University Press, 1976.

Lucas, F. L. *The Decline and Fall of the Romantic Ideal*. London: Cambridge University Press, 1963.

Taylor, Richard H., ed. *The Personal Notebooks of Thomas Hardy*. New York: Columbia University Press, 1979.

Woolf, Virginia. 'The novels of Thomas Hardy.' In *The Second Common Reader*. New York: Harcourt, Brace & World, 1932.

WOLFGANG WITTKOWSKI

# VALUES AND GERMAN TRAGEDY 1770–1840

When I sat down to prepare my paper for this conference where we deal with tragedy, philosophy, and phenomenology, my attention was attracted by a recent article entitled 'Philosophical reflections on the phenomenon of the tragic.'[1] I was disappointed, however, to find that the author, a Heidegger student, represents a position which German literary criticism had disposed of ten years ago, and rightly so. Unfortunately, most of the current views which he ignores do not do justice to the phenomenon of German tragedy either.

Let me briefly summarize the older view which, in fact, represents the predominant interest of West German literary scholars after the two world wars. Typically for the whole movement, Benno von Wiese[2] speaks of tragedy as container of the tragic view. Notice the combination of the two aspects which in American studies are generally treated as separate: the traditional art form, tragedy, and the tragic view of life. Murray Krieger (*The Tragic Vision*)[3] contends that tragedy moderates, domesticates, falsifies, and covers up the existential truth of tragic nothingness. Wiese argues similarly for eighteenth-century tragedy which he calls theodicy. As he sees it, the nihilistic truth of the tragic erupts in the tragedy of the nineteenth century, and it is not until Nietzsche and the existentialism after World Wars I and II that its philosophical expression is found. This view of history as the progressive loss of religious faith turns, as Heidegger's philosophy did before, Hegel's "Weltgeist" doctrine into the negative. Indeed, Hegel understood history as tragedy and tragedy as history, thus combining them into that historical philosophy of the tragic which Peter Szondi claimed as a German specialty.[4] Since that time German literary critics in their analyses of tragedy have been using Hegel's views as well as existentialist theology and philosophy.

When ten years ago, in West Germany Existentialism and its tragic view of life came to an end, the understanding of tragedy also underwent a remarkable change. Following earlier tendencies in the East German GDR, German tragedy now is generally interpreted in two ways: (1) as a mirror and document of the ancien régime and its restauration as well as of the political failure of the bourgeoisie of that time; (2) as a prefiguration and, because of its strong educational influence also as partial cause of the citizen's



*A-T. Tymieniecka (ed.), Analecta Husserliana, Vol. XVIII*, 319–332.

political impotence and suppression up to the Hitler period and even today. This judgmental, hostile attitude toward the past and the German Classical period in particular is fostered by the universities and has taken hold of a whole generation of teachers and students. The trend becomes publicly visible on the stage where it is reflected and reinforced by many well-publicized performances. The adaptations used amount to more or less gross falsifications of the texts and reveal a biased understanding – or better, misunderstanding – of the views contained in the texts. What, then, seems to be in danger of being destroyed today is nothing less than the last great body of tragedies the western world has produced since Shaespeare and the tragédie classique.[5] Interestingly, it was the American germanist and comparatist Benjamin Bennett who recently found it necessary to recall and point to this key position of German drama between 1770 and 1840.[6] In fact, American studies on tragedy and the tragic attest to the strong influence of German philosophy since Hegel; but German tragedy itself was all but ignored.

As I take a new look at German tragedy, trying to do something for its rescue, I feel encouraged by those American studies which treat tragedy as distinct from the tragic world view. At the same time, as I am dealing with the enriching and uplifting encounters with tragedies I will refer to the phenomenology of Nicolai Hartmann. This approach may shed some light on certain striking thematic characteristics of German tragedy. It is based on the understanding that first, tragedies are not only historic documents but historic works of art which, by form as well as content, are not totally bound by their time and for that reason, speak to all times including ours. Counter to Hegelian historicism and relativism, H. J. Muller, among others, stresses this in *The Spirit of Tragedy*[7] and points out that historical and social criticism can do no more than remind us of the fact that human beings under any system will harm each other. Second, tragedy characteristically concentrates on and illuminates central elements and borderline situations of life, such as disproportionately severe punishment; the perdition of what is most worthy of preservation. In actual life, this would embitter and depress us. Presented in the form of tragedy, it does not. We see a severe, higher justice at work when the hero becomes guilty, deceives himself, and eventually invites catastrophe through precisely those steps which he took in order to prevent it and to achieve triumph instead. Such irreconcilable destruction is, to use a formula of Goethe's,[8] simultaneously associated with conciliation, albeit tragic conciliation. The consequences of his deeds, destroying the doer like an ill-fated boomerang, reveal higher justice: his nemesis. Eternal laws and forces emerge beyond man's social and moral systems with their ideologies, institutions, and authorities, and announce their power and superiority[9] even to the greatest and most powerful of heroes. The confrontation of man's highest capabilities displayed in fighting, suffering, and death with the mighty order of things is, paradoxically, what brings about the uplifting effect of tragedy.[10]

Let us take a look at these and other aspects of tragedy on the basis of Nicolai Hartmann's phenomenology.[11] What we will encounter are common anthropological traits (phenomena) which are more readily discernible in tragedy than in any other art form. In tragedy, man is under pressure and therefore behaves just as we would behave under pressure. More than in other situations, we then focus on the future – be it with confidence and courage, be it with worry, fear, and anxiety. In those highly charged situations, the emotionally receptive acts are intensified. They serve as first and most fundamental signals of the responses of others to us and our actions. The characters in tragedies are constantly searching for this information, at the same time seeking to conceal their own reactions from their antagonists. This "game" is inextricably bound up with wanting to realize certain values or not wanting to violate others – just as human existence is a chain of choices between value and value. In real life, these interactions are so small and complex that we hardly recognize them. Tragedy presents them in a form which makes them transparent, simple, large; it reveals them as elementary and yet complicated enough to convince us of their validity in life.

The sequence of expectation, reaction, and decision gives tragedy its end-oriented pathetic structure as described by Emil Staiger.[12] Not until the end does the final judgment of hero and story emerge completely. Until then it announces itself step by step through a chain of tests probing into (1) the state of affairs, (2) the positions assumed by the various parties, (3) the ethical structure and force of the characters. In this process, the initial and most subsequent stages are refuted and corrected. On another, less visible level their genuine substance is being developed. False covers fall, layer upon layer, until only the core is left. Standing out ever more clearly (and most clearly in the face of death) is what passes the test or fails, what endures – or succumbs, what is valid and what is not. It all has to do with values, however, and can only be defined that way – ethical values above all. And this again is particularly true of German tragedy.

German tragedy begins in the late eighteenth century, the period of enlightenment. It is mainly concerned with the ethics of human ideals and moral principles. In the struggle against despotic absolutism and the guardianship imposed by the authority of the Catholic church or the Lutheran orthodoxy, morality was the strongest weapon of the German bourgeoisie. Largely excluded from public life and the spheres of influence, the bourgeois in the land of the reformation and counterreformation was all the more dependent on private and public discussions of religion, ethics and art. It was in this context that he managed to express his criticism not so much of the system than of its practices. These are the circumstances that produced the three thematic spheres which also determine the structure of the historical and the domestic tragedy: ethics, religion, politics. In tragedy, the purpose of suffering and misfortune is moral improvement in political practices, social life and in man himself. God extends his help: sometimes as a model of forgiving love; sometimes, however, God or the gods interfere in human life just as arbitrarily as the absolute rulers. Therefore man in his moral autonomy may rebel against God and thus indirectly against princely despotism. That is the case in Goethe's *Iphigenie* (1787) and in Kleist's "comedy after Molière," *Amphitryon* (1807).[13]

But these are not tragedies or pure tragedies. I will deal with this protest in connection with the tragedies of Büchner and Hebbel who direct it against God and God alone, and no longer through analogy against despotism as well.

By far the most common and important form of divine intervention is that of the nemesis. While, on the one hand, if frightens the spectator, it reassures him and provides ethical encouragement as he sees how, in the struggle between good and evil, good deeds are at least recognized if not exactly supported from "on high" – and the triumph of evil is thwarted at least on the moral and religious plane.

The eighteenth-century writer takes into account the imperfect state of the public's moral strength: man has not yet reached the point where he will do what is right and good for the sake of the right and the good whether success is certain or not. The idea of a supreme moral law being physically enforced offers him metaphysical security, a comforting aid to life. To be sure this nemesis was an illusory as-if, perhaps a lie about life. Many poets, especially the main representative of the nemesis tragedy, Schiller, no longer believed in a divine, just guidance of history. In his tragedies, however, he meant it to be accepted as reality. Since most critics today share his nonbelief, they are inclined to transpose it into the tragedy itself; it has been argued that, in tragedy, the nemesis-belief is only mentioned in order to be refuted and to be demythologized.[14] Von Wiese still considered the metaphysical aspect as central. But even he viewed the divine in tragedy as becoming more and more subjective and illusory, reflecting the decline of religious belief, a tendency that becomes all the more definite the further away tragedy moves, from its beginnings in the eighteenth century, into the nineteenth century. I consider this to be a fundamental mistake. In the encounter between the public and tragedy, it is of decisive importance whether God, his intervention as judge as well as man's rebellion against his apparent cruelty do or do not possess reality in the poetic world.

Similarly misunderstood or underestimated are the two other characteristic themes of German tragedy: ethics and politics. In Germany they are currently seen in a predominantly negative light, an attitude that is related to impulses of ideas emanating from Hegel (as mentioned above). Hegel recognized the subjective conflict involved in choosing and deciding as the characteristic trait of modern man and modern tragedies. He did prefer, however, what he saw as the powerful oneness and completeness without inner conflict displayed by Shakespeare's heroes and especially those of antiquity. He also took the figures of antique tragedy as representatives of historical forces -- or of positions whose contradictions, at some future time in the historical process, would be reconciled to form a synthesis. The anticipation of such reconciliation produces a catharsis in the spectator which literally cleanses, delivers from the tragic effects of compassion and fear.

It was of great consequence that these Hegelian ideas on Greek tragedy were applied to German tragedy. This involved a shift from the viewer's emotional sphere to his cognitive mind – a shift from character to positions, forces, social classes represented, and a shift from the dualistically structured person to one unified and consistent in itself. Indeed, in Hegel's view the hero, as late in Sartre, desires his guilt.[15] Through the historical process, his actions and even his crimes attain justification and theodicy. Any sort of ethical conflict is made to look superfluous, even foolish. And so is, of course, the justice of the nemesis. The place of a timeless moral law is taken over by the inexorable law of history which is the law of the stronger and of success that sanctions any means. To explain German tragedies today thus often means to upgrade criminals and tyrants like Wallenstein, Philipp II, Elizabeth I, and Gessler. Their antagonists, the defenders of what is right and morally good are made to look questionable and ridiculous where they suffer defeat. It is argued that their illusory ethical scruples undermine the ability and competence needed to carry out the ruthless actions that in politics often seems to be the sole way to success.

All this, if I see it correctly, puts the order of values in German tragedy upside down. With the fall of the divine order of the universe, ethical values become relative and success in history becomes the ultimate judge. Along with the conflict between good and evil, between value and value, there also disappears the enlightenment postulate of a political ethics for an ethical politics and for the defense of human rights.

Now, when the values change on the stage, the spectator must also react differently. He must choose sides differently. Basically he now studies history, carrying home information and insights into the historical process. Completely out of place would be reactions such as the fear of the numinous or the metaphysical consolation of a higher form of justice or the encouragement to go to battle with its help against the forces of evil. Choosing the good makes little impression on us now, just as little as suffering and strength in the struggle for the right decision between values. There exists hardly any sympathetic interest in conflicts, tragic conflicts where the choice of one particular value inevitably means injury to another value.

I am sure you sense what is lost here. It has to do with the almost paradoxical tension of tragic effect. Schiller distinguished the aesthetic value of the force and the intensity of the hero's struggle from the many other values for which he may strive. Hartmann speaks of intended values and the values of intention, which include the most genuine motivations. For Schiller, the value of force and intensity, of man's potential to strive for moral freedom is subject to aesthetic judgement. The intended values such as loyalty and power are subject to ethical judgement.

The paradoxical tension I spoke about occurs, for instance, when the spectator sees the immensely appealing force and strength of Wallenstein's personality and, at the same time, is forced to judge the immorality of his intentions. Schiller separates these two kinds of judgement very clearly, for the sake of ethics and for aesthetic reasons. This has always posed difficulties to the recipient. Some have asked, almost in despair, whether Schiller wanted to win the public's approval of the traitor. When current criticism, in fact, takes the criminal's part or that of other dictators, the tragic tension is destroyed; justice is done neither to the aesthetic nor the ethical values.[16]

Similarly affected are (1) the dualism of the physical and the metaphysical world – the latter being of no concern any longer; (2) within the realm of values, the law which Hartmann calls the counter-current of higher values and stronger values. Higher values pertain to the individual and his set of values; stronger values pertain to the collective values of interhuman rules of duty. The tragic hero often becomes the victim of such rules, which is why they are looked upon as higher laws. They are, it is true, more fundamental and more important for the co-existence of human beings and thus are stronger. But they do not reach the value level of the individual and his ethical potential. During the age of German tragedy people became more and more sensitive to this phenomenon. It is not until the last great writer of the period, Hebbel, that we find the full awareness of the fact that the individual is sacrificed or sacrifices himself for something general and common which is of lower value. Resignation and bitterness are being intensified by the pain of knowing that here something is being destroyed which is of much higher value than that which accepts the sacrifice for its own survival. This painful split and tension between the poles in the tragic experience is lost if one, with Hegel, attributes all value to the general and grants the individual only the dignity of being its representative.

The prototype of this attitude had been the Baroque tragedy of martyrdom. Lessing, at the beginning of the period of German tragedy, rejects it because the heroes and heroines long only for entrance into heaven and recognize this as the highest value. Without ethical conflict they rush into death "willfully, unnecessarily, despising" all relatively lower "interpersonal considerations and find torture and death to be no more important than drinking a glass of water."[17] Lessing contrasts such admirable heroism with true humanity, a higher and the highest value. True humanity means a serene sensuality, hence is open to seduction and human weakness. Associated with it are the virtues of sympathy, forgiving, and compassion. Lessing, somewhat willfully, translated Aristotle's "fear, compassion and catharsis" altogether as "Mitleid": meaning compassion for oneself and others but also the education of the spectator toward the cardinal social virtue of compassion. And yet, lessing in his domestic tragedy *Emilia Galotti* (1772) is not entirely liberated from the tradition of the martyr tragedy that he had fought against. His Emilia senses that she is capable of being seduced by the prince who has just had her bridegroom killed. But in order not to succumb to temptation, she demands of her father that he should kill her rather than killing the prince and his helper. "This life," she argues (V, 7), "is really all these wicked people possess." Her moral-religious compassion for such criminal weakness underscores the difference in ethical rank between the characters. In an emotional surge which he immediately regrets, her father does what she demands. This assures him of our sympathy for his weakness and humanity, and makes it possible for us to admire him for carrying out a decision which had matured in him against his resistance. For "her virtue is dearer to him than her life" (so wrote Lessing to Nicolai on January 21, 1758). Lessing regarded this father, who accepted this kind of guilt out of love, as overwhelmingly tragic. He did not allow "the entire state to fall" (cf. above) which had been the outcome in the classical model, the killing of the Roman Virginia. No representation of that kind of action would have been printed or produced on the stage, nor could such drama have happened in real life.

Yet Lessing's contemporaries, including the princes, got the message. A reviewer of the play called out this enlightenment warning to them: "Go and acquire some understanding, you kings! Get yourselves educated, you who are the judges over this earth!"[18] So as not to appear as an agitator of the masses, the reviewer wrote these words in Latin. He, just like Odoardo Galotti and most rebels in the drama of that period, avoid under any circumstances to put himself in the wrong opposite the princely authority. This is not just a plain case of cowardice as it was recently presented on the West German theater but is meant to keep the weapon of ethics sharp. For that purpose it must remain unambigiously clear on which side dwells right and justice and on which side wrong and injustice. Odoardo surrenders to the prince's jurisdiction on earth but also points to the court of divine justice. This prospect and the death of the woman whose body the prince intended to enjoy make it clear to the evildoer that his nemesis has reached him – "the indignation over the scoundrel's good luck," as Lessing translated.[19]

This fundamental religious theme was obscured by von Wiese. He does see the theodicy of the best of all worlds, but only in the consistency of plot and character. Other than that, he sees no bridge between this world and the world of transcendence.[20] The characters know better. They recognize that the many "accidental happenings" are the work of Providence whose hand turns the consequences of evil deeds against the perpetrator. The tragedy does "silhoutte" God's creation, as Lessing demands.[21]

Misunderstood is a further element of the ethical realm in German tragedy: the principle of autonomous self-determination. Emilia wants to preserve it consciously in its purest form. The prince, she cries out, "will tear me away, will take me away: will! will! As if we had no will at all, my father!" And "Violence! violence! Who cannot defy violence? Seduction, that is the true violence" (V, 7). For "the vice of others can, against our will, make us a partner in guilt" by making us "desire to sin" (II, 6). Emilia struggles and dies in protest against the prince's attempt to change the direction of her will. This reproach is irrelevant to her father. He administers to his daughter the death she desires.

*Emilia Galotti* appeared in 1772. At that time, Goethe was working on *Faust* and reworking his *Götz*. This first drama of the German Storm and Stress glorifies the great individual as the crown of God's nature. To assure well-being, is the privilege of the government. The greed of princes and the brutality of the peasants bring to light the good nature in Götz as well as his natural will to fight for survival. Both of them bring about his downfall. God's natural world emerges as the world of Hobbes where *homo homine lupus est*. God's world, therefore, is needed also as the fundamental realm of religious transcendence where the highest values, ethical values of the individual, are unambiguous and undisputed. Gretchen, too, turns to this transcendental sphere at the end of *Faust*, part 1. Her fear of church and society has led her to kill her child, and she is partly guilty in the deaths of her mother and brother. She submits to man's justice which takes the place of divine justice (as it does in the cases of Karl Moor and Maria Stuart). From above, in the name of the One who forgives eternally, we hear the judgement "She is saved." This becomes possible only because she pleads guilty. And this, together with the tragic balance of crime and punishment enables us to endure the ordeal of the half-demented Gretchen in the dungeon. Such balance is missing earlier in the dome, where an evil spirit in a cleric's cowl, the incarnate conscience tutored by the church, tries to persuade the girl that her love for Faust that is the highest value of the individual was sinful and invoked God's revenge. Her unsophisticated mind cannot reject the argument, and as if crushed by its weight, she swoons. But never does she concur with it or take back the affirmation of her pure motives: "Yet all that drove me to it was so good, and was so dear!" Her unperturbable resistence means protest of the autonomous ethics against the social and clerical authorities attempting to dictate, against man's and woman's will, how his her natural sense of values is supposed to function.[22]

The second part of the tragedy closes on a similar note. Angels protect the dead Faust from hell although he had pledged his soul to it and although he died as an evildoer. At the same time, Faust's lack of conscience appears as the reverse of that quality which had distinguished him as superior and gained him the theatrical pardon: that is the translation into action of his unceasing striving. In the sociopolitical realm, this attitude almost inevitably violates the rights of others. According to Goethe, no doer has a conscience, only the contemplator has. Faust's ruthlessness is criticized exemplarily and not trivialized in any way. In real life, we could hardly appreciate him. On the stage and through the intimation of divine justice and love, we can do both, condemn and admire. And we may condemn because we also admire; we may admire because we also condemn. Our capacity to react transgresses our normal limits of consciousness. This parallels the difficulty, even the impossibility we encounter in trying to realise the union of "doing and thinking, thinking and doing" which Goethe calls "the sum of all wisdom."[23]

Inevitably, critics destroy the tragic tension of the polarity at the end of *Faust* whenever they replace religious and ethical judgement with the one-sided justification of evil through the theodicy of history. The problem is the same with Schiller's *Wallenstein* (1799). The hero commits high treason out of his ambition and his excessive desire for power, but also intrigued by a number of pretty good reasons, it seems. Just as in the case of Faust, we are impressed by the breadth and vigor of the man's human qualities, especially his force and superiority with which he turns away all blows of fate and all objections voiced by his conscience. His fall moves us deeply. But it may move us, too, because we perceive it as just. Indeed, we feel that divine justice, the nemesis, makes him to get caught in the net which he had put out for his enemies. He even includes the metaphysical realm in his destiny. Presumptiously he believes that the "Weltgeist", the spirit of the universe and mankind's history, protects him, and that he, the general, by reading the stars is able to tell at what hour he must act. The constellation which he interprets in that sense, however, in fact signals the hour which seals his fate.

There have been attempts to see all of this differently, and especially his faith in the stars has been taken as proof of his higher aspirations. Wiese declares that the theodicy of history, the "historiodicy" vindicates the political evildoer in Hegel's sense.[24] Hegel, indeed, more accurately deplored the fact that theodicy was missing. Fixed as he was upon peace as the most progressive goal of that historical moment, it did not occur to him that the moral order of the world, for the protection of fundamental human rights, might demand the death of the "dictator" who violates the ethical autonomy of thousands by trying to persuade them into committing breach of oath.

*Wallenstein* was written in reaction to the French Revolution of 1789 to 1892. In reaction to the European revolutions of 1830, Goethe's *Faust* concludes with strong religious elements in 1832. Hebbel and Büchner, both born, like Kierkegaard, in 1813, and the older Austrian Grillparzer, all likewise wrote dramas containing strong religious elements. The deterioration of faith did not continue in a straight line during the restauration period. It was a period that experienced a revival of religion and its problematic issues. Grillparzer's historical tragedies subject the evil deeds to divine justice, to nemesis. There is no theodicy of history. Hebbel and Büchner, on the other side, put up with the idea and opposed to it the nemesis. However, by opposing ethics to amoral politics they raised the question whether God was directing history or enforcing ethical laws.

Georg Büchner was the only active revolutionary among German writers. In *Danton's Tod* (1836) he articulates his very personal problem, the political use of force. He contemplates and dramatizes it in the sign of Christ's word, "It must needs be that offence cometh, but woe to that man by whom the offence cometh" (Matthew 18). History demands and justifies the revolution as a trial, a bath of purification, as the bloody beginning of a spiritual and moral renewal, a rebirth. But none of the French revolutionaries in the drama possesses, like Moses or the Messiah, that moral integrity which alone gives a man the right to hand out, like God, sentences of death and suffering to other men. One after the other all of them follow or will follow their victims to death under the guillotine. A spiritual song reveals that God has given to Death full authority to clean up among men whom a Schopenhauer-like will to live drives to tear each other apart because of fear and greed for sex and power.

The spectator is meant to sense that the revolution is pulling him down into its vortex, into the battle of all against all. And in a moment of truth, Robespierre muses: "Truly, the son of man is crucified in all of us: we all struggle in the garden of Gethsemane with bloody sweat, but no one redeems the other through his wounds" (end of act 1). But perhaps, as a character in another of Büchner's plays ponders, we redeem ourselves by suffering.[25] "through pain we enter God"[26] – that is what the twenty-three-year-old Büchner is supposed to have said on his death bed. That, Büchner's figures feel, is an offence, an offence committed by God who created man's nature so that is has to cause offence and thereby is delivered into God's judgement. And yet, man must accept that lot and take up his cross. If I see it correctly, Büchner incites the spectators to turn their backs on God and the religious problem, thus giving exactly the offence which evokes God's "Woe to man!" Like Kierkegaard, Büchner demands that man should do the impossible, that he should submit history and ethics to God's incomprehensible and incredibly hard will.

At first glance, Hebbel may present a similar position. *Expressis verbis* he "casts the dialectics into God himself," thus attributing the dualism of the universe to the creator himself. In his first tragedy, *Judith* (1838), the heroine offers herself to God and indeed also to Holofernes as a sexual sacrifice so as to be able to slay the enemy of God's people. After the deed, things look different. Murder is murder. Besides, Judith feels she ought to bear children to this man although he raped her and thus justified the murder. If now she would bear a son he would have to take revenge for his father's murder. To prevent that she would have herself killed beforehand. For she no longer believes that her crime ceases being a crime and might even be glorified since it was demanded and caused by God as a necessary step in the history of salvation. Judith rejects this generally accepted theodicy as immoral. She insists on her guilt, but not because she might desire being guilty as Hegel and existentialism might argue, but rather because no political-religious ideology could shake her fundamental sense of value.

Judith will not waver in her reverent responsibility toward a morally perceived "eternal order of nature which God himself is not allowed to disturb without punishment."[27] In the tragedy, God, the ruler of history, does things which he, since he also enforces the law, has to punish. The dualism of the two positions becomes more poignant by basing them on the one God. Sooner or later, however, the idea of God must dissolve behind those positions. Indeed, this happens in Hebbel's personal belief, though not in his tragedies. His protest against God as well as Büchner's which is so different mark the non plus ultra, the climax in the Enlightenment's conflict between religion and ethics. This point went unnoticed by most of their contemporaries and incidentally by most critics of our own time.[28]

One might ponder the question whether German tragedy reached its end because it was inspired by problems whose discussion and tragic dramatization could not help but become more and more reflective and difficult to understand. On the other hand, the German dramatists credited neither the masses nor the experts with the ability to really understand them. Proof and example of this could be the chorus in Schiller's *Braut von Messina*, 1803. As observer of the action, the chorus has a better over-view and knows more then the individual characters. It sees the approach of the catastrophe sooner, and the contrast between his perspective and that of the overconfident main characters strengthen the fear and pity in us. As a group taking part in the action, the members of the chorus show themselves, now as able to see, now as blind. They rebel against the princes and yet subject themselves to them. They want to avoid civil war and yet participate in it. Just like the main characters, they have at their disposal the wide and beautiful words of ethical humanity but act in the interest and within the limits of their unconscious egotistical drives. They are, however, moved by compassion in the face of the ruling family's incredible misfortune and suffering; and the chorus feels that way beyond the boundaries of social class and party. But the treasure of wisdom – especially the famous final saying "the worst of evils is . . . guilt" proves to be non-committal. It has no influence on the actions of the main characters nor on the spectators on the stage and probably also those in front of the stage.

German writers have had that experience again and again. So did the Austrian, Grillparzer. He experienced great embarrassment with the staging of one of his works in 1837. From then on until his death he would not share with the people of Vienna his most profound tragedies. The last work he published was a modest tale (*Der arme Spielmann*) in 1847.

In it we meet a highly educated narrator who is also the witness of what he narrates. He is not impressed by what is supposed to impress us. Unchanged himself, he sits down and commits to paper what he has experienced. We read it and read it again. And so it is when we read tragedies. We are hardly going to change under their influence. But we do expose ourselves again and again to their influence and their effect. That may not be much, but it is something and presumably not the worst part of us.

*State University of New York*
*at Albany*

NOTES

1 Max Müller, 'Philosophische Reflexionen über das Phänomen des Tragischen,' *Literaturwissenschaftliches Jahrbuch der Görres-Gesellschaft* 19 (1978): 3–26.

2 Benno von Wiese, *Die deutsche Tragödie von Lessing bis Hebbel*, 5th ed. (Hamburg: Hoffman and Campe, 1951).

3 Murray Krieger, *The Tragic Vision* (New York: Holt, Reinhart, and Winston, 1960).

4 Peter Szondi, *Versuch über das Tragische* (Frankfurt: Insel, 1961), p. 7.

5 Cf. Wolfgang Wittkowski, 'Zerstört das deutsche Regie-Theater die deutsche Literatur?' in *Festschrift Walter Hinck*, ed. H. D. Irmscher and W. Keller (Cologne: Böhlau, 1981).

6 Benjamin Bennett, *Modern Drama and German Classicism. Renaissance from Lessing to Brecht* (Ithaca: Cornell University Press, 1979).

7 H. J. Muller, *The Spirit of Tragedy* (New York: Knopf, 1956).

8 Johann Wolfgang Goethe, 'Nachlese zur Aristoteles' Poetik' (1827).

9 Cf. Wolfgang Wittkowski, ' "Der Übel grösstes aber ist die Schuld". Nemesis und politische Ethik in Schillers Dramen,' in *Friedrich Schiller. Ein Symposium*, ed. W. Wittkowski (Tübingen: Niemeyer, 1982).

10 Kurt von Fritz, *Antike und moderne Tragödie* (Berlin: De Gruyter, 1962); cf. Dorothea Krook, *Elements of Tragedy* (New Haven: Yale University Press, 1969).

11 Cf. Especially Nicolai Hartmann, *Ethik*, 5th ed. (Berlin: De Gruyter, 1949); *Das Problem des geistigen Seins*, 2nd ed. (Berlin: De Gruyter, 1949); *Zur Grundlegung der Ontologie*, 3rd ed. (Berlin: De Gruyter, 1949).

12 Emil Staiger, *Grundbegriffe der Poetik* (Zurich: Artemis, 1946).

13 Cf. Wolfdietrich Rasch, *Goethes 'Iphigenie auf Tauris' als Drama der Autonomie* (Munich: Beck, 1979); Wolfgang Wittkowski, 'Tradition der Antike als Tradition der Moderne: "Iphigenie" und "Die Braut von Messina," ' in *Tradition der Moderne: Festschrift Ulrich Fülleborn*, ed. Gerd Hemmerich (Munich: Fink, 1981); Wolfgang Wittkowski, *Heinrich von Kleists 'Amphitryon'* (Berlin: De Gruyter, 1978).

14 Cf. especially Walter Müller-Seidel, 'Episches im Theater der deutschen Klassik: Eine Betrachtung zu Schillers "Wallenstein." ' *Jahrbuch der deutschen Schillergesellschaft* 20 (1976): 338–86.

15 Walter Kaufmann, *Tragedy and Philosophy* (Garden City, N.Y.: Doubleday, 1968); Cf. especially Müller-Seidel (n. 14).

16 Cf. Wolfgang Wittkowski, 'Theodizee oder Nemesistragödie? Schillers "Wallenstein" zwischen Hegel und politischer Ethik,' *Jahrbuch des Freien Deutschen Hochstifts*, 1980, pp. 177–237.

17 Gotthold Ephraim Lessing, *Hamburgische Dramaturgie: Erstes Stück* (1767).

18 *Berlinische privilegirte Zeitung vom 28. März 1772*, in *Erläuterungen und Dokumente* (Stuttgart: Reclam, n.d.), p. 60.

19 Gotthold Ephraim Lessing, *Hamburgische Dramaturgie. 79. Stück* (1768).

20 N. 2, p. 39, 49.

21 N. 19.

22 Cf. Wittkowski, 'Irrestorable Destruction and Tragic Reconciliation in Goethe's *Faust,*' in *Proceedings of the Goethe Sesquicentennial at Davis, 1982* (Columbia/South Carolina: Camden House, 1984.)

23 Johann Wolfgang Goethe, *Wilhelm Meisters Wanderjahre*, bk. 2, chap. 10; cf. also *Maximen und Reflexionen*, ed. G. Müller, no. 115: "Der Handelnde ist immer gewissenlos; es hat niemand Gewissen als der Betrachtende" (Wittkowski, ' "Gedenke zu leben!" Schuld und Sorge in Goethes "Faust," ' in *Publications of the English Goethe Society* 38 [1968] : 114–45).

24 Benno von Wiese, *Friedrich Schiller*, 3rd ed. (Stuttgart: Metzler, 1963), p. 373; cf. John Neubauer, 'The Idea of History in Schiller's "Wallenstein," ' *Neophilologus* 56 (1972); 451–63.

25 Lena in *Leonce und Lena* (1836), 1:4.

26 Caroline Schulz' Tagebuchaufzeichnungen über Büchners letzte Tage (1837), in all collections of Büchner's works. Cf. Wittkowski, 'Georg Büchners Ärgernis,' *Jahrbuch der deutschen Schillergesellschaft* 17 (1973); 362–83; *Georg Büchner* (Heidelberg: Winter, 1978).

27 Dairy no. 1011. Cf. Wittkowski, 'Das Tragische in Hebbels "Judith," ' *Hebbel Jahrbuch* , 1956, pp. 7–27; *Der junge Hebbel: Zur Entstehung und zum Wesen der Tragödie Hebbels* (1955) (Berlin: De Gruyter, 1968).

28 Wolfgang Wittkowski, 'Die Aufspaltung Gottes oder das Ende der deutschen Tragödie bei Hebbel und Büchner,' in *Sprachkunst* **13** (1982), 231–243.

A. BEN CHEHIDA

# LA DESTINÉE DE LA TRAGÉDIE DANS LA CULTURE ISLAMIQUE

## POSITION DU PROBLÈME ET QUESTIONS PRÉLIMINAIRES

(a) La tragédie a-t-elle été un genre essentiel ou *accidentel* dans la culture islamique?

(b) Comment la culture arabo musulmane a-t-elle "négligé" la tragédie et par quel processus la tragédie a-t-elle fini par s'installer dans cette culture?

## LES ANTÉCÉDENTS DRAMATIQUES DANS LA CULTURE ARABOMUSULMANE

Les trois genres fondamentaux dans le patrimoine littéraire appartiennent à l'époque pré-islamique et s'expriment généralement sous la forme de:

(1) "*madh*" ou panégyrique, ce genre a pour but de mettre en évidence la bravoure, l'esprit de rescousse et la générosité de la personne louée.

(2) Le "*hija*" est une sorte de satire ou de pamphlet qui fustige l'avarice, la batardise et la couardise.

Le poète pré-islamique vise habituellement à travers la personne choisie pour cible, la tribu adverse toute entière. La fierté d'appartenance à sa "structure d'accueil" ou "kabila,"[1] est fortement soulignée. De même le devoir ancestral d'accueillir et d'héberger tout homme perdu dans le désert. Le plus grand déshonneur serait en effet de faillir à la loi sacrée de l'hospitalité en éteignant le feu dans les campements, signe de vie et guide des égarés.

(3) Le "*riths*" ou apologie consiste à chanter les vertus du défunt. La peine ressentie par le poète n'est pas seulement vécue à titre individuel, comme en temoignent les poèmes d'El-khansa qui a fixé le souvenir de la disparition cruelle de son frère Sakhr, valeureux et noble guerrier.

La fonction du "madh" et du "ritha" tend à consolider "l'esprit de corps" (el-açabiya) par l'accumulation des valeurs exemplaires dans la mémoire collective renforcée par l'organisation des joutes poétiques lors de la clôture de la foire annuelle de Okadh à Mecca.

Ainsi, compte tenu de la nature des trois genres littéraires et de leurs fonctions respectives, il y a lieu de pousser l'investigation au-delà du constat et du comparatisme litteral afin de dégager les raisons intrinsèques qui ont



*A-T. Tymieniecka (ed.), Analecta Husserliana, Vol. XVIII*, 333–340.

amené la culture islamique à ignorer le concept de la tragédie grecque, car, affirmer que cela est dû "historiquement" à une erreur de traduction[2] (laquelle erreur serait à l'origine de ce "rendez-vous" manqué entre la culture hellénistique et la culture arabo-musulmane), cela revient à ignorer le substrat culturel et les prédispositions de l'organisme récepteur. Affirmer également, "l'incapacité originelle" de la culture sémitique d'accueillir en son sein ce mode d'expression théâtral, cela suppose l'affirmation *à priori* d'un modèle de référence (le modèle grec posé, en toute évidence, comme unique mode d'expression).[3] Cette généralisation abusive tourne le dos à d'autres modes d'expression développés non seulement en Orient mais aussi en Afrique et en Asie et qu'on continue d'étudier malheureusement avec la curiosité des entomologues ou à travers les concepts douteux de "mentalité primitive" ou "d'Orientalisme." Ne faut-il pas, au-delà de ces explications géo-ethnocentriques, rappeler une vérité reconnue de tous qui considère la tragédie comme contemporaine du conflict et qui affirme tout simplement que là où il y a polarisation et coupure il y a nécessairement drame, qu'il y ait d'ailleurs théâtralisation instituée et manifeste ou dramatisation intégrée et dissoute dans le quotidien. En effet la re-présentation dramatique peut être "jouée" selon les règles codifiées, ou bien vécue et exprimée au fur et à mesure de son émergeance dans la vie de chaque jour. Les acteurs sociaux, sont alors réellement des acteurs sans masques ni paradoxes, ils improvisent *naturellement* un spectacle sans spectateurs dans lequel des êtres déterminés ne jouent pas "à être," car ils "sont".

Ce mode d'expression dramatique s'est manifesté à l'époque pré-islamique sous deux aspects:

(a) sous l'aspect du "majnoun," le fou ou "possédé par les djins" est en fait l'amoureux éperdu, considéré comme "différent," autre que le groupe d'appartenance, il n'est ni frappé d'ostracisme ni celui qui a perdu *la raison*.

Par l'association mnémonique du nom des amoureux célèbres comme Kaïs et Leïla, Kothaïr-Azza, la littérature arabo-musulmane a consacré la tragique destinée d'individus ayant choisi de s'exclure déliberément de l'activité socio-politique du "ghenos" en rompant le lien avec la tribu pour devenir "fou." A côté des "majnoun" nous voyons apparaître les poètes "thri" représentant l'amour impossible et fondamentalement chaste.[4]

(b) Et c'est avec les premiers poètes dissidents nommés les "saalik", que s'annonce un courant d'indépendance et d'égalité. Bandits aux grands coeurs, les "saalik" ont osé défier la sacro-sainte règle d'appartenance à la tribu. Ils s'attaquaient aux caravanes et répartissaient le butin sur les pauvres et les mal nantis. La tragique marginalisation des "poètes brigands" est compensée par la recherche désespérée d'un ideal de justice et de liberté.

Avec l'avènement de l'Islam, la solidarité de groupe cède la place à une solidarité plus vaste regroupant tous les membres de la nouvelle communauté ou "oumma." D'autre part, le message Coranique se caractérise par l'absence d'une tragédie originelle du pêché et ignore le dualisme générateur de conflits. Seule la transcendance et l'unicité divines sont nettement affirmées. La "mimesis" devient alors inopérante et cède la place à l'"identification" dans le discours métaphorique et mystique.

De même, la conciliation ou la concordance (tawfik) entre Dieu et la créature exclut toute forme de relation conflictuelle de type "prométhéen" par exemple.

A l'échelle individuelle, la neutralité de l'âme humaine (nafs) – tantôt attirée par le mal tantôt attirée par le bien[5] – implique donc la clarté du choix et pose le problème du libre arbitre. A ce sujet il y a lieu de rappeler la doctrine communément admise par les théologiens musulmans et les gens de la Tradition: l'âme humaine est une et indivisible et d'elle émanent les facultés dont le nombre est en fonction des actions de chacune d'elles et toutes sont rattachées au "coeur." En effet la thèse coranique soutient l'unité de l'âme tout en accordant au "coeur"[6] une fonction centrale dans la perception. Il serait à la fois le coordonnateur des perceptions sensibles, siège de la conscience (ou "sens commun" dans le sens Aristotélicien); le "coeur" joue aussi un rôle unificateur dans l'intuition. Organe essentiel il n'en est pas moins périssable alors que l'âme humaine est éternelle. Elle retournera à son Créateur dans sa totalité. C'est ainsi que la théologie musulmane résoud le difficile problème de la résurrection de l'âme et du corps réunis et évite la "dangereuse" dichotomie ré-introduite par Aristote dans son "traité de l'âme" et qui sépare la faculté rationnelle – supposée être éternelle – des autres facultés de l'âme, considérées comme périssables et corruptibles. On comprend mieux pourquoi le conflit psychologique issu du dualisme (âme-corps ou rationnel-irrationnel) n'a pas place[7] dans ce champ d'investigation. Par contre la conception islamique du monde couvre l'univers de rationalité diffuse, affirmant que tous les êtres, y compris les animaux, ont la "ratio" en partage. L'hégémonie mythique du "logos" est réduite ici à sa juste valeur. La dimension humaine apparait dans le soin constant qu'accorde la théologie musulmane à l'équilibre entre le développement physiologique des facultés humaines et le seuil de réalisation. Le "taklif" est en fonction de l'"idrak" et non en fonction du seuil d'aspiration ni du "modèle" ou "idolon." De même le critère du "syllogisme" est, par voie de conséquences, jugé insuffisant pour distinguer l'Hommc dc l'Animal. Lc "Nous" divin étant difficile à assumer par le "zoon politikon," c'est peut-être de ce côté là qu'on doit chercher les différents types de conflits entre ce que dicte la "raison" et ce que suggère le "coeur"?

Le caractère irréductible de la pensée et de la matière fortement soulignée par le platonisme ne trouve la voie de la communication qu'à travers les "modèles" ou Idées auxquelles on tend asymptotiquement au moyen de la "mimesis." La mimesis pêche par excès d'aspiration au Vrai, au Beau et au Bien. C'est la source d'inévitables conflits. Conformément à la conception païenne grecque, l'imago reste la clé de voûte des différents types de relations possibles qu'entretient l'individu avec le Créateur, ses semblables et la "polis." "Dieu a créé l'Homme à son image" disent les Chrétiens, les Musulmans considèrent, quant à eux, que "*Rien* ne peut lui être comparé" c'est pourquoi on rencontre des cas d'espèces qui ont accédé à la dimension tragique en prétendant à l'"Identité" et à la "fusion unitive" avec Dieu. Il s'agit particulièrement des mystiques "sufis"[8] dont les figures les plus représentatives sont celles d'ibn el Faerdh, de Rab$^{c}$a el Adawiya, la martyre de l'amour divin, et d'ibn Sab$^{c}$in de Murcie. Cependant El-Hallaj qui a osé dire "je suis le Vrai," "Dieu se manifeste à travers moi, je me manifeste à travers Lui": il fut (selon la formule christianisante de Louis Massignon) "le premier crucifié de l'Islam."

Il y a lieu d'expliquer comment la "mimesis" conflictuelle présuppose l'"Idée" ou le "modèle" et se résoud nécessairement par la destruction de l'un des deux termes en présence. Sur le plan social, l'enseignement du Coran et de la Tradition préconisent l'esprit de solidarité "le croyant est pour son semblable comme un édifice dont les pierres se soutiennent mutuellement" dit le prophète Mohammed. Cette solidarité sociale est renforcée par le principe du "ijma$^{c}$" ou "consensus" qui a force de règle contraignante. Le "consensus" est issu de la "jamma" (ou assemblée) celle-ci disposée en cercle, délibère et juge les affaires de la communauté. L'importance de la forme circulaire dans la pensée socio-politique islamique est notoire. En effet, la structure idéale pour la communauté "oumma" est la "jamaa" où le centre (le chef) se définit par rapport à l'assemblée qui l'entoure. Cette solidarité organique se manifeste dans la cohésion et l'ordre fonctionnel grâce à un centre qui la définit et la regroupe dans un esprit d'entr'aide et de libre consultation (choura). Le cercle est egalement la forme adoptée dans les universités islamiques (El Azhar – Kairouan – Fès – Cordoue – Baghdad, etc.) comme structure facilitant le plus la communication du savoir. Sans oublier toutefois la table basse ou "meïda" sur laquelle on mange dans les pays arabo-musulmans et autour de laquelle la famille se réuni. Conciliant l'indépendance dans la relation, le plat est déposé au centre de telle sorte qu'il soit à la portée de tous, ce qui suppose également l'estimation de ses propres besoins compte tenu de ceux des autres, sans frustration ni hiérarchie. C'est peut-être la raison pour laquelle les manifestations socio-culturelles se sont exprimées généralement dans le cadre de cette forme "halaka."[9] Qu'on évoque les cercles d'invocation (khikr) des "sufi," ou les incantations des "derviches," ou qu'on entoure "l'aède" (ou El Hakawati) pour l'ecouter, fascinés, relatant d'un souffle épique les exploits et aventures de Antar ou "Saif ibn dhi Yazin" ou "ed. Daher Bibars" ou de "Jaziya el Hilaliya." Cette forme intégre parfaitement l'individu au groupe. Cependant la première tragédie dans l'histoire de l'Islam est comtemporaine de la première grande crise qui a consacré la scission de la communauté musulmane en deux groupes: celui de la sunna et de la chiaa. La procession annuelle de la "achoura" commémore le martyr de Hussein (fils de Ali, le quatrième Khalife). La "taziyé" constitue un drame collectif vécu et théâtralise par les légitimistes qui voient en Hussein le descendant du prophète, le successeur direct de celui-ci, les légalistes par contre, conduits par Moawiya vont non seulement détourner le pouvoir en faveur du clan des "bani oumeyya", mais vont liquider le dernier représentant de la famille des "bani taleb." Moawiya instaure ainsi la dynastie omeyyade, fait prêter le serment d'allégeance" aux princes et notables de chaque contrée en faveur de son fils.

La première manifestation tragique s'explique par l'accumulation des faits suivants:

– anéantissement de la "choura" ou libre consultation recommandée par le prophète avant sa mort;

– dénaturation de l'Islam consultatif par l'instauration d'une dynastie à caractère héréditaire (ommeyyades, Abbassides . . .) le "khalifa" devient "mulk" (monarchie);

– anéantissement des aspirations de la communauté islamique non arabe (les persans) se trouvant brutalement exclus de l'exercice légitime du pouvoir: associé par les liens du mariage à la communauté persanne, Hussein, le descendant du prophète Mohamed, cristallise les aspirations des descendants de Darius. La "chia" (ou partisans de Ali et de son fils Hussein), durant plus de douze siècles n'a jamais accepté d'entériner "ce fait accompli," ce coup d'état à caractère légaliste. Depuis cette "tragédie," elle s'est structurée en ce corps islamique marginal et indépendant du pouvoir central (qu'il se nomme "successeur légitime" ou "khalifa," qu'il se nomme commandeur des croyants, sultan, ou chah).

Outre ce grand shisme, la culture islamique a vu naître et se développer différentes manifestations théâtrales issues d'interpénétrations culturelles multiples, comme le théâtre des ombres très encouragé par "Salah ed. din" (Saladin). Par ailleurs, la seule fonction de la tragédie manifestement chathartique, présentée sous la forme d'"happening," apparaît seulement au Maroc dans la "fête des toulba" (fête des étudiants) à Fès, où un étudiant est élu roi d'un jour. Ce souverain éphémère représente apparemment le rêve secret de la remise en cause et du changement.

Mais les personnages sans visages qui ont accedé à la dimension tragique demeurent des cas d'espèces ou des "précipités" culturels, témoins de l'extension de la diversité et des tensions internes au monde islamique. Parmi les curieux personnages, émerge le cas troublant de Karakouz qui serait, d'après la légende, un roi venu d'Afrique Noire et qui parcourut une partie de l'Orient et du Maghreb (où il revêt en Tunisie les traits de Bousaadiya), exposant la détresse d'un vertueux monarque déchu. Couvert de pacotilles, il est l'objet de curiosité des badauts et des enfants. Couvert de mille facettes miroitantes et d'objets hétéroclites, il danse et chante sur les places publiques. Serait-ce pour détourner les regards de son drame personnel ou pour théâtraliser l'horrible "traversée du désert" d'un homme méritant condamné à l'anonymat et à l'indifférence générale? Karakous et encore vivant même si on a oublié sa légende. Peut-être est-il le symbole de la coupure entre les citoyens et un monarque qui serait réduit à sa dimension humaine, sans fard ni "grimaces."

Quant à la destinée tragique du savant méconnu par ses concitoyens, elle s'incarne dans plusieurs "figures" du patrimoine islamique. Abul Fath d'Alexandrie dans la scène "makamet" par exemple, ou cet érudit se trouve contraint et forcé d'exploiter l'ignorance et la supertition populaire pour assurer sa subsistance.

Le personnage de Joha (ou Jha) n'accède nullement à la méchanceté mais utilise des trésors de subtilité et d'*ironie* pour éveiller la conscience de ses concitoyens. Cependant la dramatique relation entre le savant et le pouvoir est illustrée dans la communauté islamique par le "takfir" ou excommunication du célèbre cadhi-philosophe ibn Rochd (Averroès). Son projet de réforme visant à concilier la rationnalité et la législation islamique (chariaa), avec le pouvoir n'a pas bénéficié du soutien du prince. Il s'est heurté à l'opposition des théologiens de Cordoue qui ont su grâce à l'authorité du théologien-mystique "El Ghazali," réduire ce projet à néant et condamner le cadhi-philosophe à l'exil et au silence.

En résumé, la tragédie est en fait un concept moderne dans la culture islamique. Le terme même pour désigner le théâtre (masrah) signifie lieu de libération des tendances et lieu d'expression de sentiments noués et complexes. Quant au mot "maasat" ou tragédie, il désigne en général une situation ou un cas dans lequel la peine ou la douleur sont intensément vécues.[10]

Les premières représentations théâtrales dans le genre comique ou tragiques sont issues de traductions relativement modernes et datent du début du XXe siècle. La tragédie s'installe actuellement au coeur de la culture arabomusulmane, comme phénomène d'acculturation.

Dans la culture islamique, le drame est contemporain et actuel. Il réside en fait dans la superposition de deux structures antithétiques: une structure étrangère: laïque et politique (héritée du colonialisme européen ou anglo-saxon) et une structure socio-réligieuse traditionnelle et encore vivace.

La déchirure se manifeste par la destruction progressive de la "solidarité organique" et l'affirmation de l'individualisme et du nationalisme post-colonial.[11]

La tragédie culturelle s'accentue par le choix dramatique des moyens de communication, outre le langage colonial, le langage littéraire se juxtapose au dialectal.

Le conflit se situe entre l'authenticité, le changement et le modernisme, tous les deux mal définis. Il s'agit de résoudre en fait l'équation de l'acquisition des sciences et de la technologie et du bon usage des ressources naturelles et culturelles du patrimoine délaissé ou perdu, souvent évoqué dans le rétroviseur de l'Histoire à titre de consolation. En effet après avoir été le lieu de la rationalité, la culture islamique, est elle réduite actuellement à la poésie et à la mystique? La tragédie culturelle contemporaine se propose d'exprimer au grand jour ces tensions internes ignorés ou negligés délibéremment par le réductionnisme et les visions simplificatrices enveloppant le patrimoine islamique sous les concepts de "résignation" de "fatalisme" et de "soumission." La tragédie s'installe donc au sein de cette culture pour essayer de résoudre et de dépasser le dualisme entre le mythe et l'histoire, pour résoudre et dépasser l'opposition entre le droit et le fait. Tout se passe comme si l'implacable ironie qui apparait manifestement dans "le théâtre circulaire" sous le thème de l'"émigration" dans l'oeuvre de Kateb Yassine (*Mohamed, prends ta valise*), dans le drame de la nouvelle "diaspora" et l'exode des cerveaux, et qui domine actuellement la littérature arabo-musulmane, semble exprimer une nouvelle version de la tragédie du "roi Karakous."

*Tunisie/Harvard University*

## NOTES

[1] Ce nom correspond au "ghénos" grec, sa racine "kabila" signifie "accueillir" et sa forme exprime une qualité, la tribu étant ainsi la puissance réceptrice.

[2] En effet le traducteur arabe du traité de la "poétique" d'Aristote a traduit le mot "comédie" par "hija" et le mot "tragédie" par "ritha." Il a refusé apparemment d'effectuer un transfert des genres au profit d'une version adaptée.

[3] Faut-il également conclure à l'"irrationalité" des langues semitiques qui "souffrent" (?) de l'absence d'un connecteur (verbe "être" ou "estin" grec) reliant les deux termes de la proposition? Les palmes de la logique ne seraient décernées alors qu'à la proposition calquée sur le modèle grammatical grec ou indo-européen en général?

[4] Ce genre a été abusivement dénommé "amour courtois" ou "platonique."

[5] La lecture des occurences concernant le statut de l'âme dans le Coran est édifiante: "an-Nafs" y est présentée comme un topos traversé indifféremment par le bien ou par le mal.

[6] Ce concept coranique est surtout utilisé par Ghazali et plusieurs autres théologiens. Il serait utile d'engager un travail complémentaire dans la direction de recherche suivie par Mrs. Barbara R. Woshinsky autour du discours biblique et ses prolongements chez St. Augustin et Pascal. (Clark University, Massachusetts, 01610).

[7] Doit on rappeler qu'en dépit de conditions géo-économiques particulièrement dures, le taux de suicide le plus bas appartient au monde islamique.

[8] Est-ce un hasard ou un air de parenté avec le Christianisme qui a été à l'origine du grand écho qui rencontrent les "sufis" dans la societé occidentale?

[9] Dans la lignée de cette tradition, le cercle ou "halaka" est redécouvert et adopté récemment par quelques troupes de théâtre arabe. (Théâtre de la mer. Algérie).

[10] Ce sentiment semble avoir encore un lien avec le genre littéraire "du Ritha" déjà mentionné.

[11] Le drame de la superposition de l'"aumone obligatoire" (zaqat) avec l'impôt et les taxes laïques, l'institution de l'intérêt bancaire, et de l'assurance-vie, entraînent des conflits et un déséquilibre constants dans le comportement du citoyen musulman.

EUGENE KAELIN

# TOWARD A THEORY OF CONTEMPORARY TRAGEDY

## I

The recent debate among Hellenists concerning the actual intent of Aristotle's text that has come down to us in the Latin denomination *De Poetica* succeeds in throwing some light upon a number of more general controversial questions in aesthetics. Does the Greek title of Aristotle's text call for an examination of the art of making, or only of interpreting a poem? [1] Lexical information supplies the ground for the former, and yields a classification of relevant human activities from the most general – making – to the most specific – literature – via the species of imitations. For this interpretation there is both internal and external evidence. The external evidence is found in Aristotle's theoretical scientific work: definitions give the essence of things and are constructed by classifying the species to be defined – here tragic poetry – into its proximate genus – here imitations, and then indicating the specific differences between the definiendum and the other members of the same genus. And, back once again to the internal evidence of our text, what makes tragic literature or poetry different from other forms of imitation is its means (language, with pleasurable accessories, such as the dance and music of the presentation, along with the melody and diction of the poem itself), its manner (spectacle, appreciated either as represented in the poem or presented in a theater, in a dramatic rather than in a narrative form), and its object (a serious human action, i.e., in a plot containing the change of fortune in a character, along with that character's appropriate thoughts). Later on there will be added a fourth determinant, the poem's purpose – to arouse pity and fear.

Throughout our Western tradition the purported function of arousing pity and fear is to produce catharsis, which has been variously interpreted as a kind of homeopathic purging, of therapeutic purification, through an insight into the causes and effects of these emotions or, most recently, a purely intellectual clarification of the pitiable and fearful events themselves. This theory of the intellectual clarification of the incidents arousing tragic pity and fear has been entered into the debate by my colleague in classics at Florida State University, Professor Leon Golden, whose translation of



*A-T. Tymieniecka (ed.), Analecta Husserliana, Vol. XVIII*, 341–361.

Aristotle's definition of tragedy reads in part: "it is presented in dramatic, not narrative form, and achieves, through the representation of pitiable and fearful incidents, the catharsis of such pitiable and fearful incidents."[2]

Professor Golden's claim is that pity and fear are indeed felt in our reading of the poem or our seeing the play, but that the effect of these expressed feelings is nonetheless pleasant since the reader or spectator feels these basically painful emotions in connection with the direct intuition of their causes: the probable or necessary link between the incidents of the plot that arouses them. And in an earlier section of the *Poetics* Aristotle stipulates that the pleasure one gets from imitations is the intellectual knowledge one attains by experiencing them, even when they are of an evil or a painful object. But there is a second source of the tragic pleasure; it is indicated by Aristotle as being equally natural to men and as being determined by the manner or means of the imitation, indeed those "pleasurable accessories" of language used to represent the pitiable and fearful incidents.

If Professor Golden's newer interpretation of the Aristotelian catharsis is correct, the pleasure of tragedy is still mixed; it is partly determined by our intellectual response in recognizing the unavoidability of pleasures and pains under universal conditions (the intellectualist fallacy?), and partly by the pleasures of the senses occasioned by a reader's or a spectator's perception of the medial values of the linguistic medium exploited by the poet. This doctrine of catharsis therefore poses a new problem in aesthetics, viz. how to explain the summation of these two sources of pleasure into a single moment of expressiveness; for, when we experience the natural effect of tragedy, it is lived as a single, continuous, felt determination of our conscious lives. Any theory of catharsis which places all its stress on the representational value of tragedy will leave this simple aesthetic fact unilluminated. The problem is the same for any explanation given to any other representational art medium, and demands a method of interpretation appropriate to the phenomenon, one which merely describes, rather than "explaining" that phenomenon either by appealing to a set of logical first principles or by otherwise adducing the proper set of purely causal laws. Aristotle's own techniques seem to exhibit facets of both these explanatory techniques; they are partly empirical and partly deductive, partly observational and partly inferential. I shall have to return to this point later.

The ultimate gound for interpreting Aristotle's text as a set of prescriptions for making a poem is to be found in the number of references he makes to the optimal conditions for fulfilling the requirements of the best sort of plot, of the kinds of characters whose changes in fortune are the best possible for arousing pity and fear with the proper cathartic effect, and so on. Rather than being based upon the mimetic or the expressive functions of the art, however, both these considerations are formal. It is ironic therefore that the greater portion of Aristotle's audience throughout the centuries has been composed of readers looking for a better means of interpreting the various kinds of tragedies produced in their own cultures in terms of typical plots, considered as the object of representation. Plots and characters, too, are forms of expression, not objects of representation; as such, they refer to a manner of representing another idea: a metaphysical world, for the plots, and a moral ideal for the characters. As examples, consider the early Attic tragedies, which were the only ones Aristotle himself could know; those of Elizabethan England, especially those of Shakespeare; and the works of the French neo-classicists, to name only the major historical periods where this particular art form was highly developed and greatly appreciated. What expresses the tragic emotion in these different eras is the different Weltanschauung embodied in the plot and characters of the drama. Throughout the ages some of the strands within the poetic situations may remain constant, while others, more definitive of a particular culture, will differentiate one form of tragedy from another. And if this is true, there can be no one essence exhibited in every tragedy, no one set of rules for making or interpreting tragic poems.

A poetics, taken as a theoretical basis for the appreciation of literary art and as a guide to criticism, certainly has its roots in antiquity, both in the uses to which Aristotle's own poetics was eventually put, and in the later popularity of Horace's *Ars Poetica*, but represents a relatively modern phenomenon, following upon the development of aesthetics as a theory of human expressiveness.[3] Fortunately or unfortunately, what ancient and modern readers expected from their "theories of fine art" have been two different things. Besides constructing a definition of tragedy on the basis of his general methodological principles, Aristotle was concerned to show the best ways a would-be poet could construct meaningful poems, to show the derivation of the tragic genre from earlier poetic forms, such as the dithyramb and satyr plays, and to differentiate the genre from its close relatives, the epic and comedy.

It was the work of the latter day Aristotelians that turned this body of theoretical and practical knowledge into a set of hard and fast rules for interpreting the "correctness" of a poet's conception of the dramatic medium. Witness the ways in which Aristotle's rather mild and tentative suggestion of a unity of time and place, dependent upon the necessity of the overall unity of action, became hard and fast rules for French neoclassicism,[4] and how these rules were criticized in the next century. For Lessing, the most outspoken critic of French neoclassicism, it was not a matter of imitating the actions of ancient heroes in the construction of a plot that determined a tragic response, but the purification of the tragic passion as the basic emotions of pity and fear are transformed by the representation of the virtuous actions that occasion admiration in a play's spectators. The pity for a protagonist's plight merely culminated in the spectator's admiration for that character. Rather than the actions of "noble" men or heroes, the accounts of which have been prescinded from history or our mythic tradition – the French tragedians imitated the ancients in this – Lessing called for a national theatre that would reflect persons and their actions which were true to nature and contemporaneous reality, yet determined in their dramatic form by what is actually probable or necessary for a character of the type presented. Although for Lessing a dramatic poem in its action gives us an image of time (as opposed to the image of space proper to painting), his own plots were constructed on the conflict between the moral conscience and the human heart, between the ethos and the eros of the protagonist.

In spite of this difference between the imitative neoclassicism of the French tragedians and the demands of Lessing's bourgeois drama, both Corneille and Lessing maintained that Aristotle had formulated the "rules" that were to be forever valid for the creation and interpretation of tragic poems. Indeed, such has been the common error of interpretation of an otherwise innocuous text; the error has been perhaps most conspicuously committed by Lessing when he stated that *the rules* established by Aristotle's *Poetics* are as binding upon dramatists (and hermeneuts) as the *Elements* of Euclid were binding upon geometers.[5] We now can perceive the absurdity of this classicist criticism, since it has become apparent in the time since Lessing's writing that a different choice of axioms produces a different geometrical system, indeed that Euclid's is only one "kind" of geometry.

It behoves us then to return to Aristotle's text with a new interpretive aim, not to discover the immutable laws of dramatic expression, but to consider his prescriptions as determining a set of possibilities for one kind of tragic expression, precisely the only kind he himself was conversant with, the Attic Greek.

In order to judge the validity of his results, I shall begin my analysis with a consideration of Aristotle's method, and then show how alternative theories of tragedy, in particular a theory valid for our own time, is not only possible, but necessary given the differences of ideology separating us from the ancient tragedians.

II

Given the nature of the Aristotelian analysis, the question of the proper interpretation to be given the *Poetics* – whether it be an explanation of how to make poems or only of how to interpret them – would be an exercise in academic pedantry, were it not for the various uses to which the text has been put throughout the history of our Western culture. In the seventeenth century, for example, Corneille used it as a set of prescriptions for writing "acceptable" tragedies, while in the eighteenth century Lessing used it to criticize the tragedies written by Corneille and others, as well as a set of prescriptions to create the plots of his own dramas. Since there is a difference in the heroic drama of Corneille and the bourgeois drama of Lessing, they cannot both have been right on what is prescribed by Aristotle's "rules," and the original question still remains academic after considering their controversy; indeed if there is an essence of tragedy in the formal Aristotelian sense, according to which an essence is what a definition refers to, then to formulate the definition of tragedy is to do both: first, to prescribe a set of necessary conditions without which nothing could be considered of that essence or kind; and, second, to delimit those characteristics by which one would recognize a given play as an example of the genre in question. But to recognize a play as a token of a type is certainly not to criticize it. The soundness of Aristotle's procedure is to indicate that our interpretations of artworks should be based upon their manner of construction; the unsoundness, the assumption that there is anything like a "true" definition, or to put it in another way, that there is an essence corresponding to the definition we have elected to accept. Need we accept the Aristotelian definition? Let us review the argument.

There are three phases apparent in the Aristotelian treatment of tragedy. First, a set of observations on the genesis of the tragic genre from the earlier dithyrambic and satyric verse forms (while comedy is traced to earlier invective poetry which was always written in iambic meter, as opposed to the satyrical trochaic meter); second, the formulation of the definition itself, and a more detailed explanation of the "parts" of a tragic poem (plot, character, thought, melody, diction, and spectacle); and finally, an explanation of the "rules" for composing epics, a genre comparable to tragedy, the interpretation of which allows Aristotle to compare and contrast the two "serious" genres.

Since the epic is the older form, tragedy may then be interpreted as a derivation from that form via the conventional practices of serious poets writing first in dithyrambs, then in the trochaic tetrameter of the satyr plays, actually religious rites, from which tragedy has been given its name of "goat-song." It comes as no surprise, then, to read in book 4 of the *Poetics* that "It was in fact only after a long series of changes that the movement of Tragedy stopped on its attaining to its natural form,"[6] apparently in the perfections added to the form by Aeschylus (who added a second actor, limited the action of the chorus, and emphasized the dialogue), by Sophocles (who added a third actor and scenery to the representation of the plot), and by Euripides (who produced "the most tragic" kind of plots by exploiting the unhappy ending of the protagonist's situation, which was later to become the hallmark of all truly tragic plots). Yet elsewhere Aristotle wonders aloud whether this attained "natural form" could evolve still further, and finally leaves the question open: "the consideration of whether or not tragedy is by now sufficiently developed in its formal elements, judged both in regard to its essential nature and in regard to its public performances, belongs to another discussion."[7] The contradiction contained in Aristotle's reference to tragedy's having attained to its natural form in the paradigms of the art that have come down to us as the work of the Attic tragedians of the fifth and fourth centuries B.C. (here the question seems closed) and in the other citation, taken from the same chapter, which leaves the question open, seems not to bother contemporary Aristotelians.[8]

What this contradiction does reveal to the disinterested analyst, however, is that Aristotle's own method of analysis is not purely deductive-explanatory, as has been claimed,[9] but empirically descriptive as well. He proceeds from the observation that poems are a species of imitations, and imitations, of the productive arts; and he must observe the differences between poems and other forms of imitation in order to achieve his definition (given the theory of scientific definition explained above). Once that definition has been completed, as he supposes that it is in his descriptions of the "parts" of a tragedy, he would have succeeded in describing the essence of the genre – tragedy as a species of poems.

Closer examination, however, shows that Aristotle's prescriptions derive from his descriptions of the parts of a tragic poem which are not limited to those examples extant even to him; for example, plots may be simple or complex not only by the actual observation of such things, but because of the purely logical reason that if a plot is not simple, then it is complex. In the same vein, characters may be good or evil or only of middling virtue:

noble, base, or merely average. From his observation that tragedians, who are by nature noble and who therefore likewise by nature prefer noble subjects, actually do write about such subjects, and since the purpose of their writing is to produce pity and fear in an audience, it seems natural once again that each noble protagonist must possess some tragic flaw – otherwise his or her tragic suffering would appear unjust to the audience, which would respond in moral outrage rather than in pity and fear.

But what about this pity and fear? Are they a matter of observation? Or are they too deduced from an adequate idea of the moral nature of spectators? That is difficult to determine for reasons to be explained below. Whichever is the case in this particular example, Aristotle's general method of procedure is the judicious appeal to empirical and to rational evidence as the occasion of the moment demands; what guides these appeals is his previously abstracted idea that the tragedies he was familiar with could be defined as being the imitation of an action whose purpose was to produce "the tragic pleasure," i.e., to arouse pity and fear which in turn produces some kind of catharsis in the readers or spectators of the drama.

As I see it, the first principles which allow Aristotle to proceed "according to nature" (kata phusin) are not there for explaining the existence of a natural object, which would be metaphysical, but rather the various characteristics by which the writing of poetry, itself an action, could be defined, by reference to the means, manner and objects of its imitations. By his careful "kata phusin," Aristotle seems merely to be reminding us that our definitions should be couched in terms appropriate to the subject being defined. Since by observation poems are imitations (i.e., examples of productive art), they are a form of productive activities, and these activities may be most properly defined in terms of the objects depicted in these activities, as well as of the manner and the means of their depiction. This requirement is neither metaphysical nor moral, but logical. And it is precisely this logical determination of the treatment he has given poems that forces Aristotle to admit to the open nature of the "essence" of poetry. His definitional format merely sketches out a set of parameters within which to make further observations.

Consider the following epistemological claim: Aristotle claims to know the parameters within which imitative activities may be defined; he can observe how the determinants of these parameters (both of the genus and the differentia) have been developed in the history of the poetic art and how they might have been, owing to further categoreal, a prioristic, determination of each characteristic (e.g., that characters may be better than, worse than, or about equal in virtue to the average spectator; or that changes in fortune may be from the better to the worse or from the worse to the better). In every case, indeed, where Aristotle appeals to purely a priori evidence, to that which I have characterized as a categoreal determination, rather than to the observational evidence from extant tragedies, his conclusions are not necessarily of the essence in the strict sense of that term. For example, he has observed that extant tragedies imitate the actions of noble men, and that these noble men are depicted as suffering; if so, since our response to the spectacle of their suffering must be pity and fear, they must possess some tragic fault (hamartia). Part of this explanation is observational, part deductive. But the categoreal demand which is the result of a logical conclusion as set up by premises that are here moral, rather than metaphysical or purely logical, dictates that our response be pity and fear to one form of plot, outrage to another.

As a result of mixing the kinds of evidence in his analysis of tragedy, Aristotle has perpetrated a logical and vicious circle that may, nonetheless, be avoided in one of two ways, or indeed by both simultaneously. The circle is the rational explanation of the connection between the witnessing of the spectacle of a suffering hero and our natural, moral, response in pity and fear. What is the evidence for this connection? In brief, the observation that plots are the depiction of the sufferings of heroes, or men or women more noble than us. Whence, by our natural moral response to such experience, we should feel pity and fear. *Therefore*, pity and fear constitute the purpose of the poetic representation; and tragic plots must be of the nature described. To avoid this circle and its vice one need only deny one or both, that the tragic plot is necessarily of the sufferings of a subject nobler than the audience, or that the proper response even to such suffering is adequately described as pity and fear. And whether it is or not will depend upon our reflection upon our own consciousness as it attends to the expressive characteristics of tragic plots, however they are constructed.

The history of the theory of tragedy since Aristotle has been constituted either as a repetition of Aristotle's basic observations, or in a modification of his conclusions. Some of these modifications have been the denial of the nobility of the protagonist, which had been decreed as a categorial determination of tragic situations, but which in effect was the contingent result of the classical tragedians' finding their plots in the older epics; or the denial of the pity and fear, which were likewise "deduced" as natural, given the spectators' moral nature, but which were nonetheless a phenomenon merely supposed as a constituent of our aesthetic experience. But if they are, then that phenomenon should be open to an empirical, yea, phenomenological investigation.

In the following section I shall examine some of the alternative explanations of the tragic response; and this examination should permit me to approximate an alternative explanation of tragic plots. I shall then conclude this essay with an exemplification of the alternative definition.

## III

If as Aristotle claims, pity and fear are the emotions expressed by the spectacle of a suffering hero, they present two related problems of interpretation. First, both states may be characterized as negative; i.e., defined primarily in terms of pain, rather than pleasure. Pity is the emotion evoked when a person similar to ourselves faces an unavoidable and imminent pain and we are to some extent identified with that person's fate; normally we pity the suffering of others who are similar to ourselves. As witnesses to the spectacle of that pain and identified with the person suffering it, we may be said to fear the same fate for ourselves. For this reason the pity and the fear are the same emotion, occasioned by the same pain, the anticipation of suffering, either that of another person which becomes our own by virtue of our identification with that person, or of ourselves imagined to be in the same circumstances of the suffering hero. The catharsis mentioned by Aristotle is the second phase of the spectator's experience, that by which the negative effects of the pity and fear are removed – however they are – so that the entire experience is after all pleasant. In Aristotelian terms, therefore, the experience of tragic art contains an apparent paradox: that something in essence painful is experienced as a positive joy. And the doctrine of catharsis is intended as an explanation of this transmutation of pain into pleasure.

Since this transmutation of the affective values of tragedy does not take place, as far as we know, outside the realm of art, it must be the expression of these tragic values in art that accounts for the transformation. In nature we could experience pleasure at the sight of another's pain if we are sadistic; or in our identification with another's pain, if we are masochistic. But both these explanations are in themselves negative; they would not solve the paradox, at least not satisfactorily from the moral point of view. Need we be immoral, neurasthenic, or mentally disturbed to experience the "pleasures" of tragedy?

The history of tragic interpretation is an attempt to answer this question in the negative. And nowhere is the message any clearer than in the aesthetics of German romanticism from Lessing to Schopenhauer, via Kant and Schiller. Following out this history may help us in breaking the Aristotelian circle.

In the correspondence between Lessing, Moses Mendelssohn, and Friedrich Nikolai (1756–57)[10] the question turns upon the significance of the Aristotelian pity and fear. The fear need not detain us, since no one attending a theatrical performance actually loses consciousness of the fact that the performance is only a spectacle, a fictional account and not a real, natural event. The pity, when it is suddenly evoked, however, may be accompained by terror, which stands as its beginning; nor does the pity end in the contemplation of our own or of the hero's pain. If it is evoked by terror, it is ended in admiration. And the whole affective response to the tragic vision is a single continuous moment of our conscious lives: terror (suddenness of the onset of pity), pity (sympathy with the suffering hero), and final admiration for the hero's personality. The admiration felt as the culmination of the sympathy is sufficient to "purge" the original personal feelings of terror and pity.

Lessing later wrote his *Hamburgische Dramaturgie*, 104 pieces of criticism on the theory and practice of tragic composition from the early Greeks to his contemporaneous present (1767–69),[11] to defend the Aristotelian notion of catharsis. He does not adopt the primitive homeopathic interpretation of that notion, nor does he support the heroic interpretation of a protagonist's character. A character is a person considered from the moral point of view, whether the morality of that person is better than, equal to, or worse than the spectator's. What generates our sympathy for the character's plight is the moral worth of his struggle against environing circumstance; and the greater the struggle, the greater is a spectator's vicarious pain and ultimate release into the joy of admiration.

Immanuel Kant (1790) contributed to the discussion the critical analysis of the phenomenon of the sublime,[12] as an object of aesthetic judgment. An object infintely great in magnitude or in force (the mathematical or the dynamic) may arouse in its contemplator a subjective feeling of enhanced pleasure – an actual harmony between the faculties of the imagination and the understanding – so long as the unity of that object continues to be contemplated as surpassed by our subjective powers expended in the contemplation. Here too there is a dual phasing within the structure of the experience: the immediate threat to the understanding of the infintely great or to the will in the face of a real or imagined threat of destruction by an infinitely great power. Friedrich Schiller (1795, 1801)[13] identified the tragic hero as one species of the dynamically sublime object. The greater the external force expended upon the moral integrity of the hero, the greater the spectator's feeling was uplifted in recognition of his plight. The initial response of sympathy for the hero's pains is transmuted into an exhilaration that is felt in reaction to the morally sublime subject's suffering. Schopenhauer (1818)[14] treated the sublime reaction to tragedy as a special case of the dynamical sublime, which always poses a threat to the equanimity of the person contemplating it, yet which illustrates the power of the individual will to contemplate its own destruction through its identification with the hero; as a result the experience of tragedy may be said to begin in sympathy and lead to resignation in the spectator – to accept the pain of becoming individual, i.e., of developing a particular will as opposed to the primal will that is the substance of all real things.

Schopenhauer's case is a paradigm example of catharsis through purification of feelings, of the pain of willing by the understanding of the causes of that pain. A similar case is made by Hegel, the notes for whose lectures on aesthetics were composed between 1820 and 1826.[15] The tragic conflict is not so much between an individual struggling to define his own individuality and a universal force of destiny that works toward his destruction as it is between two individuals each espousing a valid and hence emotionally compelling moral value, neither of which exhausts the substance of our moral lives. Both Antigone and Creon are right from their own points of view; each embodies a rightful claim as against the other, but each of their claims is surpassed by the higher claim of an absolute right, the moral substance that must eventually transcend the differences between individual rights since it gives a foundation to all its constituent rights. Tragic heroes suffer or die, but are vindicated by their suffering or dying in the name of absolute Right.

My final case is supplied by Nietzsche. His first philosophical work, *The Birth of Tragedy*, was first published in 1872.[16] Unlike Aristotle who interpreted the acme of tragedy to be illustrated in the works of Aeschylus, Sophocles and Euripides, Nietzsche envisaged the height of the tragic experience to have been illustrated in the earlier satyr poems composed and danced in honor of Dionysus, the god of fertility, vitality, and drunken exuberation. What Aeschylus and Sophocles had done was to shape this drive to vital effusiveness into clear images of Apollonian dreamlike meditations. Where the essence of the original tragedies was the choric dance, the satyr poems generated a vision of the suffering Dionysus, the only truly tragic hero, in whose honor the vision was celebrated with appropriate action, such as music, dancing, rape, etc., which in turn produced the final mystical union with the god. In brief, tragic art was vitality given an appropriate form, a celebration of life, not a call to its renunciation.

This original religious celebration of the pains and the joys of exuberant vitality entered into decadence, according to Nietzsche, when Euripides, influenced by the essentialism of Socrates, turned the medium into an occasion of thought. The value of the experience of tragedy was not insight into the motives of a suffering heroine, nor a call for resignation in the moral purpose of the universe; it is rather a heightened feeling of our own vitality as in our experience of the sublime. Aristotle's original paradox had arisen owing to a misconception of the expressiveness of tragic poems: they do not express pity and fear felt at the sight of pains that must be avoided, they call us to accept life as it is, with all its pains that can only function as a catalyst to intensify the pleasures or joy of living. And instead of giving us something to think about in the contemplation of the goodness of life, tragic poems give us something to feel – all the more intensely, since this feeling is controlled by the clarity of the Apollonian images. Tragedy was the highest form of art, and evolved from the ancient spirit of music.

## IV

The preceding brief summary of German aesthetic theory on the subject of tragedy gives an important clue for solving the aesthetic paradox of the tragic experience, the first explanation of which is found in Aristotle's *Poetics*. Aristotle's solution to this paradox, how the painful or negative emotions of pity and fear become component parts of the positive experience of aesthetic pleasure, has indeed been debated throughout the history of our Western culture. If there is some kind of catharsis, it has been claimed, the pity and fear have been purged, or purified, or the incidents arousing these emotions have been clarified so that the pleasure of the tragedy is conceived in part in the removal of the negative emotions. Arousing a negative feeling for the purpose of purging it sounds much like giving oneself pleasure by ceasing to hit oneself with a hammer, and for this reason the purification and clarification theses have tended to supersede the former. But they are didactic; both maintain that the pleasure of tragedy stems from our understanding the natural, metaphysical or moral reasons for human suffering. If this were indeed an explanation of tragic pleasure, then the same pleasure could be experienced by our reading the proper psychological or philosophical account of these reasons, and there would be nothing specifically valuable about the poems or plays themselves.

Aristotle, of course, was aware of this problem, as he referred to the pleasures attendant upon our perception of the musical values of the verse as partially determining the tragic response. But he had no explanation of how these positive values in the first term of an expressive relation could function in transevaluating the negative values in the second. "The first term" of an expressive relation refers to the expressing vehicle – here, the surface values of the verses, while "the second term" refers to ideas expressed in the verse. And the reason for this deficiency is that Aristotle had no idea that an art object of any kind at all could be anything but an imitation of something else. An aesthetic theory couched in terms of the activity of the creative imagination that makes its objects for the values found to inhere within their structures by an appreciative consciousness had to wait upon the discovery of this phenomenon. Needed to make it clear were (1) the conception of an art object as a structual whole – so that our "pleasures" in experiencing such objects are univocally defined by the qualities of the whole and not of their discrete parts; and (2) a doctrine of aesthetic judgment that is reflective, as Kant had insisted, rather than "logical" or determinant. Thus if we can be trained to reflect upon how our reactions to aesthetic objects are controlled by our perception of those objects' internal structures, we should be able to describe those structures in such a way as to communicate the reasons for aesthetic responses, including the vaunted "tragic pleasure." And here we can be sure that recognizing a work as a tragedy becauses it possesses the necessary and sufficient conditions for being considered such has nothing to do with the aesthetic value of that work: recognition is a cognitive, rather than an aesthetic response.

In Lessing's aesthetics we find a first attempt at a phenomenological description of the definitive tragic emotion. He examines the conditions for experiencing pity and fear, and finds that the sudden onset of sympathy for a suffering hero occasions a terror that works itself out as sympathy and ends in admiration. This admiration is a positive feeling, occasioned by our perception of the positive human values exhibited by the suffering hero. But this hero need not be, by definition, more noble than us: only human, and necessarily at conflict with the external moral demands placed upon his or her life. The greater the struggle, the more powerfully felt the defeat, the greater is our admiration for the human qualities of the hero. This admiration becomes in the theories of Kant, Schopenhauer, and Schiller an introduction of the theme of the sublime.

Reduced to its simplest expression, the tragic emotion takes place in two phases, the "tragic qualm" as we experience the threat to the moral integrity of the hero and the "tragic lift"[18] as we feel ourselves raised in spirit as yet being superior to the forces that would destroy the hero. That is the experience of the sublime, and Kant expresses that experience as an active judgment of the spectator bringing into harmonious relationship the perceived threat of destruction (the object of our intuition) and the indeterminate concept of the understanding, the free moral determination of the human will. In our perception of the moral sublimity of a suffering human subject, our reaction is not one of fear, nor one of sympathy by which we vicariously assume the pains of another's suffering, but of a feeling of elation as we respond to the threat to the hero's moral integrity from the safe vantage point of our own sense of freedom. And our ability to maintain our equilibrium while contemplating the destruction of a moral value is proof enough of the "causality of freedom" in operation.

Hegel is concerned less with the tragic emotion than with the intrigue, complication and denouement of the tragic plot. Aristotle, he seems to intimate, was incorrect in separating plot from character: there is no action which is not human, and no human who does not act. A human being is defined by his actions as these are directed to the attainment of some good or the fulfillment of some right. But the pursuit of rights or goods may be one sided, and come into conflict with pursuits of others possessing an equal right. When they do there is tragic conflict, which is reduced when we perceive that each of these subjective actions are necessary for the completion of an objective ethical substance, the system of the morally right. Here we lose the felt expressiveness of the conflict, which is once more considered the phase of negativity in the tragic response, in favor of the understanding – the positive phase – of the higher claims of the absolute system of Right. So we recognize once again the turn in the determinant phases of the tragic response; but, as for Aristotle, the ultimate positive response is determined as a release from the original negativity. For this reason, Hegel does not offer a viable alternative to Aristotle.

Nor does Schopenhauer. He was capable of perceiving the function of the dynamically sublime in our conception of tragic conflict, but he chose to interpret it in terms of his own pessimistic voluntarism. Even though a spectator shows the power of the will to contemplate its own destruction, and thereby proves its own moral superiority to the forces that would destroy it, tragic plots lead us to understand the truth of the futility of human existence as an individual desire that is divorced from the primal universal Will that gives substance to all natural events, and so teaches "resignation" to the conditions of such a life.

Both Hegel and Schopenhauer give an explanation of our responses to tragedy that presupposes the truth of their respective systems of moral thought, and to this extent are defective as a universal explanation of the phenomenon under discussion. Only Nietzsche of the aestheticians surveyed in the previous section of this article meets the conditions laid down above. The Apollonian and the Dionysian elements of tragedy give us a means for exhibiting the wholistic structure of tragic expressions and for reflecting upon the manner in which our sense of vitality is enhanced as we read or view the tragic spectacle. Consider.

Whatever the source of the tragic conflict – the free, but tragically flawed individual declaring himself against Fate; individuals espousing conflicting moral values; or individuals espousing particular desires in opposition to a universal Will; the desires of the heart in conflict with a personal duty – our experience of such spectacles is the joyous affirmation of the conditions of life. This reaction is not an intellectual or cognitive judgment, although cognitive judgments are necessary for our understanding of the tragic conflicts. Concepts are a means of aesthetic expression, not its end. Our reactions to the spectacle of tragedy are the feelings of the tensions developed between the artistic means used to picture the tragic conflict – the expressing words, scenes, gestures, and the like – and the associated depth meanings of these surface phenomena – character, plot, thought, and the like, which may themselves in turn express still deeper meanings: the ontology of beings who must express themselves in action, and, in doing so, reveal the existence of a moral ideal, or the metaphysical nature of the "world" represented by the actions of some character.

Of all the most recent aestheticians that may come to mind, Roman Ingarden comes the closest to describing the internal structures of the literary work of art. They are in descending orders of depth: the phonological stratum of the words as pronounced or sung; the meanings intended as semantical units; the world of represented objects defined by the interrelationships between the semantical units; and the system of schematized aspects brought to an understanding of that world by the reader's concretizing imagery. Each of these strata is unfolding simultaneously with the others, as if the whole constituted the several voices of a contrapuntally developed harmony, each contributing its own set of aesthetically valent properties to the quality of the whole. Ingarden's image of the "polyphonic harmony" of the four determinant strata of the literary object's constitution is, of course, only a metaphor, but it is an effective one; and in two senses: first, as a means to show how "the negative values of the second term" are indeed transmutable by the positive values of the first; and second, how Nietzsche's observation that tragedy had grown, historically, out of the spirit of music is profoundly true.

Lastly, if tragic literature is composed of a polyphonic harmony of its constitutive strata, the structures of that harmony indicate how an analysis of literature, including tragedies, should be made. We may begin by reading through the strata until closure is perceived to occur: that is the pre-aesthetic formation of the literary object that comes to exist in an act of aesthetic contemplation. Once they are constituted in this way, we may then perceive the structures of the object as controlling one's aesthetic enjoyment. And, in a moment of reflection upon the values of the tensions set up and resolved within the total construct, we can then, in an attitude of post-aesthetic analysis, show how our aesthetic response to the tragedy has been controlled. Nietzsche's notion of one's aesthetic response as an enhanced vitality was conceived as a means for expressing the tensions we feel as the aesthetic object of tragic literature comes to meaningful closure in our experience of the funding strata of the literary object.

## V

As an example of the process described above I shall choose *Waiting for Godot* by Samuel Beckett, the premier tragedian of our own age. What qualifies him for that designation is the proper worldview, that any living is a process of dying – a thought that is profoundly true if only we remember that life too has its beginning, middle and end, and that the end of life is quite naturally death; and in addition thereto, the ability to express this "tragic" or pessimistic worldview in an appropriate artistic medium. Of Beckett's plays, I select *Godot* because of the way in which it does *not* fit the Aristotelian description of the essence of a tragedy, but which nonetheless produces the tragic response: the enhancement of the viewer's sense of vitality as controlled by the rhythms of the tragic revelation.

First, concerning the differences between Aristotle's theory and the qualitative experience of this play, the moral worth of the protagonists, Didi and Gogo, is evidently not determined by their being better than that of the audience; nor, since they are symbols for the human condition, are they presented as being inferior in worth to the members of any audience. They are indeed only humans already thrown into a world not of their making and in which they must achieve whatever significance there might be in accepting the challenge to live. On the surface they might even appear as "bums," subhuman, or moral derelicts, because their suffering has rendered them insensitive to anything else. And indeed they are subhuman, since each serves as a symbol for only one half of a human subject: Didi, who retains some semblance of an intellect and therefore a knowledge of causes, stands for light, the higher functions of consciousness, indeed, alone, for a disembodied spirit; Gogo, on the other hand, stinking of the feet and immersed into the feelings of his flesh, stands for the corporeal half of the traditionally defined human animal. Neither, of course, can live alone, since there is no living body without its soul, and no soul capable of expressing its feelings not embodied in some flesh. The predicament of one human being as necessarily related to another in a common action or enterprise presents a world where neither can be separated from the other, except in sleep (as occurs in the play), when they are related so closely as body and soul of a single individual.

If we continue to contemplate this relationship as Beckett presents it to us, we find all the conditions necessary for the existence of an unhappy consciousness. First, to become fully self-conscious, each of these characters – who are for us only stage presences – must find its own existence defined by the relationship it has to the other; each may in turn play the role of lord to the other's slave. The lord becomes aware of himself as lord in the subservience of the slave; but the slave himself becomes lord of the lord owing to the latter's need for a slave. As a result both will express their mutal relationship with continual bickering, which is related to us in a comic, almost senseless banter. Indeed, from one point of view the dialogue is senseless banter since both of their claims to dominance within the relationship is without ground; but from the other point of view, the banter of the conversation between the two is one of those "probable or necessary" consequences of each character's conception. Although the dialogue on the surface seems absurd, that absurdity marks the highest degree of the play's verisimiltude. Let's look once again. In the conversation between the two in which Gogo seems to be comparing his own fate with that of Christ (suffering for the sins of others), Didi objects as if to blasphemy. And Gogo retorts, not in the admission of his blasphemy, but in recognizing the difference between his own condition and the life of Christ: Christ lived in another country, and there they crucified quickly.

Gogo understands his own suffering, of course, because he feels it; that is one way, and the only authentic way to understand a feeling. What he laments is the quick death of the Savior, which, if the biblical account is correct, transpired in only three hours, from noon to three o'clock on the original "Good Friday." Gogo and Didi will not die quickly, but will go on suffering forever – as long as there shall be readers of this play. Gogo did not take himself to be the Savior; he envied Him. Jesus, of course, died on the cross; what Gogo sensed is that he too was on the cross, in the chiasmatic relationship he enjoyed with (i.e., suffered in realtionship to) his alter Ego. Since each is self-conscious only by virtue of a relationship to the other, the existence of each traces a cross with respect to that of the other. No one, even Christ himself, ever dies quickly enough to avoid the pain of dying, because the pain of dying is life itself. And where Christ's death agony lasted for only three hours, the agony of the ordinary individual must last for all his life. Gogo's musing on the suffering of the Savior is not blasphemy, but an insight into the nature and universality of human suffering.

And if there is any doubt about this matter, Lucky and Pozzo show up as mirror images of Didi and Gogo working out their destinies as master and slave. Pozzo who portrays the role of the lord in the master-slave relation is crazy, as his very name indicates. He is crazy since he thinks he is the master of his slave, whereas in reality he is dependent upon his slave's services. In this couple, Lucky the slave who thinks when he is wearing his hat is the counterpart of Didi, rather than Gogo, who appears rather slavish in his relationship to Didi. The reason is that appearances are deceiving. As David Hume has told us, reason is and by right ought to be the slave of the passions; and this is the relationship between Lucky and Pozzo. But this too is as deceiving as the relationship between Didi and Gogo; the truth of the matter is that each in turn serves as master and slave to the other. That is the condition of their crucifixion, a condition natural to all men who must live out their lives waiting for the end. And the chiasmatic tracing out of the analysis between master and slave of each reflecting couple merely repeats the image of a cross.

If the protagonists of this tragedy are not heroes, their action is a non-action. Didi and Gogo are merely waiting to die, as they surely would should Godot ever make his appearance; for any human action which achieves a prior projected goal comes to a close in the achievement of that goal. Rather than no action at all, then, the waiting of Didi and Gogo represents the condition of all human actions – the fulfillment of a temporal span of human existence. Since their being is expressed in their caring for one another, and the meaning of that being is the temporal spanning of that care, they must continue to suffer the interminable period of their waiting. Waiting, in this tragedy, then becomes the symbol for any human action at all, the most universal expression of how a human being lives its life in separation in time from its own completion; but when that course of life will have been completed, then the human being will only have died for its caring to live. According to the Heideggerian analysis of such things, waiting is an inauthentic manner of acting, acting in a deficient mode of self-determination.

This same tragic vision of life is expressed in Beckett's short televised interlude called *Breath*. Here the vehicle used to express the tragic vision of life is not composed of "words and music," but only sight and sound. In parallel sequences the televiewer perceives two sequences of intensifying and slackening moods: the visual from the darkest blackness to total brightness (from unconsciousness to consciousness) via an intermediate gray, and back again, in the systolic and diastolic movement that accompanies life, awareness, and breathing. Interspersed with this symbol of consciousness the breathing becomes audible: a baby's piercing cry, followed by a silence which changes character again as the spectators hear the beginning of a sigh drawn out into a ten second diastolic expansion, and upon reaching its highest point, descending, after a five second pause (in which the light is at its brightest, revealing a scene of consummate desolation) in a slow systolic descent into total expiration. Then the baby's cry again. The audible and visual images run in counterpoint, but in opposite "directions," tracing once again the pattern of a cross, symbol of human suffering and salvation. The whole spectacle of this "drama" lasts less than a full minute's time; and no word is spoken, although the tragic essence of the life of human consciousness is given the most tense artistic expression I know.

That tension is felt in the viewer as the visual and audile imagery conspire to evoke both the impression of the whole of a human life and an understanding of its significance. And the compression of the symbolic tension felt in a viewing of *Breath* has its counterpoint in the more conventional drama of *Waiting for Godot*. In the stage play, the tension is drawn out as the spectators are made aware image by image of the passage of time, the meaning of our human being. And the sense of loss in the arrested motion of time is only a means for the spectator's feeling the sublimity of those two "bums," each in fact only a stage presence representing characters, i.e., the moral determinations of human beings all of whom exist on a cross in their relations to others. Time separates each of the characters from himself, as the world or his relationship to the world separates himself from the other. The knowledge of this dual separation is of the essence of the unhappy consciousness. In the meantime, each lives out these conditions of separation in the consciousness of the suffering, peculiar either to a body or a mind, or to both in their most intimate relationship.

Both the play and the teleskit illustrate Ingarden's account of the structures of a literary object whose expressiveness is controlled by the tension between its constitutive strata funding into a tense moment of consciousness. And if Aristotle had understood this, or had developed a method for analyzing the phenomena of consciousness, he would not have left aesthetic theory with a paradoxical account of the values of tragedy. He had the same data as Nietzsche, all the more readily available to him, but he had not learned to make a phenomenological interpretation. His intuition told him that the intensification of human feeling that accompanies the contemplation of a negative value is a matter of aesthetic fact; but he did not realize that our awareness of this fact is the feeling itself, and needs no naturalistic explanation.

What, then, is one to think of those Aristotelians who continue to look for the proper interpretation of the tragic catharsis? If, as I have tried to show, that the response to tragedy has nothing to do with pity and fear, surely it is false to claim that these emotions are purged or purified in our experience of tragedies, or even that the incidents evoking them have only been "clarified." Apollo without Dionysus may give us art, but not tragedy.

*Florida State University*

## NOTES

[1] See Aristotle, *Poetics*, trans. Leon Golden, commentary by O. B. Hardison, Jr. (Englewood Cliffs, N.J.: Prentice-Hall, 1968), pp. 63–66; hereafter cited as Aristotle.
[2] Aristotle, p. 11.
[3] Aristotle, p. 65.
[4] See Pierre Corneille, 'Trois discours sur la tragédie,' in *Oeuvres Complètes* (Paris: Firmin Didot Frères, 1855), pp. 547–86.
[5] Gotthold Ephraim Lessing, *Werke*, Vol. 4, *Hamburgische Dramaturgie*, in *Dramaturgische Schriften*, (Munich: Carl Hanser Verlag, 1973), pp. 699–700.
[6] *Introduction to Aristotle*, ed. Richard McKeon (New York: Modern Library, 1947), p. 629.
[7] Aristotle, p. 8.
[8] Ibid., pp. 99–100.
[9] Ibid., pp. 67–68.
[10] Lessing, pp. 155–227.
[11] Ibid., pp. 230–707.
[12] Kant, *Critique of Judgement*, trans. J. H. Bernard (New York: Hafner, 1951), pp. 82 ff.
[13] Friedrich von Schiller, *Aesthetical and Philosophical Essays*, in *Complete Works*, ed. Nathan H. Dole (Boston: Aldine Publishing Co., 1910); *Letters on the Aesthetic Education of Men*, (1795); *On the Sublime* (1801).
[14] Arthur Schopenhauer, *World as Will and Idea*, trans. Haldane and Kemp (London: Routledge & Kegan Paul, 1883), bk 3, sec. 37. Original edition, 1818.

[15] Georg W. F. Hegel, *Aesthetik*, vol. 26 (Frankfurt: Europäische Verlagsantalt, n.d.); on the essence of tragedy, 546–51. Original edition, 1838.
[16] Friedrich Nietzsche, *The Birth of Tragedy*, trans. Clifton Fadiman, in *The Philosophy of Nietzsche* (New York: Random House, n.d.). On the "Socratism" of Euripides, secs. 11–14.
[17] For George Santayana's naturalistic account of the transmutation of "negative values in the second term" see his *The Sense of Beauty* (New York: Modern Library, 1955), pp. 191–240.
[18] I owe the terms "tragic qualm" and "tragic lift" to Prosser Hall Frye; see his *Romance and Tragedy* (Boston: Marshall Jones, 1922), pp. 146–50.

# B. SOME RECURRING MOMENTS OF TRAGIC SIGNIFICANCE IN THE DRAMA

BEVERLY KENNEDY

# THE RE-EMERGENCE OF TRAGEDY IN LATE MEDIEVAL ENGLAND: SIR THOMAS MALORY'S *MORTE DARTHUR*

Tragedy is not a universal phenomenon. Rather it has been specific to certain places at certain times in the development of western civilization. I should like to begin this essay by suggesting an hypothesis to explain that cultural specificity: tragedy occurs only in cultures which enjoy the expectation of social progress. In other words, men can perceive tragedy only when they have hope.[1]

In the western world there have been two kinds of hope available to man: (1) the mystic's hope for a better spiritual life for the individual soul, both in this life and after death – what may be called vertical transcendence; and (2) the humanist's hope for a better life, both spiritual and material, for the community of men in future time – what may be called horizontal transcendence.[2] The mystic's desire for vertical transcendence leads to neglect and contempt of the world, except insofar as the world offers the means necessary to achieve a purely spiritual liberation from matter and time. On the other hand, the humanist's desire for horizontal transcendence, transcendence *through* time and matter, leads to care and concern for the world. Men who hope for horizontal transcendence assume that they can, through appropriate action in the here and now, make this world a better place for future generations. Their hope implies that mankind has the capacity to vanquish evil, or, at least to avoid man-made catastrophes. It also ensures that so long as men fail to do this, their failure will be perceived as tragic.

The humanist hope for horizontal transcendence first enters Western consciousness in the culture of the Greek *polis* during the fifth century B.C. The foundation for this new hope for social progress is revealed in Aeschylus' dramatic trilogy, the *Oresteia*. The first two plays, *Agamemnon* and the *Coephoroe*, present us with a causal sequence of bloody and awful acts which threaten to destroy an entire community; the final play, the *Eumenides*, celebrates the institution of a judicial system which promises to put an end to the destructive savagery of reciprocal violence. Aeschylus, the first of the three great Athenian dramatists, is incorporating myth as well as history into his plays when he attributes this tremendous social and cultural advance to divine intervention. Indeed, all of the Athenian dramatists



*A-T. Tymieniecka (ed.), Analecta Husserliana, Vol. XVIII,* 363–378.

represent the gods as a force to be reckoned with; however, their dramatic versions of mythic history suggest in various ways that man is not helpless before the gods. Even when the gods choose not to be benevolent, Aeschylus' *Prometheus Bound* suggests that they may be tricked or forced into parting with their divine gifts – for a price. In *Oedipus the King* Sophocles leaves room for the interpretation that Oedipus might have averted his fate had he been less proud and rash. And Euripides carries this tendency much farther, suggesting that men may even be equal to the gods, certainly equal in responsibility for human misfortunes. Thus the tragic dramatists bridge the gap in Athenian Greek culture between the Homeric epics and the first literary utopia to be created in the western world. What makes their dramas tragic rather than simply historical, is the implicit belief that mankind can do better than this. The particular men and women whose histories are chronicled on stage might not have been able to do any better, given their unique character and the circumstances in which they found themselves; nevertheless, those who watch, the chorus and audience who express pity and experience fear on their behalf, can learn from their mistakes. The hope that underlies the great tragedies of classical Athens finds its fullest expression in Plato's *Republic*. Only a century later, this young man could dispense with the gods altogether and imagine that men might actually create an ideal society by their own efforts, through education and the practice of virtue. A similar literary expression of humanist idealism was not to occur again in the western world for almost two thousand years.

The Christian culture of the early Middle Ages had no idea of social progress, no hope of a better life in this world, and, as a consequence, no tragedy. Early medieval Christian theology did not allow for the possibility that man might improve the world. As a consequence of accepting Jesus of Nazareth as the Messiah, the Christian religion had translated the ancient Judaic hope of a Messianic kingdom in this world into a hope of paradise in the next. Medieval Jews might still take consolation for their sufferings in the knowledge that the Messiah would surely come to manifest God's perfect justice on earth. Medieval Christians could take consolation for their sufferings only in the hope of heaven. Following Boethius and St. Augustine, the true beliving Christian held this wretched and mutable world in contempt and regarded his life here as an arduous pilgrimage to the Eternal Jerusalem.

With the rediscovery of Greek literature and philosophy in the fourteenth and fifteenth centuries, tragedy once again became possible in the western world, provided that European humanists could find a way to assimilate the values of classical humanism into their Christian world view. It was not easy.

However, in the English literary tradition, with which I am most familiar, it is possible to distinguish two modes of assimilation which, I shall argue, did produce genuine tragedy before the end of the Middle Ages. The first mode is ironic and best illustrated in the work of Geoffrey Chaucer. As an ironist, Chaucer did not even attempt to resolve the paradox created by the juxtaposition of the values of secular humanism and Christian theology in his society and in his art. He called *Troilus and Criseyde* a "litel . . . tragedye" (5:1786), but his ironic mode of presenting this tale of Troilus' "double sorwe" has meant that critics are still debating what he meant by "tragedye."[3] The central movement of the narrative reflects the motion of Fortune's wheel as Troilus progresses from painful desire up to blissful consummation and then down again to overwhleming grief and despair after he has lost his love. However, the final event in Chaucer's narrative is Troilus' posthumous journey up to the eighth sphere, from which his spirit looks down upon this "litel spot of erthe" and laughs. The narrator interprets his laughter in light of traditional Christian wisdom and counsels his readers to eschew this world and its "feynede loves" for love of God who "nyl falsen no wight." The power of Chaucer's love story is such that most readers find it difficult to accept his concluding remarks. On the other hand, they have found it easy to identify with the lovers, given Chaucer's essentially dramatic mode of presentation;[4] easy to experience vicariously the extasy of ideal love and the tormenting awareness that a world at war does not allow such love to last. No matter what hope of vertical transcendence Chaucer may hold out to his readers at the end, they still cannot entirely avoid the feeling of regret for what might have been, had the lovers encountered each other in different circumstances. That feeling of regret, I would maintain, is fundamental to tragedy and rests ultimately upon a foundation of hope. Men do not regret the impossible. To sum up, then, insofar as Chaucer has convinced his readers that ideal love between a man and a woman is possible in this world but that it is vulnerable to the violence and treachery in human nature and society, he has offered his readers a genuine experience of tragedy: the fate of Troilus and Criseyde could not have been averted, and that is tragic; but it need not always be so.

The standard medieval definition of tragedy reflected the orthodox Christian view of the world as wretched and contemptible. That Chaucer was aware of this orthodox view is clear from the fact that he attributes it to the Monk in his *Canterbury Tales*. The Monk is, in fact, a comic illustration of Chaucer's ironic mode of juxtaposing worldly and otherworldly values in his work. In theory, the Monk has chosen to renounce this wretched world to serve God in the cloister, and the seventeen tragic tales which he tells about noble men who have fallen from "greet prosperitee" into "myserie" (7:1973–79) appear to confirm the wisdom of his choice, suggesting as they do that catastrophe is the way of the world (symbolized by Fortune's wheel) and that there is "no remedie" but renunciation. In practice, however, we know that the Monk does not confine himself to the cloister, nor does he deprive himself of any of the pleasures of the world. On the contrary, he enjoys fine clothes, fine food and fine horses, and through a series of double-entendres, Chaucer manages to suggest that he enjoys fine women, as well.

The negative response of the other pilgrims to the Monk's performance suggests that they are no more willing than he is to renounce the world wholeheartedly. The normally courteous Knight stops the Monk after his seventeenth tale, and expresses the opinion that "litel hevynesse / Is right eynogh." It is much more pleasant to hear stories of men who "clymbeth up and wexeth fortunat / And there abideth in prosperitee" (7:2767–78). By his vocation, the Knight represents the conservative medieval view of society as a stable hierarchy, the vast majority of men (the third estate) being held in check by a very small military governing elite (the second estate). Yet here he seems to be speaking for the vast majority of the pilgrims, members of the third estate who happen to be prosperous and upwardly mobile bourgeois. Their own experience has taught them that the world can be a very fine place, indeed, and may be better still. The two most sanguine and prosperous pilgrims, the Wife of Bath and the Franklin, both choose to tell tales which illustrate how the world might be improved through the exercise of human virtue. In fact, the *Franklin's Tale*, normally categorized as a romance (like *Troilus and Criseyde*), could equally well be viewed as a domestic tragedy whose predictable outcome is prevented by the idealistic and virtuous behaviour of gentle men.

The second mode of assimilating the values of classical humanism into the late medieval Christian world view blended the theoretical pessimism of the Monk with the theoretical optimism of the Franklin to create a moral view of history. It is best illustrated in the work of John Lydgate. Lydgate goes even farther than Giovanni Bocaccio in moralizing *de casibus* narratives. In *The Fall of Princes* he loses few opportunities to attribute the misfortunes of history to the sinfulness of men. Thus he manages to justify the humanist's hope by means of the Christian's faith. For, if God punishes sin with misfortune, then it is only logical to believe that he must reward virtue with good fortune. Such a train of thought was not without its difficulties, however, since it presumed the possibility of perfect justice in the world. Men who attempted to integrate humanist and Christian values in this way found themselves obliged to explain why God did not punish all sinful men, or, conversely, why He did not reward all virtuous men. The Hebrews had attempted an answer in *The Book of Job*. The pagan philosopher, Boethius, had also attempted an answer in his *Consolation of Philosophy*. Both books were immensely popular in the later Middle Ages.[5] However, late medieval Christians had one concept to bring to bear upon this question which both the ancient Hebrews and the pagan philosopher had lacked: the concept of grace.

The orthodox teaching of the medieval Church tread a fine line between Augustinian determinism and Pelagian heresy with regard to the relationship between divine grace and human will. In asserting that neither the grace of God nor the will of man was sufficient to ensure man's salvation, the Church adopted what amounted to a semi-Pelagian compromise position.[6] The will of man was free and could resist even the grace of God. Therefore, God through His grace could no more save a man against his will than a man could save himself through his own effort without the grace of God. If that teaching is then combined with the deuteronomic theology implicit in late medieval *de casibus* narratives, we have a cultural climate of thought not unlike that illustrated in the *Eumenides* of Aeschylus. There is hope that man might create the just society in cooperation with God. If all men would only seek the grace of God and exercise their free will to promote virtue, then there would be no need for a just God to punish mankind with misfortune. If the Black Death convinced men in the fourteenth century that God did indeed punish sin with dreadful calamity, then the economic progress of the fifteenth century, coupled with dramatic achievements in technology, government and the arts, must have convinced them that God also rewarded virtue with prosperity. Hope that Christian men might achieve peace and justice for all in Christendom lay behind the best work of the northern humanists, and certainly inspired the second literary utopia to be written in the western world, Sir Thomas More's *Utopia*. Hope of this kind also made genuine tragedy possible once again.

The dramatic works of Shakespeare are the high point of the renaissance of tragedy in the western world, in the English language, at least. However, I should like to devote the rest of this essay to a late fifteenth-century work which I believe was crucial to the development of tragedy in English literature between Chaucer and Shakespeare.[7] Written in 1469 and first printed in 1485, *Le Morte Darthur* balances precariously on that conventional chronological divide between medieval Christian and Renaissance humanist England. It is usually classified as a medieval work and something of an anachronism; however, if one looks at it as tragedy, then it is possible to perceive its affinity with Renaissance humanist literature. Malory's work as a whole is not cast in dramatic form. Rather it seems to be a compendium of epic, chronicle and romance. However, the final two tales, and especially the last tale in which Malory recounts the end of Arthur's life and reign, are really a series of dramatic scenes stitched together by the thinnest of authorial commentary. Malory's history is much too compendious for the stage, of course, but I would argue that it anticipates not only Shakespeare's great history plays, but also his tragedies.

Malory's work has not one, but three tragic heroes: King Arthur; his favorite nephew, Sir Gawain; and his best knight, Sir Lancelot. Each of them typifies a different attitude toward the relative importance of worldly and other-worldly concerns, and therefore a different perception of tragedy. All three attitudes were possible in fifteenth-century England, and their differences can best be illustrated by Malory's handling of the appearance of the Holy Grail at Arthur's court at the feast of Pentecost. Arthur's attitude is the most worldly. Arthur is a cultured man, and a Christian, but he is also a head of State. He thanks God for the appearance of the Grail, but he is clearly both sorrowful and angry that so many of his knights have sworn to leave him in order to go in quest of the sacred chalice. He berates his nephew for setting the others an example:

> "Alas!" seyde kynge Arthure unto sir Gawayne, "ye have nygh slayne me for the avow that ye have made, for thorow you ye have berauffte me the fayryst and the trewyst of knyghthode that ever was sene togydir in ony realme of the worlde. For whan they departe frome hense I am sure they all shall never mete more togydir in thys worlde, for they shall dye many in the queste."[8]

Arthur's primary concern in life is not the salvation of his soul in heaven, but the maintenance of his royal power on earth. He is a rationalist, a humanist and an empiricist who perceives his primary task as a ruler to be the maintenance of peace and stability in his realm. To this task religion appears to him to be irrelevant except when it threatens to deprive him of the knights he needs in order to govern effectively. Arthur is a literary reflection of late medieval kings like Edward IV and Henry VIII. He is an honourable man, a victorious general, and a successful ruler who would never allow Christian belief to interfere with necessities of state. It simply would not occur to him, anymore than it would occur to Machiavelli's prince.

Arthur's favorite nephew, Sir Gawain, responds to the Grail positively and impetuously, but without any real understanding of the spiritual perfection which is the true goal of the quest for the holy chalice. Gawain is a simple soldier and a loyal kinsman, grateful to God for his blessings, but having no idea at all that the honour of his clan and the honour of God are not one and the same, or that his relentless and cruel pursuit of the blood feud is un-Christian.[9] Nor do his experiences in the Quest enlighten him; they merely bore him. Whereas Arthur represents the education and cultural refinement of court life in the later Middle Ages, Gawain represents the less polished ways of the early medieval rural aristocracy. Gawain is just as worldly as Arthur, but his primary concern is the welfare of his family rather than the welfare of the state. To both men, then, horizontal transcendence whether in the form of perpetuation of the blood line or perpetuation of State power, is more important than the vertical transcendence represented by their Christian belief.

Malory's third tragic hero, Lancelot, Arthur's best knight, represents a stage in the evolution of European cultural consciousness mid-way between the primitive clansman and the princely head of state. Malory praises him as the model of Christian knighthood. Unlike Gawain, Lancelot understood fully what the quest for the Holy Grail implied, and persevered unto the end; however, he did not choose to renounce the world at that time in order to seek the higher perfection attainable in the monastic life, as Galahad and Percival did. Rather, he chose to return to the world with his gift of greater grace in order to continue to serve his temporal lord. Lancelot is the literary reflection of late medieval Christian humanists like Anthony Woodville, Lord Scales, and, later, Sir Thomas More. Such men believed that it was possible to serve God by ministering to the social well-being of their fellow Christians in the world. What is more, they believed that it was selfish of the individual to abandon the care of those entrusted to him in this world, in order to seek greater spiritual perfection for himself alone.[10] Lancelot's primary service to his overlord, King Arthur, is military and judicial – to protect the citizens of the realm both from foreign attack and from domestic crime and rebellion. Because he is a French knight, who has willingly entrusted his own kingdom of Benwick to the governance of a regent in order to serve Arthur (who, as soon as he is crowned Holy Roman Emperor becomes his natural overlord), Lancelot can symbolize the Christian humanist hope for a truly just society of all men in a united Christendom. In the form of this hope, Lancelot, too, seeks, horizontal transcendence. His vision of the future is much broader than that of Gawain or Arthur, going beyond both clan and nation-state to encompass all Christian men. And it would be incorrect to say that this hope of horizontal transcendence is more important to him than the vertical transcendence represented by his Christian belief. Rather it would be correct to say that the former is truly a function of the latter. Lancelot is the only one of the three tragic heroes who ever acts as though his Christian belief were relevant to his worldly vocation and who justifies this behaviour on the grounds that God will punish sinful men with misfortune (*Works*, 270–71, 1183–84). In other words, Lancelot is the only one of Malory's heroes who perceives that God is immanent in his world and that therefore both his hope for horizontal transcendence, and his hope for vertical transcendence are in accordance with the divine will.

Malory is never a very intrusive narrator, but in the highly dramatic scenes of the final tragic tale, he disappears as effectively as any dramatist, allowing his characters to speak wholly for themselves. As each hero faces death he reveals the values which have governed his life and offers the reader a different way of understanding human tragedy.

Throughout his life Gawain has foregone vengeance only once, and then only in order to save the life and crown of his nearest and dearest kinsman, Arthur.[11] On his deathbed, however, Gawain realizes that by pursuing the blood feud against Lancelot in France, he has not only given Mordred the opportunity to seize Arthur's throne but also deprived his uncle of the one military force which might have crushed the usurper, since one hundred knights of the Round Table chose to follow Lancelot into exile. Gawain confesses his fault to his uncle in an emotionally charged scene, during which he also writes a letter to Lancelot begging him to return to England to rescue Arthur. Then, with his final breath, he charges Arthur to cherish Lancelot above all other knights.

We have here all the ingredients necessary to a moralized *de casibus* tragedy such as Lydgate might have included in his *Fall of Princes*; however, Gawain himself is not capable of drawing the moralist conclusion, that God has punished him for his "wylfulnesse" and "pryde" (*Works*, 1230). The narrator tells us that Gawain takes care to satisfy the ritual requirements for the salvation of his soul, but these nominal concessions to Christian belief – confession and extreme unction – do not provide the dramatic focus of this scene. We do not hear Gawain confess to a priest and seek forgiveness from God; rather we hear him confess to Arthur and seek forgiveness from Lancelot. Even as he lies dying Gawain's primary concern is the future well-being of his kin, and his final act – the letter to Lancelot – is an attempt to repair the damage he has done. Gawain has only two close male kinsmen left alive, his half-brother Mordred and his uncle King Arthur, and the one has become a traitor to the other. To a man like Gawain, whose primary value has always been loyalty to his kin, no prospect could be more painful than civil war between kinsmen, no crime more heinous than the revolt of Mordred against Arthur. The Christian reader may moralize Gawain's tragedy in *de casibus* fashion if he chooses – Malory has provided all the necessary elements for such an interpretation – however, Malory also makes it clear that this is not Gawain's own understanding of his tragic fall. Gawain perceives only that by acting to preserve, he has in fact lost everything he held most dear.

After his death Gawain is still trying to save his kinsman as the messenger of "Allmyghty Jesu." Thanks to "Hys speciall grace" Gawain is able to appear to Arthur in a dream to warn him of his fate if he should do battle with Mordred and to convince him to arrange a truce so that Lancelot will have sufficient time to come "with all hys noble knyghtes, and rescow you worshypfully, and sle sir Mordred and all that ever wyll holde wyth hym" (*Works*, 1234). Arthur believes his nephew and tries to negotiate a truce but his efforts are undone by the chance appearance of an "addir" whose sudden sting causes one soldier to raise his sword and thus, unwittingly, give the signal for attack to both armies. At the end of the day's slaughter, only Arthur, two of his knights, and the traitor Mordred are left alive. In Malory's version, one of Arthur' knights reminds him of his dream at this point in an effort to persuade him to leave Mordred alone; however, Arthur is so enraged by the sight of his fallen knights that he cannot forbear taking vengeance into his own hands, whatever the price he must pay.

Here we have all the ingredients necessary to a humanist tragedy such as Shakespeare might have written, even including the ghostly messenger. Of course, Arthur does not take nearly so long to believe his ghost as Hamlet to believe his, but that fact merely points to a fundamental difference in the character of the two princes. To borrow Macchiavelli's terminology, Hamlet is a prince who tends by nature to be circumspect, whereas Arthur is a prince who tends by nature to be impetuous. Princes prosper so long as circumstances require the sort of behaviour dictated by their character; but "when there is a clash they fail."[12] Arthur's rash and impetuous behaviour has got him into trouble before, as, for example, when he ordered the slaughter of all children born on the first day of May in order to destroy his incestuously begotten son (*Works*, 55). And, one could argue that, had he listened to Gawain's better counsel and refrained from judging the queen to summary execution (*Works*, 1175), thus forcing Lancelot to become his enemy in order to rescue her, he might have prevented the entire catastrophe which ensued. However, as Arthur lies waiting for the barge which will take him to Avalon, he does not reflect upon his past mistakes, nor does he seem to feel any need to confess his past sins. Rather, he confronts failure and death with very much the same attitude as Shakespeare's Hamlet. In fact, in Arthur's case, death is protrayed literally as a journey to an "undiscovered country, from whose bourn / No traveller returns" (*Hamlet*, 3.1.79–80). As in the case of Gawain, the Christian reader may moralize Arthur's fall as a divine punishment for his sins, whether incest or murder; however, this is clearly not the way Arthur himself understands his tragedy. He neither fears damnation nor hopes for heaven, and his parting request for prayers seems almost like an afterthought. What most troubles Arthur at the moment of his departure from this world is that he must say to his one remaining knight of the Round Table, "Comforte thyselff, . . . and do as well as thou mayste, for in me ys no truste for to truste in" (*Works*, 1240). Like Gawain, Arthur has been denied the one form of transcendence which had most meaning for him. He has lost his royal power and he has failed to ensure the succession. His only son is dead, a traitor; and he has no way of knowing whether or not the distant kinsman he formerly designated as his heir, Constantine of Cornwall, will be able to pick up the pieces of this catastrophe.

Lancelot returns to England with an army, too late to save his lord, but he accepts that loss with proper Christian humility, as one who knows that "ayenste deth may no man rebell" (*Works*, 1251). Then he goes in quest of his lady, only to discover that she has taken the veil and therefore is not "dysposed" to go back with him to France as his queen. This loss he accepts with much more difficulty in yet another emotionally charged scene in which he explains to Guinevere that he, too, would have forsaken the world for the cloister long ago during the quest for the Holy Grail, "had nat youre love bene" (*Works*, 1253). This reading, taken from the Winchester manuscript, makes it appear that Lancelot refused the religious life earlier only on account of his love for the queen. However, the Caxton reading, "had nat youre *lord* bene," reminds us that Lancelot has always been as devoted to Arthur as he was to the queen. Indeed, his willingness to come now to the rescue of the man who so recently and cruelly made war upon him in his own lands is remarkable proof of his devotion. It is this single-minded determination to serve Arthur, his overlord as Holy Roman Emperor, which makes the French knight and king such an appropriate embodiment of the Christian humanist hope for the future of mankind. That his values reflect the late medieval ideal of knighthood as ordained by God to ensure the well-being of men in society, shows to what extent the values of Christian Humanism were anticipated earlier in the medieval period.[13] Thomas More was a knight as well as a saint. In order to make Lancelot a worthy representative of the Christian humanist ideals which we now associate with Thomas More, Malory had to alter his character a good deal, particularly in his relationship with the queen. His revelation in this scene that Lancelot would have married Guinevere, had she been so "dysposed," epitomizes his view of him as her "trew" lover.[14] Lancelot's commitment to her, like his commitment to Arthur, represents the Christian humanist's commitment to this world and his hope for horizontal transcendence.

Lancelot seems to forsake this hope along with the commitment to the world when he decides to embrace the religious life, becoming first a monk and then later a priest. However, when he is called upon to celebrate the requiem mass for his lady and then to bury her by the side of her husband at Glastonbury, we are allowed to perceive that he has never really given up hope for the world. As we might expect, the sight of his lady alone does not move him so much as the sight of the tomb which encloses her and Arthur. The Glastonbury hermit chides him, arguing that his "sorow-makyng" is displeasing to God, but Lancelot is not dismayed. His reply pierces to the heart of the Christian-humanist paradox:

> "Truly," sayd syr Launcelot, "I trust I do not dysplese God, for He knoweth myn entente: for my sorow was not, nor is not, for ony rejoysyng of synne, but my sorow may never have ende. For whan I remembre of hir beaulté and of hir noblesse, that was both wyth hyr kyng and wyth hyr, so whan I sawe his corps and hir corps so lye togyders, truly myn herte wold not serve to susteyne my careful body. Also whan I remembre me how by my defaute and myn orgule and my pryde that they were bothe layed ful lowe, that were pereles that ever was lyvyng of Cristen people, wyt you wel," sayd syr Launcelot, "this remembred, of their kyndenes and myn unkyndenes, sanke so to myn herte that I myght not susteyne myself." (*Works*, 1256)

Lancelot grieves because he knows that he is to blame for the loss of these two people and all they represent by way of human beauty and goodness. His sorrow "may never have end" because what has been lost to the world and the future of "Cristen people" is irreplaceable. And yet he is certain that his sorrow does not displease God because it is not sorrow for the loss of sinful pleasure; rather it is sorrow for his failure to cooperate more perfectly with the grace of God so that Arthurian society might more nearly have achieved and sustained the Christian-humanist hope of a truly just society.

In Malory's three tragic heroes we can trace the development of tragedy in English literature from Chaucer to Shakespeare. Gawain can be seen as the protagonist of a *de casibus* tragedy such as Chaucer's monk might have told. Malory does not explicitly moralize Gawain's tragedy, as John Lydgate would have done, but, given Gawain's confession of willfulness and pride, and given his reported reception of the sacraments of penance and extreme unction, the reader can easily supply the missing inference. On the other hand, Arthur is the protagonist of a Renaissance humanist tragedy, of the sort Shakespeare brought to perfection in *Hamlet*. Lancelot occupies a place midway between these two tragic heroes in my developmental model. He is the protagonist of a Christian humanist tragedy in the tradition of Chaucer's Troilus, with this crucial difference: Malory is not essentially an ironist. He does not take his reader with Lancelot on the posthumous journey to paradise and then leave him with the unresolved paradox of two possible and radically contrary points of view towards earthly misfortune. Rather, Malory leaves his reader firmly on the ground and tries to resolve the paradox from that point of view. As though he were one of Lancelot's final companions, the reader observes him groveling and weeping on the tomb of Arthur and Guinevere, growing weaker and thinner day by day, week by week. Then the reader hears Lancelot ask the Bishop for his last rites as a Christian, and, finally, in the middle of the night, he rushes to his bedside to find him "starke ded; and he laye as he had smyled, and the swettest savour aboute hym that ever they felte" (*Works*, 1258). Lancelot has died, literally in the odour of sanctity, and therefore the reader must believe in the truth of the Bishop's dream. Nevertheless, neither the Bishop's joyful laughter as he dreamed that Lancelot was assumpted into heaven by flights of angels, nor Lancelot's own smiling face, can stop the tears of those he leaves behind: "Than was there wepynge and wryngyng of handes, and the grettest dole they made that ever made men" (*Works*, 1258). It may be true that Christian tragedy is not possible. But Christian-humanist tragedy is possible, because the vertical transcendence of the individual cannot compensate for his loss to society. As Lancelot wept for Arthur and Guinevere, so Sir Bedwer and Sir Bors weep for him.

I began this essay by suggesting an hypothesis to explain why tragedy has occurred only in certain places at certain times and only in the Western world. I have argued that tragedy is possible only in a culture which has the idea of social progress, a culture in which it is possible to believe that men can better their world either by their own efforts, or in cooperation with the divinity. Tragedy must be perceived first in life before it can be reflected in literature, and I would argue that in the tragedies of Gawain, Arthur and Lancelot, Malory is reflecting the perceptions of his age. England in the third quarter of the fifteenth century was experiencing prosperity and a rising standard of living; it was also enjoying a cultural renaissance reflected in art and architecture as well as in improved education, higher literacy and more and more humanist literary activity. Equally important as a precondition for the perception of tragedy, Englishmen were becoming much more nationalistic, proud of their unique law and institutions, proud of their history. King Arthur was the focus of much of this growing interest in English history; indeed, the reputed brilliance of his reign became the model for imitation by Edward IV and Henry VIII. In short, Englishmen were becoming proud of their achievements as a nation, and I would argue that where there is pride in the economic, political or cultural achievements of an entire society, there is the implicit awareness of the possibility of continued achievement – or progress – just as there was in the Greek city states of the fifth century B.C.[15]

If the hope of horizontal transcendence is what makes tragedy possible, or perceptible, in the first place then that hope, if it can be revived after misfortune, is surely the best sort of consolation for tragic misfortune. There can be no earthly consolation for Malory's tragic heroes; each of them, looking back, can see how he might have prevented the catastrophe if he had acted differently – it is that sense of having failed to do what might have been done which is the core of the tragic experience both for the protagonists and for the audience. Each of them can also see that, given his innate character and the circumstances in which he found himself, he could not have acted differently. That sense of the inevitability of what has happenened is equally crucial to the tragic experience, and is what makes acceptance possible. But what of hope? Hope underlies the first response, for one cannot regret what might have been, unless one believes that what-might-have-been was truly possible. Resignation underlies the second response, however, unless one has some grounds for hope other than the innate character of men. The Christian humanist has such grounds in the concept of grace, and in the belief that if man will cooperate with the grace of God, he can vanquish evil and put an end to misfortune on earth.

Malory closes his great work with a reference to that Christian humanist hope in the form of an epilogue which recounts the fate of Lancelot's companions:

For the Frensshe book maketh mencyon – and is auctorysed – that syr Bors, syr Ector, syr Blamour and syr Bleoberis wente into the Holy Lande, thereas Jesu Cryst was quycke

> and deed. And anone as they had stablysshed theyr londes, for, the book saith, so syr Launcelot commaunded them for to do or ever he passyd oute of thys world, there these foure knyghtes dyd many bataylles upon the myscreantes, or Turkes. And there they dyed upon a Good Fryday for Goddes sake. (*Works*, 1260)

At the time Malory wrote this passage, Pius II's great crusade against the Turk to liberate Constantinople had just failed miserably with the death of the Pope at Ancona in 1464. There is no reference in any extant "Frensshe book" to such crusading activity on the part of Lancelot's followers, but I would argue that Malory wanted his invention to have the authority of history because he felt it was an essential poetic truth. After Ancona there was never to be another Crusade; however, Malory could not have known this, and it is not, finally, of any importance. Malory chose to have Lancelot's companions end their days on crusade against the heathen because there was no other image available to him which could so clearly convey the ultimate Christian humanist goal: the creation of a truly just society within a universal Christendom. Whatever one thinks, from the vantage point of our more modern and enlightened perspective, of the actual practice of knights on Crusades, or even of the theory that it was alright to kill in the name of God, the Crusades were the most effective symbol in the later Middle Ages of a cooperative undertaking of God and man – man fighting, protected by the grace of God, to liberate the earth from the reign of Satan. (After all, even with our more modern and enlightened perspective, we are not immune to such modes of rationalization; witness our prolonged attempt to "liberate" Vietnam of the Devil of Communism in the name of God and Democracy.)

There can be no doubt that fifteenth-century Englishmen regarded the fall of Arthur as a great historical tragedy, certainly of a magnitude equal to the fall of Agamemnon in Greek history. Just as there is the suggestion in Aeschylus' *Oresteia* that Athenians may learn from their tragic history, so in Malory's *Morte Darthur* there is the suggestion that Englishmen may learn from the tragic history of King Arthur and his knights of the Round Table. Malory goes as far as he can in transforming this tragedy into a divine comedy for the reader. He gives to each of his heroes some form of vertical transcendence: Gawain goes to heaven, Lancelot also, as a saint; and Arthur becomes a living legend, whose return is still eagerly awaited by Englishmen as a kind of secular Second Coming.[16] He also provides within his text the horizontal transcendence which Gawain and Arthur most desired when he informs the reader that their kinsman, Constantine of Cornwall, was chosen King of England and ruled "worshypfully" (*Works*, 1259). Finally, although he cannot provide within his text the kind of horizontal transcendence which

Lancelot most desired, he does try to rekindle within his reader the Christian-humanist hope which Lancelot represented. Insofar as he has succeeded in doing this, he has provided his readers with a consolation far beyond the aesthetic pleasure afforded by his art; he has, in fact, provided them with a moral purpose for their own lives.

*Marianopolis, Quebec*

NOTES

1 To my knowledge there is no work on tragedy which attempts such a cultural definition. George Steiner defines tragedy rather narrowly as "predestined or inevitable disaster" and on that basis argues that the Judaic and Marxist and Christian world views are all anti-tragic, at most admitting of "partial tragedy," (*The Death of Tragedy* [New York: Knopf, 1961], p. 335). At the other extreme, Northrup Frye defines tragedy so broadly, as a kind of isolation, or alienation, or death, that he finds it endemic in life, a condition to which neither men nor gods are immune; see his *Anatomy of Criticism: Four Essays* (Princeton: Princeton University Press, 1971), esp. pp. 35–43. In developing his sociogenetic theory of religion, René Girard discusses tragedy as an important stage in the displacement and secularization of sacrifice. He regards Greek tragedy as a reenactment of the sacrificial crisis which achieved a "partial demolition of . . . myth," (*Violence and the Sacred*, trans. Patrick Gregory [Baltimore: Johns Hopkins University Press, 1977], p. 84).

2 In the phrase "horizontal transcendence" I am using the word "transcendence" in much the same way as it has been used in German philosophy, most recently by those of the Frankfurt School; see Theodor W. Adorno's essay, 'Cultural Criticism and Society,' in *Prisms*, trans. Samuel and Sherry Weber (London: Neville Spearman, 1967).

3 On Chaucer's ironic presentation of the tragedy of Troilus, see Donald Howard, *The Three Temptations: Medieval Man In Search of the World* (Princeton: Princeton University Press, 1966), pp. 146–60. Mona McAlpine has recently offered a Boethian interpretation of *Troilus* which concludes that the poem is "the comedy of Troilus and the tragedy of Criseyde," in *The Genre of Troilus and Criseyde* (Ithaca: Cornell University Press, 1978).

4 Francis Lee Utley, 'Scene Division in Chaucer's *Troilus and Criseyde*,' in *Studies in Medieval Literature in Honor of Albert Croll Baugh*, ed. MacEdward Leach (Philadelphia: University of Pennsylvania Press, 1961), pp. 109–38.

5 See Howard Patch, *The Tradition of Boethius: A Study of His Importance in Medieval Culture* (New York: Oxford University Press, 1935), and Lawrence Besserman, *The Legend of Job in the Middle Ages* (Cambridge: Harvard University Press, 1979).

6 See R. G. Parsons, 'Pelagianism and semi-Pelagianism,' *Encyclopedia of Religion and Ethics* (Edinburgh, 1917), 9:703–11.

7 Although he devotes a good deal of attention to medieval drama and to medieval narrative, or *de casibus*, tragedy, Willard Farnham does not even mention Malory's work in *The Medieval Heritage of Elizabethan Drama* (Berkeley: University of California Press,

1936). I cannot explain his omission except as part of the general tendency among scholars and critics of English literature to ignore the *Morte Darthur* until after the appearance of Vinaver's edition in 1947.

8 *The Works of Sir Thomas Malory*, ed. Eugène Vinaver, 2d ed. rev. (Oxford: Clarendon Press, 1974), p. 866. All subsequent references to the *Morte Darthur* will be to this edition and will be made in the text.

9 Gawain's ethics are remarkably like those of the Mediterranean clan-based societies described in *Honour and Shame: the Values of Mediterranean Society*, ed. J. G. Peristiany (Chicago: University of Chicago Press, 1966).

10 The best, and possibly the earliest English statement of this Christian humanist attitude is to be found in the work of the mystic, Walter Hilton. See his 'Epistle to a worldli lord on mixed life,' in *Yorkshire Writers*, ed. C. Horstman (London: Sonnenschein, 1895), 1:264–92. Hilton's little treatise enjoyed great popularity in the fifteenth century, to judge from the number of extant manuscripts. It was also printed three times in the early sixteenth century, in 1507, by Julian Notary, and twice by Wynkyn de Worde, in 1525 and again in 1533.

11 Gawain explicitly refused to seek revenge for the deaths of his brother, Aggravayne, and his own two sons, killed by Lancelot when they tried to trap him in the queen's chamber. He explained his refusal on the grounds that he had advised them not to set a trap for Lancelot in the first place; and, since they refused to take his advice, they were "the causars of their owne dethe" (*Works*, p. 1176). However, his real reason was the same as that for which he gave them such advice: he knew that his lord and uncle could not survive as king of England without the support of Lancelot. Civil war with Lancelot would mean the end of the Round Table and the realm (*Works*, pp. 1162, 1175).

12 *The Prince*, trans. George Bull (New York: Penguin, 1975), p. 133.

13 The classic statement of this ideal is to be found in a late thirteenth-century treatise by Ramon Lull, which was widely circulated in French manuscripts during the fourteenth and fifteenth centuries, and was thrice translated into English in the fifteenth century. William Caxton translated and printed it in 1484 (the year before he printed the *Morte Darthur*) as *The Book of the Ordre of Chyvalry*, ed. A. T. Byles (London: EETS, 1926).

14 See my article, 'Malory's Lancelot: Trewest Lover of a Synful Man,' *Viator* 12 (1981), 409–56.

15 Lynn White, Jr., has argued that "by the middle of the fifteenth century technological improvement became an aspect of Christian ethics" and that the "voluntaristic" tone of western Christianity produced an "ambiance where innovation could thrive" (*Medieval Religion and Technology: Collected Essays* [Los Angeles: University of California Press, 1978], p. xx). His argument could equally well be applied to other kinds of social and cultural advancement in the fifteenth century.

16 Helen Adolf suggests that King Arthur came near to becoming a secular saint in England, a kind of Pelagian hero in a country where the Pelagian heresy was never totally eradicated ('The Concept of Original Sin as Reflected in Arthurian Romance,' in *Schlauch Essays* [Warsaw, 1966], pp. 21–30).

MICHAEL PLATT

# TRAGICAL, COMICAL, HISTORICAL

## I

The long quarrel between poetry and philosophy is hard to judge, in part because the two quarrelers speak in different modes and if either adopts the other's mode he concedes half the quarrel. For the poet who writes a philosophic defense of poetry would seem to concede that poetry cannot defend itself and the philosopher who resorts to poetry to defend philosophy (or perhaps to show its superiority) would seem to concede that philosophy needs poetry. Moreover, when it comes to quarrels and defenses, there is an inequality in the capacities of the two. It is easier for philosophy to carry on a defense, for reasoning and giving arguments is its usual mode; poetry need not be against reason but it is not fond of the mode that pure reason or pure argument favors, namely, the treatise. Poetry is chary of arguments, even ones in its own defense, for it is the way of poetry to present or show things by imitating them.

Because of this difference in mode the interpreter and lover of poetry finds himself in an ambiguous position. What he usually writes resembles philosophy and its usual modes, the treatise, the essay, the inquiry, the article, etc., much more than it resembles poetry itself. Just by writing as he usually does he seems to concede that poetry cannot really speak for itself and that it needs to be completed by a philosophic kind of discourse. Of course, philosophers sometimes write interpretations of the works of previous philosophers but whatever their reasons (to help students, to enter into philosophic conversation with the author, to help themselves better possess the thought of the author, to correct error at its source), the mode remains philosophic. In other words, seldom do we see either poetic defenses of philosophy or poetic completions of philosophy. The implication of such a practice would be that philosophy requires the completion of poetry; only when philosophy becomes poetry will it be wise. The interpreter of poetry who believed this would be compelled to present his interpretations of poetry in a poetic mode. The fact that most interpreters of poetry do not do so would seem to indicate that they concede the insufficiency of poetry; poetry seems to require interpretation and thereby declare its insufficiency. However,



*A-T. Tymieniecka (ed.), Analecta Husserliana, Vol. XVIII*, 379–399.

it is not clear that the poets themselves agree; seldom or never do they seem compelled to interpret their own works. They seem to believe that their works speak perfectly well by themselves. Perhaps they even think that they speak better than philosophy. Perhaps Dante, for example, thought that the teachings of Thomas Aquinas could only become wisdom by becoming poetry, by becoming or changing into *The Divine Comedy*. There might then be a significant difference between the defense of poetry offered by the interpreters of poetry and the defense of poetry offered by poets themselves. It seems to me that Shakespeare, for example, does offer a *poetic* defense of poetry. In what follows I shall endeavor to uncover one part of that defense or account of poetry. While uncovering it I shall try to preserve some of the unobtrusiveness of his presentation. Not to do so would be to confess that he needs to be made obtrusive and argumentative in the manner of philosophy.[1]

## II

The part of Shakespeare's account of poetry which I shall deal with is genre. The idea that poetry comes in classes or species is one we often find in the works of those who interpret poetry. In Aristotle and his *Poetics* we find a philosopher treating poetry this way, but before I say something about Aristotle's generic approach to poetry I will try to discover what Shakespeare says about those who approach his work genericly. The first way Shakespeare would seem to speak to this matter is in the titles of his plays and in the arrangement by genre – comedy, history, and tragedy – offered by the First Folio.[2] However, we cannot be sure that this arrangement is his at all. He did not live to complete this edition. To be sure, the editors were his friends (one is mentioned in his will) and are likely to have known his views. If so, it is not insignificant that some plays change title and genre from Quarto to the Folio. For example, *Richard III* is *The Tragedy of King Richard the Third* in the Quarto of 1597; while in the First Folio of 1623 it falls under "Histories" and is retitled *The Life and Death of Richard the Third*. Shakespeare and his friends seem to have regarded generic classification rather casually. Evidence from the plays themselves suggests the same. Shakespeare seldom uses generic words in his plays. "Comic" and its cognates appear only fourteen times; "history" and its cognates appear only twenty-three times. The word "romance," the darling of many interpreters, never appears in his works (nor in the titles).[3] "Pastoral" appears only three times and "satire" five times. "Tragedy" appears most often, twenty-nine times,[4] but as A. P. Rossiter has noted,[5] less often as Shakespeare matured. These generic words are not very significant to Shakespeare. Shakespeare eschews talk about "tragedy," let alone "the tragic view," and this certainly separates him considerably from the vast majority of interpreters and philosophers.

Of course it might be that a rarely occurring word turned out to be decisive for the interpretation of a work or all his work. We must ask ourselves whether Shakespeare indicates that we are to study him for the sake of these genre terms or whether, if he does not so indicate, he nonetheless regards these terms as providing indispensable guidance in the study of his works.

In all of Shakespeare's work there is one moment where genre terms are thematic and one place in which the highest concentration of genre terms occurs.

Hamlet: My Lord, I have news to tell you.
When Roscius was an actor in Rome –
Polonius: The actors are come hither, my lord.
Hamlet: Buzz, buzz.
Polonius: Upon my honor –
Hamlet: Then came each actor on his ass –
Polonius: The best actors in the world, either for tragedy, comedy, history, pastoral, pastoral-comical, historical-pastoral, tragical-historical, tragical-comical-historical-pastoral; scene individable, or poem unlimited. Seneca cannot be too heavy, nor Plautus too light. For the law of writ and the liberty, these are the only men.

(*Hamlet*, 2.2.381–92)

Genre terms only become thematic when a garrulous university man who loves neoclassic things ("I did enact Julius Caesar." 2.2.99), who fancies himself a literary critic (" 'Mobled queen' is good" 2.2.492), and who adheres to neoclassic rules makes his appearance as a character. Such a man is a butt of laughter and disdain. Fancying himself critic of plots by genre, he is in reality a willing tool in the ungeneric and bloody plots of villains like Claudius.

But to understand the full significance of the fun Shakespeare pokes at Polonius, the man who thinks he knows things by genre and by sententiae, we must recall the words of another critic and university man, Francis Meres, who in *Palladis Tamia: Wits Treasury* (1598), only a few years before the writing of *Hamlet*, offered his praise of a young poet and dramatist:

As the soule of *Euphorbus* was thought to liue in *Pythagoras*: so the sweete wittie soule of *Ouid* liues in mellifluous & hony-tongued *Shakespeare*, witnes his. . . .

As *Plautus* and *Seneca* are accounted the best for Comedy and Tragedy among the Latines: so *Shakespeare* among the English is the most excellent in both kinds for the stage: for Comedy, witnes his. . . .

As *Epuis Stolo* said, that the Muses would speake with *Plautus* tongue, if they would speak Latin: so I say that the Muses would speak with *Shakespeares* fine filed phrase, if they would speak English. . . .[6]

Neoclassic minds are alike; Shakespeare gives the same comparisons to Plautus and Seneca to his Polonius. What Shakespeare thought of Meres's mind and the mind of genre critics is sufficiently, and briefly, and wittily indicated by his transfer of Meres's conceptions to Polonius.

Nevertheless, Shakespeare adds a wrinkle in his account of those who see things in genres. Polonius hyphenates the genre terms and by such inflation makes them into an unintended joke. Once he begins thinking about genres he cannot keep them apart; they begin to mingle and to produce young. His opening sentence builds from small units to large and complicated structures. He ends with a great four-masted genre: "tragical-comical-historical-pastoral." Such wanton mixing of pure kinds would, I suggest, offend a purist. In the figure of Polonius Shakespeare shows the absurdity of genre critics.

He views them with a critical, if genial superiority. His drama outstrips the canons of antiquity with its insistence upon separation of styles and purity of genres.[7] If you try à la Polonius to understand his drama with genre terminology you will either become a French neoclassicist (and declare him no correct poet for he allows mice to be mentioned in the first scene of *Hamlet*), or surrender your allegiance to genre, or become something in between, a Polonial fool. I venture to guess that when Shakespeare had done the thought about human, divine, and natural things which *Hamlet* manifests, talk of "tragic," "comic," "pastoral," and "historic," seemed weary, ancient, and unprofitable. Shakespeare's portrait of Polonius advances the indirect question: "Is it not paltry to follow Polonius when you can follow Shakespeare?"

Yet how are we to follow Shakespeare? After Polonius how can we talk of genre at all? Doesn't Shakespeare's portrait of him forbid us to pay attention to genre? Pay attention to Hamlet and his problems, not to tragedy. Hamlet doesn't pay attention to tragedy, why do you? Pay attention to Rosalind and her graces, not comedy. Does she pay attention to comedy?

Then why do you? This is good advice: we learn from poetry only by learning from particular poems and no knowledge of the essence of a genre helps us. No such knowledge of comedy would long satisfy us for the loss of our favorite comedy. No knowledge of the essence of tragedy is more precious to us than the knowledge of a particular tragedy. Give us *Lear*, not "tragic vision." If God had *Lear* in one hand and the essence of tragedy in the other, it is clear which we should have him choose.[8]

Shakespeare shows us that Polonius with his genre talk is really immune to poetry. " 'Mobled queen' is good" (2.2.492), he opines. Polonius is ever judging and never facing poetry; he is never delighted, startled, thrilled, perplexed, provoked, abashed, or taught by poetry and never will be. He will never "mind true things by what their mockeries be." He sees poetry in the manner he does, from behind an arras, an arras of genre. Even the flustered Gertrude manages to face the player Queen and utter something perceptive ("The lady doth protest too much" [3.2.222]). Even the seedy Claudius is genuinely arrested and caught by the representation of his crimes. Not so Polonius.

Now behind Polonius and his concern with genre is the infinitely greater mind of Aristotle whose treatment of tragedy in his *Poetics* is surely generic. Much can be learned from what he has to say about poetry and tragedy, but one cannot really say that he ever reads any tragedies. There is no remark in the *Poetics* which comes close to the perceptiveness of Gertrude's remark about the Player Queen. Aristotle never speaks in the face of some tragedy. He never seems to read them or one of them as if it might tell him something he does not already know. Hence he is always writing *about*, viewing them like an army in profile rather than like friends in face to face conversation. Thought provoking as his comments about tragedy may be, they do not suggest that he ever read poetry to learn something or that he encourages us to. Instead he endeavors to instruct us on how to judge poetry. Of course, it may be that his implied view that a philosopher has nothing to learn from poetry is quite justified. It is surely a long question, only to be resolved by long interpretations of poetry that succeed in showing Aristotle wrong or, by failing to, show him right. On this point the opposition between Aristotle and Shakespeare is too absolute to permit a short inquiry to bear fruit.[9]

## III

Considering the absurdity of Polonius and the deficiency of Aristotle, should we ask about genre in Shakespeare? It seems to me that we should. For when one has appreciated the criticism that Shakespeare offers us in his ridicule of Polonius, when one has avowed that single works are what we learn from, not classes of works, there still remains the stubborn sense that Shakespeare's works do fall into classes, or rather families. A few plays may bear the title tragedy in one place and history in the other, but, we notice, no comedy is ever titled a tragedy and no tragedy is ever titled a comedy. There may not be the strict separation of styles, which was one of the principle distinguishing features of ancient drama, but there is a feeling of a dominate tone. If we were to open our Shakespeare at random and read for a while, without looking at titles, we would soon be able to say whether we were reading a comedy, tragedy, or history. To be sure, there would be some border line scenes. We might even say that, for example, the first half of the *Winter's Tale* is a tragedy, but in general as we read we soon come to feel the generic differences in Shakespeare's work and if we do not seek a greater precision than the things will allow perhaps we can talk sensibly about these "families" in Shakespeare. Indeed, is not Shakespeare himself very precise when he allows Polonius the absurdity of "tragical-comical-historical-pastoral": does this not describe the generically imprecise *Hamlet* precisely?

There is another consideration which urges the same conclusion. Our protection. The fact that Shakespeare could take the same plot and make of it either a tragedy (*Romeo and Juliet*) or a comedy (*Midsummer Night's Dream*, with the plot of *Romeo and Juliet* repeated in the "Pyramus and Thisbe" play) ought to make us pause. The fact that he can take the same plot and make either our eyes weep or our hearts leap should make us wonder whether he is a a magician. He makes us weep, he makes us laugh; ought we to trust him? Or ought we to say, "You, a teacher? Why do you stylize human life so? Life is neither a comedy, tragedy, nor history. These are abstractions. They are lenses of perception. Yes, you grind those lenses finer than any before you, but what does that show, that truth shows herself naked in your words? Hardly. You are a virtuoso. You love play too much. Your very loveliness tempts us, wantonly, to forget that you artists deform life with your forms. If your plays are so true, so faithful, so instructive, how come they can be grouped in families? Do human things arrange themselves this way? And if your plays are a precious expression of the truth of human things, if they faithfully represent our lives, then how come they always resemble you? They always have your mark? We read you for just a moment and we say 'that's Shakespeare.' How then can your works be true?"

Both because observation tells us that Shakespeare's works arrange themselves in families and because this fact might warn us that he falsifies what his lovers claim he represents, we should turn to the question of genre in his work. Here we will do well not only to observe it in his work but also to ask his works what he understands by tragedy and by comedy and also their relation. The last is especially important and to bring out its significance I shall turn for a moment to what Plato tells us about comedy, tragedy, and their relation in the *Drinking Party* (or Symposium).

At the end of the dialogue only Socrates is awake; to his drowsy companions, the nearly asleep Aristophanes and the sleepy Agathon, he asserts that the same man can have the knowledge required for writing comedy and tragedy, "that the fully skilled tragedian could be the comedian as well." This remarkable assertion fails to stir either the comedian Aristophanes or the tragedian Agathon to reply. "While they were being driven to this (conclusion), they began to nod; first Aristophanes dropped into a slumber, and then as day began to dawn Agathon also. When Socrates had seen them comfortable, he rose and went away."[10] Socrates' assertion is to be understood in the light of the action that accompanies it: the comic poet falls asleep before the tragic poet. This sequence accords with something precise in what Socrates says about comedy and tragedy. He says that the fully skilled tragedian could be a comedian; he does not assert the converse. The tragedian seems to include the comedian and not vice versa. To understand this we must recall what has transpired in the dialogue. The comic poet makes us laugh at almost all things; in his speech Aristophanes made fun of men and also of the gods. Not so the tragic poet; Agathon speaks of *eros*, the assigned topic, as a beautiful god. What he deifies and Aristophanes had laughted at, as all too human, Socrates proves to be human, and not ignobly so. The superiority of the tragic poet according to Socrates and to Plato seems to rest upon the superiority of enchantment over disenchantment; the tragedian enchants his audience with the beautiful gods; the comic poet only disenchants. However, since he who enchants must himself be disenchanted the tragic poet includes the comic.[11] In this sense life is a comedy to those who think and a tragedy to those who feel. Or putting it another way, we have tragedy so that we will not perish of comedy. At least this seems to be the implication of what Socrates says when viewed in the light of the comparative sleepiness of his two companions.[12]

For proof that what Socrates said in the dawn was true, the world had to wait nearly two thousand years. It had to wait for Shakespeare. Yet today the fact that Shakespeare wrote both tragedy and comedy is as unremarked as it is undoubted. That it was and is remarkable I shall try to show in what follows. What it signifies I shall also try to understand.

The fact that there could be a Shakespeare, a man who wrote both comedy and tragedy, was known to Socrates. Nevertheless, the way in which Shakespeare confirms Socrates would cause Socrates to wonder. Not only does Shakespeare's tragedy differ from ancient tragedy, and his comedy from ancient comedy, but the relation between them differs as well. To put the difference between ancient drama, both comedy and tragedy, and Shakespeare most briefly, I would say: in place of the divine you find woman. Let me elaborate.

## IV

Shakespeare the tragedian can be described as one who takes the best human beings and breaks them; he fashions good, strong, noble men and then he watches them suffer and die. So did the ancient tragedians. A decisive difference remains. As these men of Shakespeare fall toward dusty death, there is no assurance of a "world elsewhere" to comfort them or us. No gods or God promises them a felicity they could not find in this world. No trapeze net of supernatural reward or supernatural interpretation catches their fall. No Olympic gods are there to beautify the ugliness of corpses. The chief difference between ancient tragedy and Shakespearean tragedy is the presence of the gods in the former and their absence in the latter.[13] This difference is not unconnected with woman and Shakespeare's representation of woman. To understand this, let us turn to comedy.

Shakespeare the comedian can be described as one who arranges marriages. He fashions lovely, posed and gentle women and then he marries them. Arranging such marriages is the great theme and final happy chord of his mature comedies. By the power invested in him alone do his lovers find their soul's joy in another. Or their soul's perdition. Arranging marriages is a tricky thing. Shakespeare could be like a pander or like a tyrannical father or like a good friend who wins the lasting gratitude of both parties.[14] Doing it rightly requires that the lovers themselves think it was all their own idea. As a tragedian Shakespeare (or some part of him) must be willful and cruel, like fate. As a comedian Shakespeare (or some part of him) must be teacherly and gentle, something like freedom. Here too Shakespeare differs from antiquity, both its drama and its philosophy. In ancient comedy there is nothing enchanting; what is enchanting, the gods, appears in tragedy. Shakespeare's women are not gods, yet they are the most enchanting beings in all his drama; they tempt us to believe what no ancient would, that human things can be worthy of awe. It is in comedy that they primarily do so.

For the ancient poets, especially the tragedians, the enchanting thing was the gods (and when the ancient philosophers deposed Zeus and put in his place *nous* they only went further in the same direction, away from earth). Between Shakespeare and these ancients lies Christianity and for the Christians the truly enchanting thing is the Holy Creator God, like Zeus and *nous* divine, but also creative, and also "Holy," mysterious, unfathomable (as YHWH-Elohim tells Job). In never presenting either the gods or this God Shakespeare seems to depart from both antiquity and Christianity; however, in doing so, Shekespeare seems to benefit from Christianity whose God comes down from the heavens, walks the earth, talks with men, seems a man, suffers death, and yet remains enchanting. It is only that in Shakespeare what was god to the Greeks and God to the Christians becomes woman. It is likely that the intermediary step between the worship of God become man and the love of woman is the worship of the Virgin. The Shakespearean result is Rosalind, Beatrice, Kate, Miranda, Perdita, and Cordelia.

As a whole, Shakespeare's women are certainly more intelligent than his men; they are also more lovely. It is an unconfessed secret of Shakespeare's sway over each generation that he makes the men and boys in his audience fall in love with his women. He does this without, however, inspiring envy among the women in his audience. For when they gaze at his women they fancy they know that Shakespeare loved women. Here is the testimony of Mrs. Mary Chesnut, writing in Richmond in the last days of our Civil War:

> There is love-making and love-making in this world. What a time the sweethearts of that wretch, young Shakespeare, must have had. What experiences of life's delights he must have had before he summoned up the Romeo and Juliet business from his own internal consciousness. Also, that delicious Beatrice, and Rosalind. The poor creature to whom he left his second-best bedstead came in second best all the time, no doubt, and she hardly deserved more. Fancy people wondering that Shakespeare and his kind leave no progeny like themselves. Shakespeare's children were half his, only; the other half was only the second-best bedstead's. What could you expect from that commingling of materials?

I cannot resist quoting a little more:

> Shakespeare knew all the forms and phases of true love. Straight to one's heart he goes, whether in tragedy or comedy. He never misses fire. He had been there (as the slang goes). No doubt the man's bare presence gave pleasure to the female world, and he saw them at their best. But he effaced himself; he told no tales of his own life.[15]

No doubt Shakespeare would have appreciated Mary Chesnut. Her unashamed reasoning from work to worker, which might offend a literary theorist, is rewarded with an insight that seldom comes to those who think without affection. Shakespeare should reward her with those lovely greeting words of Othello, "my soul's joy!"

Shyly we seize Othello's words to express our own affection. They seem to say, better than we could, what we have wanted to say or hear said. They and the sentiment they express come naturally, if shyly, to us. What could be more natural than to greet one's beloved this way? Yet we must acknowledge that no ancient and Christian would do so.[16] An ancient might greet a friend in this manner; we might hear such words escape the lips of Socrates after one of his long trances; a Christian might greet Christ this way or someone from whom he had received Christ's teaching, but none of these would greet a woman this way. The passionate affection of Romeo and Juliet speaks immediately to us, without the prop of footnotes or the help of interpretation, yet we know that it would not speak to Odysseus or to Caeser or to Abraham. Nor would it speak immediately and unambiguously to Paul or Augustine. We are forced to admit that this affection, which seems so immediate and self-evident to us, is the daughter of time. It was born in Provence in the early Renaissance; the sonnet attended its nativity; the city states of the Italian peninsula adopted it; their poets escorted it North of the Alps, and today it is welcome everywhere.

As a consequence of honoring love we barely regulate it. As modern lovers, modern parents, and modern citizens we have as our creed "that government is best which governs love least." Ancient peoples were more sober and so was their drama. So too is Shakespeare. In *Romeo and Juliet* he portrays this beautiful and immoderate affection in its first home, the city state of Renaissance Italy, and decks it out with the rhymes of its birthplace, Provence; for the lovers meet with a sonnet as well as a kiss. As Shakespeare fills us with the beauty of love, so much so that we forget to censure the suicides at the end (as we should according to Christian teaching), he does not fail to note the problems which this affection poses for the city. When he loves, Romeo feels that he and his beloved lose their family names; love leaps over all walls; it gentles everyone it touches. Romeo in love is tempted to imitate the loving Christ by telling all warriors to put up their bright swords, with disastrous results, as Shakespeare notes; it is his meddling that leads directly, if unintentionally, to his cousin's death. How dangerous to families and cities and friends are those who behave as if there need not be such distinctions as nation, region, family, friend!

At about the same time Shakespeare became a mature tragedian by writing *Romeo and Juliet*, he also, by writing *A Midsummer Night's Dream*, became the man Socrates in the dawn spoke of. In this comedy Shakespeare shows what the phenomena of love looks like in an ancient setting. It is far less dangerous. Lysander is no Romeo. Love is a tragedy in modern Verona and a comedy in ancient athens; the Pyramus and Thisbe play presented at the climax of the drama is the Romeo and Juliet story told by a lovable idiot, cause of mirth not woe. It is selected by the prince who, Plutarch tells us, first united Athens into a city and it is selected for his marriage to Hippolyta. Perhaps his task of sorting out the lovers and through marriage knitting up his city in civil peace is made easier by the nature of love in ancient Athens; would Lysander ever say "my soul's joy?" Love is not so spiritualized in Athens. The divine does not reside in a woman. On the other hand, we must notice that in Athens princely and priestly powers are united in Duke Theseus, while in Verona the footstamping prince can never quite get together with the slow-footed Friar. Perhaps what Athens had and Verona lacks is the unification of civil and ecclesiastical power.

That this is so is suggested by what Duke Vincentio does in *Measure for Measure*; he is never more a Duke than when he puts on a cowl; he is never more a prince than when he is a priest; he is never more a comedian than when he is making tragedies for his people. His problems are the saints and the sinners, those who prefer a cloistered virtue and those who prefer no virtue at all; the two groups, represented by Isabella and by Pompey, abhor each other while gaining a certain justification for their way of life from each other. Both disdain public things, of which marriage is the most pertinent, the one because they are against the flesh, the other because they wallow in it. The crafty, cowled, hidden Prince must first stir them with an incipient tragedy, the death of Claudio, and then reform them with a comedy, marriage. By so doing the Prince reforms Vienna, the center of the Holy Roman Empire. While his reformation is certainly Imperial and probably Roman, it cannot be said to be especially Holy. (While the whore houses are closed, so too are the cloisters.) It is a solution which he himself partakes in; he marries. Indeed one may wonder if all his calculated tragedies have not been partly moved by the hope of this comedy.

## V

In speaking of comedy and tragedy, of woman, and of love we have naturally come to speak of politics. It is now time to observe a certain connection, never made entirely explicit by Shakespeare and perhaps not capable of being made precise, between the drama of Shakespeare and the momentous changes wrought by Henry VIII in the English regime. It was under this regime, with its religious adjustment, and under the rule of Henry's daughter that Shakespeare and his company flourished. Henry VIII did two things which were fundamental to all that followed. He destroyed the monastic life of the Church; with the resulting wealth he endowed a new nobility, loyal to himself, the now supreme head of the no longer Catholic Church. As St. Thomas More suspected, these changes could not but diminish the persuasive force of Christian teaching, one consequence of which is the possibility of a tragedy such as Shakespeare's in which the dread of something after life is no longer a secure Pole star to guide human life but has become, at best, a terrible, bewildering, and harmful intrusion into human life, much like the Ghost against which Hamlet so mightily strives and over which he triumphs, only to find life deprived of all vital savor. Santayana lamented that Shakespeare was a poet utterly without religion; his tragedies certainly do not present the truly enchanting God; nor do they show us men adoring, marveling at, or worshipping this God. As Santayana remarks, there is not one good prayer in Shakespeare.[17] It is doubtful that Shakespeare would have been permitted to present human life so free from the Biblical interpretation without Henry VIII's innovation.

Henry VIII did a second thing and it prepared the way for Shakespeare's comedy. Whether for reasons of state, as some maintain, or for reasons of ungoverned appetite, as others allege, Henry saw to it that he became supreme in the matter of marriage. He was a prince who would decide for himself whom he would marry (and marry and marry and marry and yet again marry). To be the prince he wished to be, he found he had also to be a priest and so he severed England from the mainland of Papal authority and made himself *pontifex maximus*. What is so unhappily divided in the Verona of *Romeo and Juliet*, civil and ecclesiastical power, he united. The union never brought the fretful tyrant happiness, but it is among the conditions that contributed to Shakespearean comedy. In them it is Shakespeare who unites these powers. To his characters he is prince and priest. It is he who arranges all marriages. It is through his self-invested power that his amiable young men come together with his witty lasses.

As a tragedian Shakespeare resembles a tyrant for he takes the best men, breaks them on some fatality in themselves or in human things, and sends them to death ("be absolute for death"); as a comedian, he resembles a good, fatherly prince who sees to it that his beloved daughter falls spontaneously and freely in love with a prince whom he, in his better wisdom, knows will be best for both her and the realm as well. As a tragedian Shakespeare is always preparing his men and women for death; as a comedian he is always preparing them for marriage.

## VI

In order to begin to understand the relation between these two preparations and therefore the relation between tragedy and comedy according to Shakespeare we will now turn to *The Tempest*. Its first scene is a little tragedy, an epitome of every Shakespearean tragedy, whereas the whole is a comedy, the most complete of all Shakespeare's plays. Perhaps the relation between his tragedy and his comedy is suggested by the fact that this comedy contains a tragedy. Never do we find the converse pattern. *Hamlet* has a tragedy within a tragedy. The plot of *Romeo and Juliet* is contained within *A Midsummer Night's Dream*, not *vice versa*. The tragedy that constitutes the first part of *The Winter's Tale* is reversed and redeemed by the comedy of the second part. Never do we find a comedy rounded out with a tragedy, or a comedy within a tragedy. The single candidate is the comedy of Gloucester's miraculous salvation from his fall over Dover Cliffs; from his point of view it is a tragedy that ends in comedy. Not for us. We see altogether too transparently the falsehood of the gods that his son enchants him with to be ourselves enchanted, even for a minute. It seems that for Shakespeare comedy embraces tragedy.

To understand this we must consider the opening scene of the *Tempest*. All is tempest and, strive as they do or wish to, none of the seafarers can resist it. In a storm the man with genuine skill rules and so it is here. It is the Boatswain, rather than any mainland prince, who gives the orders. Still, those orders make no difference. Since he cannot command the "roarers," he must prepare for death. Even as he strives against the tempest, he tells others to give thanks that they have lived so long. His very last words,

> What, must our mouths be cold? (1.1.49)

tell us what we have most to thank life for, for warm bodies and for speech. We are a combination of life and speech. With his warm mouth the Boatswain bids life farewell, blessing it, not clinging to it. There is no better preparation for death in all of Shakespeare. The Boatswain "dies" so as to fill us with wonder at the miracle of this combination of warmth and speech which we are. In this tragedy it is not the gods that we find enchanting. The truly enchanting things are humans. It is a wonder that there should be creatures with warm lips. That these lips will be cold, that human life is also mortal life, ought to inspire us not so much with fear (the Boatswain fears not) but with wonder and affection.

That this scene and this teaching epitomize Shakespearean tragedy is suggested by the fact that Lear too ends looking at cold lips. "Never, never, never, never, never," he says over the slain Cordelia and dies looking at her lips.

> And my poor fool is hanged: no, no, no life?
> Why should a dog, a horse, a rat, have life,
> And thou no breath at all? Thou'lt come no more,
> Never, never, never, never, never.
> Pray you undo this button. Thank you, sir.
> Do you see this? Look on her! Look her lips,
> Look there, look there – *He dies* (5.3.306–312)

Here is our obstinate human love. The deepest fear in *King Lear* is a species of the fear of nothing; it is the fear that once gone, we shall never, never come again. This fear presupposes a sweet, even fierce attachment to here. But here would not be here for Lear without his beloved, wronged, beloved Cordelia. Here is not here without the beloved. Thus it is that while Lear asks why there is no life in his poor fool, he refers to her corpse as "Thou." Obstinate love indeed. Five times he then repeats the terrible "never." First indignantly, finally blankly. Turning away from what will never come again, he asks to be unbuttoned for he would spare others the labor of undressing his corpse. So stricken as he is, he yet maintains courtesy. Suddenly he bids the man look upon Cordelia. Where is Cordelia? Her eyes are closed. Look to her lips. Why lips? Lips don't close like eyes. Through lips pass words that soothe – "her voice was ever soft, gentle and low – an excellent thing in woman" – and through lips the words that undo, as Cordelia's words undid all Lear's cunning yet tender plans. Yet "love and be silent" were Cordelia's first words. Now she is silent. It would comfort Albany to know that here or elsewhere we are rewarded for our patient suffering; not Lear. Bidding others to look at Cordelia's lips, he knows that we must love and be silent. Lear dies bidding us look. He loves and is then silent. What are we humans? The Boatswain's farewell tells us we are a union of warmth and speech. To this Lear's attending to Cordelia's lips adds something; warm and speaking as our lips are, they also love, even when they do not speak. Indeed, we are the being who love and then fall silent.

What is the relation of this tragic teaching to the comedy of the *Tempest* which follows and embraces it? To understand this better we should turn to the masque Prospero puts on for his daughter and her beloved and ask: what is enchanting? Into this scene Shakespeare has poured his superabundant powers. To see the splendid antique deities treading the earth once more many a Renaissance humanist would have given his soul (Faust did). They come to bless this marriage. But they nearly upset it, for the impressionable Ferdinand avers that he would like to stay on the Isle. When he says "Let me live here ever!" (4.1.122), we may wonder if he is not somewhat more impressed with these spirits than a bridegroom nearing his wedding day should be. Fortunately these enchanting spirits disappear abruptly.

Through the masque and its abrupt disappearance Ferdinand will learn to love well his mortal wife and prefer her to these immortal spirits. In this regard he is led by her for she is not enchanted by them. In order to become disenchanted it helps to be next to some one who is unenchanted. What Miranda finds enchanting ("oh brave new world") are human beings, not immortal spirits. The spirit of Shakespeare's comedy is the same as his tragedy. In preparing for death and for marriage we are to love well the human things we must leave ere long. The truly enchanting things are not the gods or *nous* or God but human beings. Ancient tragedy, comedy, and philosophy agreed: it is a fearful thing to love what death can touch. Love what is above the human. Shakespeare is the comedian he is, the tragedian he is, and hence the dramatist he is, because he replies: it is a fearful thing to love what death can touch. Only among the dying things will you find what is lovable. Look at Miranda. Look at Cordelia. Cherish them.[18]

## VII

What Shakespeare would say to Socrates about tragedy and comedy is sufficiently clear. What he would say to Aristotle is embedded in it. *The Tempest* teaches us to wonder. According to Aristotle tragedy should teach us to pity and fear. It is only in the *Metaphysics* that he speaks of wonder and distinguishes poetic wonder from the wonder in which philosophy originates (982b12–983a23). When the poet wonders, he is satisfied to remain in wonder and does not try to replace his ignorance with knowledge; and of course he is satisfied to let others, his audience, also remain in wonder. Chief among the things that the poets deem wonderful are the gods. Where the poets end, the philosophers begin. When the philosopher wonders he wishes to replace his ignorance with knowledge. (However, when it is knowledge of Being which he desires it is not clear that he ever leaves wonder behind; it seems that Being is only available to man in this wonder.)[19] Since

Shakespeare's Prospero only presents the wonderful gods to withdraw them and fix his son-in-law's attention upon Miranda, it seems that Shakespeare cannot be easily classed with the poets, for those whom he teaches to wonder, Ferdinand, Miranda, and us, are encouraged to know what they first wonder at. Ferdinand and Miranda will know each other in marriage. Ferdinand must not only wonder at his father's survival but come to know his crimes. Miranda must not only wonder at the mainlanders but come to distinguish an Antonio from the rest. So it seems that the wonder which Shakespeare's art encourages is "philosophic" in Aristotle's sense. With this corollory: that it is by looking at Miranda, looking at Cordelia, looking with wonder at them, gazing breathless, so caught up in looking that upon finishing a play one begins again ("is that Cordelia?" then "once more"), that we come to wonder at Being.[20]

In doing so Shakespeare never encourages us to so wonder at Being that we do not fail to care for beings, beings such as Lear, Edgar, and Cordelia. Beings who do not care for beings do not wonder at Being. And if you care for Cordelia, if you want her "once more," then you will be ready to cry "hold your hand" to any villain approaching her. And therefore ready to kill any being who has slain her, as Lear does.

## VIII

Considering the care for human things that Shakespeare's wonder entails should remind us of something we have overlooked until now: Shakespeare wrote not only comedy and tragedy but history.[21] Indeed his histories are the most obvious instance of his care for human beings, which would not be genuine if it did not include care for his fellow Englishmen. Perhaps the greatest gift ever given to a nation by a poet, Shakespeare's histories are crowned with a portrait of the best king England ever had. No man in all of Shakespeare pits his intelligence against adversity with more success than Henry V. He is intelligent care in action and the active image of his maker.[22] That Shakespeare, the solitary poet of Sonnet 94, created and esteemed the active Henry V is not as extraordinary as it at first seems.[23] The "natural riches" of Henry V set both the Carlisles and the Machiavels to school.

Ignorance of first causes uncovers the human good. The absence of providential intervention in human affairs sets the stage for the appearance of the human virtues, both by calling us to set things aright for ourselves and by permitting us to do so without those helps that would render our virtues calculating and ignoble. "History" is a mess that would melt into anarchy or harden into tyranny if it were not for human virtue; without courage we could prevail against no fear, without moderation against no excess, and without justice against no corruption; and without practical wisdom our virtues themselves would not be able to secure or preserve the good things of this earth. No final victory is, however, to be hoped for. Things are always about to slide from mess to calamity to chaos. Human virtue is always required to thrust itself "once more" into the breach. The character of Shakespeare's ideal English Prince is summed up by the maxim; in philosophy, ignorance; in life, courage; in public, practical wisdom; in private, solitude. The night is "more than we know" but our watchfires cast enough light to know ourselves by.

If female virtue shines brightest in Shakespeare's Comedies, in a Rosalind or a Perdita, male virtue shines brightest in the Histories, in a Bastard, a Talbot, and in a Henry V. In the marriage of Henry V and Catherine of France manly intelligence and feminine wit seem at last matched. Not accidentally the play in which Henry's active wisdom shines includes both a tragedy, the night before Agincourt, and a comedy, the marriage after Agincourt. In the night before, we see Henry and his soldiers prepare for death; in the days after, we see Henry and his bride, Catherine of France, prepare for marriage. That this crown of the histories includes a comedy that completes a tragedy suggests that Shakespeare is fundamentally an historian and hence something entirely unprecedented. Not only did no Greek dramatist write both comedy and tragedy, none wrote history. Nor did Socrates in the dawn imagine the possibility. Shakespeare is not only the English Sophocles and the English Aristophanes but the English Thucydides and more, for he is all of these at once, a Homer. Because he was the tragedian he was, the comedian he was, and the historian he was, Shakespeare was Shakespeare.

## IX

The long quarrel between poetry and philosophy is hard to judge but perhaps one part of it is now, after this inquiry, somewhat easier. Since Shakespeare ceased to write, two philosophers have written about tragedy, without we think taking sufficient account of his work. Although his work is designed, like Hegel's, to comprehend all history, his sense of history is very different from Hegel's. Like his own Henry V, Shakespeare faces the scene of human troubles and pleasures entirely without the comfort of "progress." So long as there are humans there will be troubles and there will be pleasures; where there are men there will be vices and where there are men there must be virtues. History is a "breach" into which we must ever and again "once more" commit our lives, our fortunes and our virtues. The virtue that seeks and finds the good in the particular circumstances and that chooses the best possible although it loves the best far more, we may call practical wisdom, after Aristotle. In his *Phenomenology of Mind* Hegel tries to replace such φρονεσις with conscience, but conscience is inarticulate, like Billy Budd hitting Clagett. It is true that Shakespeare gives no apology for practical wisdom in the articulate way of Aristotle's *Ethics*, yet his way of showing practical wisdom embodied in action and in speech, in a Henry V, is superior in one respect: It is more vividly and widely instructive. To the ruled it shows what kind of ruler to prefer, to rulers it shows what kind of ruler to be. Great men sometimes need philosophers, to protect their virtues from bad theories; peoples always need poets, to protect them from false images. Although the English-speaking peoples have never had a philosopher worthy of their virtues as Socrates, Plato and Aristotle were worthy of the people of Marathon, Salamis and Thermopylae, they have had a poet.

The second philosopher to write of tragedy, after Shakespeare but without we think taking sufficient account of him, is Nietzsche. Nietzsche not only wrote of tragedy like Hegel but is the only thinker ever to champion tragedy or call his own thought tragic. Although he maintained that the best want to rule, contrary to the Platonic Socrates, and admired Caesar, calling the Übermensch "Roman Caesar with the soul of Christ", nevertheless Nietzsche's substitution of solitude for moderation in the Platonic-Socratic quartet of virtues and his characterization of prudence as knowledge of the ways and means of protecting one's solitude indicate how far the life of a statesman was from his experience or imagination.[24] Nietzsche was suited to understand a Coriolanus, a Brutus, a Cordelia and even a Falstaff, but not a Henry V. He did not appreciate the truth that sometimes when natural right – or what Nietzsche calls "an order of rank" – is diluted and even betrayed by conventional right it is naturally right that it be so. With his "once more" song he tries to suppress the nausea that noble spirits naturally feel for men made by convention alone, men made by a tailor, as Shakespeare has his Kent say. Yet Nietzsche was often nauseous and often cries *Pereat humanus et fiat superhumanus*.

Nietzsche is on the whole very different from Marx, whose ideal state, in which man loafs from morning to evening, is the same as the worst state for Nietzsche, the state of the Last Man, without aspiration, delight or self-knowledge; yet Nietzsche shares with Marx (and also the author of

Revelations) a delight in catastrophe. They await the end of humanity as we have known it with more hope than fear, and Marx awaits it with faith and *Schadenfreude*. In celebrating such a ruler as Henry V, for whom the good is not an ideal, Shakespeare shows that his sympathies lie elsewhere. In *The Tempest* he has portrayed a demi-creator willing to make men suffer. However, Prospero fits the suffering to the men and also does so with no intention of creating "brave new men." He wishes only to regain his kingdom, find a husband for his daughter and live without the mastery of nature. It is significant that he is tempted to drown the brother he cannot reform and refrains. Shakespeare understood the intention to new-create humanity by first destroying it, in a tempest, in a deluge, or in a concentration camp, as wicked, and in Gonzalo's dream (2.1.143 ff.) he has portrayed the ideal of communism as silly.[25]

In conclusion then, we offer the thesis that modern philosophy has more to learn from Shakespeare than he has to learn from it.[26]

*University of Dallas*

NOTES

1 To most of the points in this section there are significant exceptions. The works of both Plato and of Nietzsche employ a poetic mode of presentation in order to seek wisdom. Heidegger is a philosopher who seems to regard poetry as higher than philosophy, for he reads Hölderlin as if he were coauthored by Being. Among poets, we may note that Virgil is known to have been a student of Plato and Goethe appreciated Spinoza exceedingly.

2 All references to Shakespeare will be to the *Complete Works*, ed. by Alfred Harbage (Baltimore: Pelican Books, 1969).

3 Despite this fact and despite the fact that both print a photo of the *First Folio* table of contents, both the *Pelican Shakespeare* and the *Riverside Shakespeare* ed. G. Blakemore Evans (Boston: Houghton Mifflin, 1974) arrange some of the plays under the title 'Romances.' Only editor Harbage of the *Pelican* offers his reasons.

4 I have checked these counts with Marvin Spevack *A Complete and Systematic Concordance to the Works of Shakespeare*, 6 vols. (Hildeshein: Georg Olms, 1969).

5 *Angel With Horns*, ed. by Graham Story (New York: Theatre Arts Press, 1961), pp. 253 ff.

6 *Four Centuries of Shakespeare Criticism*, ed. Frank Kermode (New York: Avon Books, 1965), p. 31.

7 See Erich Auerbach's sensitive comments on Shakespeare in the light of the ancient separation of high and low, tragic and comic styles, in *Mimesis: The Representation of Reality in Western Literature*, trans. Willard Trask (Garden City, N.Y.: Doubleday, 1953), pp. 274 ff.

[8] I want to thank my colleague Eileen Gregory for her vigorous opposition to my views on genre, which prompted me to gather my thoughts on comedy and tragedy together for the first time.

[9] One way to address it would be to offer an interpretation of *Othello*, for Iago like Aristotle (*Poetics*, 60a26 and following) suggests that "incredible possibilities," e.g., that Venetian Desdemona should love a Moor, cannot be. By saying, as it were, "I know these Venetian women" (3.2.200 ff.) and thereby subsuming the rare incredible Desdemona under a class, a species, a genre, he does persuade Othello that Desdemona is unfaithful. But he does not persuade Shakespeare and he does not persuade us. It may be incredible but Desdemona is true. What Aristotle forbids the poet, Shakespeare presents.

[10] *Drinking Party*, 223d. There are good comments on this passage in Stanley Rosen's *Plato's Symposium* (New Haven: Yale University Press, 1968), pp. 325–27.

[11] By carefully stating that the tragedian can be the comedian, and not vice versa, Socrates implies that tragedy and its gods are mere enchantment. When he reduces Agathon and his speech to nothing with a few questions, Socrates disenchants; when he reports Diotima's speech, he enchants. Perhaps Socrates knows that the same man could write both comedy and tragedy because Plato knows . . . himself.

[12] While interpreting the *Drinking Party* we must keep in mind the stylistic difference between ancient comedy and tragedy, the one suited to low, private, daily things, the other suited to high, public, and splendid things.

[13] If I am correct, *Lear* suggests that if there is a creator God, then no creature of His can have evidence of His existence for such evidence would compromise the uncalculating character of all true virtue. If you know of a world elsewhere, where your Creator rewards your patience virtue, then your virtue is tainted with calculation. Creation means separation. In other words, we cannot conclude from the absence of God in Shakespeare that he is an atheist; he may be an unusual Christian. Everything else about him is rare, so this should not surprise us.

[14] Shakespeare did win such gratitude from a couple whom he brought together while he lived with the Montjoys in London. See S. Schoenbaum *William Shakespeare: A Documentary Life* (Oxford: Clarendon Press, 1975), pp. 211–13.

[15] Both quotations are from Mrs. Chesnut's diary entry for March 12, 1864; see Mary Boykin Chesnut, *A Diary from Dixie*, ed. Ben Ames Williams (Boston: Houghton Mifflin, 1949), pp. 391–92.

[16] Antony in Shakespeare's *Antony and Cleopatra* just might greet Cleopatra with these words. However, that only proves our point, for Shakespeare differs most from Plutarch in endowing his Cleopatra with an unancient attraction.

[17] 'The Absence of Religion in Shakespeare' in Kermode (cited above), pp. 158–69. However, see Note 13 above and consider *Cymbeline* 5.4.9–29.

[18] The comic ending of *The Tempest* is the work of father Prospero. The tragic Lear wanted to give up his kingdom, marry his best daughter well, and prepare for death. Comedy seems not only to be more inclusive than tragedy but higher; it shows us higher sorts of human beings.

[19] I owe this account to Seth Benardete's 'On Wisdom and Philosophy: The First Two Chapters of Aristotle's *Metaphysics* Alpha,' *Review of Metaphysics* **32** (1978), 205–215.

[20] This special relation between human beings and Being in Shakespeare resembles the relation in Biblical religion between creatures, especially man, and the Creator. Since

man is said to be in the image of God, it ought to be possible to study man so as to know God. In wondering at the one comes to wonder at the other. We find the same in the study of Shakespeare; from wondering at the works we come to wonder at their maker. See my 'Shakespearean Wisdom?' in *Shakespeare as Political Thinker* ed. John Alvis and Thomas G. West (Carolina Academic Press, 1981).

[21] Since Shakespeare there seems to be only one poet who has written both comedy and tragedy: Kleist.

[22] For further discussion of the histories see my *Shakespeare's English Prince* (Washington: University Press of America, 1984), *passim.*

[23] That in 'Sonnet 94' Shakespeare has described himself is the main idea of my essay 'Shakespearean Wisdom?' cited above.

[24] *Zarathustra* III, 'Von alten und neuen Tafeln,' 21; *Der Wille zur Macht* 983; *Jenseits von Gut und Böse*, 284; *Zarathustra* II, 'Von der Menschen-Klugheit.'

[25] For a complementary account of Tempest 1.1 see my essay 'Shakespeare's Apology for Poetic Wisdom' in *Shakespeare and the Arts*, ed. Cecile Williamson Cary and Henry S. Limouze (Washington: University Press of America, 1982), pp. 280–293.

[26] I am very grateful to the Alexander von Humboldt Foundation for the leisure to revise this essay.

FRANÇOISE RAVAUX

# THE DENIAL OF TRAGEDY: THE SELF-REFLEXIVE PROCESS OF THE CREATIVE ACTIVITY AND THE FRENCH NEW NOVEL

In 'Nature, Humanisme et Tragédie,'[1] Alain Robbe-Grillet requested that denial of tragedy be the first and necessary commitment made by the contemporary artist.

For Alain Robbe-Grillet, tragedy is the "sublimation of a difference." It is defined as an attempt to transcend the distance existing between man and the world. Sartre and Camus have denounced the romantic conception of communion between man and the world and acknowledged the distance existing between them. However, they have "recuperated" this distance in describing it as an abandonment of man by the universe and as a motivation for rebellion and action. Robbe-Grillet vehemently rejects this new anthropocentric perception and the anthropomorphic qualities it invests in the world.

The systematic tragification of the universe being felt as the result of a deliberate act of volition, Robbe-Grillet cannot help doubting any proposition positing tragedy as natural and definite and, consequently, any contemporary art forms deriving from the same apperception or perpetrating the misconception.

Since doubt has manifested itself, Robbe-Grillet acknowledges the necessity for his "looking further," i.e., for specific and refined investigations on the subject. He directs his research on the nature of the literary work of art, positing the distance existing between man and the world as a point of departure and using the narrative form as a field and a vehicle for his investigations. Although it appears as a perpetual movement, tragedy freezes man and the universe in a vertical relationship since the distance existing between them is felt as sublime and necessary. It prevents any possibility for movement or progression. Research, exploration, take more naturally the form of a novel in their horizontal progression, whereas sublimity, verticality, would adopt more easily the stable and fixed structures of tragedy.

Robbe-Grillet's novels are not the end-products of his investigation. They represent the search itself and its various steps. The reader is invited to participate and has in fact an active and essential part to perform. He is asked first to free himself, along with the author, from ready-made ideas about literature and life (p. 149).

New novels are presented as intentional entities. The author makes very



*A-T. Tymieniecka (ed.), Analecta Husserliana, Vol. XVIII*, 401–406.

clear in his theoretical as well as fictional writings that his novels are not intended to reproduce, nor to express, the external spatiotemporal world. They are works of creation and acknowledge themselves as such. Posited as creations, they negate the reality of their physical object, maintain this reality at the required distance. It is a break from the naturalist illusion that conveniently entertained the confusion between the subjective entities of fiction and the physical spatiotemporal world they refer to.

Robbe-Grillet's search has therefore its own specific field of investigation, its own objects taken in their actual context and in their true nature.

In this self-regulated world of fiction which does not comply with the rules of the natural world nor with the accepted compromises of the naturalistic illusion, the traditional elements of the novel attain a specific new status. (1) Fictional time, for instance, becomes the actual duration of the unfolding of the novel. (2) Characters are poles of organization of elements. (3) Described objects are not important for themselves but rather for the actual movement of the description. (4) Metaphors lose their traditional characteristics of synthetical expressive devises to become structural articulatons of the text. Whereas expressive traditional metaphors serve as convenient bridges between the world of fiction and the spatiotemporal world, structural metaphors are common hinges allowing the communication of different levels or cells of a same text. Traditional metaphors are vertical relationships whereas structural metaphors are horizontal relations. (5) Words exhibit their dual and reversible nature of material entities and producers of meaning.

Meaning is less important as content than as mechanism of production.

Plots therefore multiply and often contradict themselves. Contradiction however is perceived only in relationships with the categories, rules and logic of the natural world. Polysemy does not infer any inconsistency in the logic of the fictional world, but is only multiplicity of meanings.

This systematic reduction to the world of the novel has an ultimate consequence: the fiction could refer only to itself. Intentional and dynamic entity, the novel refers to the dynamic progress that informs it, i.e., the creative process. As Jean Ricardou stated, the novel is not anymore the writing of an adventure but the adventure of writing.

Robbe-Grillet's novels are phenomenological investigations on the nature of the literary work of art. The author has transported the investigator in the fictional world and made the investigation in progress, as well as the various theoretical aperceptions on the nature of fiction, the very subjects of the novels. If, for instance, fiction is a narrative form which is experienced as unfolding noematically on time, the author sees to it that the reader actually *experiences* the unfolding consciously, knowingly.

The novel, *Un Régicide*,[2] has been chosen to illustrate the nature and the complexity of Robbe-Grillet's investigations.

Published in 1978, the book had been written in 1949 and is actually the author's first novel.

At first, the novel seems particularly hermetical, developing several unrelated plots, alluding to characters belonging to different places and times. It appears as a very complex intricacy of contradictory and unconnected events where the narration, for instance, switches from present to past, where confusions and fusions occur between agents and receivers, causes and consequences, objects and subjects.

However, specific structures become progressively visible. Characteristics of duality and reversibility are emerging. Little by little, three contexts are unfolding centered around three "characters." It becomes more and more apparent that there is somebody who is writing and who then dies in a room; there is also a first person narrator who is relating a story happening to him on an island, remote in space and time; there is a hero, acknowledged as such, called Boris, who lives in a contemporary and large city very similar to that of our spatiotemporal world.

The story of the islander is written in such a way that it allows for the unfolding of two different interpretations, on the part of the reader. The description of the islander's wanderings on the island, his difficult progression through high water and quick sands, reverse into the relation of the overflowing and restless activity of the creative process. The multiple and forceful solicitation of meaning overcomes, for instance, the scriptor of the text as water sweeps away the struggling man of the island.

The same text therefore generates two different readings: the struggles on the island and the creative process endeavor. Moreover, one reading is always exclusive of the other. When one is perceived, the other is not. The attention of the reader is entirely solicited by one of the contexts. But this attention can be switched, stolen, for the benefit of the other. This is the reason why one story is actually reversing into the other. Moving back and forth from one context to the other, the reader *experiences* the dual and reversible nature of the text. The text is perceived in its materiality and in its quality of illusory world of meaning. The characteristic of textuality and of referentiality of language is therefore directly apprehended by the reader. Moreover, the reversing of one story into the other presents further consequences.

The creative activity is performed by the scriptor locked in his room. The story emerging from this activity is the story of the narrator/islander. The story of the islander is unveiling therefore as the creation of the scriptor. The quality of duality and reversibility of the two contexts is a manifestation of the relationships between the creating agent and the created material. The activity of the scriptor can be compared to the noetic act and the story of the island to the noematic content, as the two sides of a same intentional act.

Like the scriptor and the narrator, the narrator and the hero entertain specific relationships. The first person narrator relates the story of the island in the present tense and describes his island dissolving in the midst of hazes and rains, or disintegrating under the burning sun. It is a mysterious place, out of time, difficult to locate in space, inhabited by strange and legendary characters as mermaids for instance, or heroes of old sailor's songs. It evokes the mysterious realm of dreams and the imagery used discloses the obscure and disquieting regions of subconsciousness.

The story of the hero recounts, in the past tense – the literary French *passé simple* – the adventures of Boris, acknowledged as "the hero," who lives in a modern and busy large city, and plans to kill the king of the country.

The same characteristic of duality and reversibility observed before between scriptor's and narrator's stories will be perceived between the reality-like story of Boris and the consciousness-like story of the narrator. Repetitively, one of the stories will suddenly reverse into the other. For instance, the description of the rhythm of the waves of the island shows through the depiction of the crowd cheering and acclaiming the king and vice-versa (pp. 178–79). The description of the hazy, humid, almost nauseous atmosphere of the island suddenly takes the place of the stream of Boris' consciousness, benumbed by the inactivity and boredom of a typical Sartrian Sunday morning (pp. 22–23).

Boris' fiction evoking situations of the spatiotemporal world progressively unfolds as a product of the creative narrative activity of the islander/narrator. Fiction and narrator are the two sides of the same intentional act, the narrator being, this time, associated with the noetic activity when Boris' fiction is the noematic content. Thus again the characteristic of duality and reversibility of the two contexts as the two sides of a same intentional entity but also their specific nature. The subjective operating entity is described as a consciousness, the narrator's, whereas the intentional structured content is described as the fictional world of Boris.

The novel, therefore, discloses three levels corresponding to three specific entities; the scriptor, the narrator and the hero. The scriptor's context is that of language and creative process in progress; the narrator's, that of a mind at the limits of the subconscious; the hero's, a world of fiction. The three entities are also the extensions of the same intentional act, the hero and his world being the creation of the narrator whose self and world are the creation of the scriptor whose principle and world are the creation of the scriptor whose principle and world are created by the author. The multilayered structure of the novel is similar to a series of encased Chinese boxes.

The book ends with the death of the scriptor, the organizing principle, and enunciator of the text. By his death, the scriptor puts a simple stop to all activities in progress. The death takes place in a room located at the same time on the island and in the city, narrator and hero having returned to their common point of origin: the scriptor from whom they had extended.

Same and different at the same time, scriptor, narrator and hero, all die from a similar kind of death but, *in* their own context, and *from* their own context. The scriptor is suffocated by unrestrained language; the narrator following his own path and self, cause of his own fate, meets his progressive mental dissolution; the hero Boris becomes his own murderer, the agent of his own physical death.

The novel is presented as being at the same time a laboratory of language with the struggling scriptor, a mental landscape with the wandering narrator and a world of meaning, of fiction, with the hero Boris. Language, psychic, meaning are all parts of the composite substance of the narrative entity.

Made of this substance of tripartite composition, the novel has appeared at first as a juxtaposition of three different, independent stories. They have progressively unfolded as double reverses of one another. They are, in fact, "equivalents," i.e., similar but belonging to different contexts. Superimposed on each other, imprinted or molded in different substances, each of them, belonging to one of the three layers of the tripartite substance – language, mental and meaning. They are consubstantial with one another.

The construction of the novel presents such precision and complexity that should the book be opened at any page, it's three layered structure and tripartite substance would soon unfold.

Moreover, the three layered substance is perceived by the reader in its nature and in its dynamics and vitality through the constant reversing of one layer into the other. Since the three layers are exclusive of one another and equally strong in their successive solicitation of the reader's attention, the literary substance is actually perceived as an animated entity.

Minutely constructed and elaborated, revealing its own functioning, unfolding its different components, animated by its own movement, self-contained, perfectly tight from any intrusion from the natural world, a novel as *Un Régicide* is an actual literary object. With its content tightly folded, its mechanisms set, it is a specific and separate entity awaiting its use by the reader.

Offering an actual apperception of the narrative form as an animated entity and in the complexity of its different components, this experience represents an actual encounter with the narrative literary work of art in its essentiality; this apperception being equated, moreover with the aesthetic experience.

Instead of presenting his novels as some devices permitting access to the physical, external world, Alain Robbe-Grillet offers them to the reader as essential entities, providing a true enlightenment on the nature of literarity. The author opens a new field of phenomenological investigation of the nature of the literary work of art. This investigation is presented as an exploration following an horizontal progression. Each step of this exploration takes the form of a literary creation, a novel. The advance of Robbe-Grillet's exploration materializes into the progression of the creative production of the author.

Modern man has realized that he has failed in his attempt to entrap the natural world in his systems of references. He has lost his faith in the anthropomorphic qualities he had invested in his world. He has understood that perception is a creative activity in continual progress and that his contact with his world exists only through a rapport of parallelism and horizontality.

It is understandable therefore that contemporary artists as Robbe-Grillet reject the concept of tragedy, since its characteristics of staticity, sublimity, verticality do not correspond to man's apperception of his relationships with his world nor its stable and fixed structure to their formulation.

It is through the forms that man creates and especially through the perception of their true nature that modern man can finally bring some signification to his world.

Contemporary artists have therefore chosen to explore and to analyze this process of creative activity. Robbe-Grillet and his followers have adopted the novel as the literary form the most congruous to man's perception and exploration, that is to say, the new novel – the self-reflexive process of the creative activity.

*University of Richmond*

NOTES

[1] Alain Robbe-Grillet, *Pour un nouveau roman* (Paris: Gallimard, 1968).
[2] Alain Robbe-Grillet, *Un Régicide* (Paris: Editions de Minuit, 1978).

## C. THE MODES OF THE TRAGIC

JOHN LYONS

# TRAGIC CLOSURE AND THE CORNELIAN WAGER

Cornelian tragedy challenges many of our modern conceptions of the genre. It is tragedy without the tragic,[1] a conception which may be authentically classical[2] but which is far from the definite fall we frequently associate with the genre.[3] Attention has been drawn towards the opposition of Cornelian and Racinian tragedy primarily in terms of the optimistic aristocratic spirit versus the pessimistic Jansenist view of the world, or towards the structure of plays based in one case on suspenseful plot (Corneille) and in the other on a sense of complete certainty of the final outcome (Racine).

I would like to reconsider the Cornelian tragedy with regard to these two traditional descriptions, the questions of suspense and optimism, and to suggest ways in which they may be related. In doing so I will draw attention to two aspects of the ending of Corneille's tragedies.

Georges May, in *Tragédie cornélienne, tragédie racinienne* (1948),[4] analyzed at length the mechanisms by which Corneille created and maintained suspense, which May describes as the fundamental emotion in Corneille's theater. Thus pity and fear, the traditional emotions provoked by tragedy, are not major elements of Corneille's theory or practice: "la crainte sinon sous la forme adultérée et intensifiée de curiosité, ni de la pitié sinon sous la forme de surprise" (p. 22). As May describes the playwright's aims they are "essentiellement de retenir l'attention du public d'éviter qu'il ne soit indifférent à l'action, et, par conséquent, ne s'en détache . . ." (p. 61). In plot construction the consequence of this approach is to introduce many unforeseen obstacles or changes, to interrupt the performance precisely at moments when a new obstacle arises, and to delay the *dénouement* until the extreme end of the play. The end of the fourth act of Corneille's works is particularly suspenseful. An obstacle is encountered just before the end of the act, and the spectators have to spend the intermission in suspense. Corneille favored the playing of violins during the intermission to maintain this mood of some impending disaster (May, pp. 79–84). Then, in the fifth act the "catastrophe" or resolution is kept for the very last moment. In his first *Discourse* (1660), Corneille wrote of the fifth act: "Il faut, s'il se peut, lui réserver toute la catastrophe et même la reculer vers la fin, autant qu'il est possible. Plus on la diffère, plus les esprits demeurent suspendus. . . ."[5]



*A-T. Tymieniecka (ed.), Analecta Husserliana, Vol. XVIII*, 409–415.

These characteristics of the Cornelian theatre are well known. I mention them only as preliminary to consideration of the subject that has received relatively little attention except perhaps from Corneille himself: the problematic closure of the great Cornelian tragedies. It is notorious that *Le Cid* (1637) provoked one of the great quarrels of French literary history, producing no less than thirty-six pamphlets in favor of the play or against it. The primary defect of the work seems to have been the apparent reconciliation of Chimène and Rodrigue. Should Chimène marry Rodrigue, henceforth called Cid, or should she forever refuse such a marriage on the grounds that her respect for her own honor obliged her to abhor such an acceptance of the man who killed her father? In this debate at least three approaches present themselves: the ethical ("What is Chimène's duty?"), the dramatic ("What is necessary to complete the action of a tragedy?"), and the purely hermeneutic ("What does happen at the end of the play? Does she or doesn't she accept the marriage ordered by the king?).

Corneille himself, writing twenty-seven years after the first performance, addresses these three questions. He suggests that Chimène's honor requires that she refuse the marriage and that Rodrigue must be saved from the risk of death: "Il est vrai que dans ce sujet il faut se contenter de tirer Rodrigue de péril, sans le pousser jusqu'à son mariage avec Chimène" (p. 219). But when he reaches the question of what does happen at the end of the play, Corneille indulges in a certain equivocation. Of the marriage he writes, "Pour ne pas contredire l'histoire, j'ai cru ne me pouvoir dispenser d'en jeter quelque idée, mais avec incertitude de l'effet; et ce n'était que par là que je pouvais accorder la bienséance du théatre avec la vérité de l'événement" (p. 219). Thus, in Corneille's own words (but the Corneille of 1660, who is, like us, critic and observer) *Le Cid* ends on a note of uncertainty, an uncertainty that is desirable as a means of reconciling historical and dramatic requirements. The ending of the tragedy is a compromise which permits the spectators to believe what they want. If they wish to accentuate the moral duty of Chimène, they can follow Corneille's interpretation of Chimène's silence: "Je sais bien que le silence passe d'ordinaire pour une marque de consentement; mais quand les rois parlent, c'en est une de contradiction: on ne manque jamais à leur applaudir quand on entre dans leurs sentiments . . ." (p. 219). On the other hand, if the spectators wish to reconcile dramatic representation with historical belief, they may content themselves with the knowledge that the king has ordered a marriage and that no explicit verbal statement within the play indicates that this marriage will not take place: "Cet Hymen différé ne rompt point une loi / Qui sans marquer de temps, lui destine ta foi. / Prends un an, si tu veux, pour essuyer tes larmes" (vv. 1819–21). Without Chimène's signified consent, a marriage has been prescribed by the authority of law.

Silence and uncertainty are the keys to this *dénouement*, poised on the threshold between the dynamics of the plot and the "facts" of history. The plot and history will not enter into a collision at the end of the work. Instead, by an ingenious disposition of silences, or "holes", they will blend or mesh in different ways according the interpretations of different spectators and of subsequent historical periods in the play's career. The marriage is not the only event that is thus left in suspense. Its institutional contrary, so to speak, the funeral of Chimène's father, slips into silence: "Le moindre mot que j'en eusse laissé dire," wrote Corneille, "pour en prendre soin, eût rompu toute la chaleur de l'attention et remplie l'auditeur d'une facheuse idée. J'ai cru plus à propos de les dérober à son imagination par mon silence . . ." (p. 220). Hence, in Corneille's own description of *Le Cid*, the ending is characterized by a tension between the apparently total conclusion of the plot, the end of the perils to which the protagonist is exposed, and, on the other hand, a number of factors which weigh against such a sense of ending. The particular brilliance of Corneille's construction is to maintain the suspense and thus the public's attention until the very last scene. This type of construction, which May points to as being typical of Corneille's plots after *Horace* (e.g., *Hèraclius*), is actually right here in his first great tragedy. But the price of this postponement of the resolution is that the dynamism of the conflict is never effectively arrested but only concealed; Corneille's two uses of "silence" in his description of the *dénouement* are revelatory of this effort to conceal a certain number of ongoing problems at the end of the play.

Let us look at another Cornelian tragedy which has a closure that is, at first glance, radically different, *Cinna ou la Clémence d'Auguste* (1642). In this play the suspense is maintained by the uncertainty of the fate of Cinna which depends on the decision that Auguste will make in the last scene. The emperor has been, during the entire length of the play, the object of a conspiracy mounted by Cinna and Emilie. Both have been protected and favored by Auguste, but Cinna has resolved to make himself the instrument of Emilie's vengeance on Auguste, who had killed her father. Maxime, also trusted by Auguste, has become involved in this conspiracy out of a desire for justice. In the last act Auguste, during an interrogation of Cinna, learns suddenly of Emilie's part in the conspiracy and then Maxime's. Since Auguste had already shown considerable impatience at Livie's counsels of clemency in act 4, how can the evidence of spreading disaffection change the emperor's mind and argue for mercy? The outcome of this act is, however, the pardon of all the conspirators. Again the last scene is based on the imposition *in extremis* of a royal or imperial decision, a decision against which the plot itself had furnished ample argument. The acceleration of avowals of betrayal during the final scene does, in one sense, seem to demonstrate the fragility of human gratitude. The final decision to pardon all the conspirators seems therefore to be much less a demonstration of political acumen than a direct display of conquest of self. Auguste's control is not necessarily reinforced over the others, but it *is* increased for himself. The result is an ending which again leaves a large problem open. Will Auguste's decision lead to the end of his reign, will there be more conspirators, or will the clemency instead lead to a previously unknown peace? The answer is clear. We know, historically, that the reign of Augustus Caesar was not brought down by conspiracy. Yet the play itself has offered little ground for the assumption that past wrongs can be forgotten (e.g., Emilie) or that Auguste can live without exercising a tyranny. The moments after the play, the unforeseen future of this dramatic world, is projected, by Livie's speech that comes within six verses of the final curtain, as a reversed image of the world contained within the play: "Après cette action vous n'avez rien à craindre: / On portera le joug désormais sans se plaindre; / Et les plus indomptés, renversant leurs projets, / Mettront toute leur gloire à mourir vos sujects ..." (vv. 1757–60). On the other hand, no indications within the play show clemency as a policy capable of producing such a future, for gratitude stifles neither filial vendetta, nor desire, nor the principled love of freedom. By accumulating the revelations of conspiracy in the last act, Corneille maintains suspense but also creates a pattern of betrayal against which Auguste's decision stands as particularly surprising and heroic. In terms of the dynamics of the plot nothing justifies Livie's optimistic prediction. Auguste's act stands, therefore, as a *wager*. It leaves open the question of the political effect of the decision.

In *Horace*, another king makes another last-minute decision. Here, as in *Le Cid* and *Cinna*, the authority which restrains the conflicts that are still active until within a few verses of the end does so in the absence of assurance that the causes of the conflict will cease. The accused murderer and greatest hero are one person. Horace, having killed his sister Camille for her insulting accusations of brutality and barbarism (he has just won his glory by killing her lover), submits to royal justice. He agrees that the king may order his death; in fact, he begs for death. Yet, his reasons betray an unresolved difference in values: "La mort seule aujourd'hui peut conserver ma gloire. / Encore la fallait-il sitôt que j'eus vaincu, / Puisque pour mon honneur j'ai déjà trop vécu. / Un homme tel que moi voit sa gloire ternie, / Quand il tombe en péril de quelque ignominie . . ." (vv. 1580–84). Although Horace is willing to admit that he has fallen into disgrace, he does not repent of his act, nor does he see it as an abberration. It is rather a consequence of his being what he is, "un homme tel que moi." There is no reason then to assume that the death of Horace will guarantee Rome against future bloodshed of the same sort, should the circumstances arise. Horace's claim that Tulle can execute him without danger to the state is double-edged: "Rome ne manque point de généreux guerriers, / Assez d'autres sans moi soutiendront vos lauriers" (vv. 1589–90). If there are many warriors like Horace there are also many potential assassins of those who criticise them. The king chooses not to lose Horace ("Ta vertu met ta gloire au-dessus de ton crime, / Sa chaleur généreuse a produit ton forfait" (vv. 1760–61), but he concludes with an allusion, hopeful but not without uncertainty, to the unresolved conflicts: "Mais nous devons aux Dieux demain un sacrifice, / Et nous aurions le ciel à nos voeux mal propice, / Si nos prêtres, avant que de sacrifier, / Ne trouvaient les moyens de le purifier. / Son père en prendra soin, il lui sera facile / D'apaiser tout d'un temps les mânes de Camille" (vv. 1771–76). There is, within what precedes, no assurance that the *mânes* are easily appeased, no indication that Horace has assumed a new attitude towards his glory, nor that the fundamental code leading to the tragic event has been altered.

In these three works, Corneille in various ways makes the imposition of the *dénouement* a decision which does not entirely arrest the ongoing problematic of the plays, but merely attempts to fasten the spectator's attention on an apparently authoritative gesture of conciliation. In doing so, he is conscious, at least in *Le Cid*, that the forces preventing resolution earlier in the play are not calmed by such a gesture, or are calmed only by the will of the king to the extent that that will identifies itself with dramatic necessity. In *Le Cid*, the dramatically necessary saving of Rodrigue is in conflict with the equally necessary preservation of the honor of Chimène. Corneille deals with the problem by rejecting it into the post-dramaturgic world of history, with a statement from the king which is not ratified from within the world of the tragedy. In the case of *Cinna* a similar prophetic statement is made by Livie. Then political authority as incarnate in Auguste assumes that statement and imposes it by will: "J'en accepte l'augure, et j'ose l'espérer: / Ainsi toujours les dieux vous daignent inspirer! / . . . / Et que vos conjurés entendent publier / Qu'Auguste a tout appris et veut tout oublier" (vv. 1775–76, 1779–80). The union of prophecy and will in the king in *Le Cid* is here broken in two. The political power "dares to accept" the augury, thereby revealing its own potential disappearance as authority. At this juncture, the threshold between drama and post-dramaturgic history, the latter fills the gap of tragedy. The procedure of *Le Cid* is reversed, for in the case of *Cinna* we are expected to believe in the lesson of history (Auguste will prosper in peace) where in *Le Cid* we were expected to suspend the historical account in favor of the defiant silence of Chimène. In *Horace* we know also that the future towards which the play opens as the curtain falls will not be marked by further rebellion of the triumphant but humiliated hero. Yet we also know that historical reassurance is necessary for us to walk away from the theatre with such certainty.

I mentioned earlier that Cornelian suspense is, in my view, linked to Cornelian optimism. By this I do not mean optimism in regard to any individual or any role (e.g., the "hero") nor even in regard to humanity, conceived as a species with the capacity for self-regulation and improvement. Instead, Cornelian optimism displays itself as an opening towards the future in the belief that the future itself will or must bring the resolution of conflicts for which no remedy is apparent within the world of the play. This trust in the future does not mean that Corneille has a particular faith in history. The playwright's comments on *Le Cid* reveal, on the contrary, an opposition between Corneille's reading of the implications of the dramatic conclusion and the historical reality which he concedes. But the effect of the closure is to produce in the spectator the *desire* to trust in the conciliatory gesture of authority. Such closure emphasizes the coincidence of the ethical and the dramatic while allowing the hermeneutic to pass into the background. The desire to see the hero removed from peril – the Cornelian requirement for the end of tragedy – and the desire for the triumph of some ethical principle are sufficient to make the spectator cling to the gesture by which the play ends. But this gesture only functions by assuring us that the problems raised by the plot will somehow be resolved in the future. The last words of the king in *Le Cid* are significant for all three plays: "Laisse faire le temps, ta vaillance et ton roi." Time, heroic obedience, and royal authority are conjugated here in a way that removes suspense, at last, but only at the cost of a blind confidence not in true historical time but in time as it is projected by the authority of the play itself. One of the traditional comparisons between Racine and Corneille holds that Racine shows mankind as it is while Corneille shows mankind as it should be. Or, in the words of my colleague Lawrence Harvey, "In Racine the gap between the real and the ideal cannot be bridged. In Corneille it can, but at a cost, which is higher in some plays than in others."[6] This "cost" is not only within the fable of the play (for instance, accepting the death of Camille) but in the structure, for the ideal world of the play is never entirely at peace with historical reality to which we return in leaving the theatre. This gap, this opening, this wager constitutes one of the important characteristics of Cornelian tragedy. It suggests that suspense is not merely a mechanism for retaining the attention of the spectator (without comparable means Racine succeeds in holding the audience) but an aspect of the Cornelian world-view. There are conflicts between the ideal (ethical, dramatic) world and the world of experience. Cornelian tragedy depends, for its type of closure, on finding a public that is willing to "pay the cost" of confidence in the future.

*Dartmouth College*

## NOTES

[1] Distinction made by Henri Gouhier in 'Tragique et transcendance' (1962), quoted in T. J. Reiss, *Tragedy and Truth* (New Haven: Yale University Press, 1980), p. 1. The absence of the tragic from Corneille's plays has been affirmed by R. J. Nelson in *Corneille: His Heroes and their Worlds* (Philadelphia: University of Pennsylvania Press, 1963).
[2] As we are reminded by Lucas in *Tragedy: Serious Drama in Relation to Aristotle's Poetics*, rev. ed. (New York, 1962) and, of course by any rereading of the *Poetics*, section 53 b 1.
[3] R. B. Sewall, *The Vision of Tragedy* (New Haven: Yale University Press, 1959), p. 31.
[4] Urbana: University of Illinois Press, 1948. See also J. Boorsch, 'Remarques sur la technique dramatique de Corneille,' *Yale Romanic Studies* 18 (1941), 101–62.
[5] Pierre Corneille, *Oeuvres complètes*, ed. André Stegmann (Paris: Le Seuil, Collection 'L'Intégrale,' 1963), p. 829. All quotations from Corneille are from this edition.
[6] Lawrence E. Harvey, reviewing R. J. Nelson, *Corneille*, in *MLN* 82, no. 3 (May 1967): 378.

BARBARA WOSHINSKY

# INTUITION IN *BRITANNICUS*

This study represents a stage of my research into the epistemology of French classical tragedy: the relation in classical literature between words and truth.[1] Alongside this theoretical focus, I am methodologically and temperamentally committed to the close analysis of texts: my work attempts to unite literary theory with practical criticism. In this paper, I will begin by briefly outlining the hypothesis of the classical *episteme* as it has been developed by Michel Foucault, Louis Marin,[2] and others; then I will discuss the place of tragedy within this theory of classical discourse, referring to Timothy Reiss' *Tragedy and Truth*. In Reiss' book, which is an intelligent extension of Foucault's theories to literature, he argues that French classical tragedy is central to, and indeed helped constitute, the truth-centered discourse of the seventeenth century.

The central part of my paper consists of a study of Racine's *Britannicus*, through a close reading of one passage which forms, in my view, the poetic center of the play. This reading has led me to certain conclusions about the truth or truths contained in Racine's text. In addition, I think the examination of this one specific text may help us to clarify the notion of literary truth in general.

According to Michel Foucault (in *Les Mots et les choses*) the classical era was marked by a dramatic change in linguistic views. No longer a concealer of mysteries, language becomes a conveyer of knowledge. During the Middle Ages and the Renaissance, language was considered a gift of God to man, which still retained some relation to the divine, through magic, prayer, and incantation. The study of ancient languages and etymologies was seen as another possible access to divine truth. Greek, for example, was thought more authentic than French because it was closer in time to that original language spoken in the Garden of Eden, when man and God were still communicating freely. This concept of language also affected the status of what we now call literature: a written work was thought to contain a secret meaning, wrapped in cipher, allegory, or emblem, which the reader must strive to decode. Thus the act of reading becomes a hunt for a secret treasure buried at the heart of the text – at the heart of language.



*A-T. Tymieniecka (ed.), Analecta Husserliana, Vol. XVIII*, 417–423.

According to Foucault, with the onset of the classical age, this concept of language changes. Words are no secret any more; they no longer participate in divine mystery; rather, they are a human creation designed to express exclusively human thoughts. The Port-Royal *Grammar*, published in 1660, begins: "Parlez, est expliquer ses pensées par des signes, que les hommes ont inventez à ce dessin."[3] This view of language also carries with it a particular notion of temporality: thought moves through time in a linear sequence. According to Hobbes in the *Leviathan*, "The general use of speech is to transfer our mental discourse into verbal; or the train of our thoughts into a train of words."[4] Language represents (transfers) thought; speech is linear because thought is linear.

What is the relation of these philosophical concepts to classical tragedy? According to Timothy Reiss, tragedy participates in a network of discourses (both scientific and non-scientific) which mesh together to create the "reality" of the classical *episteme*. With some obvious and necessary differences, he sees the development of classical tragedy as analogous to that of the experimental sciences. The culmination of this development is manifest in the plays of Racine, where "the character becomes a *person* endowed with psychological depth, and tragedy is taken as a container of certain knowledge showing the moral, psychological and so forth functioning of the individual and society" (*Truth and Tragedy*, p. 5). Reiss expresses his fundamental thesis in the dense formula, a play is a "thought experiment" (p. 219). This formula implies that a work of literature presumably shares the same epistemological status as a scientific treatise.

Does classical tragedy indeed form an integral, undifferentiated part of the philosophical discourse of its time? Or to put it another way, do the philosophers really seek the same truth as the poets? I will hazard an answer to these questions after taking a close look at Racine's *Britannicus. Britannicus* is probably not Racine's best, and certainly not his best-known, tragedy; but it is an interesting one for our purposes, for historical as well as textual reasons. *Britannicus* was Racine's second major play; after the success of *Andromaque*, his contemporaries expected great things of him. But when *Britannicus* was produced in 1669, reactions varied from disappointment to consternation. We will return to the question of audience reception after examining the internal workings of the play.

*Britannicus* takes place in Rome, near the beginning of Nero's reign. At the time the play begins, Nero has been a model emperor for three years. But portents are not good. Agrippina fears her son is beginning to resent her authority over him and the empire. The night before the action begins, Nero has performed a decisive act, perhaps the first of his career: he has ordered his guards to carry off Junia. Through this abduction, Racine poses the central dramatic question of the play: what direction will Nero take? Will he continue in the path of virtue, or is this criminal act a first step on the road to tyranny? At the beginning of Act II, in a famous passage, Nero himself answers some of these questions. He tells us he is in love with Junia, and wishes to divorce his wife in order to marry her. He goes on to describe his first sight of Junia and the thoughts and feelings it aroused in him. Nero's monologue is a luminous example of what one writer has called "charged speech";[5] its poetic energy has already attracted the attention of many perceptive critics.[6] In my own analysis of the passage, I will focus on the issues I raised earlier concerning the relation between poetic language and truth.

Excité d'un désir curieux,
Cette nuit je l'ai vue arriver en ces lieux,
Triste, levant au ciel ses yeux mouillés de larmes,
Qui brillaient au travers des flambeaux et des armes,
Belle, sans ornement, dans le simple appareil
D'une beauté qu'on vient d'arracher au sommeil.
Que veux-tu? Je ne sais si cette négligence,
Les ombres, les flambeaux, les cris et le silence,
Et le farouche aspect de ses fiers ravisseurs,
Relevaient de ses yeux les timides douceurs.
Quoi qu'il en soit, ravi d'une si belle vue,
J'ai voulu lui parler, et ma voix s'est perdue:
Immobile, saisi d'un long étonnement,
Je l'ai laissé passer dans son appartement.
J'ai passé dans le mien. C'est là que, solitaire,
De son image en vain j'ai voulu me distraire
Trop présente à mes yeux je croyais lui parler;
J'aimais jusqu'à ses pleurs que je faisais couler.
Quelquefois, mais trop tard, je lui demandais grâce:
J'employais les soupirs, et meme la menace.
Voila comme, occupé de mon nouvel amour,
Mes yeux, sans se fermer, ont attendu le jour.
Mais je m'en fais peut-être une trop belle image:
Elle m'est apparue avec trop d'avantage:
Narcisse, qu'en dis-tu?

(*Britannicus*, II, 2)[7]

This monologue encompasses the recital of two visions: the initial depiction of the "real" Junia, as she appeared to Nero the night before, later gives way to Nero's compelling visual fantasy of her. However, this simple distinction between fantasy and reality does not adequately characterize the passage. The woman, Junia, never exists independently; both visions are "fictions" created in Nero's mind and evoked through words. Junia's fictional status is clear in the syntax of the passage. Only once, at the end of the text, is Junia allowed to become a grammatical subject (*Elle*, 1.24). Everywhere else, she is the direct object, the *l'*, the silent *e* of Nero's gaze (1.2). Similarly, two lines begin with "disembodied adjectives": [*Triste* (3) and *Belle* (5)] which have no substantive referent in the passage. Lines 3 and 5 also reinforce another aspect of the text: its static, *tableau*-like quality. The capture scene, occupying the first half (up to line 15), is structured on a series of oppositions which cancel one another out to produce an effect of stillness. For example, the perfectly balanced line 8, "Les ombres, les flambeaux, les cris et le silence," creates an effect reminiscent of classical painting – a Poussin "Rape of the Sabine Women." On a psychological level, this painterly representation reflects the essential coldness of Nero's erotic response: like Soames Forsythe in *A Man of Property*, Nero has the collector's eye.

Yet there is a more than esthetic distance separating beholder from beheld. While in her presence, Nero could not bring himself to speak to Junia (1.12). He lets the "real" woman disappear in order to transform her, in the second moment of his vision, from sight into image (*vue*, 11; *image*, 16) – an image which takes on a near-hallucinatory power. Alone in his room, Nero carries on whole conversations (or rather monologues) with the image he has created: "trop présente à mes yeux je croyais lui parler; / J'aimais jusqu'à ses pleurs que je faisais couler. / Quelquefois, mais trop tard, je lui demandais grâce: / J'employais les soupirs, et même la menace" (16–20). These lines contain the imperfect, a tense denoting in French repeated, continued and, in this case, obsessive activity. But their temporal referent is ambiguous: is Nero recalling or imagining? are the tears he caused to flow Junia's real tears, her imagined tears, or even, perhaps, the tears she will shed later on? All these allusions are present, in an abolition of time. For, as I will show, Nero's monolog contains not only a recollection of the past, but a preview of coming events. For example, in the very next scene, Nero uses both *soupirs* and *menaces* in an effort to win Junia. He sighs at her feet (*respirer à vos pieds*) but also warns her that if she turns him down she will regret it. Racine's language is naturally more elegant: "Et ne préférez point à la solide gloire / Des honneurs dont Cesar prétend vous revêtir, / La gloire d'un refus sujet au repentir" (II, 3, p. 257). When Britannicus appears in act 3, Nero's threats become more open: "Si vous n'avez appris à vous laisser conduire, / Vous êtes jeune encore, et l'on peut vous instruire" (III, 8, p. 271).

As all these examples indicate, Nero's monolog performs a function of "poetic pre-enactment": it serves as a blueprint for future dramatic action. Once one realizes this, another puzzling feature of the play becomes clarified. In act 5, scene 6, a scene which is no longer performed, Junia comes back on stage after Britannicus' death, and Nero tries to console her "par tous les soins que la tendresse inspire" (290). Nero's speech is interrupted by the arrival of his mother, who literally steps between her son and Junia, bringing the *tête-à-tête* and the scene to a close. Racine reports in his preface that spectators found this short scene odd (*étrange*): "Mais, disaient-ils, cela ne valait pas la peine de la faire revenir" (p. 234). So the scene was cut in subsequent performances. But our analysis of II, 2 makes it clear why Junia must come back in V, 6: to fulfil the poetic prophecy contained in line 19 ("Quelquefois, *Mais trop tard*, je lui demandais grâce.") When Nero asks Junia's forgiveness, it is certainly too late – Britannicus has been murdered at Nero's orders – but he must ask in order to show it is too late, and thereby close the fatal poetic circle opened in act 2.

I use the phrase "poetic circle" advisedly. For if one looks at the basic dramatic conception of *Britannicus*, one sees it is not linear, but circular: the action goes nowhere. One reason for this circularity is that the play embodies a fixed concept of dramatic character. Racine wrote of Nero in his preface, "Il a *toujours* été un très méchant homme. . . . je l'ai toujours regardé comme un monstre" (p. 233). In the play itself, Racine has Agrippina say to her son: "Vous êtes un ingrat, vous le futes toujours" (p. 278). Character does not develop; it simply *is*. On a linguistic level, the static nature of the play is underscored by the continual echoing of words, phrases and whole verses in the mouths of different characters. Racine calls our attention to this feature in II, 3 when Nero almost parodies to Junia the words Narcissus had spoken to him just moments before: Narcissus' "Que présage à mes yeux cette tristesse obscure" (p. 249) becomes Nero's "Lisez-vous dans mes yeux quelque triste présage?" (p. 254). In fact, everyone's future is mirrored in Nero's eyes. This mirroring effect is reinforced by the borrowing of lines from Nero's monolog. Nero says of Britannicus, "Je me fais de sa peine une image charmante" (II, 8, p. 261), a clear echo of his reference to Junia, "Je m'en fais peut-être une trop belle image" (II, 2, 1.23). Junia also echoes Nero's lines without having heard them, saying to Britannicus: "Votre image sans cesse est présente à mon âme" (III, 7, p. 268). Cp. Neron: "De son image en vain j'ai voulu me distraire. / Trop présente à mes yeux je croyais lui parler" (II, 2, 11.16–17). Agrippina's speeches also contain variations on Nero's monolog. For example, "J'essaierai tour à tour la force et la douceur" (266) paraphrases his "j'employais les soupirs et même la menace." As Racine remarked in the *Seconde Préface* to *Britannicus*: "Voici celle de mes tragédies que je puis dire que j'ai le plus travaillée" (p. 236). Indeed, it is difficult to see these echoes and repetitions as other than deliberate; they point to Nero's monolog as a poetic center from which everything else in the play arises.

To return, in conclusion, to our original question – whether the philosophers seek the same truth as the poets – the answer, for *Britannicus*, is yes and no. On the one hand, our analysis of the play's structure reveals little of the analytical-referential model of classical discourse. In the classical, Cartesian model, one arrives at knowledge through deduction – etymologically, the leading-out of concepts into a rational sequential order. In *Britannicus*, deduction is replaced by intuition – etymologically, seeing-in. In a moment of illumination, Nero, like the divinely crazed seer of Greek tragedy, both perceives the future and reveals it in his gaze. But this latter notion of truth is, by necessity, non-dramatic; discourse reveals truth, but cannot produce it. To make *Britannicus* a drama, Racine must create the impression of plot development, of freedom, of unfolding time – all of what he does, on the whole, quite competently. Nevertheless, as Racine's audience correctly sensed, the play is "out of line" with the orderly unrolling of a dramatic plot. According to one viewer's account, "The first act promises something wonderful (*quelque chose de fort beau*); the second maintains this promise; but by the third act, it seems as though the Author has given up; the fourth act does not help us forget how bored we were during the third; and the fifth is saved only by the pitiful end of Britannicus and the retreat of Junia *chez les vestales*."[8] From act 2 on, time in *Britannicus* resembles what Timothy Reiss calls, in his study of *Phèdre*, the "fabulous no-time" of myth and fable (*Truth and Tragedy*, p. 277). On the other hand, if we agree with Michael Black that poetic language attempts to bypass some of the processes of conceptualization in order to "put us in contact, as direct as can be, with the movement of other minds,"[9] *Britannicus* does fulfill the classical imperative to express thought: it represents a poet's mind at work. But this mental process takes a very different form from Hobbes' train of thought, rolling down its narrow gauge track. Perhaps a more apt metaphor for Racine's achievement is suggested by the mathematical-mystic Pascal, who defined eloquence as "un portrait de la pensée."

*University of Miami*

## NOTES

[1] Parts of this article appeared as, 'Truth in Literature: The Little Man Who Wasn't There,' Bunting Institute of Radcliffe College 'Working Paper,' 1980.

[2] Cf. Foucault, *Les Mots et les choses* (Paris: Gallimard, 1966), and Marin, *La Critique du discours* (Paris: Editions de Minuit, 1975).

[3] Reprint of 1660 edition (Menston, England: Scolar Press Limited, 1967), p. 5.

[4] Chap. 4, 'Of Speech' (Oxford: Clarendon Press, 1967), p. 24.

[5] Michael Black, *Poetic Drama as a Mirror of the Will* (New York: Harper and Row, 1977), p. 10.

[6] Cf. for example Serge Doubrovsky, 'L'arrivée de Junie dans *Britannicus*: La Tragédie d'une scène à l'autre,' *Littérature* 32: 27–54; Jules Brody, 'Les Yeux de César,' in *Studies in Seventeenth Century French Literature*, ed. J.-J. Demorest (Garden City, N.Y.: Anchor Books, 1966), pp. 185–200; Jean Starobinski, 'Racine et la poétique du regard,' in *L'Oeil vivant* (Paris: Gallimard, 1961), pp. 69–90.

[7] Racine, *Théâtre complet*, ed. Maurice Rat (Paris: Garnier, 1960), p. 239. Further references to *Britannicus* will appear by page number in the body of the text.

[8] Edmé Boursault, *Artemise et Poliante* (Paris, 1670), p. 12.

[9] *Poetic Drama as a Mirror of the Will*, p. 16.

MARILYN STEWART

# MYTH AND TRAGIC ACTION IN *LA CELESTINA* AND *ROMEO AND JULIET*

The fact that literature is understood in the modern world to refer primarily to written texts may blind us to the link between literary genre and memory. The poet inherits genre as he inherits a language; he discovers artistic ways of seeing in the conventions and compositional forms of his art as "integral, materially expressed inner acts of man's orientation in reality."[1] The created world of tragedy, with its own space and time, is not so closely bound to the reality of everyday life as it is bound to the memory of a tragic way of seeing and representing existence. The esthetic "distance" in space and time so necessary to the genre of tragedy was discovered by the ancients, who linked the insights of tragic art to myth; this distance between mythic action and events in everyday life is recreated and "remembered" in tragic works of art throughout the course of literary history. When a poet begins to see and to express a fiction in a particular way, his work moves within a tragic orientation.

The seductive power of art is such that particularly memorable characters, Romeo and Juliet for example, seem to take on a reality of their own. We speak of their "personalities" as if they were individuals and analyze their behavior as if to find a way their deaths might have been averted. Yet, of course, Romeo and Juliet are not human beings; they are artistic creations, abstractions or extractions from human life given an esthetic form or shape. It is the whole "action" of Shakespeare's tragedy, *Romeo and Juliet*, that is analogous to human life; and hence the generic orientation of such a poetic work, not its verisimilitude, gives shape to dramatic characters. Dramatic action has a certain integrity in performance (and reading may be a kind of performance) that any analysis after the fact may overlook.

The curious fact is that what may appear to be an identical action – the death of young lovers, for example – may be dramatized from several different perspectives. In *A Midsummer Night's Dream*, Shakespeare parodies the very plot which serves as the "soul" of the tragedy of *Romeo and Juliet*. In this festive comedy, a group of citizens dramatizes the story of the mythic lovers Pyramus and Thisbe in dubious honor of the wedding of their King. What would be obstacles from the point of view of tragedy – the wall between the lovers' houses, the mis-timing of the lovers' meeting, and the



*A-T. Tymieniecka (ed.), Analecta Husserliana, Vol. XVIII*, 425–433.

suicide of the lovers – become, in the antics of Bottom and company, laughable incongruities, which nevertheless celebrate, as befits comic action, the joy of a wedding feast. Thus, it cannot be myth in the sense of "basic plot" that generates tragedy.

The basic plot of star-crossed lovers is at least as old as Ovid's *Metamorphoses*; it was, in fact, a favorite tale of Renaissance poets.[2] Despite the age and popularity of the story it tells, however, Shakespeare's *Romeo and Juliet* is the first tragedy to dramatize such a tale. A brief comparison of *Romeo and Juliet* with a Spanish drama published a century earlier, in 1499, indicates the usefulness of understanding genre as esthetic perspective.

Indeed, Fernando de Rojas' *La Celestina* bears a striking resemblance to *Romeo and Juliet*, though it is highly unlikely that Rojas' version of this story served as Shakespeare's source. *La Celestina* probably did, on the other hand, influence Shakespeare's contemporary Miguel de Cervantes. It is worth noting that it was in printed form rather than in performance that Rojas' play achieved popularity; in fact, some scholars recognize its comic orientation by calling it a "picaresque novel in dialogue."[3] The hero of *La Celestina*, like Romeo, falls in love at first sight with a beautiful young lady of noble birth, but the love is forbidden and must be kept from public knowledge. Like Romeo, Calisto risks his honor to scale a garden wall surrounding his mistress' house. In both dramas, the lady is promised by her unsuspecting parents to another suitor, and in both the lovers' union is fostered by a feminine go-between who garrulously affirms the delights of passion. Each young heroine takes her life to join her lover in death. However, though the theme of both dramas obviously concerns the idealization of passionate love, each play gives a very different perspective of meaning to this ideal. *La Celestina* is a comedy and *Romeo and Juliet* a tragedy; the difference lies in the attitude or perspective within which the action is dramatized.

To be sure, Rojas sees his plot seriously; he proclaims in the prologue that *La Celestina* is "composed as a lesson for those crazy lovers who are so carried away by their unruly passions that they worship their mistresses as divinities. . . ."[4] This prologue serves an esthetic function similar to that of a chorus: it embodies the practical wisdom of social man. The prologue adds that the tale should also serve as ". . . a warning against the wiles of procuresses and the lies of wicked servants." It should be noted that, whereas the choral introduction of *Romeo and Juliet* directs attention to the strife of warring families and to the role of Fate in the "misadventured piteous overthrow" of lovers' dreams, here attention is called to those human traits which everyman sees as the cause of disorder and disruption in the business of daily life: lust, greed, and deceit.

Rojas' young lovers resemble characters from medieval romance: Calisto, a noble youth, meets the beautiful but arrogant lady Melibea by accident and falls instantly in love. He prides himself on his suffering, following the conventions which, in a medieval romance, elevate passion: he isolates himself, composes poetry, and reflects on his "tempestuous grief." In the space and time of this comic world, however, such passion descends to the level of a practical inconvenience. Calisto's servant Sempronio, who cannot see the esthetic satisfaction of his master's action, advises him to come down to earth:

> Stop your high-flown language, sir. Let's have no more of your poetry. Speech that's not common to all, that's not shared by all, and that all do not understand is not good speech. (p. 135)

In *La Celestina*, as in comedy in general, "good speech" is well-mannered but prosaic, often enlivened by the proverbial wisdom of folk sayings.[5] Parmeno, another servant, hesitates to be disloyal to his master, but comes to the practical decision to follow "the way of the world":

> Since traitors are called wise and loyal men fools, I'll go with the crowd. . . . Let him break and destroy, let him smash and spoil, and give his property to procuresses, I'll take my share! There's good fishing in troubled waters, they say. (p. 60)

Romeo's companions Benvolio and Mercutio, of course, also find a lover's suffering ridiculous if not incomprehensible, but they remain loyal to Romeo. Calisto's servants, however, decide to put their master's passion to their own material advantage by making a deal with Celestina, the local bawd.

These servants have love affairs of their own which parody the passions of their betters. The motifs of the angry and impatient young mistress (Elicia), love at first sight (Parmeno and Areusa), and the encounter of love with death (Sempronio and Parmeno are killed by the Night Watch) serve to degrade the potential nobility of the passion of Calisto and Melibea. Actually, Melibea's counterparts, who are both prostitutes, take what is, from within the perspective of *La Celestina*, a more sensible course of action in the face of their lovers' deaths: they mourn but not excessively. Areusa's response to Elicia's news of their lovers' deaths exhibits good common sense:

> O Parmeno, my love! What sorrow I feel for his death! I am sorry I fell so deeply in love with him so quickly, since our love was to be so short. But since the bad news has come,

since the disaster has happened, and since their lives are lost and can't be restored by tears, don't be so distressed or you'll go blind with weeping. For I think your sorrow is only a little greater than mine, and you see how patiently I bear my loss. (p. 207)

The two "ladies" honor the memory of Celestina by continuing to go about their "business."

Rojas' original title was *The Comedy or Tragi-Comedy of Calisto and Melibea*. That the work became known as *La Celestina* probably reflects the popularity of the intriguing bawd. Though the initial scenes of the play depict Calisto's lovesickness and his servants' disloyalty, once Celestina comes on stage she dominates the action. As Gerald Brenan says,

Celestina's philosophy is that of the person who lives for and by sexual love. Everything erotic charms her: the approaches, the intrigues, the confidences, the deceptions, the singing and dancing and feasting, the act itself. For her, love-making takes precedent over all other things. . . .[6]

The ill-fated passion of Rojas' lovers serves to underscore the powerful eroticism of Celestina's image; whereas Juliet eventually scorns the advice of her Nurse, Melibea entirely trusts herself to Celestina's "wisdom." Lesley Byrd Simpson goes so far as to assert that Melibea's "inevitable conquest by the old witch and her terrible death are necessary devices for displaying Celestina's gifts to the full."[7]

The deaths of Calisto and Melibea, then, are more ridiculous than sublime. Calisto, fearful he has been discovered by the Night Watch, falls clumsily from a ladder as he climbs over Melibea's garden-wall; his brains are splattered, we are told, on the pavement. Melibea's response is to summon her parents, blame Fortune for her dishonor, and prepare to leap dramatically to a similar fate. Her proud proclamation of her intentions reveals the extremity of her madness rather than the nobility of loving self-sacrifice: "His death calls for mine. It invites me to be swift and allows me no delay. It summons me to fall also, so that I may follow him in everything" (p. 241). When Celestina hoards the payment Calisto has given her for her assistance, her just reward, from the play's perspective, is to be murdered by the servants with whom she has conspired. Here is the traditional comic motif of out-tricking the trickster.

In contrast to these ridiculous and overblown postures, the tragic poet creates what Ortega y Gasset calls a "noble heroic fiction," which "rises above the inertia of reality through the greatest exertion; it lives by aspiration."[8] As Shakespeare expresses it in *Romeo and Juliet*, the story of star-crossed love becomes such a "noble heroic fiction," though Rojas' critique of the heroic ideal of love is no less convincing than Shakespeare's affirmation. *La Celestina* clearly fulfills the artistic intention of degrading not actual human passion but the "noble heroic fiction" we are wont to make of it.

*Romeo and Juliet* initially provides a comic perspective on the heroic ideal of love.[9] The Romeo of the first scene of the play closely resembles Calisto. Sick with love for the fair Rosaline, he complains to his friend Benvolio in much the same way that Calisto disputes with his servant Sempronio about the meaning of a lover's passion:

> Love is a smoke made with the fume of sighs,
> Being purg'd, a fire sparkling in lovers' eyes,
> Being vex'd a sea nourished with loving tears.
> What is it else? a madness most discreet,
> A choking gall and a preserving sweet. (I.i.193–97)[10]

Benvolio's response, unlike Sempronio's attempt to profit from love, is to resolve to help Romeo forget the lady. For Shakespeare, as for Rojas, passion is "unruly" and open to ridicule whenever it simply complements *amour propre*. But Romeo's lovesickness dissolves the moment he meets Juliet: "Did my heart love till now? forswear it, Sight! / For I ne'er saw true beauty till this night" (I.iv.168–69). From this initial encounter, the lovers bring to a brief tragic life the ideal of human love as a noble aspiration, divinely inspired and well worth the sacrifice of mere selfish needs and desires. In the mythic action of Shakespeare's drama, death is not an escape or a punishment, as it is in *La Celestina*, but a witness to the integrity of the lovers' union.

Nevertheless, Shakespeare's poetic idealization of love retains its link to sensual passion. Spoken just before the delicate lyricism of the garden scene, Mercutio's ribald remarks call to mind the frank sensuality of Celestina:

> If love be blind love cannot hit the mark.
> Nor will he sit under a medlar tree
> And wish his mistress were that kind of fruit
> As maids call medlars when they laugh along.
> O Romeo, that she were – O that she were
> An open-arse or thou a pop'rin pear. (II.i.33–38)

A comic earthiness certainly marks the character of Juliet's Nurse, but in this drama the "way of the world" is elevated by a tragic perspective. The very language of the characters removes them from a comic world of prosaic business. The senseless violence of quarreling and revenge which permeates *La Celestina* may be seen in the feuding of Capulet and Montague, with this important difference: that the violence in *Romeo and Juliet* is motivated by family honor, not by envy and lust for money.

Though the love of Romeo and Juliet is unmistakably sensual, it moves from the enclosure of the lovers' world through the limits of physical or material reality to the communal ideal of overflowing love. Juliet says,

> My bounty is a boundless as the sea,
> My love as deep – the more I give to thee
> The more I have, for both are infinite. (II.i.175–77)

The individual characters of the young lovers are transformed by the world in which Shakespeare brings them to life. They are united by the orientation of this world, not alienated from it. Edgar Stoll describes Shakespeare's lovers in a manner that defines as well their dissimilarity to Rojas' characters:

> Their struggle is not with each other, nor within themselves, but only with their quarreling families, against the stars. They do not misunderstand or deceive, allure or elude, suspect or tantalize, turn, naturally or unnaturally, from love to hatred or wreak themselves upon each other, as both in drama and in life lovers not uncommonly do. . . . There is not even the veil of a disguise between them. . . .[11]

They desire marriage as well as sexual union and contrive to have the ceremony rightly performed by the Friar, a character who has no analogue in the world of *La Celestina*. Together Shakespeare's young lovers mourn the death of Tybalt, and Romeo can even summon compassion for his rival Paris in the presence of what they both believe to be Juliet's corpse. Unlike Melibea's dramatic suicide, Juliet's response to her lover's death does not call for a rambling complaint against Fate. She gently chides her lover for leaving her:

> O churl, drunk all and left no friendly drop
> To help me after? I will kiss thy lips,
> Haply some poison yet doth hang on them
> To make me die with a restorative.
> Thy lips are warm. . . .
> Yea, noise? then I'll be brief. O happy dagger,
> This is thy sheath – there rust and let me die.
> (V.iii.163–67, 169–70)

Juliet is not driven to seek death; she freely chooses to sacrifice her life as if the sublimity of faithful love makes death an entirely proper response.

In tragedy, the fall of the heroic character upholds and affirms the living presence of human aspiration to some noble ideal, whereas in a comic world of action what would be noble succumbs to the pressure of a practical reality more closely allied to the demands of every day life. A comedy may reverse the perspective of tragedy by viewing the act of defeat or loss as a mistake which might be avoided. Tragedy, in contrast, validates a perspective of reality that cannot find expression in daily life yet carries with it an inner necessity. Oedipus must discover who he is and blind himself and leave Thebes; Orestes must slay his mother, be pursued by Furies, and acquitted by Athena; Hamlet must hear the ghost, catch the conscience of Claudius, and die in taking his revenge; Macbeth must listen to witches, plunge Scotland into chaos, and die at Macduff's hands; Romeo and Juliet must meet, love instantly, and die in one another's arms. This is a mythic world; the world of actual day-to-day experience does not recognize such heroism or such absolute, unchangeable necessity.

The esthetic perspective of tragedy is impersonal; tragic action occurs upon a cosmic stage sanctioned by myth and removed from the space and time of actual existence. Comedy, likewise impersonal, is bound less to mythic necessity than to the whims of Fortune; it requires not so much a happy ending as a triumph of practical communal wisdom. In the space and time of *La Celestina*, passion remains unruly, destructive not merely to individuals, but to the orderly progress of every day affairs. The nobility of the tragic world, always dependent upon idealism, is absorbed by a comic vision of matter-of-fact reality. The solitary lament of Melibea's father, which concludes the play, shows him to be as alienated from the communal wisdom which has prevailed as are the young lovers whose death serves as a "warning" against disorder:

> O world, world! Many men have said many things of you, many have inveighed against you for your deceits. . . . I believe in my youth that you and your deeds were governed by some order. But now . . . I find in you only a maze of error, a frightful desert, a den of wild beasts, a game in which men go round in circles, a lake full of mud, a country choked with thorns, . . . a river of tears, a sea of miseries, labour without profit, sweet poison, vain hope, false joy, and real sorrow. (p. 244)

The "real sorrow" of the father is private, his alone, shared neither by artist nor audience. In sharp contrast is the final scene of Shakespeare's tragedy, marked by the reconciliation of Capulet and Montague:

A glooming peace this morning with it brings,
The Sun for sorrow will not show his head.
Go hence to have more talk of these sad things:
Some shall be pardoned and some punished;
For never was a story of more woe
Than this of Juliet and her Romeo. (V.iii.305–10)

The "glooming peace" which concludes *Romeo and Juliet* is shared by Nature itself: "The Sun for sorrow will not show his head." The Prince of Verona, the human ruler of this tragic world, proclaims its uniqueness: "never was a story of more woe / Than this of Juliet and her Romeo." Like Thebes at Oedipus' exile, or Denmark at Hamlet's death, the community senses its loss – and so does its audience. The paradox is that in this sense of loss and waste is the discovery of the value of the heroic. In *Romeo and Juliet*, this heroic value is bound to *caritas*: love as communion or love as the ideal perfected form of Eros.

The idealized forms of the good, the true, the beautiful, the heroic – as tragedy shapes them – are not determined and cannot be measured by the facts of daily experience, because "our time," the temporal present in which we live, does not lend itself to tragic poetry.[12] We live perforce in an unstable world of "becoming," whereas the tragic world is a stable world of "being" whose action was well described by Aristotle as "serious, complete, and of a certain magnitude." The genre of tragedy must be understood as a living attitude, embodied in a culture, an orientation in reality which awaits expression and extension by the artist:

Shakespeare, like any artist, did not build his works with dead elements or with bricks, but with forms already laden and overflowing with meaning. . . . For the artisan writer, genre is a simple template, while the great author awakens the potential meaning embedded in the form.[13]

It is important to recognize that a concern for genre carries with it a concern for history: an acknowledgement that creativity itself is not an isolated event of an individual consciousness but an implicit dialogue with history and culture. The phenomenon of tragedy, grounded in ancient artistic forms and recreated in *Romeo and Juliet*, orients us towards a "potential meaning" in human experience, a unique and irreplaceable expression of ideal love.

*Southern Methodist University*

## NOTES

[1] Mikhail Bakhtin, *The Formal Method in Literary Scholarship*, trans. A. J. Wehrle (Baltimore: Johns Hopkins University Press), p. 134.

[2] Shakespeare's probable source was Arthur Brooke's translation from Italian of *The Tragical History of Romeus and Juliet*, p. 1562. For interesting discussions of the various versions of the tale of young lovers, see the *New Variorum Romeo and Juliet*, ed. H. H. Furness (New York: Dover, 1963), pp. 397–408.

[3] See H. Warner Allen, introduction to *La Celestina*, trans. James Mabbe (1631), ed. H. Warner Allen (New York: Dutton, 1923).

[4] *La Celestina: The Spanish Bawd*, trans. A. J. Cohen (New York: New York University Press, 1966), p. 17. All subsequent quotations from the play are taken from this translation.

[5] On the esthetic function of such "commonplaces," see Stephen Gilman, *The Art of 'La Celestina'* (Madison: University of Wisconsin Press, 1956), pp. 120–25.

[6] Gerald Brenan, *The Literature of the Spanish People* (Cambridge: Cambridge University Press, 1963), p. 133.

[7] Introduction to *The Celestina: A Novel in Dialogue*, trans. Lesley Byrd Simpson (Berkeley: University of California Press, 1971), p. vii.

[8] José Ortega y Gasset, *Meditations on Quixote*, trans. Evelyn Rugg and Diego Marin (New York: Norton, 1961), pp. 157–58.

[9] Susan Snyder ('*Romeo and Juliet*: Comedy into Tragedy,' *Essays in Criticism* 20 (1970):391–402) argues that the play begins as a comedy then shifts genre to become a tragedy.

[10] The text of *Romeo and Juliet* cited throughout is the edition of Richard Hosley (New Haven: Yale University Press, 1954).

[11] Elmer Edgar Stoll, *Shakespeare's Young Lovers* (New York: Oxford University Press, 1937), p. 4.

[12] See Mikhail Bakhtin, 'Epic and Novel,' in *The Dialogic Imagination*, trans. Caryl Emerson and Michael Holquist (Austin: University of Texas Press, 1981), pp. 3–40, for an illuminating discussion of the distinctive time and space of the "older genres" of epic and tragedy.

[13] Mikhail Bakhtin, '*Realizing Potentials More Boldy*,' trans. William Mandell, *Soviet Studies in Literature* 7 (Summer 1971):201.

BERNADETTE LINTZ MURPHY

# DU DÉSORDRE À L'ORDRE: LE RÔLE DE LA VIOLENCE DANS *HORACE*

Après *Le Cid*, *Horace* surprend par la violence qui se déploie dans la pièce. Cette violence est d'autant plus frappante qu'elle touche des êtres soudés les uns aux autres par des liens très étroits. A la violence qui oppose les deux villes Rome et Albe se substitue celle qui fait se massacrer les Horace et les Curiace. La violence atteint son point culminant dans le fratricide d'Horace. Pourtant, à la fin de la pièce, le meurtrier est absout et la vie reprend son cours normal. Comment sommes-nous passés de la menace d'extermination qui pèse sur les deux villes dans les premiers vers, à l'unité et à l'affirmation de la vie qui s'imposent dans les derniers vers de la pièce? Quel est le rôle de cette violence qui éclate à plusieurs reprises? Pour essayer de répondre à ces questions, nous nous servirons des recherches faites par René Girard sur le phénomène de la violence qu'il analyse dans deux de ses ouvrages: *La violence et le sacré*, et *Des choses cachées depuis la fondation du monde.*

*Horace* met en scène la crise qui oppose deux villes, Rome et Albe. C'est Sabine qui, dans les premiers vers de la pièce, fait le bilan de la "guerre ouverte"[1] entre les deux villes et annonce la bataille décisive qui doit se dérouler le jour même. Cette guerre qui déchire les deux villes est vue par Sabine comme un matricide. En se dressant contre Albe, sa ville d'origine puisqu'elle a donné naissance à Romulus, Rome se révolte contre sa propre mère:

> Albe est ton origine; arrête et considère
> Que tu portes le fer dans le sein de ta mère. (55–6)

La lutte entre les deux villes constitue un affrontement entre deux parties d'un même peuple, comme l'affirme le dictateur d'Albe:

> Nous ne sommes qu'un sang et qu'un peuple en deux villes:
> Pourquoi nous déchirer par des guerres civiles? (291–2)

C'est ce caractère de lutte intestine, d'affrontement au sein d'un même peuple, qui justifie l'emploi du terme "parricides":

> Ils étonnent comment leurs mains, de sang avides,
> Volaient, sans y penser, à tant de parricides. (319–20)



*A-T. Tymieniecka (ed.), Analecta Husserliana, Vol. XVIII*, 435–447.

Matricide ou parricide, la guerre entre Albe et Rome est ainsi comparée avec insistance à un meurtre familial.[2]

Dans le cadre de cette lutte familiale, les deux villes constituent ce que René Girard appelle des "doubles", que rien ne permet de différencier et qui sont "dominés par une réciprocité violente."[3] D'après Girard, la crise se déclenche lorsque les différences entre les membres d'une communauté s'effacent et font place au désordre et à l'indifférenciation violente:

> L'ordre, la paix et la fécondité reposent sur les différences culturelles. Ce ne sont pas les différences mais leur perte qui entraînent la rivalité démente, la lutte à outrance entre les hommes d'une même famille ou d'une même société.[4]

Les causes de cette guerre des doubles sont tour à tour évoquées par Sabine et Curiace. Sabine attribue l'initiative de la violence à Rome qui, par désir d'expansion, aurait jeté son dévolu sur sa ville-soeur:

> Je sais que ton Etat, encore en sa naissance,
> Ne saurait, sans la guerre, affermir sa puissance. . .
> Que les dieux t'ont promis l'empire de la terre. (39–40; 43)

Curiace, quant à lui, se contente de rapporter les paroles que le dictateur d'Albe a prononcées sur le champ de bataille. L'ambition et le désir de domination animent également les deux villes:

> Que si l'ambition de commander aux autres
> Fait marcher également vos troupes et les nôtres. (303–4)

Les deux villes sont en proie au "désir mimétique"[5] qui les pousse à s'exterminer. Soucieux d'éviter cette catastrophe, le dictateur d'Albe fait appel à la raison qui doit mettre un frein à ce désir:

> . . . Que faisons-nous, Romains,
> Dit-il, et quel démon nous fait venir aux mains?
> Souffrons que la raison éclaire enfin nos âmes. . .
> Pourquoi nous déchirer par des guerres civiles,
> Où la mort des vaincus affaiblit les vainqueurs. . .
> Nos ennemis communs attendent avec joie
> Qu'un des partis défait leur donne l'autre en proie.
> (285–87; 292–3; 295–6)

En mentionnant ces "ennemis communs", le dictateur élargit le tableau et présente un conflit encore plus vaste que celui qui oppose les deux villes. L'introduction de ce troisième élément s'avère décisive. Pour conjurer la menace extérieure, il faut que les deux villes allient leurs efforts et opposent un front commun à l'adversaire:

> Ils ont assez longtemps joui de nos divorces;
> Contre eux dorénavant joignons toutes nos forces. (299–300)

Si le rassemblement des forces contre l'ennemi est fortement souhaité par les deux camps, il reste à résoudre la crise intérieure, à ressouder les deux membres rivaux du même corps. Le dictateur suggère la solution:

> Que si l'ambition de commander aux autres
> Fait marcher aujourd'hui vos troupes et les nôtres,
> Pourvu qu'à moins de sang nous voulions l'apaiser,
> Elle nous unira, loin de nous diviser. (303–6)

La mention des mots "sang" et "apaiser" évoque l'idée de sacrifice. Il s'agit d'éviter que la communauté entière ne périsse en sacrifiant seulement une partie infime de la population, sacrifice qui ramènera l'unité. René Girard a longuement parlé du sacrifice et de sa fonction au sein d'une communauté en crise. D'après lui, le sacrifice sert à détourner la violence de la communauté en la concentrant sur une victime choisie à l'unanimité. L'expulsion de cette victime chargée de la violence doit rétablir l'ordre et les différences. Dans *Horace*, la solution sacrificielle prendra la forme d'un combat entre des guerriers de chaque camp:

> Nommons des combattants pour la cause commune:
> Que chaque peuple aux siens attache sa fortune. (307–8)

Les combattants ainsi choisis vont reproduire au niveau individuel la lutte qui déchire la collectivité. L'expulsion de ces victimes sacrificielles devra refaire l'unité de la communauté car, comme l'affirme le dictateur:

> Et, suivant ce que d'eux ordonnera le sort,
> Que le faible parti prenne loi du plus fort. . .
> Ainsi nos deux Etats ne feront qu'un empire. (309–10; 315)

L'annonce de la solution fait immédiatement tomber les armes et rétablit la concorde entre les deux camps. La rivalité conflictuelle s'évanouit. Elle est transférée sur ces victimes qui polarisent toute la violence: "Trois combattront pour tous . . . (325). Du plan collectif nous passons au plan individuel puisque c'est à la lutte des doubles et à la mise en oeuvre de la solution sacrificielle que nous allons assister.

L'acte II est construit sur la révélation du nom des combattants et sur l'accueil qui est réservé au choix fait par les deux chefs. La nomination des deux familles Horace et Curiace est révélatrice. Comme le dit Horace au début de l'acte II, Rome aurait pu nommer des guerriers dans diverses familles romaines, tout comme Albe aurait pu arrêter son choix sur d'autres que les Curiace. Mais ce choix ne doit pas nous surprendre puisqu'il ne fait que substituer aux deux villes les deux familles. La résolution de la crise doit passer par l'affrontement de ces deux familles et par l'extermination de l'une d'entre elles. Le combat devra mettre fin au rapport de double qui existe entre les Horace et les Curiace, si intimement liés par "l'amitié, l'alliance et l'amour" (418).

Evoquant les doubles célèbres tels que Caïn et Abel, Romulus et Rémus, René Girard affirme que "l'un des deux frères doit mourir pour que les doubles disparaissent, c'est-à-dire pour que la différence renaisse et que la cité soit fondée. Le meurtrier est unique mais il représente la communauté tout entière en tant qu'elle échappe au rapport de "double".[6]

Comment le choix de Rome et d'Albe est-il accueilli par les combattants? Pour Horace, ce combat offre l'occasion de se forger un sort exceptionnel, d'atteindre la "gloire" (378, 400, 492) qu'il désire si ardemment, et de se rendre "immortel" (406):

> La gloire de ce choix m'enfle d'un juste orgueil;
> Mon esprit en conçoit une mâle assurance. (378–9)

Une analyse des termes dont Horace se sert pour exprimer son aspiration à l'héroïsme révèle la fréquence des mots "gloire", "orgueil", "vaincre", "renommée". Horace se fait une certaine image du héros qui se dessine dans la troisième scène de l'acte II (431–452). Pour mériter le nom de héros il faut s'élever au-dessus de la masse, il faut accomplir un fait de bravoure exceptionnel et inégalé, il faut se mettre "hors de l'ordre commun" (436). Au niveau de la pièce, le désir d'héroïsme apparaît comme un désir contre nature et exige la mise à mort d'une partie de soi-même:

> Mais vouloir au public immoler ce qu'on aime,
> S'attacher au combat contre un autre soi-même,
> Attaquer un parti qui prend pour défenseur
> Le frère d'une femme et l'amant d'une soeur,

> Et, rompant tous ces noeuds, s'armer pour la patrie
> Contre un sang qu'on voudrait racheter de sa vie,
> Une telle vertu n'appartenait qu'à nous. (443–9)

D'emblée, Horace éteint en lui toute vie affective, et coupe les amarres qui le rattachent à sa condition:

> Avec une allégresse aussi pleine et sincère
> Que j'épousai la soeur, je combattrai le frère;
> Et, pour trancher enfin ces discours superflus,
> Albe vous a nommé, je ne vous connais plus. (499–502)

A la détermination d'Horace qui s'exprime en maximes: "Qui veut mourir ou vaincre est vaincu rarement (385)" s'oppose la résignation désespérée de Curiace:

> Je vais comme au supplice à cet illustre emploi. . .
> Rigoureuse vertu dont je suis la victime,
> Ne peux-tu résister sans le secours d'un crime? (537;595–6)

D'entrée de jeu, les deux groupes de guerriers représentés par Horace et Curiace se divisent en sacrificateur et victime. L'idée du sacrifice est également évoquée par Sabine qui veut qu'Horace ou Curiace l'immole afin de rendre la lutte entre les deux familles légitime puisque le combat servirait à venger sa mort:

> Commencez par Sabine à faire de vos vies
> Un digne sacrifice à vos chères patries. (643–4)

L'acte III introduit une péripétie dans le déroulement des événements. Le spectacle de ces deux familles si intimement liées et qui sont condamnées à s'entre-tuer apparaît révoltant aux deux camps:

> Tous accusent leurs chefs, tous détestent leur choix;
> Et ne pouvant souffrir un combat si barbare,
> On s'écrie, on s'avance, enfin on les sépare. (790–2)

Pour calmer l'opposition violente des soldats et empêcher que les hostilités ne reprennent, le roi de Rome s'en remet à la décision des dieux. Un sacrifice est fait qui doit révéler la volonté divine. En sanctionnant le combat, les dieux lui donnent un caractère de légalité et le sacralisent. Cette péripétie permet à Corneille de définir le rapport qui existe entre les dieux, le roi et le peuple. Tout l'acte III, et plus particulièrement la scène 3 qui occupe le centre de la pièce, est consacré à explorer ce rapport. Alors que Sabine croit, comme la plupart des personnages de la pièce, que les dieux tiennent compte dans leurs décisions des désirs et des "sentiments" (862) des hommes pris individuellement,

> Les dieux n'avoueront point un combat plein de crimes;
> J'en espère beaucoup, puisqu'il est différé,
> Et je commence à voir ce que j'ai désiré. (828–830)

Camille, elle, écarte résolument une telle conception:

> Le ciel agit sans nous en ces événements,
> Et ne les règle point dessus nos sentiments. (861–62)

Loin de se mettre à l'écoute des simples mortels, les dieux se contentent d'inspirer les rois, leurs représentants sur la terre:

> Disons plutôt, ma soeur, qu'en vain on les consulte.
> Ces mêmes dieux à Tulle ont inspiré ce choix;
> Et la voix du public n'est pas toujours leur voix;
> Ils descendent bien moins dans de si bas étages,
> Que dans l'âme des rois, leurs vivantes images,
> De qui l'indépendante et sainte autorité
> Est un rayon secret de leur divinité. (840–846)

Monarque de droit divin, le roi est chargé de révéler aux hommes la volonté du ciel et de veiller à l'exécution du plan divin. Une hiérarchie se dessine qui place les dieux au-dessus du roi et le roi au-dessus des hommes. En faisant porter tout l'éclairage sur le rôle crucial du roi, cette péripétie prépare le terrain pour l'intervention de Tulle dans l'acte V.

Le récit du combat entre les Horace et les Curiace est confié à Valère. Ayant perdu ses deux frères, Horace se replie et attend chacun des trois adversaires qu'il tue l'un après l'autre. Les paroles qu'il prononce au moment de tuer le dernier Curiace sont celles d'un sacrificateur:

> J'en viens d'immoler deux aux mânes de mes frères;
> Rome aura le dernier des mes trois adversaires;
> C'est à ses intérêts que je vais l'immoler. (1131–33)

Par ailleurs, le texte établit une comparaison entre le dernier Curiace et une victime sacrificielle:

L'Albain percé de coups ne se trainaît qu'à peine
Et, comme une victime aux marches de l'autel,
Il semblait présenter sa gorge au coup mortel. (1136–38)

Il est à remarquer que de ces trois exécutions, deux sont des vengeances familiales alors que seule la dernière se conforme aux conditions du combat et sert, dans l'esprit d'Horace, à assurer la victoire de Rome.[7] Tout comme son père, Horace est possédé par l'esprit de vengeance et l'exécution des deux Curiace n'en est que la première manifestation. S'étant rendu maître de la violence, Horace est désormais appelé "héros" (1129). En mettant fin à la guerre des doubles, il ramène l'ordre et la paix au sein du peuple. Son acte est considéré comme un acte de violence purificatrice et sera suivi "D'un sacrifice aux dieux pour un bonheur si rare" (1152).

Pourtant, si Horace se rend maître de la violence, il s'en rend en même temps esclave, comme le montre l'affrontement entre lui et Camille. Les analyses du désir mimétique faites par Girard vont nous aider à comprendre cette crise violente qui se termine par le meurtre de Camille. Pourquoi, en effet, ce déchaînement de la violence entre le frère et la soeur? Avant l'arrivée d'Horace sur la scène, Camille a déjà résolu de braver son frère et de chercher sa propre gloire:

Dégénérons, mon coeur, d'un si vertueux père;
Soyons indigne soeur d'un si généreux frère:
C'est gloire de passer pour un coeur abattu,
Quant la brutalité fait la haute vertu. . .
Pour ce cruel vainqueur n'ayez point de respect;
Loin d'éviter ses yeux, croissez à son aspect. (1239–42; 1245–6)

D'emblée, Camille se pose en rivale de son frère et laisse éclater son désir et sa colère. Elle réduit à néant les arguments d'Horace en les prenant à son propre compte. Dès qu'il apparaît, Horace se pose en vengeur:

Ma soeur, voici le bras qui venge nos deux frères. . .
Quand la perte est vengée, on n'a plus rien perdu. (1251; 1261)

Se saisissant de l'argument de la vengeance, Camille le retourne contre Horace:

Et j'oublierai leur mort que vous avez vengée;
Mais qui me vengera de celle d'un amant
Pour me faire oublier sa perte en ce moment? (1264–66)

Sur son chemin vers la gloire, Horace butte contre l'obstacle que constitute le désir rival de Camille. Dans un premier temps, il cherche à lui faire lâcher prise: "Suis moins ta passion, règle mieux tes désirs" (1273). Ayant accompli l'exploit qui le met au-dessus de l'ordre commun et qui lui donne ce statut quasi-divin si ardemment désiré, Horace veut servir de modèle aux autres. Tout comme il le demande à Camille,

> Tes flammes désormais doivent être étouffées;
> Bannis-les de ton âme, et songe à mes trophées;
> Qu'ils soient dorénavant ton unique entretien, (1275–77)

il l'exige de sa femme: ". . . te réglant sur moi/Fais-toi de mon exemple une immuable loi (1361–2).

Mais Camille ne cède pas. Au contraire, elle attaque son frère par son côté le plus sensible qui est son orgueil et sa gloire. Elle veut le faire tomber des hauteurs où l'a porté son action et s'élever elle-même à la même occasion:

> Puissent tant de malheurs accompagner ta vie,
> Que tu tombes au point de me porter envie!
> Et toi bientôt souiller par quelque lâcheté
> Cette gloire si chère à ta brutalité! (1291–4)

Elle lui assène un dernier coup en lançant sa malédiction contre Rome. Il est intéressant de noter qu'un souffle épique anime ces quelques vers qui décrivent la chute apocalyptique de Rome, souhaitée par la jeune fille. En se posant comme l'auteur de ces calamités, Camille s'accorde un statut égal sinon supérieur à celui du héros Horace. Ce défi héroïque qu'elle lui lance, constitue une dangereuse concurrence pour Horace. Il reproduit, au sein de la même famille, le rapport de double qui avait existé aussi bien au niveau de la collectivité qu'entre les deux familles Horace et Curiace et qui se caractérise par la réciprocité violente.

En effet, Camille imite la violence de son frère. Si Horace est un "tigre altéré de sang" elle sera "une furie attachée" à ses "pas" (1287; 1285). Nous assistons à un véritable déchaînement de la violence réciproque qui se traduit par les imprécations et les malédictions que Camille lance contre son frère et contre Rome et qui culmine dans le meurtre de la jeune fille. Ainsi, Horace succombe à cette violence qu'il s'est appropriée lors du combat contre les Curiace. Comment justifie-t-il son acte? A Procule qui l'interroge, il répond qu'il a fait "un acte de justice" (1324), qu'il a puni la faute de Camille par "un châtiment soudain" (1321). Horace s'arroge le droit à la justice et se pose en vengeur légitime de l'affront fait à sa famille et à son pays: "La plus prompte vengeance en est plus légitime . . . (1232).

En se servant à la fois des deux termes "justice" et "vengeance", Horace essaie de faire passer son crime pour un acte légal. Car cette violence à laquelle il vient de succomber n'est pas, comme dans le cas de l'exécution du troisième Curiace, un acte de violence purificatrice, agréé par les dieux. C'est bien d'une vengeance, d'un acte de violence impure qu'il s'agit. Même le vieil Horace reconnaît le caractère criminel de cet acte:

> Je te plains . . . / . . . d'avoir par sa mort
> déshonoré ta main . . . / . . . son crime . . . /
> Etait mieux impuni que puni par ton bras.
> (1412; 1414; 1417; 1418)

Celui qui s'arroge le droit à la vengeance risque de mettre en route un cycle de violence interminable. Comme le montre Girard, c'est le système judiciaire qui, en "rationalisant" la vengeance, en la limitant, l'empêche de devenir immanente et en écarte la menace pour la communauté.[8] Dans le réquisitoire qu'il adresse au roi, Valère, le porte-parole de "tous les gens de bien" (1482), dénonce cet acte de vengeance commis par Horace et expose, avec une logique implacable, les conséquences qui en découlent pour la collectivité. En prenant la justice dans ses propres mains, Horace a mis en mouvement un processus de violence qui peut se répandre et décimer la communauté:

> Quel sang épargnera ce barbare vainqueur,
> Qui ne pardonne pas à celui de sa soeur? (1501–2)

Il souligne par ailleurs la menace qu'un tel acte constitue pour le roi qu'il dépouille de son autorité et de ses prérogatives:

> Arrêtez sa fureur, et sauvez de ses mains,
> Si vous voulez régner, le reste des Romains. (1489–90)

C'est donc au roi Tulle qu'il revient, à la fin de la pièce, de régler le sort d'Horace.

Avant de rendre son verdict, le roi écoute les requêtes des divers personnages. Alors que Valère réclame la mort pour Horace, Sabine, une fois de plus, offre sa vie en sacrifice, cette fois-ci pour expier le crime de son époux. Le vieil Horace, quant à lui, essaie de neutraliser les arguments de Valère en soulignant l'aspect strictement familial de l'acte violent commis par son fils:

Il ne prend intérêt qu'aux crimes de sa race:
Qui n'est point de son sang ne peut faire d'affront
Aux lauriers immortels qui lui ceignent le front. (1676–8)

Appelé par le roi à se défendre, Horace passe sous silence la mort de sa soeur et ne se préoccupe que des moyens de "sauver" sa "gloire" (1554). C'est pour cela qu'il demande la mort. Au lieu de la ressentir comme une punition, il la recherche pour immobiliser sa gloire, pour empêcher la détérioration de sa renommée. Dans un long développement, il affirme que celui qui a atteint les sommets de l'héroïsme ne peut s'y tenir éternellement parce que l'opinion publique exige un renouvellement constant des actes héroïques. C'est pour soustraire sa gloire aux vicissitudes temporelles qu'il demande sa mort au roi:

Si bien que, pour laisser une illustre mémoire,
La mort seule aujourd'hui peut conserver ma gloire. . .
Permettez, ô grand roi, que de ce bras vainqueur
Je m'immole à ma gloire, et non pas à ma soeur. (1579–80; 1593–4)

L'intérêt de l'Etat – il pourrait après tout continuer à défendre et à servir Rome – lui est moins cher que sa gloire et sa renommée.

Face à des requêtes aussi diverses, quel sera le verdict de Tulle? Dans un premier temps, il affirme le caractère criminel du fratricide d'Horace, punissable de mort. La condamnation véhémente de "cette énorme action", qui "outrage la nature et blesse jusqu'aux dieux" (1733–34), fait ressortir l'impératif de l'interdit. Au sortir de la crise, il s'avère nécessaire de restaurer l'interdit contre la violence afin de protéger la communauté. Pourtant, à côté de cet acte de violence impure et destructrice et parti du même bras, il y a l'acte de violence unificatrice dont l'importance rachète le crime contre le soeur:

Vis donc, Horace, vis, guerrier trop magnanime:
Ta vertu met ta gloire au-dessus de ton crime;
Sa chaleur généreuse a produit ton forfait;
D'une cause si belle il faut souffrir l'effet. (1759–62)

La nature des services rendus au roi et à l'Etat met Horace au-dessus des lois:

De pareils serviteurs sont les forces des rois,
Et de pareils aussi sont au-dessus des lois. (1753–54)

S'il ne doit pas payer de sa vie le crime commis, il lui faut néanmoins vivre dans le monde et se plier au code de la majorité. Cette affectivité qu'il avait supprimée en lui et qu'il avait remplacée par la violence, il doit lui redonner toute son importance:

> Vis pour servir l'Etat; vis, mais aime Valère:
> Qu'il ne reste entre vous ni haine ni colère. (1763–4)

L'impératif de l'interdit se double d'un véritable appel à l'amour que Tulle lance à Horace et à Valère, amour nécessaire au bon fonctionnement de la communauté. Ainsi les prétentions d'Horace à l'héroïsme se trouvent ruinées et c'est en tant que "guerrier" au service de l'Etat qu'il devra continuer à vivre.

La remarque centrale de cette dernière tirade prononcée par le roi ne manque pas d'attirer notre attention: en parlant des lois qui exigent la punition du crime d'Horace, Tulle dit:

> Qu'elles se taisent donc; que Rome dissimule
> Ce que dès sa naissance elle vit en Romule;
> Elle peut bien souffrir en son libérateur
> Ce qu'elle a bien souffert en son premier auteur. (1755–8)

Au matricide évoqué par Sabine au début de la pièce, s'ajoute le fratricide qui, pour Romulus comme pour Horace, accompagne la fondation de l'Etat. Alors que le combat glorieux qui assure la renaissance de l'Etat romain agrandi mérite de passer à la postérité, le crime familial doit être escamoté.

Comment s'y prendra-t-on pour dissimuler cet acte de violence impure, ce que S. Doubrovsky désigne du nom de "péché originel?"[9] Une fois de plus, c'est au religieux qu'il sera fait appel, et une fois de plus, c'est le roi qui présente la solution. La purification d'Horace, "souillé" (1426; 1293; 1528) par la violence, sera obtenue par les rites préparatoires au sacrifice d'action de grâces décrété par le roi. Purificateurs et unificateurs, ces rites permettront de réintégrer Horace dans la communauté réconciliée autour du sacrifice, et "d'apaiser . . . les mânes de Camille" (1776). Les deux amants morts seront réunis dans un même tombeau, réunion qui symbolise celle des deux villes. A l'aube de l'ère nouvelle qui s'ouvre pour Rome, la vie jaillit de la mort des victimes et la communauté s'édifie à partir du tombeau. C'est au roi, l'exécuteur de la justice divine,[10] qu'il revient de rétablir l'ordre et les différences, en assignant à chacun son rang et sa fonction au sein de la communauté.

Une lecture girardienne d'*Horace* permet ainsi de déchiffrer dans cette tragédie un discours sur la violence qui s'articule en plusieurs étapes. La révélation de la vérité de la violence vers laquelle la pièce progresse inexorablement passe par deux événements tragiques, le combat et le fratricide, tous deux également nécessaires à la mise en lumière du désir mimétique qui fonde les rapports humains et de ses conséquences, tant au niveau individuel que collectif. Chemin faisant, la pièce expose la violence réciproque, le mécanisme de la victime sacrificielle, ainsi que les moyens mis en oeuvre pour dissimuler la vérité de la violence. La dernière tirade du roi met en évidence la restauration et le renforcement des interdits qui s'imposent au terme de la crise mimétique, de même que le rôle unificateur et producteur de paix du rituel. Ainsi, *Horace* propose à la fois une progression de la crise à sa résolution et une régression de l'histoire d'Horace à celle de Romulus et, partant, à toute crise fondatrice productrice de culture.

*Colgate University*

## NOTES

[1] Pierre Corneille, *Théâtre choisi* (Paris: Garnier Frères, 1974). Toutes nos citations renvoient à cette édition.

[2] C'est ce qu'ont très bien relevé des critiques comme Lawrence E. Harvey dans 'Corneille's *Horace*: A Study in Tragic and Artistic Ambivalence' in *Studies in Seventeenth-Century French Literature Presented to Morris Bishop*, ed. J.-J. Demorest (Ithaca, N.Y.: Cornell University Press, 1962), p. 86, et Serge Doubrovsky dans *Corneille et la dialectique du héros* (Paris: Gallimard, 1963), p. 152.

[3] René Girard, *Des Choses cachées depuis la fondation du monde* (Paris: Grasset, 1978), p. 324.

[4] René Girard, *La Violence et le sacré* (Paris: Grasset, 1972), p. 77.

[5] Dans le désir mimétique, le sujet modèle son désir sur celui de l'autre et convoite l'objet que lui désigne le désir de l'autre. Les deux désirs, qui élisent le même objet, ne peuvent manquer d'entrer en conflit. Il s'instaure alors une rivalité entre le sujet et le modèle qui, en s'intensifiant, aboutit à la violence destructrice. Au paroxysme de la crise mimétique, cette violence se substitue à l'objet initial et devient "à la fois l'instrument, l'objet et le sujet universel de tous les désirs." Pour une explication plus détaillée du désir mimétique, voir Girard, *La Violence et le sacré*, pp. 203–208.

[6] Girard, Des *Choses cachées*. . . , p. 47.

[7] Nous tenons à remercier le Professeur Madeleine Alcover pour cette suggestion précieuse.

[8] Girard, *La Violence et le sacré*, p. 40.

[9] Doubrovsky, op. cit., p. 181.

10 Cf. la déclaration de Tulle:

> "... je ferai justice:
> J'aime à la rendre à tous, à toute heure, en tout lieu
> C'est par elle qu'un roi se fait un demi-dieu (1476–78)."

# PART FOUR

# LITERARY SIGNIFICANCE AND MAN'S SELF-INTERPRETATION WITHIN THE LIFE-WORLD

## LITERARY THEORY AND PRACTICE

# A. LITERARY EXPRESSION
# AND MAN'S BEING-IN-THE-WORLD

JACQUES GARELLI

# THE ACT OF WRITING AS AN APPREHENSION OF THE ENIGMA OF BEING-IN-THE-WORLD

> this senseless game of writing is to arrogate to oneself, by virtue of a doubt – the drop of ink related to the sublime night – some obligation to recreate it all, with reminiscences, to prove that one is indeed there where one ought to be (because, allow me to express this apprehension, an uncertainty remains.) (Mallarmé)[1]

> instead of saying that I am in time and in space, or that I am nowhere, why not rather say that I am everywhere, always, by being at this moment and at this place? (Merleau-Ponty)[2]

The preceding quotations lead directly to our primary objective: to show that the enigma of "Being-in-the-World," such as it is presented in Merleau-Ponty's last work, *The Visible and the Invisible*, is the very one which brings to life Mallarmé's poetic thought.[3] This is a most fertile convergence, for it affords sight of the phenomenological experience of what we will call a "thinking place"[4] in its concrete constitution as a poem, that is, as a movement, captured in words, of a temporalizing intentional projection of a "being-there" directed toward the world.

The logical difficulties raised by Merleau-Ponty's statement notwithstanding, it is the ontological formula, "I am everywhere, always, by being at this moment and at this place," that we wish to investigate here, for this statement poses the question of the ubiquity of the "visible seer," situated in the middle of the nondelineable order of things, from where all seeing is affirmed and all questioning emerges. This is a problem, it must be noted, that provokes an outrage, for as soon as it is articulated, it is repressed as though arising from madness, set aside as useless, or qualified, not without irony, as "misplaced"! Mallarmé, in fact, has claimed, with regard to this question of the inaugural act of creation, that his difficulty is so formidable that "He who accomplishes it, fully, withdraws from it."[5] And Artaud, haunted by the question of the ontological "overflow" of man's body onto his surroundings – that which raises the so little understood problem of the ubiquity of being – attributes the lack of comprehension of this question to an aborted conception of the notion of "World":



*A-T. Tymieniecka (ed.), Analecta Husserliana, Vol. XVIII*, 451–477.

And from where does this degradation of filth come? / Answer. / From the world's having not yet been constituted, / or from man's having but a narrow idea of the world / and his forever wanting to keep it? / It comes from man's *having stopped* / one fine day / the idea of world. / Two routes were offered to him: / That of the outside infinite, / that of the lowly inner life.[6]

Let us now follow the itinerary of a philosopher who did not "stop the idea of world" and who, faced with two possible routes, chose, lucidly, that of the "outside infinite," leaving to representational thought, sequestered in the specters of subjectivity, that of the "lowly inner life." In fact, Merleau-Ponty articulates with extreme precision that enigma which arises – for he who dares to let himself go – from the viewing of art works, as from the profound contemplation of a landscape, from, in other words, the visibility of their presence. To deal accurately with this question, however, it must be reinstated within its site, the field of "Being-in-the-World". What is paradoxical is that this particular problem is cast aside in the name of an interpretation which does not take into account the ontological level on which it rests. In fact, if Merleau-Ponty were to suggest that "me," this entity of flesh, I am, as an entity, the mountain which stands at a distance, between two blinks of my eyes, the assertion would stem from meaninglessness as it would contradict a principle of logic according to which the same entity cannot be both itself and another. What this implies, correlatively, is that it cannot be here and there, in the same time and in two different spatial forms. Considering the question more closely, however, it is clear that the logical contradiction and absurdity appear only at the center of an ontic interpretation of the situation. And in no way does Merleau-Ponty pose the question of the ubiquity of the entity as such, but rather that of the ubiquity of the being of this particular entity, the "being-there," which lives out its relation to things on the "explosive," thus dilated and, as we will see, "bifurcated" mode of "Being-in-the-World." This is a situation which effects at the heart of the question of Being an ontologico-existential structure, the sign of our localized and timed belonging to this earth. Conversely, a discourse which, out of a concern for purity, would strive to no longer take into account the necessarily existential insertion of all its questions – including that of Being – could only ignore the mystery of this problem. But, as Merleau-Ponty asks, "Do we or do we not have a body?" Yes or no, has anyone ever gone out of it to utter from this "nonplace" and this "nontime" some universal statement on Being? Moreover, could a philosophy of language stand alone, in the air so to speak, without a localized and temporalized embodiment to in some way give it life? If we do not wish to play a game of thoughts disembodied from essences, we must accept the necessity of situating our meditation on the order of the "here and now," where, whether it be remembered or not, it makes itself heard. And it is precisely because the question of Being *extends beyond* that of the localization of the human being, *while anchoring itself within it*, that the problem of ontological ubiquity comes into question.

This is a problem of which neither philosophers of essence nor those of language, who take the realm of discourse for a field of autonomous meanings, can have any real sense, for it is the simultaneous belonging of the "being-there" to both its body and to the open field of Being, the two irreducible and inalienable, which leads the human entity to raise the question of his ontological ubiquity. To grasp this situation, however, it must also be understood that this structure of Being is neither "an immediate given of consciousness" nor a mere fact of existing. It is not "a reality in itself" but rather, as an "opening" and a transcendence, it is an experience to be effected, an effort to be made. In a word, it is a structure of the World to order and to conquer for the "being-there" can well live in the oblivion of Being, subjected to the violence of a crushing space mapped out by a mechanized time. The difficulty of the problem as regards this dual structure stems from our tendency to think of what is really a possible double dimension of Being of the human – which can either get caught up in the "here" and the "now" or "open onto" the World within the activated experience of the transcendency of Being – to think of this double dimension according to the logical equation of two heterogeneous, self-sufficient realities.

It is in an effort to reroute the primordial experience of the "World" along paths foreign to those followed by subjective and objective thought that Merleau-Ponty unveils – within a "site" whose position can be conceived of only in terms of a "thinking locus" – the enigmatic unity of Being which defines the ontological belonging of the "existent." This is an attitude identical to that effecting what is essential to Mallarmé's poetic "voice." In this perspective, the Merleau-Pontian terms of "visible-present," "seer," and "thinking visibility" in no way obviate the notions of space and time but rather lead to their conception well beyond a realistic interpretation which would regard them, on the one hand, as the "real framework" incorporating the "existent" and, on the other hand, as "a priori forms of feeling" of a subject withdrawn into himself, applying its structuring forms from the outside while looking down on a reality "in-itself," henceforth reduced to the objective condition of a grand spectacle "enacted on the private stage of consciousness." It is in this respect that the following remark of

Merleau-Ponty concerning the spatiotemporal situation of the visible present can best be understood:

> For the visible present is not *in* time and space, nor of course, *outside* of them: there is nothing before it, after it, about it, that could compete with its visibility. And yet it is not alone, it is not everything. To put it precisely, it stops up my view, that is, time and space extend beyond the visible present, and at the same time they are *behind* it, in depth, in hiding.[7]

The question of the site or locus, therefore, discloses the problem of a reciprocal envelopment and overflow of Being and thought, as soon as one agrees to contemplate this question outside of the Cartesian and post-Cartesian theory of representation, which constitutes our modern and contemporary cultural a priori. It thus follows that the metamorphosing interplay between the "ego" and the "World" leads to a total restructuring of the "Topos" such that the "World" thinks itself through the emerging perimeters of things from within its "Fold." What this implies is the substitution of a progressively constituted "locus" which conquers itself over itself while revealing itself directly on the invested space, the substitution, in other words, of this "thinking visibility" for the representative pulverization of entities by a subjectivity enclosed within the silence of its consciousness. Merleau-Ponty has written: "The visible can thus fill me and occupy me only because I who see it do not see it from the depths of nothingness, but from the midst of itself; I the seer am also visible."[8] The visibility of the seer attests to his belonging to the open field of the visible which implies, correlatively, his structural insertion within what Merleau-Ponty calls, in accordance with Artaud's terminology, the "flesh" of the world. Far from proclaiming itself from outside a spectacle, the thinking vision unfolds within the perceptually invested heart of the world which, from then on, in its thickness of Being, reveals itself as a *thinking locus*. It is at the horizon of this situation that we can read that text, *The Visible and the Invisible*, which is fundamental to our understanding of the authentic experience of "Being-in-the-World," the experience, in other words, of the mortal grasping hold of himself in the very movement in which he is revealed to himself, over there, bursting forth onto his surroundings. The entire issue of the ubiquity of Being is founded on this visibility and, in the words of Merleau-Ponty,

> What makes the weight, the thickness, the flesh of each color, of each sound, of each tactile texture, of the present, and of the world is the fact that he who grasps them feels himself emerge from them by a sort of coiling up or redoubling, fundamentally homogeneous with them; he feels *that he is the sensible itself coming to itself and*

*that in return the sensible is in his eyes* as it were his double or an extension of his own flesh.[9]

What is this "sensible itself coming to itself" – this emergence outside of the depths of a world in accord with a "sort of coiling up" and a "redoubling" which bears witness to the homogeneity of the thought of the unfolding of things – if not the constitution of a situation according to which the circumferences of a place unfold their being in a "thinking omnipresence" characteristic of the situation of the ontological ubiquity of the "Being-in-the-World." This is a topos which manifests itself according to its dimension of logos, a totality one of whose thousands of folds, radiating in the distance, is activated by mankind. And so it is that the thought of space and time leads to a grasping of the bifurcated conquest of the "mortal fold" unfolding within the perceptual investment or endowment of things. This is the unwinding of a game – the very game of ontological ubiquity – which follows from the spatiotemporal metamorphoses of the World. And it is on this perspective that the entire weight of Merleau-Ponty's reflection depends:

The space, time of the things are shreds of himself, of his own spatialization, of his own temporalization, are no longer a multiplicity of individuals synchronically and diachronically distributed, but a relief of the simultaneous and of the successive, a spatial and temporal pulp where the individuals are formed by differentiation.[10]

What must be noted, from this point on, is the emergence of the individual outside of an englobing primordial unity, the differentiation from a multiplicity whose number is reflected in that of physical humans, the proliferation of objects and people at the horizon of a structural totality where mankind takes hold of himself in the very movement in which the World opens up in a thinking visibility. The ontological solidarity, as yet hardly fathomed, which unites in an indecomposable totality the "being-there" and the World, forbids the conception of things in the objective strangeness of the "in-itself." Rather, it leads to their integration within the field of a co-belonging which establishes in the same unity place and word, time and space, thought and world.

What the philosopher is contemplating here is nothing other than the thought of the Mallarméen "place" which, in the *Coup de Dès* and *Igitur*, substitutes for an externally objectivated spectacle the radiant unfolding of a site thinking itself from within itself. This is a place or locus endowed with an "ego" which conquers itself while letting the movement occur which absorbs it over its limits and, in this sense, abolishes it as a private individuality. From then on the order of fact can no longer be separated from that of essence, for it is in speaking of things that the locus, in speaking of itself, constitutes itself.

Fact and essence can no longer be distinguished, not because, mixed up in our experience, they in their purity would be inaccessible and would subsist as limit-ideas beyond our experience, but because – Being no longer *before me*, but surrounding me and in a sense traversing me, and my vision of Being not forming itself from elsewhere, but from the midst of Being – the alleged facts, the spatio-temporal individuals, are from the first mounted on the axes, the pivots, the dimensions, the generality of my body, and the ideas are therefore already encrusted in its joints.[11]

For the meditations pulverized on the illusions of a thinking subjectivity which aims to assimilate the otherness of the object to itself through an impossible representational identification of the presence to itself is henceforth substituted the meditation on a site where "thinking" unfolds in the "leaving-extended-in front" of its Being whose word gathers up the surroundings into a World. Thus Merleau-Ponty is led to the question of the ontological overflow of the limits of the "being-there," a situation which requires the revision of those realistic conceptions – which appear a matter of course – concerning the absolute and, in a sense, abrupt limits of the body. And it is the concept of "flesh" that reveals the most intimate comprehension of the "being-there" in that investment which both the thinking vision and touch effect, near and far, on things which, thereafter, reveal their own outline. Through this movement of Being and thought mankind bursts out of what are seemingly the limits of his "restricted body" to expand over the World. It is in this perspective that Merleau-Ponty asks: "Where are we to put the limit between the body and the world, since the world is flesh? Where in the body are we to put the seer, since, evidently there is in the body only 'shadows stuffed with organs,' that is, more of the visible?"[12] What Merleau-Ponty calls "this strange adhesion of the seer and the visible"[13] is such that neither the body nor the World can be thought of on a factual level, for beyond all objectivity, all reality in-itself and, furthermore, all ideality – since one cannot reduce the World and the body to simple linguistic meanings – there is a "reciprocal insertion and intertwining" of the seeing body and the visible. This is a situation which requires the effort of thinking, in accordance with a new ontological formula, the vision and visibility which unfold in the World.

There is vision, touch, when a certain visible, a certain tangible, turns back upon the whole of the visible, the whole of the tangible, of which it is a part, or when suddenly it finds itself *surrounded* by them, or when between it and them, and through their commerce, is formed a Visibility, a Tangible in itself, which belong properly neither to the body qua fact nor to the world qua fact. . . . *Thus since the seer is caught up in what he sees, it is still himself he sees. . . .*[14]

Merleau-Ponty could not emphasize any more strongly the indissoluble connection between the "being-there" and the unifying and englobing field of the "Being-in-the-World" as it is explicitly stated *that in seeing the World, the "being-there" sees himself in the distance*. Reciprocally, however, through this "self-visibility" which manifests itself outside of the limits of the restricted body, it is the World itself which, from one of its folds, opens onto itself and, in so doing, sees itself in the very act in which the "being-there" both see *it* and *itself*. An ontic psychological realism would not fail to reject this outrageous situation by reducing it to an act of schizophrenia when, in truth, it stems from the question of the ubiquity of Being characteristic of the thinking visibility at work. This ubiquity of Being is what both englobes and traverses the unique "being-there" and is what, hereafter, can only grasp itself in the quadruple unifying "narcissistic" passion whose hierarchic order can be soberly stated in the passions of the World for the World, the World for the "being-there," the "being-there" for the World and the "being-there" for itself. This is so to the extent that the "being-there" "haunts itself" within the irruptive movement which projects it onto the entities offered by the englobing presence of its own surroundings.

For the rationalist pantheism conceived within the analytic abstraction games of substance and its modes – such as it was bequeathed to Spinoza and those Hegelians ideally absorbed in the dialectical course of history by the Cartesian realist severing of the *res cogitans* and the *res extensa* – Merleau-Ponty has substituted not the narcissism of the individual enclosed within his body, but the self-affection of the World for the World whose vision of the "being-there" activates, from one of its thousands of folds, its inextinguishable opening. This reciprocity of action and passion effected through the exercise of touch and seeing establishes the ubiquity of the "being-there" in the opening and opened passion of the "There is" and "the World." And it is in this sense that the passage immediately following the preceding quotation is to be understood:

there is a fundamental narcissism of all vision. And thus, for the same reason, the vision he[15] exercises, he also undergoes from the things, such that, as many painters have said, I feel myself looked at by the things, my activity is equally passivity – which is the second and more profound sense of the narcissism: not to see in the outside, as the others see it, the contour of a body one inhabits, but especially to be seen by the outside, *to exist within it, to emigrate into it, to be seduced, captivated, alienated by the phantom, so that the seer and the visible reciprocate one another and we no longer know which sees and which is seen.*[16]

For he who agrees to set his thinking on that course which opens the "being-there" onto things, a nondelineable situation marking our condition as mortals avowed to this world, there is, therefore, an "ontology of the visible" at the center of which is posed the enigma of the ubiquity of Being. This attitude necessarily leads to the substitution of the open field of presence of a "thinking locus" for the representational thought of the World enclosed like an image within the secret of subjectivity. Indeed, the "being-there" comes out of it and it is this *of it* which raises, in all the simplicity of the term, the perpetually repressed enigma of the ubiquity of the "Being-in-the-World."

Here we must note that if the thought of this "locus" which is revealed in the World's "coming into itself," a movement which carries within it the mystery of the "Verweltlichung" or "worldification," if this implies an ontology, a somewhat narcissistic ontology of vision or looking, in no way is this because, by some coincidence, sight would see itself seeing. This would be entirely foreign to Merleau-Ponty's thinking, for it is but a tributary of a subjectivity's reflexive thought and, precisely, one that is challenged in *The Visible and the Invisible*. Rather, what must be understood is that seeing the world, in the act of viewing, are both the "being-there" and the World which see themselves and each other. The dictates of conscious vision must not be sought, therefore, within the empty digression of being's practise of annihilation characteristic of the scheme of nonthetic consciousness which, while escaping itself, gives way to a thetic consciousness of itself. This is a scheme, it must be said, that arises exclusively from the subjectivist thinking of the "For-itself" as opposed to an "In-itself" which constitutes a forceful and solid antipathy. To speak of a sight of seeing, even in terms of the annihilation of Being, would be to enclose this problematic of vision within the illusions of subjectivity while Merleau-Ponty's entire effort is to open it up into the "haunting" of the World. To attain this level of Being and thinking which does not arise from dialectics, intuition, or reflection and which requires inventing new methods of meditation, of which the notions of "intertwining" and "chiasm" permit a preliminary conception, an inquiry into the "Being-in-the-World" must be initiated, one which considers the question of the union of the "topos" and "logos" in the opened ubiquity of the "thinking place."

In implicating the identification of thinking with Being, this conception relates to arguments of Parmenides, the domain of which is restored by contemporary ontology. And it becomes a question of meditating the well-known fragments 3 and 8 of Parmenides' *Poem*, not within the framework of a philosophy committed to the course of absolute idealism as witnessed by history, but within our very connection with the entity. This means that the identification thinking with Being must be conceived as that eruptive movement of the "being-there" which unfolds its perceptual space while endowing those things shaping its horizon which are revealed as constituting its very limits. This movement sets up a site and opens out a place. And it is in this fold that Being and thinking are one and the same. The identification of thinking with Being is, in other words, the eruption of a "thinking locus" which "comes into itself" as it opens out into the endowed certainty of its surroundings. The "being-there" emanates from within it to the extent that it kindles the narcissistic passion in the lived enigma of its ubiquity of Being.

If to think is to be identified with Being and if the act of thinking unfolds within the *perceptual* eruption of the "being-there" projected onto entities, two essential conclusions are indicated: the first is that the Being of the entity which perceives identifies with the perceptual act which projects it onto things. This implies, as a necessary correlative, the second conclusion: that man's "being-there" is "*outside*," having burst forth straight from the perceptual endowment of the World. And it is this which constitutes the *ontological field of the thinking locus*.

These are formidable conclusions disturbing not only to conventional subjectivist thinking, but to the most profound conception of both man's limits and the power of his sight and language which, by the overlapping of their self-constitutional interrelating, commit him to being himself only through his endowment of the World. Such is the meaning which must be given to the opposition between the routes of the "outside infinite" and the "lowly inner life." A close reading of the two passages from Parmenides will offer proof of the validity of these conclusions.

Our attention is first drawn to the well-known section 3: "Τὸ *γὰρ αὐτὸ νοεῖν ἐστίν τε χαὶ εἶναι.*" This is an enigmatic statement which has inspired numerous translations. We will not follow that of Andre Préau who, in his translation of Heidegger's essay "Moira" writes: "Car la même chose sont pensée et être"[17] ("For thought and being are the same thing"). This gives the term "*νοεῖν*" a nominal value when it is important to preserve the infinitive value of the Greek verbal form. For this reason, the translation by Jean-Jacques Rinieri would appear preferable: "Le même, lui, est à la fois pen*ser* et être."[18] It is the same, at once, *to think* and to be. The latter interpretation conforms to Heidegger's German which follows that of W. Kranz: "Denn dasselbe ist Denken und Sein." This question of translation is essential, for it prevents any idealist assimilation of Being into thought at the same time as it emphasizes the co-belonging of thinking and Being. In fact, Parmenides' phrase does not in any way reduce Being to the order of thought, as was ultimately the case in the idealist interpretations of Berkeley and Hegel. Rather, in a radically different perspective, the order of thinking is conceived in terms of Being. In fact, for Parmenides, *to think* is already part of *Being*. Thus, it is not the thought, the concept or the idea which, as such, allows for an understanding of Being. No idealist assimilation is possible as regards this point. More accurately, it is the dimension of the act of thinking – manifest in the infinitive value of to think – which reveals, underlying the idea, the operational movement of Being which establishes it. This is why, in Parmenides, what is referred is the "act of thinking," *νοεῖν*, and what is referred to is Being: "*εἶναι*." The act of thinking, in other words, is brought over or "conferred" to Being while, in Berkeley, in the formula "esse est Percipi," what is referred is "esse" and what is referred to is "Percipi," which bears witness to the referral of Being to and its actual absorption within thought. This is an attitude characteristic of absolute idealism.

Three contemporary interpretations confirm this claim. First, as Heidegger notes in *Moira*:

> In the two forms which are given him [section 3 and 8, 34], Parmenides constructs his statement in such a way that each time *νοεῖν* (thinking) precedes *εἶναι* (being). Berkeley, on the other hand, names esse (being) before percipi (thinking). This appears to indicate that Parmenides harbors a preference for thought and Berkeley for being. And yet it is the opposite that is true. Parmenides confers thought to being. Berkeley accords being to thought. To respond to the Greek phrase and to identify in a way with it, the modern thesis should read: percipi = esse.[19]

Second, Beauffret declares in the same persepctive:

> In the well-known line which proclaims the co-belonging of being and thinking in the dimension of the Same, he does not, in fact, opt for *being*, as in the *Critique of Pure Reason*, but rather for *thought*. It is not said of being that its determination must, in the last resort, respond to the security of the *awareness of self*, but rather that "thought" itself is only there to respond to the opening of the domain of the Open which is the very domain of being.[20]

And third, Henri Birault emphasizes that contrary to Berkeleyian and Hegelian idealism, which reduces being to thought, Heidegger, in accord with Parmenides literal expression, of which he inspires a profound meditation, restores the order of thinking to the field of being:

The formula for this identification is Esse = Percipi. It is the formula for all modern idealism, for that of Hegel as much as, or for more than, for that of Berkeley. . . .

The Heideggerian interpretation of Parmenides' word procedes, however, from a mode of thought entirely different from that of Hegel. The formula in question names first the *Esse* and appears, therefore, to favor being over thought. Parmenides' expression names first the νοεῖν and thus appears to refer the εἶναι back to the νοεῖν. And yet the opposite is true. Parmenides returns thinking to being, Berkeley restores being to thought.[21]

The reversed attitude, characteristic of the "moderns" corresponds to the passing from a thought of being to a philosophy of the representation of the object and of presence to itself. This is characteristic of Cartesian and post-Cartesian reflection, whose correlates are the *Ego Cogito* and the *res*, the thing reduced to the abstract universality of "the object." And thus it marks the change from an ontological[22] meditation to a metaphysical theory of knowledge. While Parmenides attributes thinking – which, when understood in its broadest sense, incorporates within it the *perception* of the world – to the open field of presence, modern idealism reduces being to thought, defined as the intellectual representation of the object:

The modern thesis states something about being in the sense of an objectivity constituted by a grasping representation. The Greek phrase attributes thought, understood as a gathering perception, to being, that is to say, to presence.[23]

It is this reduction of being to thought which leads us astray in the illusions of absolute idealism: "This is why any interpretation of the Greek phrase which drifts in the direction of modern thought goes astray from the start."[24] And thus there remains both the task of maintaining reflection on the level established by Greek thought and, therefore, that of probing what is literally affirmed as an identification of thinking with being. Without a doubt, the identification of the Tó *αὐτὸ* remains enigmatic as long as one considers the unity of the "same" in terms of logical or mathematical relations, in terms, in other words, of a sign outside of the two terms it unites. In fact, when one sets up the arithmetic equation 6–4 = 2, the relationship of equality does not identify with either of the two terms that it unites. It remains outside of them. More accurately, however, is there really any question of a relationship? Must the Tὸ *αὐτὸ* be conceptually modeled after a sign?

Thus stated, the problem is unsolvable, for a logical or mathematical equation cannot unite two heterogeneous orders: in this case, the *νοεὶν* and the *εἶναι*. But this impossibility relates strictly to this mode of thinking. Can it not be asked if the Tὸ *αὐτὸ* would disclose, rather than a logical relation, the dimension of the same, or sameness, characteristic of the belonging of thinking to being which requires consideration, as such, in its unity? After all, why would the expression "the same" necessarily designate a relation between two heterogeneous orders and not the disclosure of a co-belonging to one single dimension? Why would the evidence of the ontological dimension and, therefore, of the "weight of being" fundamental to the act of thinking, *νοεῖν*, which implies knowing, not manifest outside of the "representational" field the identification of thinking and being? For what is an act of thinking which would not be? And what is thinking outside of the act which utters it? This would be a thinking deprived of being, which is inconceivable, for to think is, from all evidence, to already be: this is a certainty located not on the Cartesian subjectivist level of the *Cogito sum*. But the ontological Neutrality of the "There is," the "Es gibt," the "Il y a." How then can we not precisely state, as does Parmenides, "For the same is thinking and being" and thus reveal the ontological dimension of the *νοεῖν* which, as an act of thinking, unfolds within the field of being's belonging? What must be investigated, at this point, is this mode of the unfolding of "thinking," which is a mode of being and which, while it reveals the entity in the light of knowing, realizes the ontological movement of its accomplishment, a temporalizing dimension of being. We know how frequently a fair number of minds turn away from this style of meditation in the name of some so-called tautological weakness which, it must be said, has yet to be demonstrated. The tautology, in fact, is an affirmation of the "sameness" in the field of the entity, one that is supported by its connection to a logical equation. And yet, Parmenides' claim is not an effort to reveal the equality of two entities! In a radically different perspective it discloses the dimension of being within the entity "thinking," the ontological dimension, in other words, within a structure of knowing which, at first sight, is ontic. In reality, for he who thinks, being is prescribed within the thinking act.

If the Cartesian experience of the *Ego Cogito* and the development of modern idealism do not allow for the localization of the ontological site in which Parmenides claim unfolds, it must be noted that since antiquity the Platonic theory of ideas has barely contributed to the understanding of this issue. In fact, for Plato, only the *νοεῖν* abstracted from the perceptual field allows for a contemplation of ideas in the purity of their detachment from the senses. In this perspective, being arises exclusively from the order of the "beyond perception" and it is in the neo-Platonic tradition that

Plotinus interprets Parmenides' claim. Being, in this regard, arises from the domain of the *νοητά* which, in their ideal purity, are cut off from the field of the *αἰσθητά*. As Heidegger notes,

Plotinus interprets Parmenides' claim in the Platonic sense according to which Parmenides is saying that being is something non-perceptual. The weight of this claim falls on thought, although in a sense other than that understood by modern philosophy. Being is characterized by the non-perceptual mode of thinking.[25]

A philosophy which, out of a concern for purity, refuses to restore Parmenides' meditation to the very heart of the concrete perception of entities *εόντα* is plagued by the tendency to essentialize the *νοεῖν* and, in short, the order of being, *εἶναι*. Ultimately, it is on this level that section 8, which completes section 3, which we shall now consider, is located:

*Ταὐτὸν δ' ἐστὶ νοεῖν τε καὶ οὕνεχεν ἔστι νόημα. Οὐ γὰρ ἄνευ τοῦ ἐόντος, ἐν ᾧ πεφατισμένον ἐτσίν, εὑρήσεις, τὸ νοεῖν · οὐδ' ἦν, γὰρ ⟨ἢ⟩ ἔστιν ἢ ἔσται ἄλλο πάρεξ τοῦ ἐόντος, ἐπεὶ τό γε Μοῖρ' ἐπέδησεν οὖλον ἀκίνητόν τ' ἔμεναι' τῷ πάντ' ὄνομ' ἔσται, ὅσσα βροτοὶ κατέθεντο πεποιθότες εἶναι ἀληθῆ, γίγνεσθαί τε καὶ ὄλλυσθαι, εἶναί τε καὶ ουχί καὶ τόπον ἀλλάσσειν διά τε χρόα φανὸν ἀμείϑειν.*[26]

Here, Andre Préau's translation, which agrees with that of Kranz adopted by Heidegger, must be closely followed for, in so far as the *νοεῖν* of Parmenides is fundamentally constituted through the contact with the *ἐόντα*, in other words, entities, it restores Parmenides' meditation to the very heart of the world:

To think and the thought that "is" is/are the same thing *for without the entity in which it resides as an uttered thing, you would not know how to find the thought. Indeed, there is nothing, or there will be nothing, outside of the entity*, since the Moira caused it to be a whole and immobile. All that mankind, convinced of its truth, had so fixed, therefore, would be but a name: "to become" as well as "to perish," "to be" as well as "to not be," "to change places" as well as "to pass from one bright color to another."[27]

We wish to insist on both the

*Οὐ γὰρ ἄνευ τοῦ ἐόντος, ἐν ᾧ πεφατισμένον ἐστίν, εὑρήσεις τὸ νοεῖν . . .*

and the

*. . . οὐδ' ἦν γὰρ ⟨ἢ⟩ ἔστιν ἢ ἔσται ἄλλο πάρεξ τοῦ ἐόντος,*

in order to emphasize first, that according to Parmenides, the "thinking," in its dimension *of being*, proclaims itself at the very center of a meditation on the entity, establishing ontology, once and for all, at the very heart of the *perception of things* and second, that this ontological meditation could in no way be constituted outside of the *perceptual field* of the entity. In the words of Parmenides, "Without the entity in which it resides as an uttered thing, you would not know how to find the thought," a statement which is ultimately confirmed by the following assertion: "Indeed, there is nothing, or there will be nothing, outside of the entity."

Now, it is strange that this insurpassable, nondelineable reference to the *entity*, as the basis for the unfolding of the being of thinking, does not appear in Jean-Jacques Rinieri's translation where ["étant"] "entity" is translated by ["être"] "being." Far from ontologizing Parmenides text, this essentializes it as, cut off from all contact with things of the world, being then asserts itself in an empty abstraction unfilled by the notion of "word" when deprived of all connection with the entity. As Parmenides claimed: "For never without the being in which it becomes word, will you find thinking; for nothing else was nor is or will be next to and outside of being.[28] Two consecutive times, in lines 36 and 37, Parmenides writes *του ἐόντος* which clearly introduces the ontological meditation into the field of the perception of the world of entities. At this point, a meditation on being can not be abstracted into a reflection on language, purified of all contact with things. This attitude of "flight" beyond the other of the entity is characteristic of the neo-Platonic interpretation of Parmenides' text, and it is this attitude which appears to have recently found its way into those contemporary interpretations seeking to constitute a philosophy of language cut off from all contemplation of the world. This tendency conforms neither to Parmenides' thought nor to Heidegger's interpretation of it in his last works. In *Moira*, Heidegger underlines the way in which thinking is approached as an act of perceiving. How is this subtlety articulated? Very simply in the following terms:

> Instead of *ειναι*, Parmenides now says *ἐόν*, entity, which, by its double meaning, designates the Fold. But the *νοετν*, in its own right, is now called *νόημα*, that which is taken in the attentive act of perceiving with respect to what it perceives.

Now, the note on the bottom of Préau's translated page clarifies: "Das in die Acht genommene eines achtenden Vernehmens." The perceptual act *νοεῖν* concerns what it perceives and is attentive to it. It respects its freedom by letting it be, leaving it "extended-in front."[29] Thus it is from the "fold," which causes the entity to appear, that thinking unfolds its being, as the *ἐόν* is determined as "the reason the thought is present."[30] But this unfolding of "thinking" in the fold of being from where the entity radiates is nothing less than the mark of the λέγειν which proclaims itself as a "leaving-extended-in front." It is this structure of being which must now be contemplated in order for an understanding of how "thinking," from the unique fact that it thinks, unfolds its presence of being which opens it onto the entity. And it is within this structure as well that the locus – which is place and which takes place – is established, creating a situation which, while difficult to accept, is evident in the following analyses. Heidegger states:

The νοεῖν, whose belonging to the ἐόν we would like to consider, is founded in the λέγειν and unfolds its being from it. There, in the λέγειν this "leaving-extended-in-front" takes place leaving the thing spread out in its presence.[31]

It is the meaning of this movement which allows us to understand the poetic utterance's structure of being whose unfolding causes to appear, within the presence of the place, the movement of the emergence of things present. Again in *Moira*, we read:

We must learn to think the being of language, beginning with its expression, and think of this expression as the "leaving-extended-in-front" (λόγος) and as the causing to appear (φάσις).[32]

What does the expressive act of uttering cause to appear in the very unfolding of its being? Precisely, and this beyond any tautology, that "*perceptual* place" in which the presence of present things unfolds. "The uttering of language, which both reigns within the fold and manifests it, is the gathering together of that presence in whose appearance present things can appear. This φάσις, which Parmenides thinks, Heraclitus names λόγος, the 'leaving-extended-in-front' which gathers together."[33]

If the νοεῖν discloses, while leaving-extended-in-front the entities that it promotes in the illumination of presence,[34] it is this movement of being and "Co-naissance," knowing, which, at the same time, unfolds the νοεῖν in the world and, in so doing, establishes the place where the unfolding takes place, that which reveals itself as "being-in-the-world," fully being, in other words, itself, the same, and yet, over there, spread out in the shelter from which the things promoted to understanding emerge. The act of perceiving is therefore an "ontological overflowing." It is not an essence to categorize among tools of knowledge, but a movement which establishes the ubiquity of being – whose site is the "World" – in its worldifying expansiveness.

If, nevertheless, the disclosure grants the illumination of presence, it is in implementing both – in the case where present things must appear – a "leaving-extended-in-front" and a perceiving[35] and, in this way, bringing them into play, while retaining thought in its belonging to the Fold.[36]

Thus the work of art can be conceived of only as a dispensation of this fold as its "poetic" function is to effect the illumination of the entity – whether it arise from vision, hearing or the word – onto the unveiling presence of the place: "the multiple appearing of presence emerging, itself, from the unfolded Fold."[37]

The fold must be thought of, therefore, in the unity of the *νοεῖν* and the *εἶναι* while being conceived in the overflowing movement which projects being . . . there, perceiving and thinking, alive, onto the surroundings of the world. This is what connects the meditation on thinking with the apprehension of a "thinking locus." It is a structure of being essential to the comprehension of any work of art:

The Fold shelters within it the *νοεῖν* and that which it thinks *νόημα* as a proclaimed thing. But what is perceived in the thought *is the presence of present things*. The uttering which thinks and which corresponds to the Fold is the *λέγειν* in as much as it leaves the presence extended-in front.[38]

This locus of the world is also a metamorphosis of the locus where shield and unfold themselves some of the modeling forms, that which – out of the act of appearing – levels with the surface of things.

There, what merits being thought shelters itself causing us to consider as an entity the relation between thought and being, the truth of being in the sense of the unfolding of the Fold, the reserve observed by the Fold, (the *μὴ εον*), when present things predominate (*τὰ ἐόντα, τα δοχοῦντα*).[39]

For both that thought which, shut within the secret of a closed subjectivity, seeks to assimilate the otherness of the object into itself in the impossible quest for the presence to itself and for the study of a vitalized portion of space, reduced to the broadly sketched outlines of a body, is now substituted the meditation of a locus, where "thinking" unfolds within the "leaving-extended-in-front" of its being, of which mankind's speech opens out the fold while assembling his surroundings in a World. This is a situation which constitutes, in the open field of his belonging to things, mankind's dimension of being, while proclaiming and promoting the life of the place: it is the motion of a *Logos*, in other words, which unfolds as *Topos* and, in so doing, effects "the dispensation of all that is dispensed,"[40] that which "leaves extended in front in the light where a thing is held by precisely that which gives it a name."[41]

This reciprocity between the human who speaks and the place which unfolds necessarily leads the thinker to reconsider the question of the ontological dimension of his situation as a "being-there" avowed to the World. This is what poets have always done, for writing, by the movement of its expression, disturbs the field of the "being-there's" belonging to the World while the expressive act unfolds the passing of nonbeing to being.

Thus Mallarmé can soberly pronounce, in a style which recalls the glaring self-evidence of Heraclitus' and Parmenides' passages:

Nothing. . . .
Will have taken place. . . .
But the place. . . .
Except. . . .
Perhaps. . . .
A constellation.[42]

It is this place that thinking unfolds. In it the figures of the world venture. And yet, must it not be emphasized, in contrast, that which Mallarmé's text encloses, in the spacing on the page, a spacing which discloses the shattered, splintered, fragmented scansion of the "thinking Topos," is nothing other than a correlating annihilation of the human "personality," this

. . . memorable crisis
or the event [which]
might have been completed with no possible result in view human[43]

whose progressive disappearance is formulated as

. . . an ordinary swell [disclosing] the absence. . . .[44]

This "absence," however, is not a total dissolution. This "no possible result human" does not mark the inception of a "barbaric" order. Rather, it establishes the open field of the metamorphoses where the "fold," recalled long ago by mankind, is accomplished in the "games of the World." The place remains. And that passage of Mallarmé's text situated precisely within the "fold" of the book, a passage whose rhythm we can now reinstate, affirms this:

NOTHING
of the memorable crisis
or the event
might have been completed with no possible result in view
human
WILL HAVE TAKEN PLACE
an ordinary swell discloses the absence
BUT THE PLACE
any mediocre plashing as if to disperse the empty act
abruptly which otherwise
by its lie
would have justified
the perdition
in these parts
of the void
in which all reality is dissolved
EXCEPT
in the heights
PERHAPS
at so distant a place
that it fuses with infinity
above human interest
as pointed out to him
in general
by such slant by such slope
of lights
toward
what should be
the Septentrion or North
A CONSTELLATION
cold from neglect and disuse
yet not so much
that it does not count
on some empty and superior plane
the next collision
siderally
of a final reckoning in the making
watching
doubting

revolving
blazing and meditating
before it halts
at some final point which consecrates it
All Thought emits a Throw of the Dice[45]

Every interpretation of this well-known passage from Mallarmé, which either conceives of man in the form of a subjectivity recoiled in on itself or, most recently, attempts to reduce the text to a game of repetitive tropes and abstract meanings forming the thereafter purifed field of an essentialized writing, has worn thin. This passage can only be understood within the ontological perspective of the "Being-in-the-World" which, to be grasped, requires a meditation on the "thinking place" whose perimeters are endowed with the Logos of the human being. In as much as this "thinking place" is Being and act of thinking, it is precisely that which "holds-itself-extended-in-front-in-the-distance." In fact, that which poses within its enigma the reinforced presence of the capital letters is nothing other than the "place of the World" in its celestial topology, that place in which the temporary, as mortal, act of thinking is invested. Here the Word, in its sacred dogmatism, predicts itself in the future perfect tense. This is a strictly formal grammatical clue which would be superfluous if it were not a temporal projection conceived as already accomplished, allowing the "being-there" to foreshadow its death. Emerged from the field of a World which is "always-already-there" and which grasps its dispersed being only in folding over its own limits, the nothingness which erupts through the investment of its borders acknowledges the eternity of the place where, provisionally, as temporarily projected forward, in this abyss, in the very forming of the constellations, it thinks itself. Mankind scans the drift of this movement through his combined act of living and thinking, an act resolved by Mallarmé in the "next collision / sidereally / of a final reckoning in the making / watching / doubting / revolving / blazing and meditating" before abolishing itself in the death which eternalizes it, a death which the poet calls a "halting / at some final point which consecrates it." Now, this drift thinks itself and thereby accedes temporarily to being only through the fusion of mankind with a distant place, a fusion which projects him, makes him burst forth while, simultaneously, revealing him, outside of his nothingness.

It is this situation that is considered in the following passage:

any mediocre plashing as if to disperse the empty act / abruptly which otherwise / by its lie / would have justified / the perdition / in these parts / of the void / in which all reality is dissolved / EXCEPT / in the heights / PERHAPS / at so distant a place / that it fuses with infinity. . . .[46]

The thought of mankind's burst of being in the distance is, in its own right, so related to that of the ubiquity of the "Being-in-the-World," a Logos which reveals itself as a Topos, that Mallarmé goes so far as to name the possible localization of its celestial topology:

above human interest / as pointed out to him / in general / by such slant by such slope / of lights / toward / what should be / the Septentrion or North / A CONSTELLATION. . . .[47]

If this structure of death forms a star and if it preserves, beyond forgetting and emptiness, the trace of what were thought and life, as its deceased support, then the living person cannot think himself detached from this place which shows itself to be both his revealer and his receptacle. This implies a return, in full circle, to the following passage:

cold from neglect and disuse / yet not so much / that it does not count / on some empty and superior plane / the next collision / sidereally / of a final reckoning in the making. . . .[48]

It is this situation which, above and beyond all tragedy, requires a renewed interpretation of "Un Coup de Dés" ("A Throw of the Dice"), for if this text assumes as its fate the unpredictability of chance, it unveils itself as a moment of the eternal world game, of which the mortal fold of man, contingent upon an astral localization, temporarily reclaims the throw. This is a bursting, a breaking into pieces, an attraction and a diversion scanned – in their movement of calling and in their "gravity" – by words precisely spaced on the white of a page.

It is not only in the intuition of the celestial world that Mallarmé conceives of the destiny of mankind, in accord with the terms of a "thinking place," but, more intimately, in the inhabited microcosm of a room as well. In fact, what is witnessed in the unfolding of the itineraries marking Igitur's coming into the world, in Mallarmé's text of that name, is – in function of the diminishing identification of the self with the self – the subsistance of a place, at once Being and thinking, which, beyong all individuality, sets itself within the temporally haunted presence of a World. How can this event, which withdraws all individuality from humans, be better revealed than in the neutrality of a name which, at once, "depersonalizes" its foundation and manifests the relation of the thought, "*Igitur*," with the enigmatic emergence of a place. This is a situation which establishes within the same constitutional unity of Being both "Logos" and "Topos," for it is a word which names and manifests its presence of being while a place sets itself up around a personality which, in its individuality, tears itself down. To the extent that the "Ego" fades in self-annihilation into the hangings and furnishings of the room, the place thinks itself in that movement which constitutes it as a World. It is from within that it opens itself out. It is on its surroundings that it takes hold of itself. It is on things that it establishes itself. The effort to situate this "tale," as Mallarmé calls the text "Igitur," through a qualification of the place as textual, would be meaningless for in giving the illusion of an answer, the problem would only be eluded. The question of the mode of being of the place would still remain, and it is this enigma, one that we must now decipher, which is manifested in the well-known line from "Igitur" that reads, "A MIDNIGHT presence certainly is upheld."[49]

What then is the being of this "presence," what is the being of "upheld" and what is the being of this time which, above and beyond any closed-in individuality, unfolds as a "World"? For Mallarmé, as for Parmenides long ago, it is not a question of disassociating the dimension of being from thinking, a dimension affirmed here in the uttered haunting of a room. Rather, it is the Logos, as that which "leaves-extended-in-front," that allows for the arrangement of the "Topos," in this case, Igitur's room. And this is a situation to be comprehended through the contemplation of the act of "*ὄνομα*," beginning with "*λέγειν*," for, as Heidegger claims, in his exploration of Heraclitus' meditation, "*λέγειν* means at once 'to say,' 'to speak,' 'to name,' but it is intelligible only when thought of in its most proper acceptation, that of 'to extend' and 'to gather up.' "[50] What must be understood, therefore, is how the persistance of "thinking," which is both Word and Being, unfolds before it the spacing of the room. And understanding this allows for an understanding of the profound meaning of naming for, in the words of Heidegger,

> To name means: in calling, to make appear. What is gathered up and deposited in the name finds itself, from the very fact of this positing, being-extended-in front and appearing. To name *ὄνομα*, if one thinks of this act beginning with *λέγειν*, is not to express the meaning of a word, but to leave extended in front in the light where a thing is upheld by that by which it has a name.[51]

He who does not contemplate in the entirety of its implications the double ontological dimension of the *λέγειν*, "unfolder of space," cannot come close to the thought of "Being-in-the-World" and the manifestation of this structure in the effecting of a work of art. Furthermore, he who does not accept the risk entailed in encountering the inordinate, what appears in its excising phase as Strange, can not come close to the thought of the "place"!

That this struggle of the mortal who "comes down the stairwell, bound for the bottom of things, in the role of the 'absolute' "[52] in an effort to conquer the unfolding of his "Being-in-the-World," that this struggle motivates a quest for infinity which differs radically from that sought by mathematical thought is clearly shown in the segment of *Igitur* entitled "The Argument": "The infinity springs from chance, which you've denied. You, dead mathematicians: *I project myself absolute*."[53] Now, this conquest implies both the abolition of the individual, in his restricted life, as well as an old conception of space:

Simple word and gesture. Touching upon what I tell you by way of explaining my life. *Nothing will remain of you*, the infinite finally escaping the family its suffered in, the old space, without chance. *It had good reason to deny its* life and make a try for the absolute.[54]

The unfolding of the ontological ubiqutiy of the place, which thinks itself through the presence of things, in the section of Mallarmé's text entitled "Midnight," absorbed all human individuality to the extent that this individuality is proclaimed but one time, in the first person singular, as a memory: "I recollect that its gold went etching into the absence. . . .[55] whereas the grammatical and signifying subjects of each statement which cover the totality of the six sequences of text 1 are affirmed successively within the neutrality of the dreamed presence of an hour (paragraphs 1, 3, 4, and 6), the neutrality of the contingency of celestial conjunctions (paragraph 2) and the neutrality of the unfolding of a setting which is reflected in an idea (paragraph 5). Each instance is the anonymous opening of a place which takes hold of itself, dreams itself, and thus thinks itself through the *ἐόντα* constituting, in this way, the decor of the room. It is this situation, entirely new in French letters, that effects the identification of Thinking with Being within the perceptual unfolding of a Place into which the "Being-in-the-World" of a "dreamer" has been absorbed. Both our earlier reflections on the "intertwining" and "chiasms" of Merleau-Ponty's "thinking visibility" and our investigation of the unity of the *τόπος* and the *Λόγος* in the identification of the *νοεῖν* and the *εἶναι*, in Parmenides, now allow us to understand the topology of the Mallarméan room where Igitur is totally absorbed, Being and thinking, in the dreamed unfolding of his setting. And it is in the light of these considerations that the following six paragraphs, in their successive order, can be understood:

"A MIDNIGHT presence certainly is upheld."[56]

"Developer of midnight, [it] [the infinite chanciness of its conjunctions] [has] never yet pointed to its exact conjunction. . . ."[57]

"And the Midnight presence lives in the vision of a room in time. . . ."[58]

"It's the pure dream of a Midnight vanishing into itself. . . ."

"Dead for so long, an ancient idea sees itself as it were by the light of the monster in which its dream has suffered. . . ."[59]

"From the finished furnishings that will settle into the darkness like the hangings . . . useless . . . the hour draws itself into this echo, at the threshold of the panels opened by its act in the Night. . . ."[60]

For the most part, the difficulty of this text arises from the representational orientation of reader's seeking to interpret its meaning within the framework of a subjective thinking while, in fact, its statements are not contingent upon the representation of an *Ego Cogito* to endorse them. In an entirely different way, Mallarmé "allows" the "always-already-there" of a World whose site unfolds in the distance, to utter itself, through the midnight, thus igniting – through this spatial movement – the spark of a thought grasped upon the things over which it extends. In fact, in each of the text's passages, the space opens itself out in a fold which endows its own perimeters as it unfolds the setting of a Topos which transforms into Logos in the articulated presence of its limits. Dispersed over the world, the act of the *νοεῖν*, whose eminent perceptuality we proclaimed in our consideration of section 8 of the *Poem* by Parmenides, discovers its Being through its contact with the *ἐόντα* which reveals to it its thinking ubiquity in the traversed space of the room. This results in a total reciprocity wherein it is the room which thinks itself through its hangings, its consoles and its tables in the endowed presence of midnight. Thus, "mysterious furniture detains a vague vibration of a thought,"[61] a movement characteristic of the revelation of *νοεῖν* on the perceptual contact with its surroundings. Due to the resistance of the *ἐόντα* on which its opening catches and breaks, the act of thinking, from this point on, "metamorphoses" in things seen and, in this way, thinks itself in that strangeness which both enlightens and "deletes!" Accordingly, we can now restore the thought presence of midnight to the perceptual totality of the encountered setting where it installs itself as things:

And the Midnight presence lives in the vision of a *room in time* where mysterious furniture *detains a vague vibration of a thought, the luminous breakaway from the recurrence of the waves and their first enlargement; at the same time (with limited movement) the room prior to the striking fall of the hour is immobilized in a narcotic calm of pure "ego" dreamed – of for a long time, whose time however is resolved in the hangings which stop it*, finishing them with its splendor, *its shudders* deadened by an oblivion like languishing long hair around a face illuminated by mystery, with void eyes equal *to the inmate's* in the mirror, *stripped of all meaning but presence itself.*[62]

Here the Being of thinking is set by Mallarmé within the spatializing extension of a Place which he qualifies as a "room in time" and which radiates as "luminous waves." Now, this flow of waves, in its material expanse, is nothing other than the boomerang of a "vibration of a thought," "extensio-animi" stopped by the *ἐούτα* constituting the furnishings of this "moving limit" which, from then on, merit the title of a "room in time". The "distentio animi" of the " 'ego' dreamed-of for a long time" is profoundly pondered in its temporalizing dimension as a creator of space, its unfolding on the *ἐόντα*, which reveals its so precisely formulated thought metamorphosed in hangings, that the poet can write of an " 'ego' dreamed-of for a long time, *whose time however is resolved in the hangings which stop it*, finishing them with its splendor, *its shudders deadened* by an oblivion like languishing long hair around a face illuminated by mystery, with void eyes equal *to the inmate's* in the mirror, *stripped of all meaning but presence itself.*" The ontological primacy of the presence of Being over meaning correlates with that by which the *νοεῖν* establishes itself in the *εἶναι* while the strangeness of its self-discovery on the furnishings, here, the hangings, installs its Presence in the 'Beyond-meaning.' This is the route of the "outside Infinite" which is substituted for the subjective representation of the "lowly inner life" wherein the interplays of meaning dissipate. Furthermore, the "being-there's" movement of the opening of time implies the abolition of a form a individuality at the center of the unfolding of the entities of the World which is what establishes the presence of the Being of the dream on the articulated surroundings of the room:

It is the pure dream of a Midnight vanishing into itself, whose grateful Splendor, living alone in the breast of its accomplishment plunged in shadows, sums up its sterility via the pale complexion of an open book, which is presented by the table: the page and ordinary decor of the Night, except that there is still hovering some silence of an ancient word uttered by it, in which this Midnight evokes its limited and emptied ghost by the following words: I am the hour that must make myself pure.[63]

At this point, Mallarmé's meditation on the unity of *τόπος* and the *Λόγος* is so concise that the presence of the hour is uttered in one word which metamorphoses – an echo and a fragment of light – between silence and darkness in the immutable permanence of the decor:

From the finished furnishings that will settle into the darkness like the hangings, which already have been dulled into a permanent shape of always, useless; while the spark of virtuality produced by its own apparition in the mirroring of the obscurity scintillates the pure fire of the diamond of the clock, sole surviving jewel of the endless Night – the hour draws itself into this echo, at the threshold of the panels opened by its act in the Night: "Goodbye, night; as I was your tomb, now a surviving shadow will metamorphose into Eternity."[64]

This movement extended through the irrecusable presence of the word which took possession of the room is effected by a language act which reveals itself as "act in the Night," an intentional process, in other words, while also a movement of the *Verweltlichung*. It is in this way that the generative force of language, through its constitutive spatializing power, unfolds through the *ἐόντα*, the ontological ubiquity of the "being-there," expanded in the distance, from the very fact that its resplendence is procclaimed from one of the many thousand of the World's folds. If thinking implies Being's reclaiming of the human "being-there," then this revindication unfolds in the thought investment of a place. Thus it is a new conception of those Sites haunted by Mankind – sites whose constitutive field are the Poet's word, the Musician's score and the Painter's canvas – which requires meditation within the framework of the awareness of the ontological ubiquity of the "Being-in-the-World," the locus in which unfolds the immemorial Mystery of the work of art.

*Université d'Amiens*

Translated by Dr. Lois Oppenheim

NOTES

[1] Stéphane Mallarmé, 'Villiers de l'isle Adam' in 'Quelques médaillons et portraits en pied,' *Oeuvres complètes* (Paris: Ed. la Pléiade, 1945), p. 481.
[2] Maurice Merleau-Ponty, *Le Visible et l'invisible* (Paris: Gallimard, 1964), p. 152. [Eng. trans. by Alphonso *Lingis, The Visible and the Invisible*, ed. Claude Lefort (Evanston: Northwestern Univ. Pr., 1968), p. 113.]
[3] Translator's Note: Unless otherwise noted, the translation of quotations is ours. In those passages cited by Garelli, the effort was to remain as close as possible to the French with few, if any, alterations made, even for the sake of clarity.

[4] Translator: *Lieu*, a term recurring throughout the text, is variously rendered as "place," "locus" or "site," depending upon the particular context.

[5] Mallarmé, 'Villiers de l'isle Adam,' p. 481.

[6] Antonin Artaud, 'La Recherche de la Fécalité', in *Pour en Finir avec le Jugement de Dieu, Oeuvres complètes*, Tome XII (Paris: Gallimard, 1974).

[7] Merleau-Ponty, p. 152. [Eng. trans. Lingis, p. 113.]

[8] Merleau-Ponty, pp. 152–153. [Lingis, p. 113.]

[9] Merleau-Ponty, p. 153 (Underlined by us). [Lingis, pp. 113–114.]

[10] Merleau-Ponty, p. 153. [Lingis, p. 114.]

[11] Merleau-Ponty, pp. 153–154. [Lingis, p. 114.]

[12] Merleau-Ponty, p. 182. [Lingis, p. 138.]

[13] Merleau-Ponty, p. 183. [Lingis, p. 139.]

[14] Merleau-Ponty, p. 183. [Lingis, p. 139.] (The first underlining is Merleau-Ponty's while the second is ours.)

[15] "the seer . . . caught up in what he sees".

[16] Merleau-Ponty, p. 183. [Lingis, p. 139.]

[17] 'Moira' in *Essais et conférences* (Paris: Gallimard, 1958), p. 279.

[18] *Le poèm de Parménide* (Paris: P.U.F., 1955), p. 79.

[19] Heidegger, p. 286.

[20] 'Introduction à une lecture du poème de Parmenide,' *Le poème de Parménide* (Paris: P.U.F., 1955), pp. 69–70.

[21] *Heidegger et l'expérience de la pensée* (Paris: Gallimard, 1978), p. 383.

[22] We accord the term 'ontological' its dual value of a meditation on being *εἶναι* and on *λέγειν* such as it was established by Heidegger and Parmenides. (Cf. the followings pages.)

[23] *Moira*, p. 286.

[24] *Moira*, p. 286.

[25] *Moira*, p. 287.

[26] The remarks already made concerning the translation of the infinitive, *νοεῖν*, remain the same as regards this Section as well.

[27] *Moira*, pp. 279–280. Section VIII. 34–41. (Underlined by us.)

[28] *Le poème de Parménide*, p. 87.

[29] *Moira*, p. 292.

[30] *Moira*, p. 293.

[31] *Moira*, p. 294.

[32] *Moira*, p. 297.

[33] *Moira*, p. 299.

[34] *Moira*, p. 303.

[35] The note on p. 303 of *Moira* states: "A *λέγειν*, an expression, and a *νοεῖν* a 'knowing' " which attests to the necessity for contemplating the act of "perceiving" as being and as thinking. A fair number of contemporary attitudes, absorbed in the game of language meanings, appear to have forgotten this.

[36] *Moira*, p. 303.

[37] *Moira*, p. 306.

[38] *Moira*, pp. 306–307. (Underlined by us.)

[39] *Moira*, p. 309.

[40] *Logos* in *Essais et Conférences*, p. 271. "Das Geschick alles Geschicklichen."

[41] *Logos*, p. 271.

[42] Mallarmé, 'Un coup de dés' in *Oeuvres Complètes*, pp. 474–477. [Eng. trans. *Daisy Alden*, (Tiber Press, n.d., n.p.)]
[43] "Un coup de dés," pp. 474–475. [Eng. trans. Aldan.]
[44] "Un coup de dés," pp. 474–475. [Eng. trans. Alden.]
[45] "Un coup de dés," pp. 474–477. [Eng. trans. Alden.]
[46] "Un coup de dés," p. 475. [Eng. trans. Alden.]
[47] "Un coup de dés," p. 477. [Eng. trans. Alden.]
[48] "Un coup de dés," p. 477. [Eng. trans. Alden.]
[49] 'Le Minuit,' in *Igitur ou la folie d'Elbehnon* (Paris: la Pléiade, 1945), p. 435. [Eng. trans. Jack Hirschman (Los Angeles: Pegacycle Lady, 1974), p. 9.]
[50] *Logos*, p. 270.
[51] *Logos*, pp. 270–271.
[52] *Igitur*, 'Argument', p. 434. [Eng. trans., Hirschman, p. 7.]
[53] *Igitur*, p. 434. [Hirschman, p. 7.] (Underlined by us.)
[54] *Igitur*, p. 434. [Hirschman, pp. 7–8.] (Underlined by us.)
[55] *Igitur*, p. 435. [Hirschman, p. 9.]
[56] *Igitur*, 'Le Minuit,' p. 436. [Hirschman, pp. 9–10.]
[57] *Ibid.*
[58] *Ibid.*
[59] *Ibid.*
[60] *Ibid.*
[61] *Igitur*, 'Le Minuit,' p. 436. [Hirschman, p. 9.]
[62] *Igitur*, 'Le Minuit,' p. 436. Hirschman, p. 9. (Underlined by us.)
[63] *Igitur*, 'Le Minuit,' p. 435. [Hirschman, p. 10.]
[64] *Igitur*, 'Le Minuit,' p. 436. [Hirschman, p. 10.]

MICHAEL D. RILEY

# THE TRUTH OF THE BODY: MERLEAU-PONTY ON PERCEPTION, LANGUAGE, AND LITERATURE

## I

When Maurice Merleau-Ponty died in 1961, the world of professional philosophy was well aware that one of the Twentieth Century's most significant philosophic voices had been lost. Less obvious at the time, and I suspect still less than fully appreciated now, was the loss incurred by theorists of language and literature. Yet, as Ronald Bruzina points out, one of Merleau-Ponty's most valuable advances beyond the great inaugural work of Husserl was precisely in the phenomenology of language.[1] It is unlikely that the space of a brief essay can do much to remedy that situation. I would like to attempt, however, an interpretative summary of Merleau-Ponty's major insights regarding the body, perception, and language, and to suggest the relevance of those ideas to selected issues affecting literary theory. The emphasis throughout will be on the integrative or synthetic tendency of Merleau-Ponty's thought, a tendency which leads him to attack dualistic thought in all its forms. This synthetic tendency is especially evident in three insights with special relation to language issues: the inherence of meaning, the inherence of ambiguity, and the inherence of process to human existence.

Merleau-Ponty's entire phenomenology has sometimes been called, in de Waehlen's phrase, a "philosophy of ambiguity," and the description is apt as long as it remains free of pejorative connotations. For Merleau-Ponty, the term "ambiguity" indicates the complexity of lived experience, and ambiguity can be observed in all areas of that experience. Perception itself immerses us in ambiguity: "The phenomenon of perspective reveals a synthesis of immanence and transcendence [i.e., presence and absence] which contains the ambiguity of perception. For the perceived thing exists only insofar as I perceive it, and yet its being is never exhausted by the view I have of it. It is this simultaneous presence and absence that is required for 'something' to be perceived at all."[2] Two things are significant here. When I perceive anything – this page in front of me, for example – it is always in perspectival terms of foreground and background. The price I pay for the clarity with which I perceive the top of this particular page, if you will, is the relative fuzziness of the rest of my perceptual field. Yet that field is



*A-T. Tymieniecka (ed.), Analecta Husserliana, Vol. XVIII*, 479–493.

not nonexistent for me; by a shift of my attention, I could retrieve more precisely the ashtray on my desk, rather than this page, but that change would necessitate the relative absence of this page. This is one form of perceptual ambiguity. A second and more important form is contained in the realization that even the foreground of my perception is ambiguous; this page, even if it were blank, would transcend my grasp of it. That is, although I can certainly perceive a blank page of paper – and more, understand essentially what it is – I can never exhaust its possibilities or empty it of meaning. It always offers me more; there is always more to be known. As Merleau-Ponty says in regard to the common object mentioned above, "If I try to verify my perception of [an] ashtray, my task will be endless, for this perception takes for granted more than I can know in an explicit way" (*PP*, 383).

What is true of perception is equally true of my general sense of myself and of my activities. As to my sense of myself, "What I understand never quite tallies with my living experience; in short, I am never quite at one with myself. Such is the lot of a being who is born, that is, who once and for all has been given to himself as something to be understood" (*PP*, 347). As to my activities, "All human projects are contradictory because they simultaneously attract and repel their realization" (*S-NS*, 40). My urge to transcend myself through love, for example, is always matched by the urge of my self-centeredness to maintain itself. In other words, "The act which draws together at the same time takes away and holds at a distance, so that I touch myself only by escaping from myself" (*PP*, 408). Insofar as I lose myself in love, it may be that I discover what I really am; insofar as I attempt to lose myself in love and fail, I may discover my isolation with trebled force. "This ambiguity is not some imperfection of consciousness or existence, but the definition of them" (*PP*, 332).

Our second insight, and the greatest manifestation of our ambiguity, is the specifically human upsurge of meaning in existence. Again we must begin with the body: "Through its 'sensory fields' and its whole organization the body is, so to speak, predestined to model itself on the natural aspects of the world. But as an active body capable of gestures, of expression, and finally of language, it turns back on the world to signify it" (*Prim P*, 7). As Herbert Spiegelberg reminds us in his translation of "etre-au-monde," Merleau-Ponty's description of the primary phenomenon of human existence, we are not merely "in" the world but "at" it.[3] And it is precisely the intentional activity of the body and perception, the fact that both are expressions in action, that establishes meaning as inherent in human existence.

Indeed, it is precisely the inherence of meaning in existence that gives point to our equally inherent ambiguity: We can never disentangle ourselves either from meaning or from existence, for the two are synonymous. Simply because we are in the world, as Merleau-Ponty says in a shrewd turn on Sartre's famous phrase, "We are *condemned to meaning*, and we cannot do or say anything without its acquiring a name in history" (*PP*, XIX). As embodied subjects, we are resonant vessels of expression, and what we express is the indissoluble union of our being and Being. We are "condemned" to receive *and* create, to observe *and* express, to act *and* reflect – and never quite to sort out the boundaries between any of them.

It is crucial to recognize that the true ambiguity of our state derives from the equal presence and absence of meaning: We can no more say that we are not in contact with meaning than we can say that we are in contact with absolute meaning. Merleau-Ponty stands firm with Husserl in recognizing at the very heart of phenomenology what Husserl called the "intuition of essences," which means that "grasp of meaning is essential knowledge, knowledge of essence," that "there is a true object and there is a true grasp of it."[4] At the same time, as in the individual perception, our essential knowledge, while quite real, is never complete. We can never persuasively explain or reflectively encompass the world, for it always transcends us, escapes our attempts to finalize it, despite the fact that throughout our daily lives meaning is everywhere, so naturally omnipresent that we take it utterly for granted. As we have seen, this ambiguity defines us. And it is the inherence of meaning that defines our proper ambiguity.

Our final insight, the inherence of process, is merely the temporal term necessitated by the inherence of meaning and ambiguity. The inherence of process, however, should remind us of the positive side of the preceding analysis, for process is the agitation set up by a world which both defines us and is defined by us, an agitation which continually propels us onward in the project of growth itself. While our reflective understanding, for example, never keeps pace with the fullness of our lives, we do attain meaning and retain it as the ever-increasing foundation of still further knowledge. This process, this project, is life itself; truth is an unfolding, and it unfolds in us, beginning with the primary phenomenon of the body's perceptual inherence in the world, and extending from there through all the realms of symbolic activity.

The premier example of that symbolic activity is of course language itself, the keystone of the cultural world, and we can observe within it the same pattern of meaning, ambiguity, and process already noted. We must again turn first to the body, however, for language is not only as native to man as sight and hearing, it is as much a part of bodily experience as well, most obviously in the spontaneity of speech. Indeed, the connection between the body and language as primordial expression is crucial to the consistency found in all of Merleau-Ponty's further ideas on language:

> The body is our general medium for having a world. Sometimes it is restricted to the actions necessary for the conservation of life, and accordingly it posits around us a biological world; at other times, elaborating upon these primary actions and moving from the literal to a figurative meaning, it manifests through them a core of new significance: This is true of motor habits such as dancing. Sometimes, finally, the meaning aimed at cannot be achieved by the body's natural means; it must then build itself an instrument, and it projects thereby around itself a cultural world (*PP*, 146).

In short, the phenomenal body itself guarantees the active, expressive, creative (and symbolic) nature of "being-in-and-at-the-world"; it guarantees further that such expressive existence is inherently charged with meaning, a fact fully evident in the very first of all cultural objects, "the body of the other person as the vehicle of a form of behaviour" (*PP*, 348). This composite "style" of expressive being is the living presence of the person himself within the perceived world; it is not the "sign" of that person's presence, but that person himself. Moreover, that "style" of being is seen, shared, understood, for the world is an intersubjective world. We must admit, therefore, that "Expression is not one of the curiosities that the mind may propose to examine but is its existence in act" (*PP*, 79). Furthermore, "[The] primordial human gesture contains in seed all the characteristics of more sophisticated acts of expression" (*PP*, 186) – acts such as language and art.

For Merleau-Ponty, then, there is a direct connection between bodily gesture and linguistic gesture: "Words have meaning as behavior has meaning, for the spoken word, for Merleau-Ponty, is a veritable type of behavior and a true gesture. . . . [The] word as gesture is not simply a heuristic device but the demarcation of the nature of the spoken word."[5] Throughout his observations on language, Merleau-Ponty insists on the vital importance of the gestural meaning inherent in authentic speech and writing as the only way to understand the actual contact language has with being – the being of the speaker, listener, and the world they mutually inhabit and depend upon. Only as the *lived inscription* of embodied subjects can authentic language be seen to retain the force it actually imposes on our lives.

Merleau-Ponty's anti-Cartesianism is evident here, for his view of the primordially gestural potential of language is in part a response to the failure of either empiricism or intellectualism adequately to confront the appearance of meaning at all, especially of new meaning. As he says in one memorable sentence, "We refute both intellectualism and empiricism by simply saying that *the word has a meaning*" (*PP*, 177). The richness of this observation involves the fact that neither alternative comes close to the phenomenal *fact* of meaning: empiricism can never get convincingly from its dance of atoms and molecules, rods and cones, hammer and stirrup, neurons and synapses, to the sheer presence of *significance*; and intellectualism can never get convincingly from a subjectivity solely responsible, ultimately, for all the meaning it attains, to the hosts of *new significances* which irrevocably alter and inform that subjectivity. As always, the great strength of Merleau-Ponty's middle way resides in its conformity to the fullness and complexity of our actual experience.

Of the ambiguities inherent to the phenomenon of language, two closely related ones deserve particular mention: the relation between thought and language and the relation between perception and language. Neither relation permits a clear-cut differentiation. As Gillan reminds us, it is the unique genius of language to mediate "between the horizon of ideality and the horizon of perception."[6] But this boundary is anything but plain, for the ambiguity already inherent in perception finds itself articulated in the symbols of language. Pregnant with things themselves (the perceptual world) and thus truly gestural ("existence in act"), language also responds to and brings to being the range of *significance* found in that world as language slides through its chromatic scale from specific to general, from concrete to abstract, from syntactic structure to living presence. Language both complicates *and* clarifies the perceptual world. The ambiguity peculiar to language, while based on the ambiguity of perception itself, extends beyond sensory ambiguity to the full range of symbolic ambiguity which in fact expresses the specifically human fecundity of lived experience.

We can appreciate this interdependence in an even broader way by realizing that we cannot finally separate thought itself from authentic language: "It is through expression that we make [thought] our own. The denomination of objects does not follow upon recognition; it is itself recognition" (*PP*, 177). But the validity of such recognition still rests upon perceptual contact. Indeed, "It is the office of language to cause essences to exist in a state of separation which is in fact merely apparent, since through language they still rest upon the ante-predicative life of consciousness" (*PP*, XV).

The silence of primary consciousness and the upsurge of articulate meaning, therefore, co-exist in a relationship of necessary ambiguity. While not always strictly identical, neither can they ever be truly separated; they exist as desire and fulfillment, call and answer. "What then does language express, if it does not express thought? It presents or rather it *is* the subject's taking up of a position in the world of his meanings" (*PP*, 193). If we once allow language to become irrevocably sundered from thought, we not only trivialize language but also deny the validity of perception which language touches and completes. Communication – in fact, knowledge itself – is thereby rendered impossible, for the dance of the code has no necessary relation to the codifier. Ultimately, only some form of post-Cartesian nihilism remains, in which *en soi* and *pour soi*, body and consciousness, I and Thou reside in a state of sterile separation.

A final pair of related ambiguities deserves mention, the first of which again echoes our perceptual circumstances. Just as perception inhabits, in Husserl's words, a transcendent world which "is always already there," so also does language, even the language of primary gestural expression. Thus, our expressive activity can never be absolute. The world on which our primary expressions depend transcends our linguistic grasp; although we grasp essential truths, we never grasp them entirely. Therefore, absolute expression is impossible. And yet our primary expressions, precisely because they reveal essential truths, however incompletely, likewise outrun our interpretative grasp; although we share their inherent meanings, we can never exhaust their potential. Consequently, absolute interpretation is equally impossible. But primary expressions are not the sole inhabitants of our linguistic world. Precisely because the linguistic world is in large part "always already there," a related temporal ambiguity also arises.

In language, as in perception, past, present, and future merge; what *has been* said, collectively or personally, becomes either repeated or transformed in what *is being* said, which is itself only a portion of what *remains to be* said. "What has been said" Merleau-Ponty calls "sedimented" language to distinguish it from "creative" language, the authentic gestural language we have been discussing. But the union here is an intimate one. Sedimented language is far from dead – though it may indeed degenerate to mere mindless repetition – for it always exists as infinitely transformable, vitally potential. Indeed, except for the comparatively rare instances when new language is actually created, authentic expression ordinarily engages in just such a reactivation of the already given. Moreover, all language which is presently sedimented was at one time, however distant, creative; and all creative language, however innovative or unique, is itself destined to sediment, thus becoming available for still future transformations. This necessary interplay, then, is another of the vital ambiguities of language.

This same interplay also makes clear the inherence of process – of history – to the phenomenon of language. Repetition and creation, expression and interpretation: each calls forth the other in necessary partnership in the endless project of language which has as its object no less than the elucidation of a world:

> Each act of philosophical or literary expression contributes to fulfilling the vow to retrieve the world taken with the first appearance of a language, that is, with the first appearance of a finite system of signs which claimed to be capable in principle of winning by a sort of ruse any being which might present itself. Each act of expression realizes for its own part a portion of this project, and by opening a new field of truths, further extends the contract which has just expired. (*S*, 95)

Merleau-Ponty's refusal to simplify the nature of our experience, his insistence that philosophy conform to the actual complexity of that experience, with all its frayed ends of ambiguity and temporal uncertainty intact, might appear to some the outgrowth of fundamentally negative impulses. Nothing could be farther from the truth. For Merleau-Ponty, truth and meaning – however fraught with hazards and however partial their satisfaction of our omnivorous desires – not only exist, they inhabit every corner of our lives. From the silences of Being itself they are called forth, and call us forth, preeminently in the activities of perception and language. Merleau-Ponty expresses quite beautifully the essence of his own intellectual project, as well as its principal merit, in the following words: "It is easy to strip language and actions of all meaning and to make them seem absurd, if only one looks at them from far enough away. . . . But that other miracle, the fact that, in an absurd world, language and behavior do have meaning for those who speak and act, remains to be understood" (*S-NS*, 39).

## II

For Merleau-Ponty, art is the very clothing of the phenomenal body because "[it] is related above all to the perceived world with which we are in living contact."[7] Each of the arts accomplishes in its own unique fashion the greatest extension of significance possible of the body's sensory fields, a density of significance so great that the richness of the work of art most resembles the transcendent nature of Being itself, the "world" on which perception ultimately depends. Despite their individual variations, therefore, each of the arts arises within the context of meaning, ambiguity, and process inherent in the perceptual presence of the phenomenal body to the world itself. The core of all the arts resides in the same human center – our expressive "existence in act."

Although Merleau-Ponty ranges freely among all the arts, literature holds an especially valued position in his pantheon – not surprisingly, given the importance language assumes in his phenomenology. Literature is a "conquering language" which not only partakes of being, like all authentic language, but which installs itself in being. It thus acquires an enormous power. "The process of expression, when it is successful, does not merely leave for the reader and the writer himself a kind of reminder, it brings the meaning into existence as a thing at the very heart of the text, it brings it to life in an organism of words, establishing it in the writer or the reader as a new sense organ, opening a new field or a new dimension to our experiences" (*PP*, 182). Such a statement will appear far-fetched only to the intellectualist for whom words are merely beads on a symbolic abacus. His problem, however, is to explain how words so fraught with "artifice" can instruct, move, and even change him, at times so completely and irrevocably.

The price literature pays for this power, this inauguration of a conquering meaning, is its ambiguity. But ambiguity is in fact that very density which enables literature to appear so uncannily like perceived experience itself:

> What is irreplaceable in the work of art? What makes it far more a voice of the spirit, whose analogue is found in all productive philosophic or political thought, than a means to pleasure? The fact that it contains, better than ideas, *matrices of ideas* – the fact that it provides us with symbols whose meaning we never stop developing. Precisely because it dwells and makes us dwell in a world we do not have the key to, the work of art teaches us to see and ultimately gives us something to think about as no analytical work can. (*S*, 77)

Thus the very nature of artistic expression guarantees the existence of criticism. Art calls criticism forth as inevitably as an interlocutor calls forth a spoken conversation. But just as the artist is transcended by the world and the language he inhabits, even while he touches both essentially, so the critic is transcended by the world of the text and the language it transforms, even while he also reaches it essentially. "Creation thus begins and ends in human experience. . . . The expression is never complete because there is always a further call to interpretation, and each individual interpretation in turn makes its appeal to reflective criticism which, for Merleau-Ponty, is the very essence of philosophy itself."[8] Perception and bodily gesture, speech and listening, writing and reading, art and criticism: all are "project" – the unfolding, the pressing forward of meaning toward further clarification whose progress can neither be denied nor absolutely finished.

At its best, criticism will facilitate the authentic communication of the primary text, even though it can never replace the reader's direct experience of that text. Since "the joy of art lies in its showing how something takes on meaning" (*S-NS*, 58), what Merleau-Ponty most applauds is a revelatory criticism that assists at the birth of meaning and which therefore makes that meaning more accessible, a criticism which will make manifest the complexities of expression, rather than limit, simplify, or categorize them, and which will keep its own pretensions to absoluteness or closure modestly in check – an ideal hardly surprising, since it defines equally well the philosophical task phenomenology sets for itself.

Several recent *contretemps* in that world of literary criticism and theory suggest the usefulness of Merleau-Ponty's mediating, integrative perspective. One such issue is the question of whether a given literary text has any "stable meaning" worthy of discussion. Another issue, nearly aligned with the first, involves the increasingly stark dichotomy drawn between the world of the text as it is subjectively created in the mind of the solitary reader and the seemingly autonomous operation of linguistic codes or structures as a virtually deterministic force in the worlds inhabited by both author and reader. The starkest disagreement in both cases often arises by the presumption of an overly exclusive subject-object separation, a separation which runs counter to the thrust of Merleau-Ponty's entire philosophy.

Sufficient stability for transfer must be recognized, in both perception and language, if meaning is to exist at all – as it manifestly does – but that meaning must not be viewed as exhaustive if the transcendence, the necessary ambiguity, of the object is to be retained. We must say of the text what Merleau-Ponty says of the perceptual world itself (because, as we have seen, language partakes essentially of that world): "The world remains the same world throughout my life, because it is that permanent being within which I make all corrections to my knowledge, a world which in its unity remains unaffected by those corrections, and the self-evidence of which attracts my activity towards the truth through appearance and error" (*PP*, 328). We can no more hope to overmaster the text than the author could hope to overmaster his experience by composing it; yet we can no more deny its inherent meaning than the author can deny the work's inherent sense of formal closure upon its completion or the gestural imprint which makes it his own. The reader, fellow-traveler in an intersubjective world, apprehends this original accent precisely *because* it occurs against a pre-given context (syntactic, semantic, generic, cultural, personal, perceptual); the reader's experience of the text is directly analogous to his perceptual experience, in which the freshness of immediacy always arises within the context of memory:

> A telling utterance or a good book impose their meaning upon us. Thus they carry it within them in a certain way. . . . Speech is, therefore, that paradoxical operation through which, by using words of a given sense, and already available meanings, we try to follow up an intention which necessarily outstrips, modifies, and itself, in the last analysis, stabilizes the meanings of the words which translate it (*PP*, 388–89).

Similarly, neither the reader's uniqueness of response to the linguistic artifact nor the seeming autonomy of the artifact (or of language itself) can be emphasized without violence being done to the true nature of the experience. For one thing, the reader never merely "takes in" a text; he "opens out" to receive it. Meaning arises, in fact, because of this necessary *conjunction*. Furthermore, as the articulated voice of man's relation to meaning, language can never be separated convincingly from the living voices that have shaped, adapted, and reshaped language continuously; language can only be made into a "thing" when man also becomes a "thing." The sense of "absence" or distance language impresses upon us at times (as when a single word is stared at for a great length of time) is no more definitive than its remarkable "presence" – the expressiveness and commonality language by its very nature presumes and seeks to perfect. This interacting presence and absence suggests, quite exactly, the never-ending project which language *is*, whether considered culturally or personally, for writer or reader.

Moreover, neither emphasis on the subjective consciousness of the reader nor emphasis on the supposedly objective formative structure of language can account for how the resultant gap is ever bridged, a bridge which "knowing," by any definition, regularly constructs:

> We would deprive [language] of its distinctive characteristic by realizing the result of expressive operations in advance in a heaven of ideas [or, we might add, in a hell of deterministic structures]; we would lose sight of the leap these operations take from already available significations to those we are in the process of constructing and acquiring. And the intelligible substitute we would try to base them on would not exempt us from understanding how our knowing apparatus expands to the point of understanding what it does not contain (*S*, 95).

Thus, from neither the subject nor object perspective can thought and meaning be denied a necessary partnership with language: "Language is not meaning's servant, and yet it does not govern meaning. There is no subordination between them. Here no one commands and no one obeys. What we *mean* is not before us, outside all speech, as sheer signification. It is only the excess of what we live over what has already been said" (*S*, 83). And one great attraction of Merleau-Ponty's views of language is that they preserve a balance which actually conforms to the complexity of our linguistic experience. Language is weighted neither in the direction of abstract, suprapersonal patterns in whose grooves we merely trot along like sheep; nor is it wholly plastic and subjective, a private theater in which we sit like god-like directors, projecting isolated fantasies to which we forlornly cling as the only reality we have. As our own experience of meaning assures us, language and literature embrace – complexly, ambiguously, transcendently – both extremes.

Many of criticism's most hallowed staples can also be considerably enriched by means of Merleau-Ponty's phenomenological perspective, particularly the function of figurative language and irony, and the integrity of style and context.

Since *all* speech and writing involve the taking over of available language resources in pursuit of a transformed significance, literature can be seen as the most complete accomplishment of that activity, one in which the resources of language come closest to full realization. While it would be a mistake to credit figurative language with all of the victories of this "conquering language," it would be equally difficult to minimize its importance. Figurative language thrives at the outermost boundaries of thought, language, and perception, asserting a fitness between man and the world at first unlooked-for and thus inherently creative. Indeed, figurative language depends upon the implicit assumption that the world is an endless source of potential significance whose meanings emerge most dramatically when the signifying mind finds its most daring flights held aloft by the world itself.

Furthermore, figurative language is frequently animistic, affirming the necessary interrelation between the specifically human world and the world of things. As William Carlos Williams says in his short poem "A Sort of a Song," the nature of metaphor is to "reconcile the people and the stones." Even when figures of speech do not specifically refer to the human world, the imaginative displacement of expected perception and expected language attests to the abiding presence of mind – much as connotation (an important part of what Merleau-Ponty designates the "gestural" sense of language) inevitably attests the abiding presence of human history at the heart of language, a history inclusive of emotion and sensation as well as objectivity.

Uniquely saturated with humanized – i.e., symbolic – significance, therefore, figurative language as a whole manifests to a special degree man's ambiguous stance as equally the discoverer and the creator of meaning and establishes that line of necessary ambiguity directly from the perceptual world to the linguistic world. To plumb the world's meanings is simultaneously to plumb the potentialities of the mind, its language, and the other. The writer creates the idiom of his work, endowing that work with his own accent; yet the reader encountering the comparison recognizes its fitness, wondering perhaps that he had not noticed it for himself. Creation and discovery, symbol and thing, writer and reader, particularity and universality intermingle in the service of meaning and the intersubjective community meaning presumes. And the world is enriched by the reacquisition of one of its abated truths.

Symbolism, in its usual literary sense, shares the operations of metaphor but asserts even more strongly the inherence of meaning in the worlds of perception and discourse, and the unlimited potential for the increase of that meaning. From a symbol, meaning radiates in many directions, but without ever sacrificing its existential integrity. The literary symbol thus provides a perfect instance of the immanence and transcendence of meaning, and of the necessary ambiguity that defines that relationship – all of which becomes manifest as a process of unfolding, whether in my own multiple readings or in the history of criticism itself.

In some respects the ambiguity inherent in existence finds its most inevitable vehicle in ironic apprehension and expression. Language, and the world it refers to, folds back upon itself in order to reveal itself. Yet for a word, a scene, a picture to resonate with irony, it must presume a known (though never a completed) world. So it is with imagination itself: meaning must inhere before it can be challenged, complicated, and transformed.

Because for Merleau-Ponty all primary linguistic expression, like our expressive bodily behavior, touches being itself, he always insisted that the integrity of the entire expression must be maintained. Thus another well-traveled critical concept reappears with renewed force: "[The] principle of Plato and Aristotle so dear to classical aestheticians as 'organic unity' is no mere metaphor. In the presence of this living thought, the hearer or reader comes to possess a new sense organ with which to view the world; the successful literary work opens a new field or dimension of experience."[9] But it does so only in its complete integrity.

For Merleau-Ponty, "style" in its widest sense comprises this contextual richness. Style first of all must be seen to include each nuance and portion of a completed primary expression, since all the resources of language combine to create it:

The spoken or written words carry a top coating of meaning which sticks to them and which presents the thought as a style, an affective value, a piece of existential mimicry, rather than as a conceptual statement. We find here, beneath the conceptual meaning of the words, an existential meaning which is not only rendered by them, but which inhabits them, and is inseparable from them (*PP*, 182).

Language can only appear arbitrary – a mere matter of endings and structure – when we limit it to its conceptual or objective potentialities alone:

[It] would no longer appear so if we took into account the emotional content of the word, which we have called above its 'gestural' sense, which is all important in poetry, for example. It would then be found that the words, vowels and phonemes are so many ways of 'singing' the world, and that their function is to represent things . . . because they extract, and literally express, their emotional essence (*PP*, 187).

Style is therefore the inevitable signature of authentic expression, for it results from the specifically human engagement with all the resources of language in purusit of truth, a mode of being not merely caught, but still alive as a virtual presence. When his world truly inhabits his expression, then the writer himself inhabits that expression to the same degree; and it is the resultant style or accent of vision and experience which touches and alters the reader most dramatically, rendering as it does the fullest possible communication.

Students of literature have of course always appreciated this, recognizing clearly that "the power of a writer resides more in his style, in his manner of presentation, and his interlacing of words, than in the communication of 'objective truths'."[10] But to say this is to say simultaneously that objective truth touches but narrowly either the richness of language or the even greater richness of being which language seeks to elucidate, a richness which depends for its continued unfolding on the endless creativity of perception and mind united with an endlessly creative vehicle.

Indeed, insofar as knowledge itself depends upon a perceptible pattern within which my faculties feel relatively "at home," other people and the world itself may be considered stylistic vehicles. However great the disparities and inequalities in the world of my experience, equally apparent is a fitness and a "sense" to which my faculties respond as one portion of being to another, a pattern which makes knowledge itself possible:

A style is a certain manner of dealing with situations, which I identify or understand in an individual or in a writer, by taking over that manner myself in a sort of imitative way, even though I may be quite unable to define it; and in any case a definition, correct though it may be, never provides an exact equivalent, and is never of interest to any but

> those who have already had the actual experience. I experience the unity of the world as I recognize a style (*PP*, 327).

At one point in *Signs* Merleau-Ponty reflects with considerable satisfaction on the intellectual conditions of his day. At no time like the present, he says, have science and politics so intently questioned their own primary assumptions; at no time like the present has literature been so "philosophical," so ready to reflect on the nature of language, truth, and its own expressive activity. Were philosophers themselves to weaken in their resolve, he says, all of these other forces would be ready to call them back to philosophy. Then he closes with this thought:

> Unless this uneasiness consumes itself, and the world destroys itself in experiencing itself, much can be expected of an age which no longer believes in philosophy triumphant but is through its difficulties a permanent appeal to rigor, criticism, universality, and philosophy militant (*S*, 157–58).

Clearly, the force of these remarks has only been intensified by the passage of two decades. It is altogether likely, for example, that we are now witnessing a great age of linguistic and literary analysis, one in which philosophy is indeed being pressed to its task from many directions at once. Yet the result often resembles nothing so much as an Eastern bazaar, with something indeed for everyone, but roped off in countless narrow and competing stalls and set off by the cacophony of endless haggling and recrimination. One may justly wonder whether the "rigor" and "universality" Merleau-Ponty looked for has often enough appeared; the "militancy" we have had in full measure.

The abundant presence of both rigor and universality in Merleau-Ponty's own work attests strongly to its permanent value. Most important perhaps is his refusal to consign human experience to narrowly subjectivist or objectivist definitions, closely followed in importance by his subtle and careful grounding of all experience in the operations of the body and perception. His recognition that human experience is inherently expressive, creative, and meaningful sustains both knowledge and communication, even while the full force of inherent ambiguity and temporality is unhesitatingly faced. While its special characteristics receive extensive and sensitive analysis, from the realm of spontaneous speech to that of the great literary artifact, language as a primordial phenomenon is housed in the fullness of human experience, not separated but indissolubly implicated in a wider definition.

Nor is too much claimed. Merleau-Ponty's phenomenology is generative rather than closed, like all primary expressions. The miracle, the "silence" at the heart of existence is not so much a problem to be solved as it is a call requiring an answer, a need requiring a voice. With luck, we shall attend most carefully to those voices as rigorous, universal, and humane as that of Merleau-Ponty himself. For in the end, we are what we say we are.

*The Pennsylvania State University, Berks Campus*

## NOTES

[1] *Logos and Eidos: The Concept in Phenomenology* (The Hague: Mouton, 1970), p. 10. References to Merleau-Ponty are from the following editions, with the abbreviations used in the text: *Phenomenology of Perception*, trans. Colin Smith (London: Routledge and Kegan Paul, 1962): *PP; The Primacy of Perception and Other Essays*, ed. James M. Edie (Evanston, Ill.: Northwestern University Press, 1964): *Prim P; Sense and Non-Sense*, trans. Hubert L. and Patricia A. Dreyfus (Evanston, Ill.: Northwestern University Press, 1964): *S-NS; Signs*, trans. Richard C. McCleary (Evanston, Ill.: Northwestern University Press, 1964): *S; The Visible and Invisible*, trans. Alphonso Lingis (Evanston, Ill.: Northwestern University Press, 1968): *VI.*

[2] John O'Neill, *Perception, Expression, and History: The Social Phenomenology of Maurice Merleau-Ponty* (Evanston, Ill.: Northwestern University Press, 1970), p. 16.

[3] *The Phenomenological Movement: A Historical Introduction*, vol. 2 (The Hague: Martinus Nijhoff, 1969), p. 522, n. 1.

[4] Bruzina, pp. 39–40.

[5] Garth Gillan, 'In the Folds of the Flesh: Philosophy and Language,' in *The Horizons of the Flesh: Critical Perspectives on the Thought of Merleau-Ponty*, ed. Garth Gillan (Carbondale, Ill.: Southern Illinois University Press, 1972), p. 6.

[6] Ibid., p. 8.

[7] Albert Rabil, Jr., *Merleau-Ponty: Existentialist of the Social World* (New York: Columbia University Press, 1967), p. 212.

[8] Eugene F. Kaelin, *An Existentialist Aesthetic: The Theories of Sartre and Merleau-Ponty* (Madison: The University of Wisconsin Press, 1962), p. 304.

[9] Kaelin, p. 289.

[10] Laurie Spurling, *Phenomenology and the Social World: The Philosophy of Merleau-Ponty and its Relation to the Social Sciences* (London: Routledge and Kegan Paul, 1977), p. 55.

# B. LITERARY INTERPRETATION OF "REALITY"

FÉLIX MARTÍNEZ-BONATI

# FICTION AND THE TRANSPOSITION OF PRESENCE

It is characteristic of Husserlian phenomenology [1] that it methodically retracts from the direct description of the given object back to the reflective description of the acts of consciousness that give constitution to the object. In a sense, what is hidden in the immediate intuition of the object is brought to light by the *intentio obliqua* of reflective intuition or of the transcendental reduction. There is a basic duality here, the duality of transcendence and immanence, of world and subject, a duality replicated *within* transcendental consciousness by the distinction of *noema* and *noesis*. Leaving aside very significant philosophical differences, I dare say that, topologically speaking (that is, considering only the *position* of the concepts in the overall system of the theory), this duality corresponds to the traditional opposition of the represented and the representation, the object and its image.[2] What is hidden in the object and is brought to light by the examination of the image as such is the structure of the image *and* the formal constitution of the object – insofar as the object is given to the subject, that is, as phenomenon. By *structure*, in a wide Structuralist sense, I mean the system of rules that determine the formation of the image and the set of components articulated by these rules.

Up to a certain point – which I will indicate – I can term my approach *phenomenological* in the Husserlian sense (as I did in the subtitle of my book) or, less restrictively, *transcendental* in the Kantian sense, because I try to grasp the essence of the phenomenon by reflecting upon constitutive principles of objectivity which determine the subject's activity of knowledge. These subjective principles, according to the philosophical tradition I am evoking, are not private or particular, but universal, and they are the very ground of objectivity. (Whether this universality is absolute, as Husserl thinks, anthropological, as Kant thinks, or ethnocentric, as many think today, is not my present concern.) Not only the real world we live in and the languages we speak, but also the fictions of art and literature are objectively, that is, intersubjectively constituted. That we can meaningfully talk, agreeing and disagreeing, about the imaginary beings of literature is proof of their objectivity.

My general theme here is the phenomenon of literature, the nature of the



*A-T. Tymieniecka (ed.), Analecta Husserliana, Vol. XVIII*, 495–504.

object that we experience in the everyday-reading of novels, poems, or plays. I prefer to begin by limiting my considerations to narrative literature, especially to fiction and to the monological standard discourse of an authorial narrator. The following remarks explore some aspects of the constitution of literary fiction. I must begin, however, with a few notes on communication in general.

It is a fact deserving attention in the present intellectual circumstance that whenever we see written signs we proceed to understand them as being ultimately the product of a singular act of expression or communication. Somebody, at some place and moment, and for some reason, inscribed the signs (or caused others to do it for him). Indeed we assimilate written signs to oral utterances insofar as we understand both of them as instruments of a human action, thus bound to a personal responsibility and to a singular situation or origin. The meaning of the written as well as of the oral sign can be of merely private or of public relevance, the pertinent circumstances momentary or durable. "Pick up the package at six," in a piece of paper, differs in this respect enormously from "Know Thyself" engraved in stone over the entrance of the temple. In the first case the message is essentially incomplete and worthless, ineffective, if we cannot retrieve the circumstances of origin (that is, decipher all the implicit and explicit deictic references). In the second case, not knowing who was the individual whose decision led to the engraving, or the exact time and minor circumstances of his/her act, is almost entirely irrelevant for the meaning and value of the sentence. However, the original cultural circumstance is relevant for its meaning. What was meant at that time by *knowing*, by *self*? Whose authority endorsed the imperative? What form of life is addressing itself with these words?

The narrow or wide circumstance of its effective use corresponds in the sign to the narrow or wide conceptual sphere of its intrinsic meaning. We can provisorily limit ourselves to the assumption that the widest possible circumstance of an utterance's appropriate, meaningful use is the life of a culture, the narrowest, an instant of one individual's life. Differences in the scope of the pertinent circumstance explain why we tend to engrave in stone *dicta* such as "Know Thyself" and not messages such as "Pick up the package at six."

But what should we say if somebody in his right mind did engrave in stone or print in many copies the sentence "Pick up the package at six"? We would probably take it as a quotation, the quotation, say, of a famous verbal utterance made in circumstances and by somebody of historical prominence; something like MacArthur's "I shall return." Dicta of universal content, such as "Know Thyself," belong directly to a public and durable context. Narrowly occasional utterances can accede only indirectly to such meaning, that is, by means of the public imaginary reproduction and fixation of the formerly private circumstance, so that it becomes an item of the public world. Historiographical (as well as journalistic) imagination reproduces the single event as a repeatable image. It does not really *reproduce* the unretrievable past event. It only "represents" it. It represents the past event in a way that constitutively subordinates the present representation or image to the past event – that is, to its traces (documents, pieces of evidence, the memory of witnesses). The quotation of a historical utterance is one of the forms of historiographical representation.

If we press a little harder on this matter we see that reading a written text (or listening to the recording of a speech) is always similar to understanding a quotation, because the actual experience of reading or listening cannot be taken as the original circumstance of the utterance in question: we read here and now an utterance made at another time and place. And, of course, we have to imaginarily transpose the utterance to its original circumstance. Thus, we observe the date of the newspaper, the date of the first publication of the book and the background of his author, and so on. Recorded or read sentences are essentially reproductions of internally or externally spoken sentences, or of originally written sentences (written necessarily at some moment for the first time). There is, then, an important difference between reproduced (written and recorded) and originally produced sentences. And this is a difference in the way we, readers and listeners, take them and constitute them as signs.

However, this difference might not be essential from a more abstract point of view. One could maintain that even in the spoken exchange of words we have always to transpose the sentence we hear in our circumstance to the speaker's act of producing it in his circumstance – which will never be totally identical with ours. Is, then, not all understanding some form of understanding "quotations," reproduced sentences, representations of the ungraspable original act? Are not all utterances a kind of writing? The difference indicated above seems to be, then, not essential, but quantitative: the leap from the present reproduction to the absent original is bigger and therefore noticeable in written sentences, while minimal and disappearing in spoken exchange.

Yet, a quotation proper will be structurally different from an originally said sentence. In the case of quotation we transpose the sentence heard by us first to the speaker's oral utterance, and, then, again to another, the original source. Otherwise we misunderstand the quotation.

The fact that in linguistic communication we are always confronted with reproduction and transposition of circumstance, that is, with representation (with a kind of writing, if your want to fashionably say so), is not more important than the fact that we are always collapsing the difference between the present sign and the absent original. We cannot understand signs unless we see them as if they were (part of) the act of the original subject. There might never be such a thing as the direct experience of the presence of the other's expressive act (or of one's own for that matter), but if we want to understand the sign we perceive we have to construct the imaginary experience of that presence. We have to relive through representation the event of the other's internal and external expressive act. Presence and voice of the other are "imaginary" experiences but nonetheless necessary and constitutive moments of all understanding. The original expressive action (as physical event and as intentional act) appears, then, as a kind of regulative idea of communicative experience ("regulative idea" in a Kantian sense, that is, something that lies beyond possible experience but is a necessary reference to construct the order of experience). That is why, as we said at the beginning, we always take written signs as the expression of an original and singular event of intentional communication. We automatically convert the letter to a vague image of sound (mostly internalized, "mute," temporal sequences of an individual's expression).

Our understanding of our life seems, thus, to be determined by an ontology of events (some of which are communicative acts). Unrecoverable, unduplicable events, forming part of unrepeatable situations. As Husserl (who, by the way, does not accept the view of representation I am sustaining) indicates, facts have essences, and the singular event carries in itself a universal form. Hence, reproduction of the individual *form* allows for the representation and the imaginary (indeed illusory) retrieval of the past or absent event. We cannot retrieve the event itself, but we can (really) actualize its nature or essence, and thus create an image that will illusorily evoke the presence of the absent event. Understanding goes always against the current of time imaginarily placing present images where past events are supposed to have been.

The transposition of the present image is the way we construct our world, spatially and temporally. The search for the origin of any sign is not a metaphysical quest but a fundamental mechanism of understanding, even more, of all experience. To read signs without connecting them to their origins (if at all possible as a meaningful operation, which I doubt) would imply the repression of meaning and world in favor of casual associative excursions in the field of language, or a quasisolipsistic contraction to the series of impressions and sensations. To read signs without connecting them to their origins is like having retinal impressions not seen as objects in space. We can reasonably accept that there never is immediate presence of objects as naive realism would think it, or as Brentano holds (and also Husserl does in certain passages).[3] This rejection of realism corresponds, of course, to the old physico-physiological critical doctrine of knowledge and perception, partly based on Newtonian optics, that leads to Kant's epistemology. Presence belongs to the ungraspable *thing in itself*. But we cannot live without projecting presences in the modes of perception, memory, and imagination proper. Without the imagination of presences there would be no world and no consciousness.

Now, how is this transpositive operation performed when reading a novel appropriately? Reading, or listening to, a fictional narration is a modified act, based on the act of listening to or reading a "true" narration. If we open, say, Albert Speer's *Memoirs* and read a paragraph or two, we imagine, very vaguely, but positively, an authorial presence making those assertions. If we have reasons to think some of the author's most important assertions as factually inaccurate and many of his underlying assumptions as morally wrong (or, to the contrary, his narrative true to fact and his attitudes fair-minded) we will judge of his personality accordingly. Those are *his* words, his communicative deeds. We have transposed the read sentences to an origin in the author of the text, and, moreover, we imagine them as mostly, if not always, serious acts of assertion which are consistent with his durable convictions.[4] In other words, these assertions define by their implications the identity of the author, and they are determined, at the same time, by the author's personal and historical circumstances. They are addressed to a defined public, confronting previous assumptions and views on the subject matter and on the way to deal with it.

Let us now consider a fictional narrative such as Thomas Mann's *Bekenntnisse des Hochstaplers Felix Krull*. We read the sentences transposing them to an origin that, at the beginning, is radically more vague than the imagined author of Speer's memoirs. We do *not* imagine Thomas Mann speaking (as we would if we were reading his published letters). We imagine a speaker whose person and personal circumstances are not a part of the events that really have taken place in the world. We imagine, certainly, that these are this fictive person's words, his deeds, but these words and deeds are as fictive as the person who utters and performs them.

But did not Mann himself write these sentences? In a sense, yes, of course. But he was not writing *his* (mostly serious) assertions – those consistent with his durable convictions. He was imagining the assertions of another speaker or writer whom he was inventing along with the assertions. Mann's literary, fictional words are, then, not direct speech or linguistic communication, but icons of a speech never really intended by anybody as an act of language in the world. In my book I term "pseudosentences" the really performed, spoken and heard or written and read, icons of an absent speech, and "authentic, but merely imaginary sentences" the absent speech itself when it is a fictive one. I would term "absent or represented *real* and authentic sentences" the signifying acts of the author of a nonfictional text. Directly heard or, generally, perceived original expressions can be termed, in this theoretical context, "real and authentic sentences."

The conception of the representative nature of experience in general – that I have assumed – should not serve to dilute the differences between perception of original speech, as a relatively direct form of knowledge, and reading, or listening to recordings, as a markedly transpositive operation. More important for our theme is that we should not overlook the difference between transposing read sentences to an original authorial act of speech, and transposing read sentences to an original act of speech that is *ipso facto* imagined as being fictitious, merely imaginary. Literature in a strict sense is defined by this latter operation of reading. Literature is fictive discourse. This means: literary discourse is imagined as emerging from a person who never existed. The (real) author of the text may use the fictive speaker to express the author's views; the style of the fiction may be not different from the one the author displays in his own real assertions (letters, essays, etc.). The fiction may resemble the author's personal attributes. But this does not deny the difference I am pointing to.

A nonfictional text is bound to its circumstances in a way different from the poetic text. The truth of a historiographic or journalistic narrative is – at least in a fundamental part – a matter of facts, of singular facts. The truth of a novel, if any, is not the truth of the narrator's account of the pertinent story. The nonfictional narrative text is contextually connected with other texts referring to the same singular events. This is not the case of the fictional narrative, whose contextuality might be archetypal, ideological, stylistic, generic, or thematic in an abstract sense, but will never be determined by (nonexistent) common narrated singular events. The nonfictional text reflects explicitly the views of the author. The fictional one may be *entirely* ironic, that is, the display, perhaps merely tentative, of a view not held by the author within his durable convictions. Such ironic creation, certainly, also expresses the mind of the author and is his responsibility, but the connection thus necessarily made is not identical with the responsibility of a speaker to his speech or a writer to his straight statement.

Fictional speech has a fictional audience, not only not identical with the real readership but occasionally very dissimilar from it. Historiography may also address an original public different from subsequent readers. This difference, however, is subjected to limitations not binding fiction – where, for example, the narrative discourse can be addressed to an assembly of animals, ghosts, Martians, and so on.

The repeated claim of the autonomy of literature is certainly wrong if it is taken in an absolute sense (seldom, if ever, found in its proponents). However, there should be little doubt that literary discourse is one essential step removed from the relation to real circumstance of any form of real discourse. Fictional imagination relates more indirectly to reality than historiographical or journalistic imagination.

An important implication of the fictivity of literary discourse is that, as all fictional imaginations, it can assume fantastic properties, that is, escape the laws of reality. As we have superhuman heros in fiction, we have also superhuman discourses – I mean discourses imagined as spontaneous utterances of a human being which no human being can be expected to produce: say, the sonnet spoken in a moment of passion by a character in a play; also the omniscient narrator's discourse in realistic novels is a fantastic discourse implying supersensual powers of perception.

The basic point of my view of literature is that the discourse of poetry and fiction is fictive, has never existed as the real utterance of a real speaker. (It has existed, of course, as the real image of a speech.) To say that literary speech is fictive (or fictitious) is not the same thing as saying that it is feigned or fictional. Fictional is generally said of discourse that *refers* to fictions, without necessarily being implied that the discourse itself is a fiction. Thus, philosophers (for example Richard Rorty in a lecture given at Columbia University in 1980 with the title "Is There a Problem about Fictional Discourse?") call fictional the assertion "Sherlock Holmes lived (or lives) in Baker Street" even if made by a critic referring to Conan Doyle's creation. Clearly, the critic's sentence (or, for that matter, Conan Doyle's possible sentences *about* his works of fiction) is not fictive, but a real true-or-false assertion that can be validated or disproved by the evidence of Conan Doyle's literary works. Within his stories, however, such a sentence is not merely fictional, but fictitious, and it is, besides, a true-or-false sentence that, by the rules of fiction, is necessarily true.

As for "feigned speech" or "pretended speech," I think that the notion involved therein is inauthenticity, that is, an *intrinsic* imperfection of something that is presented as speech not being it. The image itself is here different from speech, and only deceptively similar to it. The difference between feigned or pretended speech, on the one hand, and real or fictive but authentic speech, on the other, is ontic, whereas the difference between real and fictive speech is not intrinsic but ontological.

Fictive speech is a particular speech that, like all fictions, is *known* not to exist, because it is conceived as nonexistent. In more Husserlian terms, it is constitutively intended in the ontological modality of mere imagination. Or, as Ingarden says (at the beginning of his *Das literarische Kunstwerk* in reference to the literary work in general), it is a "purely intentional object" ("ein rein intentionaler Gegenstand"). Later in the book he arrives at the concept of objects of "heteronomous being." In his *L'imaginaire*, Sartre speaks of the annihilating intentionality that sustains the imaginary. These various terms point in the same direction although they correspond, certainly, to conceptions that differ considerably. (Ingarden's theory of heteronomous being, for example, contains an explicit rejection of Husserl's transcendental idealism.) As for the view I have been suggesting, I think it possible to clarify the question of fiction within a quasi-psychologistic theory of representation. But the demonstration of such possibility belongs elsewhere.

The assertion of a nonreal ontological modality for the literary work as object of the reader's experience does not necessarily imply an appropriate insight into its merely imaginary character. Usually the notion of the merely imaginary or fictitious nature of literature does not include the linguistic base, "the text." This is the case in Ingarden's theory of the "quasi-judgmental character" of literary assertions. Fictional narrative sentences are seen by Ingarden as sentences of the author but not seriously asserted by him and not properly true or false. If they are sentences of the author (whether seriously stated or not), they have to be real acts of his, real utterances. This view of Ingarden's (prefigured in I.A. Richards' theory of poetic "pseudo-statements" and reincurred by John Searle and others)[5] is false because it postulates that something is the case (that novelists write their fiction as a series of their own but not seriously made assertions) that is not the case. Besides, this view would reduce all assertions found in fiction to one and only one logical type: the quasi-judgmental, nonseriously made, neither true nor false assertion. The fact that some assertions within fictional discourse do indeed appear to be made nonseriously (ironically, playfully, jokingly, etc.) by the narrator, and thus distinguish themselves from the rest of his discourse, is entirely inexplicable under Ingarden's assumptions. That assertions of characters (or of the narrator himself if he is of the "unreliable" type) can be definitely false within the world of the novel or with respect to it, thus contrasting with other of their assertions which have to be thought of as *true* within the world of the novel or with respect to it, is also a fact that cannot be explained if we assume that all assertions in a fictional narrative are neither true nor false.

I think that the visionary phase is the final step of the reader's poetic experience, and that intuition and concrete individual presences supersede and efface structure (the conceptual or semantic structure of words and sentences as well as structures of a higher order including all-encompassing generic forms). But the movement of the mind towards the intuition of the presence of concrete individuals within the horizon of the world – a movement constitutive of experience, both real and imaginary – has phases and develops through systematic steps. Ingarden saw the concretization of the literary work in the experience of the reader as the filling up of a schematic structure. I think that there is more to the reading operation than just adding imaginary matter to empty forms. The forms themselves are subjected to change and displacement. Reading is a transformation of given materials and structures. The structure of the ultimately experienced work is different in essence from the arrangement of elements initially given to the reader. And even this final structure is, in a sense, hidden inside the concrete presence of the poetic object. Therefore, I would say that there is a profound ambiguity in Ingarden's thesis that a literary work is a stratified construct, a multileveled compositum ("ein mehrschichtiges Gebilde"). It is a multileveled formation only insofar as we consider it as given to and processed by the reader. But it does not end that way. The teleology of aesthetic experience (as Kant indicated in his *Critique of Judgment*) puts the object beyond structure, form, and abstraction. My study of the structure of the literary work has attempted to show precisely the transformational process that leads from linguistic signs and concepts to aesthetic experience. This experience, as any, is constituted in and through subjective acts, which are determined by systems of rules, that is, are structured.

I come back to my initial remarks and conclude by observing that phenomenology is defined by reflexion on the structured intentionality of subjective acts as well as by the validation of the objective presence thus constituted. The description of the given, as it is given, falls back into the description of the constitutive forms which, at first, are hidden in its inarticulated presence. Description should return to the presence by deictically pointing to it as the inarticulated, full, and therefore ineffable limit of experience.

*Columbia University*

NOTES

[1] This paper is the text of the introductory remarks I presented for the discussion of my book *Fictive Discourse and the Structures of Literature: A Phenomenological Approach* (Ithaca: Cornell University Press, 1981) at the seminar of the World Phenomenology Institute in March 1982 in Boston.

[2] I have touched on this theme in my article 'Representation and Fiction,' *Dispositio*, 1980.

[3] See, for example, *Ideas*, bk. 1, sec. 43.

[4] Precisely because a nonfictional memoir is taken as the discourse of the real author, he can project an image of himself that is better than life. He can feign noble convictions and pretend fine gestures. The real author of a fiction does not have this possibility, except in passages where he breaks the fictional tenor.

[5] See John Searle, 'The Logical Status of Fictional Discourse,' *New Literary History*, 1975, and my critical remarks, 'The Act of Writing Fiction,' *New Literary History*, 1980.

JESSE GELLRICH

# THE STRUCTURE OF ALLEGORY

Recent reactions against American formalism (New Criticism) have shown a preference for summarizing their departure by speaking of literature as a signifying system rather than an embodiment of signified meanings. One commentator observes that in the ideal text, "the networks are many and interact . . . this text is a galaxy of signifiers, not a structure of signifieds . . . there is never a whole of the text (which would by reversion form an internal order, a reconciliation of complementary parts)."[1] Another theorist in this vein speaks of literature as an "organization of signifiers which do not designate a signified object, but instead designate *instructions* for the *production* of the signified."[2] These views of fictional discourse as a configuration of signs for producing meaning have become commonplace since the impact of semiology on literary studies. But what has not been considered sufficiently is that they were commonplace long before the structuralist activity of the last few decades.

For the mode of discourse in which the concepts of *sign*, *signifier*, and *signified* are entirely familiar dates from some of the earliest discussions of *allegory*. As a system of signification, allegory has always presented the challenge of the "dark glass," the "radiant veil," that instructs readers to supply what the discourse itself refuses to complete. Readers "made" the "text" of allegorical writing many centuries before the semiological revolution of our time. It is in every way fitting, therefore, that we are beginning to see in recent criticism the adoption of allegory as a category for explaining the relationship between texts and the history about them. This is a logical development from discussions of the so-called "allegory of reading" – the strategy of certain texts not only to signify but to mirror their own signifying gestures. For instance, Paul de Man's "allegories of reading" in Proust and Rousseau look back to older allegorical forms – the *speculum* of history or of nature familiar in medieval and Renaissance traditions – but those forms, too, mirrored the process of interpretation and turned reader into the "writer" of the text.[3]

I shall have more to say later about early allegorical writing – particularly medieval – and its relationship to modern theory. For now I wish to observe only that the recent interest in allegory is a welcome shift away from the



*A-T. Tymieniecka (ed.), Analecta Husserliana, Vol. XVIII,* 505–519.

tarnished reputation of the mode that has dominated modern criticism, and it is a shift that reveals some of the basic tenets of theory after the New Criticism. To appreciate the significance of this shift and its bearing on the understanding of allegory, I would like to turn briefly to Coleridge, who is principally responsible for the negative opinion in which allegory has been help up until the last few years. "An allegory," he says,

> is but a translation of abstract notions into a picture language, which is itself nothing but an abstraction from objects of the senses; the principal being more worthless even than its phantom proxy, both alike unsubstantial, and the former shapeless to boot. On the other hand a symbol . . . is characterized by a translucence of the special in the individual, or of the general in the special or of the universal in the general; above all by the translucence of the eternal through and in the temporal. It always partakes of the reality which it renders intelligible.[4]

Coleridge's preference for symbol over allegory is of a piece with the priority of "organic form" in his poetics. That form is "innate" because "it shapes, as it develops, itself from within, and the fullness of its development is one and the same with the perfection of its outward form."[5] It is easy to see in these formulations a metaphysics that privileges the union of object and subject as a way of making sense, and that finds in the symbol a means of harmonizing differences between man and his world, fusing divisions within the self, and investing artificial forms with natural meanings. As the *imagination*, for Coleridge, is the agency of this synthesizing, it diffuses and dissipates in order to recreate, and poetry becomes the highly symbolic form that embodies the union of natural and artificial, particular and general – the organic unity in which meaning is no longer mediated by artifice; rather the eternal is translucent in the temporal, is present in the created form. Allegory, in contrast, poses a threat for Coleridge to the natural presence of meaning. It imposes on matter a "predetermined form" and emphasizes that the bond of thought and thing, of signs and their referents, is at most arbitrary. Meaning is assigned by choice or tradition; it is no longer natural to the freshness deep down in things. The form, therefore, is arbitrary, too, the product of the "fancy," a "mechanic" or "artificial" contrivance.[6]

The profound influence of these divisions on the formalism of the New Critics is most apparent in their attention to a literary text as organic and symbolic, a heterocosm or self-enclosed object of meaning. In understanding a work, their classic distinction (as argued by Wellek and Warren) between "intrinsic" and "extrinsic" evidence favors the internal elements of style, seeing in any part a participation in the whole.[7] This structure of meaning *pars pro toto* is the essence of the Coleridgean idea of the symbol as the image that "partakes of the reality which it renders intelligible." But allegory prefers extrinsic evidence, which for Wellek and Warren is the weakest appeal in deriving the meaning of a work. By reaching outside of the boundaries of the text, allegory is perceived by the New Critics as a deterministic mode. They see allegory as the mechanical equivalence of image A in the text and referent B in the outside world. The allegorical image has no life of its own, but is determined to mean according to convention. The poet turns into a ventriloquist of tradition instead of a maker of meanings, and his poetry is little more than a "phantom proxy."[8]

These are extreme positions, but they illustrate the New Critical preoccupation with deriving meaning exclusively from the surface texture of a work. Because allegory does not respond to that expectation, the New Critical reaction was either to disparage it as a form of theology or philosophy masquerading in a fair garment ("nature to advantage dressed"), or to try to find a way of explaining it as a symbolic mode.[9] New Critics were not particularly interested in affirming allegory as symbolic, but an effort in this direction by Angus Fletcher was an attempt to move beyond New Criticial borders of the text itself and at the same time to remain within a fundamentally organicist view of literature.[10] If the allegorical image could not be adequately explained by appeal exclusively to the style of its context in a work, then one might speak of the context as it is informed by its particular literary *genre*. The apparent leaning toward Northrop Frye in this move to the symbolic concept of genre offered a way of tolerating the reliance in allegory on reference "outside" of the work, for now that "outside" was safely located within a monolithic organic system, the symbolic genre of allegory.[11] This approach opens the windows and doors of the closed New Critical text and yet reveals the prevailing organicism of modern theory. The only way, it appears, that such an intellectual milieu could tolerate allegory was to see it as symbol.

But Frye's concept of genre is not the source for the symbolization of allegory. To see more fully what is involved in the recent shift away from symbol, it ought to be noted quickly that allegory was symbolized massively throughtout the tradition of medieval language theory that begins about the time of Augustine and extends beyond Denis the Carthusian in the fifteenth century. It is true that medieval tradition is often cited by the detractors of allegory, like Coleridge, but arguments in Augustine and his followers reveal deep organicist roots. For example, the popular medieval images for explaining the relationship between a word and its meaning in the Bible are organic metaphors: within the "cortex" is the "nut" of truth;

underneath the "chaff" is the "fruit" of wisdom.[12] The books of Moses are "barley bread"; the books of the New Testament are "wheat bread."[13] Interpretation is a process of removing the hard covering of the "literal sense" (*sensus literalis*) in order to obtain the "fruit" of the "spiritual sense" (*sensus spiritualis*) underneath.[14] The "spiritual sense" of Scripture is threefold: (1) the historical continuity between Old and New Testament figures (Isaac as a type of Christ); (2) the tropological sense in which the sacrifice of Isaac illustrates Christian self-denial; and (3) the anagogical sense in which Isaac's release from death signifies the release of all souls from the bondage of sin at the Last Judgment.[15] Each of these meanings is simultaneously present in the first, the literal sense. It is essentially symbolic because it "partakes of the reality which it renders intelligible."[16] For those meanings were "written" by the "finger of God."[17] Not only can they be read in the Book of God's Word, Scripture, but also in the Book of His Work, Nature. Both books may be spoken of as "allegory," but they are clearly the symbolic "translucence of the eternal in the termporal," created in the beginning as the *Verbum Dei*.

The medieval symbolization of allegory explains a great deal about the Romantic fondness for cultural forms of the Middle Ages. They provide many organic forms for the symbolizing imagination, as Henry Adams illustrates in his Romantic account of what he sees as the order and unity of social and artistic forms apparent at Mt. St. Michel and Chartres Cathedral in the thirteenth century.[18] But the Romantics and their modern heirs, the New Critics, are not fond of medieval allegorical fictions because they are too mechanical, too arbitrary. Meaning is not retrievable from poetic expression itself. Allegorical imagery does not overflow spontaneously from recollection in tranquillity. Hence meaning is not natural to utterance and allegory is surely not the "real language of men." Wordsworth's formula for representing truth in literature has little to do with the "truth" of allegory, which is not "natural" or simultaneously "present" in its discourse.[19]

Coleridge, therefore, was correct to point out the opposition between symbol and allegory, and modern critics who would like to see it as a symbolic mode reveal more about their organicist – and medieval – leanings than they explain about the structure of allegory. However, an appreciation of the differences between the two modes is now more accessible than in the past because of the movement in contemporary theory away from the New Critical preference for symbolic meaning. I refer, for instance, to the departure from formalism in the remarks with which I began, namely, to the notion that the system of signifiers in a text does not designate a signified object, but establishes "instructions for the production of the signified." The author of this claim has also criticized formalist ideas of the text as "object" or "container" by arguing that there is a dimension in a literary work that provokes semantic transformation in order that it may be related to existing frames of reference. This argument leads to the conclusion that "the *ultimate* dimension of the text cannot be semantic. It is what we might call 'imaginary.' "[20] This slippery term is intended, I gather, to refer not to the story or fiction, but to the experience of meaning to which it points. The text itself, this argument would claim, cannot replicate or "contain" the "imaginary."

A similar departure from the objectivist position – and similarly slippery but no less useful – is the recent argument that narratives are "secret" and "dark," that these qualities invite our attention and increase, rather than forestall, the "hermeneutic potential" of a text.[21] Parabolic discourse such as we find in the New Testament or the works of Kafka illustrates well this basic feature of narrative to "reveal" but at the same time "conceal" full significance. Not quite cognizant of the reach of this quality, the first audience of Jesus' parables, the Pharisees, complained about his intolerable language, demanding full disclosure and the plain sense of meaning. Institutions of criticism have made similar Pharisaical demands by claiming privilege to the secret meaning inside a text while remaining blind to the signifying light escaping from the crevices.

The view of narratives as secret and dark is the obverse of the notion of the text as producer of meaning. I have not represented either argument completely, but only enough to indicate that they are sharp departures from formalism. If we have heard the ring of *deconstruction* in the arguments, it is most certainly evident. For those who speak of the deconstruction of form begin by questioning the priority of symbolic meaning. They argue that the objectivist concept of form assumes the presence of meaning infused throughout the surface texture or stylistic elements of a literary work. The problem here, as Paul de Man has claimed, is that language is treated as a "natural object," that sort of object which can be known by the totality of its sensory properties.[22] But in fact, says de Man, language is an "intentional object," the kind of entity that can only be understood abequately when we know the purpose or use it is to serve. By deriving meaning from the totality of sensory properties in a literary work, the objectivists bypass the intentional factor of language and consequently locate in the work the attempt at totality of the interpretive process. The movement toward totalizing and closing that characterizes understanding is hypostatized as the "form" of the work, which consequently is said to be the concretion of meaning, an "object" in its own right. The deconstruction of form would ask, I take it, that the totalizing process be recognized for what it is, a phenomenon of understanding; it would point out further that, among other things, literary form is not an object or a thing, but, as de Man puts it, "a process on the way to its completion."[23]

Some may wish, rightly or wrongly, to apply this definition of form as process to fictive discourse in general. But the one discourse that it seems to describe without much strain is allegory. For that mode, once criticized as inorganic and lacking the completeness of symbol, has certain similarities with the sense of deconstructed form just defined. Allegory makes no pretense to an illusion of totality; it is the signifying means for a sense beyond its "veil"; it is open-ended and ongoing, a "mirror" of "dark radiance," an "uncovering." These are the commonplaces of allegorical theory. But their reappearance in recent discussions of interpretation allows us to see a bit more clearly how we may understand allegory either in modern or medieval formulations. For one thing, recent inquiry points out the error in trying to assimilate allegory by symbolizing it; just as important, we are now in a position to look back at medieval allegory without trying to turn it into something else. In the remaining pages of my discussion, I would like to move to a medieval allegory – a few passages of Dante's *Inferno*. My purpose will be to inquire into the assumptions that have been made about the structure of the poem. What sense, if any, does it make to think about Dante's allegory in light of the view of deconstructed form that I have reviewed above?

If we ask what kind of allegory Dante has made, the traditional answer would document the scholarship of Charles Singleton. It would affirm that in the *Inferno*, nothing is arbitrary: all parts compose an intricate, highly controlled fabric. The model for this fabric of meaning is the so-called "allegory of theologians," the mode described above as "symbolic." The structure is symbolic because the Divine Author has inscribed all temporal things in the Book of Nature with eternal meaning and has written all sentences in the Book of His Word with threefold transcendent significance. Dante, says Singleton, has copied those great Books, has made "an imitation of God's way of writing."[24] For verification, he directs us to the 'Letter to Can Grande della Scala,' attributed to Dante; there we find that the poet himself describes the fourfold sense of Scripture and says that the *Comedy* is an allegory of the same sort. The passage begins with a citation from Psalm 114.

"When Israel came out of Egypt, and the house of Jacob from a people of strange speech, Judea became his sanctification, Israel his power." For if we inspect the letter alone the departure of the children of Israel from Egypt in the time of Moses is presented to us; if the allegory, our redemption wrought by Christ; if the moral sense, the conversion of the soul from the grief and misery of sin to the state of grace is presented to us; if the anagogical, the departure of the holy soul from the slavery of this corruption to the liberty of eternal glory is presented to us. And although these mystic senses have each their special denominations, they may all in general be called allegorical. . . . When we understand this we see clearly that the subject round which the alternative senses play must be twofold. And we must therefore consider the subject of this work as literally understood, and then its subject as allegorically intended. The subject of the whole work, then, taken in the literal sense only, is "the state of souls after death," without qualification, for the whole progress of the work hinges on it and about it. Whereas if the work be taken allegorically the subject is "man, as by good or ill deserts, in the exercise of the freedom of his choice, he becomes liable to rewarding or punishing justice."[25]

These remarks have puzzled readers no doubt since the time they were written. First, there is the question of whether Dante actually wrote the 'Letter';[26] second we are given a rather loose definition of "allegory." A medieval reader would have understood the term in its technical sense as the diachronic continuity between the Old and New Testaments;[27] but, the 'Letter' gives us something very close to Coleridge's symbolic interpenetration of the eternal in the visible form. And third, because of the symbolic coalescence of the literal into the other levels of spiritual meaning, we face yet another dilemma when Dante's poem is claimed to be a copy of God's Book, as Singleton has argued. The problem is that the spiritual senses of Scripture are sanctioned by the historicity of the literal sense: it is a documentary record of real events and people. But when Singleton follows his comparison to its conclusion, he is left insisting that Dante's literal story is so organically bound to its spiritual meaning that his "fiction" is really "history." "Let it be observed at this point," Singleton argues,

that nothing within this poem ever declares that it is a fiction. The fiction of the *Divine Comedy* is that it is not fiction. And no work ever guarded that fundamental hypothesis more carefully. Nowhere in the work is this vision of things presented as a vision or a dream. These things happened, and the poet who went that way in the flesh and experienced them is, now that he has returned, a scribe who sets them down as they had occurred.[28]

Confronted with the fact that this view of the poem makes no distinction between the story of the journey and a record of documentary facts, Singleton has retorted that all readers of the poem know, of course, that

Dante has not really taken a journey into the afterlife and has returned from the dead like Lazarus to tell us all. Yet he has continued to emphasize that the language of the poem is, as he puts it, "real" and "irreducible," like objects in the world.[29] Although this argument recognizes a separation of literal from other senses, the conceptual framework of their structure is based on interpenetration, union, and coalescence: the organic unity of the symbolic structure will not allow any other way of construing meaning, and the result is that fiction cannot be distinguished conceptually from reality. There does not appear to be any way out of this dilemma, so long as allegory is a species of symbolism.

Symbolic structure has led to another approach to the *Comedy*, equally compelling yet also at the expense of allegory. In his reading of *Inferno* canto 10, Erich Auerbach has presented the fascinating position that the allegorical reference of the figures, Cavalcante and Farinata, is so bound to their realistic portraits in the work that the reference fades from view. What we have, says Auerbach, is not a figure standing for (a "phantom proxy" of) a particular category of sin, in this case heresy; instead, reference gives place to the human form before us.[30] Dante the pilgrim is passing through a graveyard of flaming tombs when Virgil stops him.

"See there Farinata who has risen erect: from the waist upwards you will see him all."... When I was at the foot of his tomb, he looked at me for a moment, then, as if in disdain, asked me, "Who were your ancestors?" . . . Then there arose to sight alongside of him a shade, visible to the chin: I think he had raised himself on his knees. He looked round about me as though he wished to see whether someone was with me, but when his expectation was quite spent, he said, weeping, "If you go through this blind prison by reason of high genius, where is my son, and why is he not with you?" And I to him, "I come not of myself. He who waits younder, whom perhaps your Guido had in disdain, is leading me through here." . . . Suddenly straightening up, he cried, "How? Did you say 'he had'? Does he not still live? Does the sweet light not strike his eyes?"[31]

Cavalcante says nothing else. He falls back into the flames in despair of the uncertainty of whether his son Guido is living or dead. Farinata then resumes his dialogue with the pilgrim, maintaining the same attitude with which he began – defending himself as a Ghibelline leader with his stout brow and stout chest thrust forward in defiance of all Hell. His presence is so potent that he appears courageous – as one of my undergraduates described him: Farinata is the "epic hero" of Hell. Of this scene, Auerbach has said that Dante has created in the heart of the afterlife "a world of earthly beings and passions so powerful that it breaks bounds and proclaims its independence."[32] The Christian allegorical category of the sin of heresy is "eclipsed" and "obscured" by the vivid mimesis of human feelings.[33] Auerbach arrives at this position because of a bascially objectivist approach to the poem: the meaning of the scene is derived from the totality of stylistic elements as they inform each other and infuse the text with a single purpose – the pathos of Cavalcante and the courage of Farinata. This view, however, does not say much for allegory. The reading ends stressing the paradox of the realism of the journey in apparent conflict with the Christian allegorical purpose of the poem.

But the paradox, I think, may be turned to our advantage. It would be fair to say that Auerbach's reading explains the sense we have of canto 10 that the experience of pain at this stage of Hell is inexhaustible, beyond the mechanical equivalence of image and referent. Such passages compare with the many other moments in the *Comedy* when Dante tells us that the experience he has had is inexpressible; his story cannot render it completely but lags behind. This distinction between the experience of Hell as necessary penitence for the grace of religious conversion and the story of the journey through the afterlife is easily overlooked or oversimplified. For Dante, and this is Singleton's mistake, does not ask us to believe in the *story* of the quest, but only in the *experience* of conversion.[34] If that separation is closed by a symbolic structure, we will end up affirming either the historicity of the fiction (Singleton) or the irrelevance of allegory (Auerbach). Instead, it would seem reasonable to say that the *allegory* of the poem is the difference between the fiction of the journey and the real historical conversion that Dante Alighieri of Florence has had. This is to say that Dante does not create an illusion that he wants us to believe in as real. He does not make a myth. Instead, his poem knows itself and discloses itself as fiction.[35] A model for the form of the poem, then, is not a "container," the familiar medieval "cortex" and "nut," "shell" and "pearl." Its form occasions the discovery of the meaning of the grace of conversion. Allegory is not a construct enclosing truth, but a means for signifying beyond discourse the experience of conversion. The text as a "means for signifying" or a "process" occasioning religious mystery is close, I believe, to the "deconstructed form" that I have described above.

In order to consider this point further it will be worthwhile to take a look at a few more passages in the *Inferno*, particularly those in which the poet himself suggests the allegorical distance between story and experience. One of the first instances is the unforgettable sign carved in rock over the hellgate at the beginning of the descent: "Abandon all hope you who enter here." This passage joins with many others in which Dante addresses us directly as readers in order to suggest that the quest of the pilgrim is to be understood as the experience of reading a text. In addition to these allusions, there is the extended and provocative reading episode of Paolo and Francesca. Occurring near the opening of the poem, the passage illustrates the failure of adequate response and thereby traces how the larger episode is to be read. Hearing the sad lament of Francesca, Dante calls to her accross the winds of Hell and asks her to tell her story. She responds with the moving account of how she and Paolo read one day for delight the book of Lancelot and Guenevere. As they muse through the passage describing Lancelot's kiss, the moment of their reading fades irresistibly into the moment about which they read. Then, says Francesca, "We read no more that day."[36] As commentators on this famous episode have remarked, the text read by the lovers has acted like the "go-between" in popular medieval romances: they are "seduced" by the book.[37] Indeed, Dante himself makes this allusion clear when the French text is called "Galeotto" ("Gallehault").[38] But what is just as remarkable is Dante the pilgrim's response to the "text" of Francesca's story. After she records how she and Paolo were murdered by her husband in the act of adultery, Dante weeps for her and swoons to the floor the Hell, "as a dead body falls."[39] His response parallels that of the lovers. Their story has "seduced" him, and his fall repeats those of many others in this world of the fallen.

Dante's gesture, therefore, cannot signify sympathy for the damned; it signifies instead that the reader must search for what Dante the pilgrim does not understand in this account of passion. The whole passage, both the reading of Francesca and the response of Dante, illustrates the inability to read allegorically, to separate the story from the experience of passion, the world of the damned from the ongoing world of the living. The allegory of the canto is the mirror reflection of this failure to establish difference. To see the significance of this point more fully, I would like to turn now, for a final example, to one of the last episodes in the *Inferno*. It illustrates the exact opposite of the problem in canto 5, because it establishes separation with a vengeance. I refer to canto 32, in which Dante kicks the head of Bocca degli Abbati, whose body is forzen in ice from the neck down. No matter how careful we have been to resist sympathy for the damned and to separate the story from the experience of penitence in Hell, this moment always comes as a shock.

And while we were going toward the center to which all gravity collects, and I was shivering in the eternal chill, whether it was will or fate or chance I do not know, but, walking among the heads, I struck my foot hard in the face of one. Wailing he railed at me, "Why do you trample on me?" . . . Then I seized him by the hair of the nape and said, "Either you'll name yourself, or not a hair will be left on you here." Whereon he to me, "Though you strip me bald, I will not tell you or show you who I am, though you should fall a thousand times upon my head." I had already twisted his hair in my hand and had yanked out more than one tuft, he barking and with his eyes kept close down, when another cried, "What ails you, Bocca?"[40]

The passage plays powerfully on our response to realism. But as the test of response was earlier on Francesca and then on Dante in canto 5, now it is on us. The passage is teasing us to take up sides with Bocca against "the devil," Dante the pilgrim, who is tormenting him.[41] Yet we must not misperceive apparent callousness to figures in Hell. We have to know how to understand the allegory of Dante's kick, to know that it has been ordained from on High as an act of justice against a betrayer of family ties, which is Bocca's sin. As a representation of that sin – an allegorical image of it – he is trying in vain to demand human dignity, to be a person again. But he is separated from that former self by his representation as an image of betrayal, and that is why he will not tell his name, which Dante threatens to write down in his "notes" for the poem he is planning, the *Divine Comedy*. It is this final point, the naming of Bocca, that acts in the same way as the kick, for Dante tortures Bocca forcing him to say his name, and then continues to torment him further by writing his name down for posterity to remember: Dante uses Bocca's name against him, steals it from him and makes it a reference for a particular category of sin.

In the scene with Bocca, it may be tempting to repeat Dante's mistaken response to Francesca's story. But each of these incidents in their different ways deals with the separation of the story from the experience of penitential grace. Each reveals that the structure of the separating is not symbolic, is not identification and coalescence. The figures in canto 5 are "seduced" into the temptation of dissolving past into present (Francesca identifies with Guenevere, Dante with Paolo). At the other end of the tour, the opposite is true: Dante separates himself from Bocca and then splits up Bocca so viciously that we are well advised not to repudiate him for doing so. These examples suggest that the structure of reference is constituted by *time*, the old threat of the symbol as Coleridge feared. What is Francesca or Bocca but the repetition of a former self to which she or he no longer bears any resemblance.[42] The allegorical image is not constituted, so it seems, by its capacity to represent something out of the past (or future for that matter); instead, it is constituted by the fact that it *cannot* correspond to a referent, since that referent must remain lost in the past.[43]

To sum up: I have been suggesting, first of all, that we can profit to some extent by the discussion of allegory in contemporary theoretical reactions against American formalism. Next, to fault allegory – as do the New Critics – for not being a symbolic mode may now be recognized as the influence of a critical tradition preoccupied with symbolic thinking. Further, when we consider Dante's *Comedy* in light of post-formalist theoretical inquiry, we may see how symbolic assumptions lead, on the one hand, either to affirming the historicity of the fiction or, on the other, to the eclipse of the Christian allegorical purpose. These assumptions aside, Dante's poem itself seems to suggest a structure of reference that has at least one crucial point in common with recent theory – a stress on the difference and separability of a fiction from the meaning of the experience that it is striving to signify. I have suggested that the structure of this reference is based on temporality, but *not* – I should add – in the medieval sense of the continuing presence of the past. Rather, I have used the term as it appears in certain post-formalist discussions, and I would argue in Dante, too. His poem appears to break with the symbolized allegorical reading of medieval institutions. One final rejoinder: I have not been seeking to imply that Dante anticipates the attack on structure and the loose play with metaphor of contemporary writers like Derrida.[44] On the contrary, my intention has been to point out that temporality leads us to a sense of poetic form as process, not object, and I can think of no better example of that definition than Dante's medieval allegory, the *Comedy*.

*University of Michigan, Ann Arbor*

NOTES

[1] Roland Barthes, *S/Z*, trans. Richard Miller (New York: Hill and Wang, 1974), pp. 5–6.
[2] Wolfgang Iser, 'The Reality of Fiction: a Functionalist Approach to Literature,' *New Literary History* 7 (1975):18.
[3] Paul de Man, *Allegories of Reading, Figural Language in Rousseau, Nietzsche, Rilke, and Proust* (New Haven: Yale University Press, 1979). See also his seminal essay, 'The Rhetoric of Temporality,' in *Interpretation: Theory and Practice*, ed. Charles S. Singleton (Baltimore: Johns Hopkins University Press, 1969).
[4] *The Statesman's Manual*, cited in *Critical Theory Since Plato*, ed. Hazard Adams (New York: Harcourt Brace Jovanovich, 1971), p. 468.
[5] *Shakespeare's Judgement Equal to His Genius*, cited in *Critical Theory*, p. 462.
[6] See *Shakespeare's Genius* in *Critical Theory*, p. 462. For the distinction of fancy from imagination, see *Biographia Literaria*, chap. 13, in *Critical Theory*, p. 470.

[7] René Wellek and Austin Warren, *Theory of Literature* (1942; reprint, New York: Harcourt, Brace and World, 1956), pts. 3–4. See also W. K. Wimsatt and Monroe C. Beardsley, 'The Intentional Fallacy,' in *The Verbal Icon* (New York: Noonday Press, 1965), pp. 3–18; essay originally published in 1946.
[8] For a rehearsal of recent positions about allegory, see Morton Bloomfield, 'Allegory as Interpretation,' *New Literary History* 3 (1972):301–17. Convenient bibliographies on allegory may be found in *Allegories of History, Allegories of Love*, by Stephen A. Barney (Hamden, Conn.: Archon Books, 1979).
[9] Alexander Pope, 'An Essay on Criticism' (1711), l.297, in *The Poems of Alexander Pope*, ed. John Butt (1963; reprint, New Haven: Yale University Press, 1966), p. 153.
[10] *Allegory: The Theory of A Symbolic Mode* (Ithaca: Cornell University Press, 1964).
[11] *Anatomy of Criticism* (Princeton: Princeton University Press, 1957). Fletcher makes his debt to Frye explicit throughout.
[12] At the beginning of the Middle Ages, these commonplaces are discussed by Augustine in *De doctrina Christiana*, III, 12, *Patrologia Latina*, ed. J.-P. Migne (Paris, 1841), 34:72–73; late in the fourteenth century, Geoffrey Chaucer alludes to such figures for allegory, for example, at the end of the *Nun's Priest's Tale* (l.3443) in *The Works of Geoffrey Chaucer*, ed. F. N. Robinson, 2d ed. (Boston: Houghton Mifflin, 1957), p. 205.
[13] These figures from John's Gospel (6:9) are glossed elaborately by St. Jerome in the *Epistola Adversus Jovinianum* (I, 7) and incorporated by Chaucer into the *Wife of Bath's Prologue* (ll.143–46) in Robinson, p. 77 and n.; see D. W. Robertson, Jr., *A Preface to Chaucer* (Princeton: Princeton University Press, 1962), pp. 328–29.
[14] The definitive study of the several senses of medieval exegesis is by Henri de Lubac, S.J., *Exégèse médiévale, les quatre sens de L'Ecriture*, 4 vols. (Paris: Aubier, 1959–64).
[15] *Glossa ordinaria, PL* 113, 139.
[16] A thorough scholarly investigation of this claim has not yet, to my knowledge, been published.
[17] A phrase that occurs frequently in Scripture, for example, in Exodus 31:18.
[18] *Mont-Saint-Michel and Chartres* (1913; reprint, New York: New American Library, 1961).
[19] William Wordsworth, 'Preface' to the second edition of the *Lyrical Ballads* in *Critical Theory*, pp. 433–43.
[20] Wolfgang Iser, 'The Current Situation of Literary Theory: Key Concepts and the Imaginary,' *New Literary History* 11 (1979):17, Cf. the arguments about *écriture* by Jacques Derrida, *De la grammatologie* (Paris: Les editions de minuit, 1967), pt. 1 – "writing before the letter."
[21] Frank Kermode, *The Genesis of Secrecy: On the Interpretation of Narrative* (Cambridge: Harvard University Press, 1979), pp. 39–40, 43, 45.
[22] 'Form and Intent in the American New Criticism,' in *Blindness and Insight* (New York: Oxford University Press, 1971), pp. 20–35. De Man's use of "intentionality" relies, of course, not on scholastic philosophy, but on phenomenology, and particularly on Edmund Husserl's use of the term throughout his works as a modification of "acts."
[23] De Man: "What happened in American criticism could then be explained as follows: because such patient and delicate attention was paid to the reading of forms, the critics pragmatically entered into the hermeneutic circle of interpretation, mistaking it for the

organic circularity of natural processes" (p. 29). But, he continues, "the completed form never exists as a concrete aspect of the work that could coincide with a sensorial or semantic dimension of the language. It is constituted in the mind of the interpreter as the work discloses itself in response to his questioning. But this dialogue between work and interpreter is endless. The hermeneutic understanding is always, by its very nature, lagging behind" (p. 32).

[24] Charles S. Singleton, *Dante Studies I: Commedia, Elements of Structure* (Cambridge: Harvard University Press, 1954), p. 15; see Singleton's 'In Exitu Israel de Aegypto,' *Dante Studies* 78 (1960):1–24; and Robert Hollander, *Allegory in Dante's Comedy* (Princeton: Princeton University Press, 1969), pp. 9–70.

[25] 'Letter to Can Grande,' trans. P. H. Wicksteed, in *Critical Theory*, p. 122.

[26] See Helmut Hatzfield, 'Modern Literary Scholarship as Reflected in Dante Criticism,' *Comparative Literature* 3 (1951):296.

[27] See de Lubac, *Exégèse* 4 (1964), chap. 8, pp. 125–49.

[28] *Dante Studies*, 1:62.

[29] *Dante Studies*, 1:77–81. See the critique of Singleton's position by Richard Hamilton Green, 'Dante's "Allegory of the Poets" and the Mediaeval Theory of Poetic Fiction,' *Comparative Literature* 9 (1957):118–28; Singleton's rebuttal is in 9:129–35.

[30] Auerbach, *Mimesis, The Representation of Reality in Western Literature* (1946), trans. Willard Trask (New York: Doubleday, 1957): "Yet never before has this realism been carried so far; never before . . . has so much art and so much expressive power been employed to produce an almost painfully immediate impression of the earthly reality of human beings" (p. 174).

[31] Charles S. Singleton, trans., *The Divine Comedy* (Princeton: Princeton University Press, 1970–75), *Inferno*, canto 10, pp. 101–3.

[32] Auerbach: "Figure surpasses fulfillment. . . . We cannot but admire Farinata and weep with Cavalcante. What actually moves us is not that God has damned them, but that the one is unbroken and the other mourns so heart-rendingly for his son" (pp. 174–75).

[33] Auerbach: "And by virtue of this immediate and admiring sympathy with man, the principle, rooted in the divine order, of the indestructibility of the whole historical and individual man turns *against* that order, makes it subservient to its own purposes, and obscures it. The image of man eclipses the image of God. Dante's work made man's Christian-figural being a reality, and destroyed it in the very process of realizing it" (p. 176).

[34] John Freccero has stressed this distinction on several occasions; see: 'Dante's Novel of the Self,' *Christian Century* 82 (1965):216–18; Introduction to *Dante: A Collection of Critical Essays*, ed. J. Freccero (Englewood Cliffs, N.J.: Prentice-Hall, 1965), pp. 3–4.

[35] Giuseppe Mazzotta has argued a related point in stressing the "allegory of reading" in the *Comedy*; see *Dante, Poet of the Desert: History and Allegory in the 'Divine Comedy'* (Princeton: Princeton University Press, 1979).

[36] My translation: "quel giorno più non vi leggemmo avante"; Singleton, *Inferno*, V, 138 (p. 54).

[37] For example, see René Girard, 'From "The Divine Comedy" to the Sociology of the Novel,' trans. Petra Morrison, *Sociology of Literature and Drama*, ed. Elizabeth and Thomas Barns (Baltimore: Penguin Books, 1973); and Renato Poggioli, 'Paolo and Francesca,' *Dante: A Collection of Critical Essays*, ed. Freccero, pp. 61–77 (originally published, *PMLA* 72 [1957]).

[38] 'Galeotto fu 'l libro e chi lo scrisse'; Singleton, *Inferno*, V, 137 (p. 54).

[39] 'E caddi come corpo morto cade': Singleton, *Inferno*, V, 142 (pp. 56–57).

[40] Singleton, *Inferno*, XXXII, 73–106 (pp. 343–45).

[41] A provocative assessment of the allegory of this passage in relation to satire has been made by Gerald L. Bruns, 'Allegory and Satire: A Rhetorical Meditation,' *New Literary History* 11 (1979):121–32.

[42] Cf. de Man, 'The Rhetoric of Temporality': "the meaning constituted by the allegorical sign can then consist only in the repetition . . . of a previous sign with which it can never coincide, since it is of the essence of this previous sign to be pure anteriority" (p. 190); "in the world of allegory, time is the originary constitutive category" (p. 190). Freccero has suggested the relevance of de Man's insights to the structure of allegory in *Inferno*, IX; see 'Dante's *Medusa*: Allegory and Autobiography,' in *By Things Seen: Reference and Recognition in Medieval Thought*, ed. David L. Jeffrey (Ottawa: University of Ottawa Press, 1979), pp. 33–46.

[43] Cf. the penetrating approach to allegory through the psychoanalysis of Jacques Lacan argued recently by Joel Fineman: "Distanced at the beginning from its source, allegory will set out on an increasingly futile search for a signifier with which to recuperate the fracture of and at its source, and with each successive signifier the fracture and the search begin again: a structure of continual yearning, the insatiable desire of allegory"; 'The Structure of Allegorical Desire,' in *Allegory and Representation*, ed. Stephen Greenblatt (Baltimore: Johns Hopkins University Press, 1981), p. 45.

[44] It may very well be that Derrida's approach to structure has been much exaggerated by his critics; my view of allegorical structure in the *Comedy* is not unlike Derrida's position in an early essay, 'Structure, Sign, and Play in the Discourse of the Human Sciences,' in *The Structuralist Controversy: The Languages of Criticism and the Sciences of Man*, ed. Richard Macksey and Eugenio Donato (Baltimore: Johns Hopkins University Press, 1972).

MARLIES KRONEGGER

# LITERARY IMPRESSIONISM AND PHENOMENOLOGY: AFFINITIES AND CONTRASTS

When Cézanne painted the external world from the standpoint of nature itself, he stripped it of all those aesthetic associations that generally obscure its integrity and its independence of consciousness. With him, description is creative: it is disintegration and reintegration. He wishes to resituate objects in the flux, in the immediate certainty of their first appearance in the conscious mind. With Cézanne, as Merleau-Ponty points out,[1] knowledge becomes an intuitive act which appears to reveal the essence of things, or the essence of life contained in things.

There is common agreement among both writers and phenomenologists that the "being-in-the world" precedes our reflective consciousness of the world. The aim of both modern writers and phenomenologists, then, is the direct apprehension of reality. The subject, the I, constructs the reality of the world through language. For them, perception is the basis of man's reflections, and the body, anchored in the world, is the basis of perception, perceiving and being perceived, a creative structure unfolding in time. As the literary work has immanent within itself a dynamic interaction of self and world, three major questions have been of interest for writers and phenomenologists since the end of the nineteenth century: What is the essence of perception? What is the essence of consciousness? How can language express the very specific effect of the elemental sensation, the impression?

Edmund Husserl in his *Cartesian Meditations* summarized the elements of a phenomenological investigation. His method is the descriptive analysis of experience. The two main steps in this method are (1) the "epoché," the bracketing of what is termed the natural attitude: the viewer or investigator must divorce himself from the positing of any certitude until the investigation has been completed, and being divorced from such knowledge he finds himself in an epoché or a hermetic sphere from which he can procede; (2) the "reduction," paying attention only to the facts themselves. The only certitude that can be posited is that the investigator has a perception of something. The object of this perception is called the *cogitatum*, and the intentional object for consciousness, the *cogito*. When the *cogito* receives a *cogitatum*, it is the sum of many moments of perception and it is constituted as a whole ideal entity. There are many modes of perception: memory,



*A-T. Tymieniecka (ed.), Analecta Husserliana, Vol. XVIII*, 521–533.

will, interrogation, etc. The object of consciousness can be produced by direct contact with what is immediately perceptive or by recollection or projection. This method of analysis may be applied to the literary text by extension, inasmuch as the text provides both immediate sensory perception (of the totality of its signifiers) as well as a picto-grammatical production, where the text offers the possibility of an opening toward the referential world (through the totality of its signifieds). Both aspects of the sign generate phenomenological material to the reader which produces two different domains: (1) semiotic phenomenology, and (2) semantic phenomenology.

Thus, the reader is presented with a system of signs. The text is the ground for his subjective response as it is the horizon of his consciousness. He can privilege the literal meaning (and hence abolish the referential illusion) or privilege the referential meaning (and hence abolish the literal illustion) of the signs before him. If the reader privileges the referential content of the text, that is, the relation of the signs to the exterior or to their referent, then it would be possible to examine the phenomenology of the narrator's perceptive field. However, if the literal meaning is given priority, that is, the sign on the page, then it is possible to address the phenomenological production of the reader.

Authors such as Flaubert, Proust, Gide, Rilke, Joyce, and Claude Simon have demonstrated that their thought takes form only as they are writing. They adopt a phenomenological description of reality. Gerard Genette, in his recent works *Figures* and *Mimologiques*,[2] explores such texts as they find their essence in the unfolding of a *mise-en-scene* of *écriture*. Genette, and with him the new novelists, refute the Balzacian *langage-tableau*, as it is thought of as self-effacing in the process of the representation of things and society.[3] Language embodies experience, that is, the structure of language is parallel to and representative of experience.

This paper attempts to illustrate how contemporary investigations of the narrative language of impressionistic authors challenge the relationship between the narrator, the reader, and the exterior world. Does phenomenology provide a method for the description of literary texts? Can the work of the phenomenologist and impressionist author be complementary? The text is considered to be a structure which is given by the writer's use of images. These images express the writer's consciousness of the world, the meaning he gives to the world, showing us life and our experiences without explaining them. When a philosopher explores reality, he distinguishes between those perceptions that contain existential certainty and those that appear as phenomenal truths. In the latter case the investigator must deal with experience that he receives and feels yet to which he is incapable of assigning ontic value. How can we learn to see poetically? How does the sensivity of impressionist writers reveal the truths intelligence elucidates?

With the impressionist's perceptive experience, the reality of the novel changes; the traditional frozen forms of description set themselves into motion spatially. The protagonists see reality from several angles of vision at once and the objects are released without losing sight of their earlier positions. Seen at a distance, no object has any clear and detailed outline, and thus, automatically, the subject itself is subordinated to the melodious effect of colors and sounds which then can be used to evoke a particular mood. Through unity of color, tint, and tone, and blurred outlines, impressionist writers achieve musical effects as we shall see in the second part of this paper. Reality is seen as a harmony of subjects and objects, which merge together in terms of time and space, during any act of the protagonists; that is, their states melt into one another. The act of perception is more important than either the perceived or the perceiver. No longer is there *me* the narrator-protagonist, on the one hand, and that tree, on the other hand; there is only my seeing, retaining, or remembering that tree. There is no separation between the narrator and the objects: there is a narrator seeing the objects, and without objects there can be no self, and without self, there can be no objects.

Impressionist writers share with recent investigators in phenomenology (such as Husserl or Merleau-Ponty) the conviction that we cannot know reality independently of consciousness, and that we cannot know consciousness independently of reality: every act of consciousness is a consciousness of something, every act of love is a love of something. Impressionists do not return to the being of reality in itself (Kant's *noumenon*). They return to the thing which is the direct object of consciousness, to what Kant would have called appearance of reality in consciousness or *phenomenon*. The phenomenological method, as revealed in the works of Ernst Mach, Husserl, and Merleau-Ponty, outlines not only the steps which must be undertaken to arrive at the pure phenomenon, wherein is revealed the essence of appearance and that which appears, but also a description of essence as contained in a thing, the direct object of consciousness. A color, therefore, such as red, is no less a thing than a flower. For both phenomenologists and literary impressionists, the essence of the color is *in* the individual's experience of the color as we hope to illustrate.

The physicist and philosopher, Ernst Mach, diagnoses the widespread culture phenomenon of impressionism. In *Die Analyse der Empfindungen*

(1885), Mach calls spaces and times sensations, as appropriately as one does colors and sounds. The unity of space and color sensation is, for Mach and the impressionists, an interplay of the individual's consciousness and the surrounding world. The reader is intended to seize the impressionist works spatially in a moment of time, rather than as a time sequence. Space is defined as the relationships which pieces of matter, in our experience, have with one another. Time, also, is the relationship between different occasions, again within the experience of the individual. In this allusive space and time, both human mind and physical objects have exactly the same ingredients, which from the point of view of psychology are sensations, and from the point of view of physics, the constituents of things. With E. Mach and the impressionists, the antithesis of ego and world, sensation and thing, vanishes. All that exists is in contact with everything else. The German critic Hermann Bahr calls Mach's views the "philosophy of impressionism."[5]

With Mach and the impressionists, man's consciousness is the point of departure for any perception. Both impressionists and phenomenologists derive their conceptions of reality from sense impressions: it is in this context that there may be said to be an affinity between them. Thus, they do not render, for instance, a landscape, but rather the sensation produced by the landscape, asking themselves, in what sense can a scene exist or appear? Whatever is *known* is *phenomenon*, because to be known means to appear to consciousness in a special way.

With Husserl, Merleau-Ponty and the authors to be discussed in this paper, the Saussurian distinction between the signified and the referent is unsatisfactory in that it makes no allowance for the status of the latter within the reader's consciousness. Consciousness, according to Husserl in the *Cartesian Meditations*, is dependent upon the constitution of the object as *cogitatum* or intentional object in the perceiving consciousness. The object of reference is never actually present, but the mind constitutes its phenomenal reality (which is not an imagined, preconceived or imaged reality). In a semiotic system, perception of the referent may be designated more accurately as the creation of a signified in the subject's consciousness by an intentional inference of the referent as signifier. The structure of the sign remains intact at this spontaneous level, described by Husserl as a *reflexion naturelle* (natural attitude) towards reality, and serves as a base upon which further sign systems may be imposed. At a secondary level, the subject engages upon the activity of reflection as opposed to perception. This process, one of *reflexion transcendentale*, reduplicates the act of perception, but infers its signifier not from the referent but from the signified already present within the reflective consciousness.

In *Structure du Comportement*,[6] Merleau-Ponty sees consciousness as a network of significative intentions which are sometimes clear to themselves and sometimes, on the contrary, lived rather than known. Intentionality is for him an essential step for a method of description in search of essences, such as: the essence of perception or the essence of consciousness. Intentionality refers to the entire activity of engaging ourselves co-creatively with the objects in our world. Each act of consciousness is necessarily intentional, i.e., it has a reference outside itself. Man needs the world to understand reality, and the world needs man to have a meaning. For him, a wheel does not look the same to the man who does not know how to use it as it does to the craftsman himself: a wheel lying on the ground does not look the same as a wheel bearing a load. Intentionality is not only an essential step for a method of description in search of the essence of consciousness or the essence of perception: it makes language the expression of self and world for both the phenomenologist and impressionist writer.

Merleau-Ponty invokes Husserl's differentiation between the voluntary *intentionnalité d'acte* and the involuntary *intentionnalité opérante.*[7] Language as intentional act is for Merleau-Ponty a gestural action; it is not a sign for a meaning, but an embodiment, an "incarnation" of meaning. The theory of *parole* (with its non-conceptual elements: *sens émotionnel*) strikes a balance between language understood as sign and language regarded as autonomous and will help us to understand in which ways impressionist texts seriously challenge traditional notions of narrative causality. Embodiment is densest and richest in language of lyrical prose, a direct description of experience in which man's essence is resituated in the context of his existence.

The rediscovery of a naïve contact with the world, which precedes the reflective epistemological act, constitutes a primary expression which is always located within lived time and space. The return to the primordial expression of *la parole* operates a phenomenological reduction in which the self is apprehended by the lived experience only. Words are necessary to construct this transcendental consciousness which does not correspond to something existing, but on the contrary, emerges from the ambient creative silence: silence is not the opposite of speech, but the surrounding and, therefore, the setting-off element or climate of speech, which surges from, and returns to silence.[8]

For the structuralist Gérard Genette, self and world, in contrast to Merleau-Ponty, are shaped by the structure of language. While Merleau-Ponty assumes in the form of language an expression of the synoptic presence of subject and object, Genette's *Discours du récit* (*Narrative Discourse*),[9] one of the most discriminating poetics of the novel in recent years, pays hardly any attention to setting or description, to the evocation of atmosphere in which subjects and objects merge. Merleau-Ponty grasps language in its specificity as the milieu of articulation of the reality experienced by the individual:

> What is proper to the sensible (as to language) is to be representative of the whole, not by a sign-signification relation, or by the immanence of the parts in one another and in the whole, but because each part is *torn up* from the whole, comes with its roots, encroaches upon the whole, transgresses the frontiers of the others. It is thus that the parts overlap (transparency), that the present does not stop at the limits of the visible (behind my back).[10]

Both Genette and Merleau-Ponty derive their theories of language from Saussurian linguistics. When Saussure distinguished a synchronic linguistics of *la parole* and a diachronic linguistics of *la langue*, he merely juxtaposed two of his irrevocably separated perspectives. Merleau-Ponty and Genette, however, institute a dialectic of perception to bring the diachronic and synchronic views into communication. The subjective point of view envelops the objective point of view. Diachrony envelops synchrony; the dichotomy between impression and language is rejected by Merleau-Ponty who sees writing as bodily expression which is part of the ongoing primordial process of perception.

Both Genette and Merleau-Ponty do not isolate language from experience as we have seen. By experience, Genette, in contrast to Husserl and Merleau-Ponty, does not mean intentionality. He rather sees experience as grounded in autonomous language. Experience rises out of a closed system of signs which is the literary work.

To indicate Genette's argument on Proust's 'L'Age des noms,' which is also the title intended for the first part of *A la Recherche du temps perdu* followed by 'L'Age des mots,' and 'L'Age des choses,' we may note that the definition of *la parole* responds to "l'image présentée par le nom propre,"[11] being internally dynamic (*confuse, indistincte, complexe*) in that it translates itself by menas of its words. Genette proposes a linguistic concept of the *nom propre*, always linked to a geographical location, refuting *le Mot*, that is to say, ready-made meanings within the experience of language: Genette, in 'L'Age des noms,' reprinted in *Mimologiques*, accentuates the internal power of the words themselves, a verbal experience in which the word carries a meaning that conveys a thought as a style, as an emotional value, as an existential mimicry rather than a conceptual statement, concluding with Proust that "ce n'est pas des noms qu'il faut partir pour connaître les choses, mais des choses elles-mêmes":[12] the word's meaning is not in the sound, but in the modulation of existence enacted by the thing.

For Merleau-Ponty, a word or an image placed in a particular context produces a particular effect: "The writer's work is a work of language rather than of thought" as Merleau-Ponty tells us in discussing Proust. "His task is to produce a system of signs whose internal articulation reproduces the contours of experience, the reliefs and sweeping lines of these contours in turn generate a syntax in depth, a mode of composition and recital which breaks the mold of the world and everyday language and refashions it."[13]

Poe, Joyce, Baudelaire, Rimbaud, and Mallarmé have initiated a movement in fiction to adapt changes of consciousness, of feelings or sensations, to the power of words, inventing a new language which by itself will be capable of expressing the multiple nuances of the personality and feelings. Proust and Gide were the first to acknowledge the aesthetic revolution of Flaubert's *L'Education Sentimentale* in which "impression" replaces "action."[14] Detachment from the human character is a stylistic device: when Flaubert dispenses with character, naturally, he has to do away with plot too. Characters simply *appear* in their pure physical reality; they are seen before their lives and thoughts are known. The reader sees Frédéric in successive instant portraits of fragmentary and progressive images of *anéantissement, abandon, oubli, dénouement, dépouillement, désagrégation, épuisement, dépersonnalisation*, and *dissolution*. The major impression of the book is the one of distance, of detatchment, of discontinuity. Since each sensation is reflected by the mirror of the artist's consciousness , which is precisely located in space and time, no two impressions can be identical.

Flaubert's style reflects an impression that lingers for but a moment. Characteristics of this style are devices which Flaubert shares with his impressionist successors: the strong verbs are often replaced by weak auxiliaries (*être, avoir*), by colorless verbs of perception such as *paraître, sembler, glisser, osciller*, by substantivized color adjectives, by quality nouns, by impersonal synthetic pronouns, by a predominance of color over objects, by subjective impressions and onomatopoeic overtones. There is a frequent use of ellipsis to express an absence; there are adverbial expressions to suggest a certain *Stimmung*, "mood," or "atmosphere." Flaubert initiates the use of present participles to suggest an illusion and indicate simultaneity. His composition by *tableaux* is most effective when he creates distance in the way French writers of the seventeenth century would create it: with the indefinite article, adverbs and prepositions (*avec, au-loin, par-derrière*). There are inversions, crescendo effects, a lack of outline, and a predominant shifting use of the present and imperfect.

Gide transformed the narrative of *La Nouvelle Education Sentimentale*, a short fragment and the original draft of *Les Cahiers d'André Walter* (1891), into the form of a diary, the form most used by impressionist writers. Also, when Gide wrote *Les Cahiers d'André Walter*, he felt many affinities with Flaubert's *Frédéric Moreau* and even more, was deeply impressed by Flaubert's stylist devices. The major purpose of his lyrical prose, Gide tells us, is to write not in French but in music. Phenomenologically, music is anterior to spoken language, and it is in the musical structure of Gide's lyrical work that language can appear. Language is no longer understood as thought and concept, but as sensation and sound image.

In this respect, Gide's *Les Nourritures Terrestres* (1897) appears to us as almost the gospel of impressionist lyrical prose. It is a hymn of joy, of spontaneity, of instantaneousness, of discontinuity, of detachment, of dissolution. All beings are judged according to their ability to receive light. The reflection of light on cities and landscapes creates the impression of distance, remoteness, and detachment. Reality for Gide's protagonist is a synthesis of pure sensations, modulated by consciousness and changed into impressions.

We find ourselves confronted with the problem of the relationship between the thing or reality as it appears, through a single perception, and the thing or reality as it really is. The seeing or pinpointing of certain essences within the structures common to all phenomena, is the act of perception taken as a vehicle of knowing a lived-object, it is *Wesensschau, Schauen* for Rilke's Malte; it is the perceived perceiving, the speech speaking, the thought thinking as the German critic Käte Hamburger[15] explains in her recent book. Likewise, Merleau-Ponty refers to the act of perception which determines the nature of perception itself:

> L'énigme tient en ceci que mon corps est à la fois voyant et visible. Lui qui regarde toutes choses, il peut aussi se regarder, et reconnaître dans ce qu'il voit alors l' 'autre côté' de sa puissance voyante. Il se voit voyant, il se touche touchant, il est visible et sensible pour soi-même.[16]

In Rilke's *Die Aufzeichnungen des Malte Laurids Brigge*, sickness and death adopt, as Malte tells us, the characteristics of the person it attacks. He contrasts personal death with a hospital death where sickness becomes anonymous, and cannot express the person. Thoughts of death fuse with the surroundings of Malte's grandfather, Christoph Detlev, in harmony with the broken rhythm of sentences which describe the fall of objects and which lead to the major effect, the death of Chamberlain Christoph Detlev Brigge; the repetition of key words suggests, in ever louder and penetrating rhythm, death. The passage conveys the impression that death emerges from the world of broken and falling objects as a necessity. Even structure and images reflect a "coloration" of death which penetrates and dissolves both subject and object simultaneously. The crescendo of key words, obtained by an elaborate arrangement of rhythms and images, creates the effect of a musical elegy. The predominant sound-effects excite our soul, conveying the impression that these sounds live.

In Joyce's *Dubliners* and *A Portrait of the Artist*, intellectual associations are completely dispensed with whenever Joyce wishes to capture the ephemeral, translucent atmosphere of Dublin. Joyce seems to be going beyond form to seize, for a moment, the peculiarly sensual qualities of light playing havoc with the staid objects of the outside. In one passage, Stephen is journeying to Cork by the night mail train. His window is his frame on life, and outside objects flash by with kaleidoscopic effect, revealing only their essence in the repeated and momentary flaring up of tiny grains of fire. Joyce creates the impression of blurred images passing before his eyes in total silence by repetition of key words (flung), and by his use of the gerund and adjectival participles. Words take the place of objects: darkening, slipping, passing, glimmering, twinkling, etc. This blurring of images is an essential part of Joyce's technique, and so often his frame of reference is the window picture-frame, as with Flaubert who also used it with preference.[17]

*A la recherche du temps perdu* reflects the transitional process from impressionist to phenomenological vision and from geometrical to lived perspective in the act of literary creation. Both Merleau-Ponty and Genette take Proust's writing as an example of the primordial expression which precedes logical formulation. Stating that language can never be a mere purveyor of received ideas, they describe Proust's expression of ideas as an integral part of vision. Genette shows that the farther we go into Proust's *A la recherche du temps perdu* the more daringly abrupt become the uses of ellipsis, as a counterpart to the fact that the scenes in between, though they cover shorter and shorter periods of story time, contain greater and greater detail. The whole effect is of an ever-growing discontinuity between discourse time and story time. The formal structure of Proust's work reflects only a choice of interpretation which is arbitrarily imposed upon the tissue of perception. The present engages in constant dialogue with the multiplicity of individual impressions which constitute the past. Proust's rendering of reality through art is achieved by portraying the prismatic refraction of impressions which take place in the reflective consciousness, rather than by a simple reflection of the impression. The interconnecting patterns of perception create a constantly shifting prelogical perspective.

Each impression is located in a time-space continuum and is modulated by Marcel's individual point of view, the nuance of emotion renewing each memory and creating new associations. The book written by Marcel is that action of consciousness in which author and reader communicate not with each other, but with their own feeling of existence. It is a necessary correlative to self-observation in a phenomenological mode. The functions of Marcel's self-esteem seem to be taken up by a variety of interpersonal systems that operate through it. Thus the text is an exploration of both writing and reading. It is about Marcel becoming a writer, about his learning how to see his world clearly and understand it, about how to structure the vision of his world in a way that will enable him to write the book we are reading. Time arises from Marcel's relation to things. It is a network of intentionalities that are a product of spatial positions. We may be warned, though, by the German critic Wolfgang Iser: "the reader, in establishing these interrelations between past, present and future, actually ceases the text to reveal its potential multiplicity of connections."[18] A reader approaching a Proustian sentence expects a coherent whole but actually sees many fragments which he must reconstitute into a coherent whole.

In the following passage, Marcel realizes two things: first, that the way he comes to know individuals is not through a single static position in space, but through shifting and successive perceptions, and second, eternity and infinity become a physical experience in the privileged moment of heightened awareness. Writing stops Time (it gives a text a limitless duration of time) in order to allow not just a surface perception, but a unified vision, a vision which, by being written (captured or recaptured) can be perceived and comprehended by another, the reader. Words, once written down, constantly threaten to fix Marcel's existence which is of dynamic character. Marcel is a network of interrelationships, and the reader of this passage must, in order to gain access to Marcel's world, become attuned to the numerous shifts of focus, or orientation, since Marcel's phenomenal world cannot be reduced to anything else than that which appears in his/our field of perception. With Marcel, each act of perception seizes the perceived individual only in a certain respect. The perceived gives itself on and through the act of perception only by means of *profiles* (*Abschattungen*) which are correlated to a determined attitude and standpoint in Marcel, the perceiver. The passage shows the writer's struggle to describe the individual act of appearing as it is in the process, in the act of appearing. The meaning of this becomes coherent to the reader through repeated reading. Proust makes us conscious of the way he sees reality and in this process dramatically extends our field of vision, our possibility of learning to see with Marcel:

> Nous ne pourrions pas raconter nos rapports avec un être que nous avons même peu connu, sans faire se succéder les sites les plus différents de notre vie. Ainsi chaque individu – et j'étais moi-même un de ces individus – mesurait pour moi la durée par la révolution qu'il avait accomplie non seulement autour de soi-même, mais autour des autres, et notamment par les positions qu'il avait occupées successivement par rapport à moi. Et sans doute tous ces plans différents suivant lesquels le Temps, depuis que je venais de le ressaisir dans cette fête, disposait ma vie, en me faisant user, par opposition à la psychologie plane dont on use d'ordinaire, d'une sorte de psychologie dans l'espace, ajoutaient une beauté nouvelle à ces résurrections que ma mémoire opérait tant que je songeais seul dans la bibliothèque, puisque la mémoire, en introduisant le passé dans le présent sans le modifier, tel qu'il était au moment où il était le présent, supprime précisément cette grande dimension du Temps suivant laquelle la vie se réalise.[19]

With the passage we have the impression of an emerging order, of individuals in the act of appearing, viewed not in linear, but in lived perspective. To trace just a single outline of the individual would sacrifice depth, the dimension of an inexhaustible human reality. Instead, when Marcel attempts to seize the identity of others or of himself, they seem to dissolve simultaneously into a multiplicity and succession of selves in space and time. Therefore, he contrasts *psychologie plane* (two-dimensional psychology) with *psychologie dans l'espace* (three-dimensional psychology). The former is dealing with body and mind only, and the latter with body, mind, and Time. We see the depth of Time. as Gilles Deleuze states: "Time itself is plural. The great distinction in this regard is that between Time lost and Time regained."[20] The real appears to both Marcel's and the reader's field of perception constantly filled with a play of colors, noises, and fleeting tactile sensations, the background from which all acts stand out. Thus, the text has a total and sacred freedom of structure, and this evidently creates surprises, not to mention difficulties for the reader, inasmuch as the novel no longer represents a casual sequence that is coherent and that extends from one end of the book to the other. Finally, with the notions of space, time, individuality, all the conditions and limitations of ordinary life are suspended together; eternity and infinity become a physical experience in the privileged moment of heightened awareness – of writer/protagonist/reader.

Claude Simon introduces phenomenological elements into the act of writing, completing the break between primordial and secondary forms of expression. The conforting activity of reorganizing impressions upon which Marcel embarks in *Le Temps Retrouvé* is for the narrator of *Histoire* a discomforting primordial expression of his own reflective consciousness. His conception of reality becomes the expression of his *écriture* as he observes the interior dimensions of his reflective consciousness. The final words of the text bring the narrator to an awareness of his own psyche. The day has come to an end, the narrator has closed off perception of the exterior world and has confined himself as the product of his *écriture*, he is the unwilling witness of his own transformation, a Kafkaesque metamorphosis. The author's attempt at inward appraisal is defeated by the prismatic space of the *écriture* The narrator can only perceive himself phenomenologically as an unrecognizable monstrosity unpredicated by rational analysis.

We may conclude with Genette: "Critique du language, triomphe de l'écriture." The intellectual organization of the work of art is subsequent and subordinate to the impression. It is a compliment to the more immediate awareness of reality. The interest of the narrator in disengaging truths from experience leads him to accept both the phenomenological and epistemological modes of expression in *l'écriture*.

With the works of Flaubert, Gide, Proust, Joyce, Rilke and Claude Simon, we have tried to illustrate that the novel moved toward a text where the text itself is affirmed as its own reality, that is, a text concerned with representing the narrator's consciousness through which reality is filtered and by which reality is presented in narrative form. The text is not concerned with representing a reality exterior or anterior to the novel. The text has a total and sacred freedom of structure, and this evidently creates surprises, not to mention difficulties for the reader, inasmuch as the novel no longer represents a causal sequence that is coherent and that extends from one end of the book to the other.

What can we expect from both impressionist and phenomenological authors and philosophers? That in the long run they will affirm and integrate into their own approach the basic intuitions of structuralism.

*Michigan State University*

## NOTES

[1] Maurice Merleau-Ponty, 'Le doute de Cézanne,' in *Sens et Non-Sens* (Paris: Nagel, 1966), pp. 15–44.

[2] Gérard Genette, *Figures I* (Paris: Seuil, 1966); *Figures II* (Paris: Seuil, 1969); *Figures III* (Paris: Seuil, 1972); and *Mimologiques* (Paris: Seuil, 1976).

[3] Michel Butor, 'Research on the Technique of the Novel,' and 'Balzac and Reality,' in *Inventory*, ed. Richard Howard (New York: Simon and Schuster, 1968), pp. 146–84.

[4] Marcel Proust, 'Le Côté de Guermantes,' in *A la recherche du temps Perdu, II* (Paris: Pleïade, 1954), p. 406.

[5] Ernst Mach, *Die Analyse der Empfindungen*, tr. *Contributions to the Analysis of Sensations* (Chicago: Open Court, 1897).

[6] Maurice Merleau-Ponty, *Structure de comportement* (Paris: Presses Universitaires, 1949).

[7] Edmund Husserl, *Cartesianische Meditationen und Pariser Vorträge* (The Hague, 1950).

[8] Maurice Merleau-Ponty, 'Indirect Language and the Voices of Silence,' in *Signs* (Chicago: Northwestern University Press, pp. 39–97.

[9] Gérard Genette, *Discours du récit* (Paris: Seuil, 1972).

[10] Maurice Merleau-Ponty, *The visible and the invisible* (Evanston: Northwestern University Press, 1968), p. 218.

[11] 'L'Age des noms,' in *Mimologiques*, p. 315.

[12] Ibid., p. 238.

[13] Maurice Merleau-Ponty, *Phénoménologie de la perception* (Paris: Gallimard, 1945).

[14] Maria E. Kronegger, *Impressionist Literature* (New Haven: College and University Press, 1973).

[15] Käte Hamburger, 'Die phänomenologische Struktur der Dichtung Rilkes,' in *Rilke in neur Sicht* (Berlin, Köln, Mainz: Kohlhammer, 1971).

[16] Maurice Merleau-Ponty, *L'oeil et l'esprit* (Paris: Gallimard, 1964), p. 16.

[17] Maria E. Kronegger, *James Joyce and Associated Image Makers* (New Haven: College and University Press, 1968); E. F. N. Jephcott, *Proust and Rilke, The Literature of the Expanded Consciousness* (London: Chatto and Windus, 1972), pp. 155–75.

[18] Wolfgang Iser, 'The Reading Process: A Phenomenological Approach,' *New Literary History* 3 (Winter 1972):283.

[19] 'Le temps retrouvé,' in *A la recherche du temps perdu III*, p. 1031.

[20] Gilles Deleuze, *Proust and Signs* (New York: George Braziller, 1972), p. 17.

PETER STOWELL

# PHENOMENOLOGY AND LITERARY IMPRESSIONISM: THE PRISMATIC SENSIBILITY

> As far as one ventured to interpret actual science, the mind had thus far adjusted itself by an infinite series of infinitely delicate adjustments forced on it by the infinite motion of an infinite chaos of motion; dragged at one moment into the unknowable and unthinkable, then trying to scramble back within its senses and to bar the chaos out, but always assimilating bits of it, until at last, in 1900, a new avalanche of unknown forces had fallen on it, which required new mental powers to control. If this view was correct, the mind could gain nothing by flight or by fight; it must merge in its supersensual multiverse, or succumb to it. (Henry Adams, *The Education of Henry Adams*)

Henry Adams saw and understood better than most men of letters the special quality of change and upheaval that was taking place between the years 1870 and 1920. He faced squarely the issues of a fragmented, relativistic, unknowable, accidental, subjective, and accelerating world, while realizing how ill-prepared he was to absorb and adapt to such a world. His autobiography can in many ways be viewed as an historian's analogue to literary impressionism.

Impressionists found themselves caught between the ferocious revolt against the subjectivity of romanticism and the headlong dash into the anarchy of modernism. Neither literary movement or critical invention, literary impressionism was a phenomenon that existed as a set of shared assumptions, elective affinities, philosophic agreements, and common reactions to the "supersensual multiverse."

Literary impressionism sprang from neither theoretical constructs nor rigid ideology, but from reactions to the shifting intellectual and social currents of the age. This sense of change was so rapid and comprehensive that the impressionists felt compelled to express their artistic responses in highly individualistic and experimental forms. Yet their stylistic qualities emerged from the realization that since they lived in a prismatically impressionistic world, then they must recreate that world of individualized sensory perception, epistemological indeterminacy, relativism, ambiguity,



*A-T. Tymieniecka (ed.), Analecta Husserliana, Vol. XVIII,* 535–544.

fragmentation, and surfaces. They saw faith disintegrate into the confusion of chance, the chaos of unknowability, and the crisis of change. They wrote of a need to survive in this "supersensual multiverse" by having enough faith in their expanding powers of inductive perception to navigate the shoals of the unseen and unknowable.

The impressionists' first philosophical reaction was anti-mechanistic. The realists and naturalists, while rejecting romantic mechanism succumbed to the power of Darwin's scientifically observable and morally justifiable unity. The impressionists saw the contradiction in this position. Retaining empiricism, they rejected the thinly veiled moral didacticism of the realist's aesthetic omniscience. Subjective and limited sensory perception became the empirical truth for the impressionists, who believed, as William James did, that while the empiricist might attain truth, he cannot know when he has grasped it: "We still pin our faith on its [the truth's] existence, and still believe that we gain an even better position towards it by systematically continuing to roll up experiences and think." The impressionists, then, created characters who flounder through an accelerating and ambiguous world of sensory stimuli. They attempt to control, face, adapt to, and escape from such a world – while discovering that all one can ultimately do is "roll up experiences and think."

Against the background of an essentially Newtonian 19th century, the impressionists, braced by their scientific training, or at least scientific curiosity, began to intuitively anticipate the major scientific, philosophical, intellectual, and cultural changes of the coming age. To the extent that their art reflected these breakthroughs, they came to understand the new concepts in optics and photography, the wave and particle theories as applied to Bergson's *durée* and Husserl's phenomenological time, the empiricism and epistemological indeterminacy of William James, the uncertainty principle of Heisenberg, the relativity of Einstein's space and time, the phenomenologists' recognition of a reciprocal perceptual reality between subject and object, and the *Gestalt* psychologists' model of an inexplicable synthesis based upon dynamic patterns of perceived fragments. Within this framework the impressionists discovered their aesthetic goal: the fusion of form and meaning through the portrayal of human consciousness compressed into a fluid network of brilliantly concentrated sensory impressions.

Impressionism's primary tenets emerge from this flux. Its basic stylistic impulse is the shift from a description of concrete and tangible reality to a *rendering* of apperception. This recreates the perceived mood and atmosphere surrounding concrete objects – or, to paraphrase Archibald McLeish's comment about poetry, "an impressionist piece of prose doesn't mean; it is." This shift goes to the heart of impressionism's aesthetic:

> The issue is of the sensation itself and, following this hypothesis, must we see in Proust and the impressionists . . . those for whom pure sensation are wholly important, or indeed, placed within the context of the views of Sartre or Merleau-Ponty, are we present at the spontaneous transformation of the sensation by perception, with the understanding that perception coincides with the progressive creation of the perceiver?

For the impressionists, the answer to this question was clearly affirmative. They created characters who grow only as their ability to perceive the connections among fragments expands. The impressionists' belief in the primacy of perception forced the pose of the omniscient, didactic, and discursive author out of the work. He was replaced by an elusive presence who allowed all characters to perceive for themselves the ambiguous and ultimately unknowable surfaces of sensory reality.

## THE ACT OF PERCEPTION

The starting point for the act of perception in impressionist literature is the perceiver. This pinpoints the shift from the 18th and 19th century's concept of "telling" to the twentieth century's belief in "showing." The impressionists believed that, as Merleau-Ponty put it, "everything rests upon the insurpassable richness, the miraculous multiplication of perceptible being." They regarded the process of perceiving impressions as reality itself, not simply as a factor in the representation of it. "To the susceptible mind of the impressionist the surrounding world viewed at large is not simple and well ordered, but an indistinct and obscure picture made up of an irresistible flood of confused and ever changing sense impressions." At any given point in time, one impression is as valid and real as any other.

The impressionist attempts to capture the feel, texture, and consciousness of the phenomenological *tabula rasa* or, in the impressionist lexicon, "the innocence of the eye." For phenomenologists and impressionists alike everything is seen anew through fresh eyes. It is not surprising, then, that the most impressionistic early stories of Chekhov were of children experiencing a new world of sensory impressions. And James' first two impressionist novels, *The Portrait of a Lady* and *What Maisie Knew*, concerned young girls experiencing new worlds. While much impressionism filters reality through the innocent eyes of Crane's Henry Fleming or Proust's Marcel, it cannot be stressed too strongly that since perception is an act of becoming, all phenomenal perceivers view the world again, with each passing impression. Chekhov and James both understood this, so their mature impressionist works often have such older middle-aged men as Gurov in 'The Lady with the Dog' or Strether in *The Ambassadors* as their central consciousnesses. After years of anesthetized perception these characters open themselves to the sensory world. The act of perception becomes more closely connected to the act of remembering. Memory for the now active perceiving consciousness has become Proustian since the impressions of the past are held suspended in and given the texture of the present. The perceptive world is the world of the present. This concept anticipated the essence of existential phenomenology.

Tension in impressionist fiction pivots on the passive perceptual intake of sensory impressions set off against the *gestalt* synthesis of consciousness. Impressionism does not simply record the impact of pure, raw sensations on a passive receptor. If that were so, impressionist works would be little more than a compendium of impressions. Impressionism is the process by which impressions are absorbed by a perceiver and synthesized by consciousness into a *gestalt* that may lead to action or thought.

The stylistic foundation of impressionism is the creation of linguistic and rhetorical effects that render sensory impressions on eyes, ears, and consciousness. Impressionist characters rarely think causally or conceptually. Instead, if they aren't directly involved in a sensory experience, their consciousnesses are sorting, resorting, and synthesizing perceptual fragments experienced over time, but reexperienced in each present moment.

There is a dramatic and terribly important ontological shift from the verb "to be" to "seems," "appears," or "it was as if." These tropes underscore the tenuous nature of perceived reality, thrusting at the reader the textures of unknowable surfaces. The world of impressionism is the world of appearances. With the burden knowability placed firmly on perception and with the accelerating pace of sensory stimuli constantly assaulting the perceiving consciousness, it becomes possible to deal with the reality of surfaces only. There is neither the time to reach out beyond appearances toward the heights of a romantically metaphysical universe nor the analytical means to probe beneath the surfaces of a rigorously neoclassical reality. For the impressionist, one can neither penetrate nor transcend the multiple and ever-changing surfaces of the relative and ambiguous modern world – or as Henry James put it, "Life is, immensely, a matter of surface. . . ." The impressionist takes the world of appearances as reality, rather than struggle with the phantom issue of appearance versus reality. Each impression, then, is real.

But not until the impression is tempered on the forge of experience and cooled in the vat of consciousness can it be authenticated by the *gestalt*. It can then be "known."

## SUBJECT AND OBJECT

Crucial to the impressionist vision is the relationship of subject of object, "inner ego to outer world," consciousness to external reality. Naturalists and realists believed the empirical method could only focus on objectified exterior reality; symbolists and expressionists concentrated on a solipsistically encapsuled inner world. The impressionists, however, sensed, along with the phenomenologists, that, "consciousness must be consciousness of something" and that both subject and object are real. Merleau-Ponty accurately describes the impressionists' method of reciprocating mediation:

> If a friend and I are standing before a landscape, and if I attempt to show my friend something which I see and which he does not yet see, we cannot account for the situation by saying that I see something in my own world and that I attempt, by sending verbal messages, to give rise to an analogous perception in the world of my friend. There are not two numerically distinct worlds plus a mediating language which alone would bring us together. There is – and I know it very well if I become impatient with him – a kind of demand that what I see be seen by him also. And at the same time this communication is required by the very thing which I am looking at, by the reflections of sunlight upon it by its color, by its sensible evidence. The thing imposes itself not as true for every intellect, but as real for every subject who is standing where I am.

The perceiving consciousness acquires existence and meaning only in relation to the objects and events of the external world. The relationship is one of subjective objectivity where both subject and object are immersed and juxtaposed in the spatiotemporal world of perception. This idea for realizing a new relation to reality was a major breakthrough for modern literature.

Adapting this heightened consciousness to the new reality of objects means fusing the inner ego with the outer world. This phenomenological relationship underlies the entire vision and aesthetic of impressionism. It has been previously recognized as important, but its position as the nucleus of impressionism has never been fully acknowledged. No longer is there a separation between subject and object, nor is one given precedence over the other. Reality is the synthesis of perceiver and perceived – each exists and each creates meaning for the other. The new aesthetic is based upon integrated juxtaposition – a *gestalt*: the perceiver superimposes on an object actual physical qualities based upon his own memory, mood, and perspective;

the object takes on and reflects the physical properties of the surrounding environment; the object and its reflections, the entire *mise-en-scène*, simultaneously infuse the perceiver's own set of physical and psychological characteristics. This process continues *ad infinitum*, turning, spinning reacting, blending – all in the moments of durational and phenomenological time.

The stringent preoccupation with the subjective objectivity of objects that characterizes much of the *nouveau roman* was not so purely practiced by the impressionists. Instead, with the exception of Chekhov, they imbued their works with a plenitude of consciousness. Character is still a major humanizing force for the impressionists; they have not wholly removed themselves from the ties of nineteenth-century literature. Chekhov's stripped style begins to move toward the empty entropy of Beckett's silences and objects, and the camera-eye coldness of Robbe-Grillet's obsessive, object oriented perceiver. Impressionism, however, still basks in the warm light of humanism.

In Beckett and Robbe-Grillet objects emerge with crystalline clarity, whereas the impressionists find the contours indistinct. Fragmentation leads to decomposition. And this becomes as true for the impressionist characters, as foı the objects they perceive. Physical description seems lost in the welter of other objects, and metonymized through time. A character's physical characteristics are always subordinate to consciousness and scattered throughout the work. A character's psychological configuration is also so protean that it becomes difficult to envision clearly outlined character traits. Both perceiver and perceived absorb and project the tones and features of the other.

In that prismatic reality between subject and object all outlines become hazy, while each element struggles for its own identity and existence: "All distinctions are eliminated, the physical and the psychological coincide, and sensations are one and the same, the ego dissolves and everything is an eternal flux which in some places seems to stop, in others to flow swifter, everything is merely a movement of colors, sounds, temperatures, pressures, spaces and times, which on this side of the ego appears as moods, feelings, and desires."

## SPACE AND TIME

But it was in their approach to time and space that the impressionists made their greatest contribution. All the components of literary impressionism – the shift from description of objects to the rendering of atmosphere and mood, the importance of a perceiving consciousness, and the phenomenological relationship of subject and object – are, of course, inextricably woven into the impressionistic tapestry. Yet the dominant thread is the paradoxical union of Bergson's *durée* and phenomenological time and space. Arnold Hauser was really the first to notice the importance of time, believing that "impressionistic thinking finds its purest expression in the philosophy of Bergson, above all in his interpretation of time – the medium which is the vital element of impressionism." But the impressionists intuitively felt that Bergson's subjective current of human time neglected the privileged moments of spatialized presentness. His notions contributed to the continuity of consciousness, the stream of sensory impressions, the flow of human fulfillment for, according to Hauser, "Time [impressionistic, Bergsonian, Proustian] is no longer the principle of dissolution and destruction, no longer the element in which ideas and ideals lose their value, and life and mind their substance, it is rather the form in which we obtain possession and become aware of our spiritual life, our living nature, which is the antithesis of dead matter and rigid mechanics. What we are, we become not only in time but through time." It is this quality that distinguishes literary impressionism from painting. Literature must create characters responding to events through time. Most impressionists rejected concepts of time that would render change definitive: the naturalists' belief that time was solely the province of objective natural phenomena resulting finally in death or the romantic conception of transcendental moments resurrecting man from the crush of natural time. The impressionists went a step further by fusing the subjective wave of Bergsonian time with the subjectively objective quanta of perceived phenomena. Phenomenological time stresses the "moments" of events, which because they are imperceptibly framed and spatially perceived take on the literary impressionists' hallmark – temporally extended frozen moments of spatialized time that dissolve and return to the flow of durational time. Bergson provided the humanizing force of continuity, where "existence acquires actual life, movement, colour, an ideal transparency and a spiritual content from the perspective of a present that is the result of our past" and the hope of our future. Phenomenology suggested, however, "that one does not find a continuous flow of the personal life, no instants, and no intervals: but simply the consciousness of a changing world. No 'empty' moment of time between the vanished pulsation and that which is about to be produced, but rather a particular object of consciousness made up of tremblings, tensions, relaxations, in which, with my body and by my body, I recover space" (Gaston Berger).

Through their concern with time and space the impressionists discovered their basic theme – change. All writers and thinkers on the subject of time have recognized that time is change. The impressionists realized and were, in fact, compelled by the changing world in which they found themselves. Unlike historians, they translated their concerns over an accelerating, multiplying, and increasingly unknowable reality into characters who strive to hang onto the dizzying swirl of shifting consciousnesses, while responding to the changes in perceived events. The struggle, as Henry Adams so vividly saw it, was for survival and the scales held sensitive perception of the known chaos against measured faith in one's "power to guess the unseen from the seen, to trace the implication of things, to judge the whole piece by the pattern."

The movement of subject and of object and the relative movement between them, finds its most manifest expression in that figure of change, time and its inextricable companion, space. It is often argued that time itself has become the actual subject of modern literature. This is certainly true for Proust, Joyce, Woolf, Eliot, Faulkner, Robbe-Grillet, and Borges. While for centuries philosophers have been struggling with the major paradoxes of time, writers have generally accepted chronological or cosmic time – time in relation to a character's march through the hours, days, months, and years of existence or time in relation to a character's concern with death and/or eternity.

The impressionists and Bergson, however, were contemporaneously exploring the creative potential in the idea that, as Auden put it years later, "All our intuitions mock / The formal logic of the clock." The impressionists were synthesizers who did not attack the concept of chronologically absolute time, but instead found that their sympathies lay more with fusing objective and subjective time, moment with duration, and space to time. These new conceptions formed the basic structure of the impressionistically architectonic fiction that has come to dominate modern literature.

If time is change, then space is endurance. The impressionists experimented with moments of persistent, changeless time in the midst of constantly reforming space. Since they envisioned space and time together in a continually evolving relationship, it became possible for each to take on the characteristics of the other. The impressionist painters fought to capture the moment on their spatial canvases, a held instant in time, which through the vibrancy of color juxtaposition, the movement of objects, and the bursting of the frame transposed space into time. Literary impressionists, working in their temporal art, strove to spatialize time. They depended on simultaneity through the cross-cutting of scenes, the fragmentation of image complexes, and the sensory fusion of past and present. These elements formed the basis of the "privileged moment" or the mythos of "eternal repetition." Proust was then able to proclaim that "Time has assumed the form of space."

This is not to argue that time is always spatialized, for as Roger Shattuck correctly reminds us, space and time are in a constant interchange. The impressionists created moments of spatialized time in order to develop impressions, redirect the inherently temporal process of literature, alter the rhythms of prose and consciousness, and give their characters a respite from the flux of continual change. There is no immunity from change, however, because even space moves with the development of multiple perspectives. Merleau-Ponty, for instance, has observed that,

> If the observer sits in a boat and is carried by the current, we may say that he is moving downstream toward his future, but the future lies in the new landscapes which await him at the estuary, and the course of time is no longer the stream itself: it is the landscape as it rolls by for the moving observer. Time is, therefore, not a real process, not an actual succession that I am content to record. It arises from *my* relation of things.

Time is space and space is time, just as subject and object are each other in the revolutionary shift the impressionists made toward synthesis.

Duration accounts for the privileged moments and the imperceptible flow of tenses. Time – past, present and future – seems immobilized in a conjoined present, while it also slips through the divisions. But Bergson's duration could not account for the impressionists' need to create successive moments, to break up the uninterrupted *élan vital*. They sensed that the moment was a transitory fragment of presentness, not part of "one indivisible present." Bergson's *durée* led to stream-of-consciousness, yet for all its affinities with this mode (which the impressionists influenced), impressionism's distinct quality rested on the fusion of duration and succession.

The impressionists allowed their characters to experience both the unifying flow of privileged moments and the phenomenological succession of presentness. Phenomenological moments are contingent, so that those elements of the past and future that do exist are not only contained completely within the present, but in fact, *are* the present. The durational moment, on the other hand, embodies within its flowing presentness the actual entities of past, present, and future. Impressionism offers an interchange between Bergsonian duration and phenomenological succession as the only way of rendering the creative and destructive, suspended and moving, enduring and transitory, fluid and fragmented, subjective and objective qualities of truly lived time. Time for the impressionist is a subjectively experienced sense of objective events. It continues through the dynamic relationship between the quantalike fragments of successive instants that change constantly and the enduring wave of "times" that flows uninterruptedly through human consciousness.

Both the impressionist work and character are always slightly and suspensefully out of balance. Neither temporal nor spatial formations are ever complete. Time seems to fade into the background as the rush of objects and images, captured almost simultaneously, takes precedence. After the dazzling fusion of images and the overpowering bombardment of fragments, time returns to smother the spatial experiences. The relationship between time and space is dynamic and relativistic. In this respect impressionism placed itself firmly in the forefront of modern literature, anticipating the innovations of Proust, Joyce, Faulkner, and the *nouveau roman*. Space is no longer geometric; time is no longer measurable – both are apprehensive.

*Florida State University*

## C. LITERARY "READING" OF THE HUMAN DRAMA

ADNAN MOUSSALLY

# UN MODÈLE D'ANALYSE DU TEXTE DRAMATIQUE

Les postulats sur lesquels la tradition littéraire s'est appuyée pour dégager la spécificité du théâtre ne cessent d'être battus en brèche depuis l'avènement de la linguistique saussurienne. Structuralistes et sémioticiens se sont tournés vers l'agencement des éléments d'un texte littéraire afin d'en étudier le fonctionnement et d'en établir les liens entre la forme et le contenu. Pour cela, ils ont échafaudé, au cours des dernières décennies, tout un arsenal de notions, de critères et de techniques fort compliqués; ils ont occasionné des querelles de chapelles et ont développé une terminologie assez riche par ses ramifications, mais qui reste pauvre à cause de ses chevauchements ou enchevêtrements.

La masse des études produites est inégale et c'est normal, puisque tous les coups n'ont pas porté et que les tiraillements ont fait avorter certaines tentatives qui paraissaient pourtant prometteuses. En revanche, cette même dialectique a éliminé certaines données inféc ondes et en a fertilisé d'autres. Le bilan est positif puisqu'aujourd'hui, on est mieux équipé pour poser la question du rapport qu'entretient le théâtre avec les autres formes littéraires. Avec une certaine précaution dans la manipulation des instruments mis de l'avant par les théoriciens et un émondage du fouillis méthodologique qui oscille encore entre l'ouverture de la sémiologie sur d'autres disciplines et sa fermeture sur elle-même, il est possible d'en arriver à des applications pertinentes.

Comme la sémiologie est une méthode critique relativement récente dont la terminologie est quelque peu floue et la méthodologie un peu imprécise, nous avons jugé bon de présenter tout d'abord quelques données d'ordre théorique. Telle que la conçoit F. de Saussure,[1] la sémiotique c'est "la science qui étudie la vie des signes au sein de la vie sociale". Cette définition garde encore certaine valeur heuristique du fait qu'elle sert de base à des définitions ultérieurement formulées et voulant que la sémiotique soit la science qui étudie les systèmes de signes tels qu'ils se manifestent dans les langues, les codes, les signalisations. Rappelons que le signe, dont la fonction essentielle est de signifier, jouit d'un double statut:

1. Il renvoie à autre chose (relation référentielle).
2. Il renvoie à un autre signe.[2]



*A-T. Tymieniecka (ed.), Analecta Husserliana, Vol. XVIII*, 547–557.

On distingue généralement quatre acceptions du mot "signe". La première correspond à la relation référentielle de Greimas et la deuxième n'est autre que l'articulation du signe en signifiant et signifié préconisée par Saussure; les deux autres concernent sa fonction communicative. Il s'ensuit que le signe linguistique ne représente qu'un cas particulier dans le système. En admettant qu'il y ait interchangeabilité entre niveau sémiotique et niveau linguistique, nous soutenons que celui-ci est postérieur à celui-là. Au théâtre, tel décor ou tel costume sont des signes traduisibles en un langage-objet qui renvoie à un signifié ou à un ensemble de signifiés déterminables. L'auteur dramatique – ou occasionnellement, le metteur en scène – conçoit le signe à représenter par le biais de manifestations linguistiques qui servent de relais entre la conception et la concrétisation.

Patrice Pavis[3] distingue, à juste titre, deux disciplines dans le déveloopement de la sémiotique: l'une qui est axée sur la communication et l'autre qui s'intéresse à la signification. Son erreur c'est qu'il projette de poursuivre deux lièvres en même temps alors qu'il s'agit d'un choix à faire entre deux options possibles. Notre démarche se veut plus modeste, se concentrant plutôt sur la sémiologie de la communication, par le biais de l'étude des signes, tout en y distinguant les indices des signaux. Bien que le symbole ait sa place, il ne fera pas l'objet d'une analyse particulière et ce, pour diverses raisons. Tout d'abord il se confond, dans l'usage américain, avec ce que la linguistique européenne nomme "signe". Pris dans son acception psychanalytique, il aurait dévié l'intérêt vers un domaine qui échappe à notre compétence et à nos préoccupations.

Pour l'indice, nous adopterons la définition formulée par Luis Prieto voulant qu'il soit "un fait immédiatement perceptible qui nous fait connaître quelque chose à propos d'un autre qui ne l'est pas".[4] De par sa nature, l'indice nécessite donc une interprétation: la tombée des feuilles serait l'indice de l'arrivée de l'automne, par exemple.

Le signal sera pris dans le sens de tout moyen utilisé pour influencer autrui et reconnu comme tel par celui qu'on veut influencer. Cette définition implique l'existence, à un moment donné, et pour un groupe déterminé, d'un code plus ou moins socialisé: un jeune homme affichant une certaine élégance vestimentaire en une occasion particulière peut s'attirer les invitations qu'il désire. Le signal s'identifie donc à une sorte d'indice, à la seule exception qu'il est produit volontairement par l'émetteur pour manifester une intention au récepteur. Comme il repose sur une base conventionnelle, il requiert un décodage.

L'étude du procès entre des signifiants et des signifiés immédiats, c'est-à-dire de tout ce qu'on appelle la dénotation, ne permet pas la saisie globale de l'oeuvre, ni ne rend compte de sa dimension culturelle. Aussi, l'avons-nous, tant soit peu, négligé au profit d'une sémiologie qui s'appuie sur la chaîne de connotations en vue de cerner l'idéologie sous-tendant l'acte de communication. De la sorte, nous essaierons de mettre à jour les principaux mécanismes de la communication, les structures constantes ainsi que les articulations essentielles.

Dans la mesure où la communication n'est pas le simple fait de la transmission de messages construits selon un code et véhiculés par un canal quelconque, on peut la considérer comme étant l'ensemble des phénomènes qui établissent des liens entre un émetteur et un récepteur. Elle doit présupposer une volonté de susciter des réactions et, en ce sens, ses préoccupations s'approchent de celles de la sémiologie, telle que définie par Eric Buyssens, à savoir, "l'étude des moyens utilisés pour influencer autrui et reconnus comme tels par celui qu'on veut influencer".[5]

La base de la communication étant de type conventionnel, on pourrait croire que la mise à jour des principes fondamentaux de ce genre de conventions suffirait pour en comprendre le mécanisme. C'est l'erreur commise par toute analyse structurale qui préconise une sorte de code unique á l'intérieur d'une même culture ou qui n'admet pas le phénomène continuel de structuration et de restructuration que subit la convention. Pour René Thom, la théorie structuraliste est statique; on y analyse l'organisation interne dans une optique a-temporelle. Sa théorie des catastrophes pousse à une recherche structurelle dans une perspective évolutive. On s'attache à découvrir l'organisation interne dynamique des objets.

Par ailleurs, la loi du moindre effort a poussé nombre de sémiologues à transposer certaines notions du domaine de la linguistique, sans trop réfléchir sur leur valeur opératoire. C'est ainsi que le "référent" a été adopté dans l'étude des signes. Mais devant les maigres résultats qu'une telle notion a apporté à la sémiologie et les problèmes qu'elle a soulevés, il est loisible de penser que son introduction inconditionnelle a joué au détriment du champ d'investigation de la sémiotique. En tout cas, cet outil externe n'a pas su résoudre la question de l'échange communicatif.

On laissera donc à d'autres branches de la connaissance humaine, telle que la philosophie, le soin de discourir sur l'équivalence du signe et de l'objet qu'il représente. En d'autres termes, il nous a semblé facile de nous passer de la notion de référent sans nuire à l'objectif poursuivi par cette étude, souscrivant en cela aux préceptes d'Umberto Eco contenus dans le passage suivant:

> La sémiotique étudie les codes en tant que phénomènes de culture et – restant indifférente aux réalités vérifiables auxquelles les signes se réfèrent – elle doit se limiter à examiner comment à l'intérieur d'un corpus social se sont établies des régles d'équivalence entre un signifiant et un signifié (ce dernier ne pouvant être défini qu'à travers un interprétant qui le signifie au moyen d'autres signifiants) et des règles de combinaison entre les éléments du répertoire paradigmatique. Cela ne veut pas dire que le référent "n'existe pas" mais qu'il constitue l'objet d'autres sciences tandis que l'étude des systèmes de signes peut et doit se dérouler dans l'univers des conventions culturelles qui règlent l'échange communicatif.[6]

Que ce soit dans sa concetpion initiale ou au cours de sa réalisation, le projet théâtral prend à un moment donné la forme d'un texte lequel, dans sa version imprimée, est susceptible de communiquer un certain message qui reste à définir. Mais pour qu'on puisse vraiment parler d'acte de communication, il faut que le récepteur, en l'occurrence le lecteur, reconnaisse l'existence d'un tel message. Le cas échéant, le phénomène sémiologique se présente selon le schéma suivant:

CODE

EMETTEUR (DRAMATURGE) Message RECEPTEUR (LECTEUR)

Médium (TEXTE)

Le roman et le théâtre, en tant que deux grandes divisions d'un ensemble plus vaste appelé littérature, présentent certaines affinités qui pourraient mettre en doute la question des genres littéraires. En effet, le romancier et le dramaturge se servent dans leurs écrits d'une dose plus ou moins importante de dialogues, recourent aux techniques de la scène et du résumé, créent des personnages fictifs, etc. Tout comme dans le théâtre, la notion de point de vue s'atomise dans le roman moderne, surtout lorsqu'il n'y pas de narrateur unique, omniscient et omniprésent. Le lecteur fait alors face à des énoncés déceptifs qu'il est appelé à démêler par le biais de certains indices fournis soit dans la partie descriptive du roman soit à partir des didascalies que l'auteur dramatique est souvent porté à détailler.

Le texte dramatique se distingue cependant du texte romanesque du fait que dans celui-là, les didascalies ont une topographie différente de la partie dialogique alors que le romancier n'emploie pas de signes distinctifs. Mais la démarcation ne s'arrête pas là. La partie descriptive d'un roman s'adresse au lecteur pour l'aider à situer la fable et à comprendre les rapports entre le personnage, alors que les didascalies se veulent en plus une communication avec d'éventuels metteurs en scène, acteurs, décorateurs, accessoiristes, etc. La multiplicité des destinataires entraîne une certaine complicité entre le dramaturge et ceux qui le servent ou se servent de lui. Cette connivence dictée par les exigences du genre et son mode de consommation rend le théâtre intrinsèquement plus conservateur que le roman.

Et comme l'affirme si bien Michel Corvin, le théâtre s'est transformé, au cours des âges en, "un art hautement socialisé que les traditions et la culture ont doté d'un code trés précis, fruit de conventions minutieusement analysées par les critiques et les dramaturges et instinctivement détectées par les spectateurs les moins avertis." Une telle affirmation appelle une remarque concernant l'importation du théâtre chez les peuples sans tradition théâtrale. La greffe ne prend qu'après une longue et intense activité d'initiation et d'explication. Faute de procéder de la sorte, on débouche sur les constats d'échecs.[7]

En Occident, le nouveau théâtre que d'aucuns qualifient d'avantgardiste, voire de révolutionnaire, donne souvent des signes de vétusté ou traîne à la remorque des autres genres littéraires à tel point que Roland Barthes n'a pas hésité à le considérer comme un sous-produit de la bourgeoisie dont il dépend pour sa survie. Le théâtre, ne peut donc, à moins de signer son arrêt de mort, se libérer des contraintes sociales qui l'asservissent. De temps en temps, il lui arrive de lancer quelques flèches, par ci par là, pour prouver que le système est sain puisqu'il permet la critique, mais ces soubresauts sont trop isolés pour indiquer une tendance bien définie.

> Le théâtre ce n'est jamais gratuit, c'est machiné, prémédité, concerté, c'est un appareil de sédition masqué par les feux des projecteurs et les besoins de l'amusement. Si la représentation d'une pièce a du sens, c'est par la conspiration qu'il y a derrière. (Jacques Ferron, *Les Grands Soleils*.)

Tout en admettant qu'une pièce de théâtre se conçoit habituellement en vue d'être représentée devant un public, sa consommation par voie de lecture n'est pas chose rare. En outre, une bonne mise en scène découle souvent d'une lecture attentive du texte de manière à éviter de la déformer imprudemment par l'interprétation ou de minimiser sa portée. Le cas échéant, la préséance est accordée au texte.

Steen Jansen définit le texte dramatique comme un ensemble structuré des éléments dont dispose et que doit utiliser l'auteur et auxquels le lecteur reconnaît tel ou tel texte comme appartenant au théâtre. Selon Jansen, les éléments de base d'un texte dramatique se répartissent en quatre catégories, à savoir: la réplique, la régie, les personnages et le décor; les deux premières font partie du plan textuel et les deux autres du plan scénique. Il va même jusqu'à postuler qu':

> Un texte dramatique doit contenir au moins une unité de chacune des quatre catégories et [que] toutes les unités du texte doivent pouvoir être placées dans l'une ou l'autre de ces catégories.[8]

L'avantage d'une telle théorie réside dans le fait qu'elle donne droit de cité au plus grand nombre de manifestations modernes de l'écriture dramatique, même les plus insolites. Mais l'inconvénient c'est qu'elle ouvre la porte aux autres formes aussi. En plus, elle ne fait pas de distinction entre éléments implicites et explicites. La réplique, par exemple, détermine, sans qu'il soit toujours nécessaire à l'auteur dramatique de le spécifier, le ton ou le geste qui devrait l'accompagner. De même, une voix peut évoquer la présence d'un personnage caché.

Ces signes pris séparément violent l'unité du texte à moins de les considérer dans le mouvement organique par lequel la situation naît, se développe et s'achève. Nous donnerons à ce mouvement le nom d'action que nous distinguerons de l'intrigue qui consiste en l'enchevêtrement des événements au milieu desquels cette action se déroule. La situation fournit donc l'unité de l'analyse et toute description de valeur procédera en faisant ressortir les éléments diffus dans la pièce.

Toutefois la dramaticité de l'oeuvre, c'est-à-dire l'ensemble des éléments qui la font considérer comme appartenant au genre dramatique, ne se dégagera clairement, à moins d'envisager les personnages qui, par leurs gestes et paroles, matérialisent l'action et formulent les conflits. Ces personnages se grouperaient selon le rapport de sympathie ou d'antipathie qui régit leurs interréactions ou en fonction des moyens dont ils disposent pour atteindre l'objectif poursuivi.

Le théâtre, quand il vise à représenter une suite d'événements où sont engagés des êtres humains agissant et parlant, se distingue des autres genres par sa disposition à diffuser simultanément des "informations d'origine, d'intensité, de durés, de valeurs différentes proposées par décors, accessoires, costumes, bruits, éclairage, gestes, maquillage, paroles".[9] La simultanéité des signes peut envoyer un faisceau convergent d'informations. Quand il y a divergence, il faut chercher dans l'effet produit l'intention ou la signification d'un tel désaccord. Les autres genres, tel que le roman, s'emploient à diffuser des messages qui pourraient être identiques, mais ils se trouvent dans l'impossibilité de les transmettre simultanément, vu le caractére linéaire de l'écriture. Le lecteur d'un roman, pour sa part, n'éprouve pas le besoin d'en visualiser le contenu au même degré que s'il avait affaire à une pièce de théâtre.

Le théâtre est, en outre, le domaine privilégié de l'ambiguïté. Le romancier peut adopter différentes formes de narration; il se prévaut, parfois, d'une omniscience qui lui permet de se substituer à tous les personnages de son récit; il délègue, par moment, ce pouvoir au héros ou à un autre personnage qui éclairerait par son savoir les recoins cachés. Au théâtre il y a au moins une personne qui parle et l'auteur est en principe absent.[10] Le dire et, encore plus, le jugement de l'auteur ou de l'actant, selon le cas, recouvre la notion de point de vue ou système de cognition. S'il y a un narrateur unique qui ne soit ni actant, ni acteur du récit et qui refuse la délégation du savoir, il donnera à la narration une homogénéité totale et limpide. La confusion apparaît lorsque l'énonciateur disparaît ou que le récit est assuré par les acteurs d'une manière partielle ou totale.

La carence des précisions en provenance des indications scéniques rend difficile la détermination de la valeur de vérité d'un énoncé. Même si l'opposition de deux connaissants laisse planer un doute, le spectateur ou éventuellement le lecteur peut se retourner à d'autres sources d'information pour élucider la situation. Le décor, par exemple, est rarement gratuit. Il renvoie à l'origine ou à la causalité et sert à situer la problématique posée dans un cadre plus général. Les costumes et les accessoires peuvent faire pencher le jugement d'un côté plutôt que d'un autre. En dépit de l'ambiguïté inhérente et parfois soutenue par le dramaturge, on décèle dans le théâtre moderne une volonté de signifier clairement, "de construire un signifiant qui porte son signifié au devant du spectateur".[11]

L'opposition déterminé-indéterminé caractérisant le rapport entre le macrocosme théâtral et le microcosme scénique crée la tension dramatique nécessaire que vient parfois accentuer l'opposition entre lieux ouverts et fermés. Selon Etienne Souriau, la théâtralité réside dans le jeu de forces entretenu entre ces deux univers:

> Tout le caractère dramatique de ce qui passe en scène, tient à son rapport avec ce qui reste hors scène ... Si menu, si étroit, si serré sur soi que soit le monde qu'on nous présente, sans ce débordement du microcosme scénique par l'univers de l'oeuvre, il n'y a pas de théâtre.[12]

La structure des principaux thèmes s'établit à partir des relations hyper ou hyponymiques entretenues entre les éléments constitutifs; de la récurrence et de l'importance de chaque thème pris séparément. Au point de vue du langage, deux tendances se partagent le théâtre d'avantgarde qui groupe des noms aussi disparates que Ionesco, Beckett, Adamov, Audiberti et Schehadé. L'une vise à une dépréciation radicale du langage, l'autre à débarrasser ce moyen d'expression et de communication des tares qui l'infestent et qui gênent son plein épanouissement.

Comme le problème du langage se situe au centre même du débat et que les opinions véhiculées à ce sujet par les répliques lui accordent un rôle primordial par rapport aux divers aspects de la vie, il est loisible de considérer le langage comme l'axe principal autour duquel s'articulent tous les autres thèmes. L'oeuvre théâtrale demeure foncièrement une exploration de l'espace particulier du langage. La mise à jour du réseau d'obsession de tel ou tel auteur se fera à partir du repérage, dans le texte, des sujets ou des préoccupations qui se répètent sous sa plume. On comprendra alors mieux le règne de la spontanéité qui prévaut dans l'oeuvre en le rattachant à la primauté du signifiant qui explique aussi la prépondérance de la libre manifestation des idées sur toutes les formes de la pensée discursive. L'aspiration à la liberté et le besoin d'élargir le champ d'action des individus font écho au combat mené contre les formes sclérosées du langage.

En effet, la perspective théâtrale se traduit par une nette condamnation des formes sclérosées, clichés et lieux communs et par un glissement continu du rapport signifiant-signifié, tout en donnant l'initiative au signifiant, dans un souci d'affranchir la parole, "de rendre le verbe humain à son innocence et à sa vertu créatrice originelle".[13]

Une telle affirmation peut paraître tant soit peu gratuite à moins de l'étayer par de nombreux arguments tirés du vaste répertoire moderne. Là encore, les synthèses seront sujettes à caution à cause du parti pris de la démonstration laquelle est longue et difficile à établir. Une solution de rechange, plus courte et plus pratique, serait de s'inspirer de la théorie des catastrophes de René Thom pour étudier un cas limite, en l'occurrence une pantomime de Georges Schehadé, publiée en 1973. En extrapolant les conclusions, nous pouvons jeter un nouvel éclairage sur l'ensemble de la production théâtrale des dernières décennies.

*L'Habit fait le prince* est une pantomime où la réplique s'efface en faveur des gestes, du décor, des costumes, de l'éclairage, bref de tous les signes extralinguistiques. A la lecture cependant et encore plus à la représentation, on ne tarders pas à constater que cette dernière création de Schehadé ne cède en rien aux autres pièces sur le plan du langage et s'insère bien dans l'ensemble de sa dramaturgie.

Feignant respecter les lois du genre dans son dernier ouvrage qu'il se complaît à désigner sous l'appellation de "pantomime", Schehadé risque d'induire en erreur les critiques les plus chevronnés. En effet, à plusieurs reprises l'auteur intervient pour imposer à ses protagonistes le silence le plus complet et pour rappeler que dans une pantomime aucun mot n'est permis. Si l'un des personnages éternue, aucun bruit ne devra accompagner son geste: "Comme il se doit dans une pantomime" (*L'Habit fait le prince*, p. 29). De même, "faire éclater une bourse", "frapper des deux mains", "jouer de la musique", "sonner l'heure", "applaudir", "s'interpeler", le tout doit se faire dans le mutisme complet, et l'auteur de rappeler entre parenthèses: "ne l'oublions pas: c'est une pantomine" (*Ibid.*, p. 79).

Ironiquement, l'exagération dans les allusions aux contraintes du genre rend évident le doute qui plane autour de leur utilité et écarte la possibilité de les prendre au sens littéral. Une seule remarque à cet effet aurait été amplement suffisante pour indiquer la nécessité du respect des règles inhérentes au genre, tandis que la redondance dans les rappels ne peut avoir pour objet que de railler les exigences et d'en montrer l'absurdité, d'autant plus que Schehadé refuse par deux fois de s'y soumettre. Ces deux écarts significatifs méritent qu'on s'y attarde un peu.

La première dérogation se produit vers la fin du cinquième tableau lorsque l'un des joueurs impliqué dans la partie de dés, irrité par la perte qu'il vient de subir, oublie le contexte dans lequel s'inscrit le rôle qu'il joue et:

> déverse sur la table un flot d'argent qui tombe dans un grand bruit de cascade. Surgit alors sur la scène un machiniste, en bleu de chauffe, qui dit à voix haute: "Arrêtez, ça ne va pas . . . Dans la pantomime on ne fait pas de bruit. . . ." Les joueurs, figés par cette intervention, ne bronchent plus. Le machiniste ajoute à l'adresse du perdant: "Recommencez, s'il vous plaît!" Le joueur ramasse les écus bruyants, les remet dans la bourse; puis il déverse sur le tapis un sac d'écus silencieux. "C'est très bien, on continue", dit le machiniste qui, en se retirant, ajoute à l'adresse du public: "Excusez-nous" (*Ibid.*, p. 32).

En s'excusant, le machiniste s'avoue coupable et pour absoudre le personnage qui a commis une faute, il en commet une beaucoup plus grave. Son intervention à haute voix ne peut passer inaperçue, au même titre que l'inadvertance du joueur. Cette coupure brise irrémédiablement le processus d'identification et remet en cause la nature contractuelle du genre.

Rusant avec les exigences de la pantomime, Schehadé tente d'assurer une place à la parole dans une forme qui, en principe, la bannit. Pour cela il multiplie l'emploi des verbes d'énonciation, tels que: parler, dire, redire, avoir l'air de dire, sembler dire, devoir dire, chuchoter, dire un mot à l'oreille, lire, marmoner, souvent suivis de deux points et d'une réplique. Bien plus, il ne dédaigne pas d'avoir recours à la graphie, au besoin, d'où le nombre considérable d'écriteaux qui servent à concrétiser certains faits ou à critiquer certaines professions.[14]

Plus loin, une tragédie classique, en deux actes et avec réplique, est introduite. Sur un ton de parodie, cette piécette représente en creux la situation de l'intrigue dans laquelle elle est greffée. Pour la faire jouer l'auteur a recours aux quatre joueurs de dés. Ces liens font de cette "tragédie classique" une partie intégrante de la pantomime, mettant ainsi en évidence l'alternance des deux formes du jeu scénique.[15] Schehadé offre de la sorte l'envers d'un procédé dont il s'est amplement servi auparavant.

La clé de voûte de *L'Habit fait le prince* réside dans une articulation classématique du lexème "dé". Pris dans son acception d'étui de métal pour protéger le doigt poussant l'aiguille, il renvoie, par une relation métonymique, au métier de tailleur. Affublé d'un habit de prince, Armen laisse tomber un dé que Melchior ramasse discrètement et s'en sert pour dévoiler l'identité du prétendu prince. En le lui rendant, il lui signifie avoir découvert l'imposture.

Mais le dé, c'est aussi un petit cube à faces marquées de points de 1 à 6 ou de figures, et utilisé dans différents jeux où le hasard et la chance jouent un rôle prépondérant. Concrétisée par la présence des joueurs de dés et les parties qu'ils engagent, cette deuxième acception du mot "dé" s'imbrique sur la première puisque la jonction entre l'ouvrier-tailleur et les quatre joueurs rend compte de la quasi totalité des péripéties de l'intrigue.

A la fin de la "tragédie classique", les conjurés incarnés par les joueurs de dés enfoncent un dé énorme sur la tête du roi. Exploitée à fond, l'économie du langage met cette pantomime sous la dépendance d'une considération linguistique, alors que cette forme théâtrale, en particulier, se veut, en général, indépendante du langage. Dans les autres pièces, Schehadé a souvent recours à la polysémie en vue de rapprocher deux notions apparemment fort éloignées l'une de l'autre. Les lois ecclésiastiques deviennent, par le biais du mot "canon", aussi redoutables que les pièces d'artillerie: "Pauvre femme! Elle ignore que les canons de l'Eglise sont de véritables canons. (D'une voix forte:) Ils tirent!" (*Les Violettes*, p. 74.) Le recours de Monsieur Bob'le au verbe "posez" pour passer un ordre à son valet au sujet des bottes et pour solliciter la lumière de la Vierge Marie indique que l'usage fait d'une expression ne l'ennoblit ni ne la souille pour autant, pourvu qu'on s'en tienne au mot exact. "Cette possibilité de signifier, de faire signifier à deux niveaux, l'un banal, l'autre plus métaphysique, est (aussi) une caractéristique du langage beckettien et c'est ce qu'un philosophe du langage (Stanley Cavell) a appelé "hidden literality", une littéralité cachée".[16]

Ailleurs, il se crée par le biais du langage, une ambiguïté conforme à la nature du théâtre. Investies d'équivoques, les pièces de Schehadé engagent l'intelligence du spectateur, surtout s'il se livre à les dissiper. Qu'entend Alexis par "chasser la nuit"! Est-ce mettre la nuit en fuite ou poursuivre le gibier pendant la durée comprise entre le coucher et le lever du soleil? Quelques-unes des maximes du "Trémandour" frisent la franche drôlerie par la juxtaposition de termes incompatibles ou contradictoires: "Celui qui porte un chapeau doit être plus juste que les autres, car, *a priori*, il pèche contre la lune et le soleil . . ." (*Monsieur Bob'le*, p. 29). La première partie de ce précepte contredit la deuxième, d'autant plus que Bob'le s'est levé et a mis son haut de forme avant de l'énoncer.

*Collège Militarie Royal, St. Jean, Québec*

NOTES

[1] *Cf.* Ferdinand de Saussure, *Cours de linguistique générale*, notamment le chapitre III de l'Introduction.

[2] *Cf.* A. J. Greimas, *Du Sens* (Paris: Seuil), 1970.

[3] *Problèmes de sémiologie théâtrales* (Mtl: P. U. Q.), 1975, p. 4.

[4] La sémiologie *Encyclopédie La Pléiade*, volume *Le langage* (Paris: Gallimard), 1966, p. 95.

[5] *La communication et l'articulation linguistique*, Bruxelles, Université de Bruxelles, 1970. Cette définition s'assimile à celle adoptée pour le signal. Sans vouloir réduire la sémiologie à l'étude des signaux, nous insistons sur l'interchangeabilité entre indice et signal.

[6] *La structure absente* (Paris: Mercure de France), 1972, p. 306.

[7] 'Approche sémiologique d'un texte dramatique,' *Littérature* 9, (février 1973), 86–100.

[8] 'Esquisse d'une théorie de la forme dramatique,' *Langage* 12 (déc. 1968), p. 75.

[9] Cf. Michel Corvin, 'Approche sémiologique du texte dramatique', *Littérature* 9 (fév. 1973), 86–100.

[10] Dans certains cas, l'auteur dramatique a recours à un récitant qui commente l'action. Mais ce récitant peut s'assimiler à un personnage.

[11] Richard Demarcy, 'Le commerce de pain', *Travail théâtral* 4 (été 1971), 107–110.

[12] *Les Deux cent mille situations dramatiques*, p. 27.

[13] André Breton, *Entretiens*, p. 161.

[14] Déjà dans *Le Voyage*, l'écriteau de M. Strawberry joue un rôle important dans la pièce et peut être considéré comme prélude à ceux du vieux tailleur et de la rue marchande dans *L'Habit fait le prince*.

[15] A la tombée du rideau, Armen disparaît et l'auteur en profite pour rappeler qu'on est de nouveau dans la pantomime et juge nécessaire de réitérer: "l'applaudissement doit se faire sans bruit".

[16] Sina Sherzer, 'Didi, Gogo . . .', *Etudes littéraires*, (déc. 80), p. 541.

JOSEPH MARGOLIS

# THE PROBLEM OF READING, PHENOMENOLOGICALLY OR OTHERWISE

In speaking about Eugene Kaelin's speaking about Samuel Beckett's speaking about and creating a world or worlds of human self-examination,[1] we are too easily distracted from the difficulty of the undertaking and too easily persuaded of its presumption. Beckett's fictions are peculiarly definite but indeterminately located, conveyed by a cadenced language that remembers older formalities in a contemporary voice, and focused, one cannot doubt, on the drama of human resolution and irresolution and doubt and confidence and reflexive questioning and answering and awareness of same. It is easy to say that Beckett is occupied in this sense in a notably pure and persistent way with the human condition – not necessarily with its essential condition if the existential theme cannot be thus reduced, but at least with indefinitely many strands of puzzling about purposing without puzzling about purposes. In this regard, Beckett baffles by being banal – banal, that is, not in the second-order sense. So saying, we are ourselves launched almost without deliberate effort into extending the critic's labors. If it is true that Beckett's *oeuvre* is thus absorbed, then it is reasonable to suppose that all competent speakers of the language are already suitably attuned to the family of self-investigative efforts that his central characters betray and display. We need no special orientation, no clues about *genres*, idiom, learned allusions, philosophical systems, historical context, ideological allegiences, constraints on paraphrase, symbolic conventions, and the rest of the usual literary apparatus. We are not the characters of Beckett's stories and plays. But we are like them and they, like us, in generating resemblant histories in the middle of life.

We could perhaps say, with the allegedly ventriloquist narrator of *Mercier and Camier*, "The journey of Mercier and Camier is one I can tell, if I will, for I was with them all the time."[2] It is the first sentence of the story and never fails to arrest the reader's attention. Kaelin says that the "attempts of Mercier and Camier to undertake a voyage together (symbolizing a life's project) . . . [are] recounted by an omniscient, third-person narrator, who claims to have accompained them on the trip. Since [he adds] there is no justification for this claim, other than the opening statement of the narrator, one can be sure that Beckett has fallen back into the errors of his former ways, playing the ventriloquist, but allowing his lips to be seen as he breathes



*A-T. Tymieniecka (ed.), Analecta Husserliana, Vol. XVIII*, 559–568.

the words for his puppets."[3] He goes on, after a short resumé of the story, to remark:

> But if God is dead in the universe of Mercier and Camier, the narrator is not. He makes his presence felt all along the road. And that is a flaw the fledgling novelist will have to learn to control. That Beckett will eventually learn how to control his propensity to intrude upon the universe of his novels becomes clear in the collected *Stories* [that is, the 1945 *Stories*, that include *The Expelled, The Calmative*, and *The End*].[4]

Now, there are a great many difficulties here. First of all, if the narrating voice says, "I was with them all the time," then that itself is a datum favoring his having been present all the time, though it does not explain the sense in which he was present or confirm that he was indeed present all the time. Secondly, Kaelin himself remarks that the narrator "makes his presence felt all along the road"; but this means that he cannot be right, on his own say-so, that there is no justification beyond the first sentence for the narrator's claim. There are the alleged intrusions "all along the road" that confirm, in some sense, that the narrator was indeed "present." Thirdly, the narrator is not a third-person invention: he speaks in the first person; though he never speaks to Mercier and Camier; in fact, he tells the story of Mercier and Camier, including a great many details that they certainly do not tell, perhaps could not tell, surely would not be disposed to tell. "Physically, it was fairly easy going [he begins], without seas or frontiers to be crossed, through regions untormented on the whole, if desolate in parts." But Mercier and Camier themselves "did not remove from home, they had that great good fortune."[5] It is entirely plausible – but certainly not certain – that they would never have considered the narrator's own reflection. And, following the amusing failure of the two voyagers to meet at the appointed place to begin their venture, we are presented with a schedule of their arrivals and departures that looks for all the world like a train schedule and that fixes nine fifty in the morning as the moment of the accidental conjunction of Mercier and Camier. "What stink of artifice,"[6] the next line reads, which, like the schedule itself, is provided by the knowledgeable narrator – or so we suppose. Is he commenting on the artifice of the author or the mechanical features of the story or the prearranged harmony of the movements of Mercier and Camier or his own disposition to collect events as he does or his telling the story of our voyagers in the way he does having been "with them all the time"? Or are the schedule and the comment the work of a higher-order narrator?

But that's not all there is to it. For, fourthly, Kaelin takes the narrator to be omniscient. And he is clearly not, in a certain sense. For instance, the narrator speculates about what may have been in the minds of our voyagers in having selected Saint Ruth Square for their first meeting: "Mercier and Camier did not know the place. Hence no doubt their choice of it for their meeting. Certain things shall never be known for sure."[7] Why "no doubt" unless the matter was uncertain to the narrator himself? At the very least, he could not be Beckett, *if* Beckett could be said to know what he himself had produced in writing *his* story. In the same sense, the narrator muses about two dogs beaten and driven off still coitally engaged: "They must have greatly suffered," he says, an inference confirming the limitation of his own perspective. It is as if, as he says, he was with Mercier and Camier all the time; but there is a sense in which he is telling *their* story, not telling a story. Fifthly, Kaelin himself is hardly prepared to concede that an author – not merely a narrating voice – need know (or does know) the significance of all that he has produced:

> A literary work of art [he says] is an intentional object; its universe is significant only by means of its internal relations. The fictional universe was, of course, created by the writer; he may or may not be aware of its significance; but if he is, his attitude of appreciation . . . is the same as any other attentive reader's; he too must read the words to discover the structures of the universe he has created. In this sense, an author may very well have created by discovering what he meant in the first instance to say; [for instance] Molloy, the narrator, needs both Molloy the reader and Moran, his surrogate. By the dialectic of phenomenological experience the self is discovered in the other.[8]

Perhaps. But now at least no one is omniscient, not the author, not the narrating voice. But then, sixthly, it could hardly be the case, as Kaelin claims, that the author, Samuel Beckett, omnisciently intrudes into his own fictional universe – awkwardly, amateurishly, without adequate art, too obviously disruptively, by error and failure. For, for one thing, Beckett cannot know what he has wrought but must discover it like ourselves. And for a second, the fictional world is intact, complete somehow in its own system of "internal relations." Does this mean that Beckett failed to produce a fictional world or merely produced one that could have been improved? If he did intrude, then we have more than the intentional world of the fiction; and if we have that world, then he did not intrude in it.[9] There is, then, seventhly, a discrepancy between the practice of Kaelin's phenomenological critic reviewing a fiction and the precise kind of critical work required by Kaelin himself – on the assumption of Beckett's intrusion. For if a story is the *work* of an author – any story – an intentional world intended (not merely psychologically), not merely detached from its generative human source, "there" so to say for anyone's taking, intact, independent, the babble of a world anchored in the words once wrought but now no longer requiring reference to its original artificer, then there can be no viable critical procedure that refuses or ignores that originating linkage.

Here is the essential difficulty of Kaelin's conception of reading. Everything sounds scrupulous enough when we are told:

> For us, critics become self-conscious in our turn, there can only be one solution [in reading such an author as Beckett] : total respect for the universe created by the author's words, begining with their surface, i.e. their poetical chant, and proceeding along to whatever depth significance they may convey, and in as many layers as are open to clear exposition.[10]

The method Kaelin terms "structural phenomenology" or "phenomenological structuralism." It consists, he says,

> in "bracketing" out of relevance all reference to any theories, any previously established knowledge of the real world. What remains before the critical consciousness is the universe of the novel created as a unique frame of reference. . . . It implies a suspension of our beliefs in the existence of a natural world – the one we all live in, as described by the rational procedures of positive scientists and nomological theorists – in favor of an attitude of serious contemplation of the appearing phenomena, which then may be described as they appear to our attentive consciousness. The individual work of art is thus a structure of noematic elements constituted by the noetic acts of the reader.[11]

But surely, Kaelin himself repudiates – and must repudiate – his own thory by his own practice. Notice, for instance, that he claims Becket has intruded into his own fictional world: *how* could this be phenomenologically confirmed? If the fictional world is a unique intentional framework, then we cannot be sure what its internal rules must be in advance of exploring it; we cannot be sure that all fictional worlds abide by the same rules; and we cannot restrict their possibilities by reference to the constraints of the actual world.[12] *If* Beckett created the world of Mercier and Camier, then, for good or for bad, the voice of the narrator is as much a part of it as the voices of the voyagers themselves – perhaps even more so, for the story is as much the account of the narrator's telling the story as it is the account of the affairs of Mercier and Camier. Notice, also, that in summarizing the story, Kaelin remarks that the attempts of Mercier and Camier to undertake a voyage "symbolize . . . a life's project." How does he know that? One suspects that he theorizes – not at all implausibly – that the accumulating movement of Beckett's *previous* work and the apparent *continuity of the present work with that* justifies the ascription. But that means that the critic's reading *cannot* be detached from his interpretation of the *linkage* between the fictional world intended and the actual generating effort of the originating author. Why else say that Beckett failed in *Mercier and Camier* to keep from intruding into the universe of the novel but "eventually" succeeded in the *Stories*?

This, then, suggests an eighth difficulty. How does Kaelin know *what* "the individual work of art" is, what its boundaries are, where it begins and ends? And how could phenomenological bracketing possibly tell us? *Mercier and Camier* was translated by Beckett himself, but are the French and English two different stories, two versions of the same story, a translation of an original story and that original story, the bilingual parts of some unusual dual narrative, or analogous variations of the subparts of some larger and more continuous writing of which part or all of the other "works" are part? Hugh Kenner characterizes *Mercier et Camier* as "jettisoned," Beckett's "farewell to the bourgeois novel."[13] Right or wrong, the remark obliges us to link that novel with *Watt* and with Beckett's actual purpose; and besides, Kenner wrote this before Beckett published both the French and the English (1970, 1974). But also, Kenner notes the problem of translation. For instance, in the French and English of *Endgame* (*Fin de Partie*), the following symptomatic change occurs:

NAGG: (geignard): Qu'est-ce que c'est?
CLOV: C'est le biscuit classique.
NAGG: (plaintively): What is it?
CLOV: Spratt's medium.

"A desperate equivalent for the aloof precision of 'le biscuit classique,' " says Kenner.[14] Is the English, then, a translation of the French; or is it the dual of the French, within the limits and possibilities of a distinct language? "And for those of us," say Kaelin, "who have not understood this literary and dramatic achievement [that is, the later experimental plays without words or that use language in a peculiarly reduced way], or who have understood it badly, the author has given us a second chance to grasp his display of the phenomenology of the human spirit in his various translations. If the English is not clear, try the French; and for those written in French, reverse the procedure."[15] But this is either too sanguine or simply premature; and in any case, its confirmation cannot but presuppose a theoretically fashioned decision about the proper individuation of the "individual work of art," that is, about a *precondition* for the self-styled phenomenological reading that, on that assumption, is no longer purely phenomenological.

Perhaps, in Foucault's sense, "the author does not precede the works, he is a certain functional principle by which, in our culture, one limits, excludes, and chooses; in short, by which one impedes the free circulation, the free manipulation, the free composition, decomposition, and recomposition of fiction."[16] There certainly is something in this; once posited, it needs to be debated. And it is, in fact, very much the decisive clue in Kaelin's own endeavor: "We have since the beginning of this book [says Kaelin] been following Beckett's personal voyage as a search for self-consciousness in two parallel directions."[17] Was that necessary or inevitable? And does not the phenomenological pursuit of Beckett preclude the phenomenological bracketing of his work? For the first requires a linkage between the fiction and its originating author, some sense of the historically contexted endeavor of Beckett himself. But if that is true, then what the bracketing of our biographical effort could be is no longer clear.

Kaelin's most explicit statement of his theory of reading is perhaps the following:

> The signification of any sign is its reference to an object or idea outside itself, and there are as many levels of these as there are signs to be interpreted. For the whole work, there is no signification; for, the total system of signs and signified, being a closed, saturated system of linguistic relations, does not refer to anything external to itself. This is why the key to the treasure is the treasure itself. The saturation of that system is, however, felt; it is the quality of that system's wholeness, the significance that pervades the whole. . . . The hermeneutical, phenomenological critic maintains the strictness of the work's intent – that which had been isolated within consciousness in the first place by his voluntary decision to allow the work to speak for itself – by restricting his attention to the internal play of the work's linguistic structures. What is expressed is "there," in the texture of the linguistic relations; and what is there must come to expression in the critic's words, which describe the images and all the relations a thoughtful interpretation may find obtaining between them.[18]

But what are we to make of the notion of "the whole work," that has significance but not signification? And how do we know that, in the diachronic development of a crticial reading, we have achieved closure or only the reflection of our own contingent and contexted perspective? And is it the case that every literary work properly so called genuinely invites closure? Perhaps the individuation of works and the assignment of their matching authors are themselves methodically controlled by a larger theory of authors and texts that orients us in this way. The doctrine is at least mildly Kantian. But if it is a reasonable complaint, then it must be the case that the phenomenological reduction Kaelin means to practice is itself implicitly controlled by a special theory of the nature of, and relationship between, actual literary texts and actual authors. The intentional world of a story is a world subtended by a network of utterances. But what the right way is to individuate such a network, what the functional relationship of an author to such a network must be, and what the interior structure of the intentional world that network discloses is, must be open to alternative strategies that only a sustained review of human histories can help us decide – such review itself being only one among alternative ways of organizing history. So seen, structuralist phenomenology is nothing but the intention of a reader to be scrupulous, without conceding explicitly the contingently contexted and inescapably historicized understanding with which he must work – both regarding what he searches for and what he is capable of finding. In fact, it says nothing about how he means to proceed – or why.

Closure is the limiting objective of literary understanding; and works and authors and meanings are themselves in some measure defined or posited by our commitment to closure. It is not the antecedently structuring unity of an intentional world subtended that we gradually and gropingly penetrate and *reclaim*. There *is* no "treasure itself," but reading posits it – responsibly, to be sure, by reflecting on the historical reality that is the text. This is perhaps the point of Roland Barthes' well-known remarks – well-known post-structuralist remarks:

> The Text must not be thought of as a defined object . . . the Text . . . is a methodological field . . . the Text is experienced only in an activity, a production . . . the Text . . . like language . . . is structured but decentered, without closure . . . Every text, being itself the intertext of another text, belongs to the intertextual, which must not be confused with a text's origins.[19]

To admit intertextuality is to admit, at one and the same time, the contingency of historical interpretation and the impossibility of liberating *any* scrupulous reading of a text from that text's own presumed actual history. There are no bracketed appearances of a fictive world to which we can turn: there are only situated readers and situated texts, and a compelling interpretive hit is merely honored in being thought essential or timeless. The peculiarity of Beckett clearly lies with the extraordinarily spare (and at the same time florid) intensity with which he displays the vagaries of the human mind wondering and wandering over its purposing (regardless of purpose) and its wondering about such reflection. In an elementary sense, existential phenomenology *is* the philosophical enterprise most congruent with this *oeuvre*; and in a certain sense, therefore, Beckett *is* the phenomenological author par excellence. But to say this is not quite to agree with Kaelin:

Samuel Beckett is one of the most successful philosophical literary artists of our times. He had made his point by bringing his various fictional worlds to closure on the basis of a philosophical idea. In no *oeuvre* known to this writer has there been such a persistent attempt to come to grips with the understanding of personal existence, no work in which it has been shown how consciousness becomes fully aware of itself. That indeed was the project of the phenomenology of mind, which must itself define the meaning of "spirit," human if not absolute.[20]

What is the difference, for instance, between Beckett's having been philosophically informed in producing his work and his having actually worked as a "phenomenological structuralist"?[21] And why should we suppose that the "dramaturgical" (the second-order, philosophical) reflections conveyed, say, in *Cascando* provide an adequate clue either to the interpretation of that play or to the whole of the subsequent *oeuvre*?[22]

There is a curious but telling clue to the plausibility of Kaelin's *non sequitur*. It lies in his having assimilating Beckett to Heidegger. The plausibility rests on what, quite reasonably, we take to be the related projects of Beckett's entire output. It is precisely the peculiarly existentialized reflection of the stories and plays that at one and the same time invites the application of phenomenological theory to Beckett's purpose and work, the presumption of the adequacy of so-called phenomenological criticism in the reading of his various pieces, *and* the appeal to the putatively bracketed responsiveness of every sensitive reader to the pure texts themselves. The *non sequitur* rests on concluding, from the congruence of the phenomenological schema of interpretation – necessarily, unfalsifiably, congruent – with the structure of *any* communicative pattern actually represented, that that pattern must somehow *intend* (not psychologically, perhaps) that very schema. So for example Kaelin observes that, in *How It Is*, there are variously permuted exchanges involving Kram, Bom, Pim, and Krim. In all, "there is [he says] a triple layer of hermeneutical expansion." The represented exchanges are circular, but they convey a deeper circularity, "in that no one can make an interpretation of anything, be it an experience or its written record, without having been acquainted with the experience or the record. Implicit, then, is the knowledge of the context to be interpreted."[23] Apparently, this provides sufficient grounds on which to invoke Heidegger's doctrine of *Vor-Struktur*. In Heidegger's own terms, meaning concerns the disclosure of *Dasein's* Being to itself: "*Meaning is the 'upon-which' of a projection in terms of which something becomes intelligible as something; it gets its structure from a fore-having, a fore-sight, and a fore-conception*." The process involves a circle indeed, but it is not a vicious circle since understanding assimilates, historically so to say, "the existential *fore-structure* of Dasein itself."[24] In Kaelin's terms, the actual exchanges of the characters of *How It Is* rather blandly (but suggestively) signify that "no one can make an interpretation of anything, be it an experience or its written record, without having been acquainted with the experience or the record"; that is, acquainted *at least* in that way in which a "fore-having" (a *Vorhabe*) obtains. But in the phenomenological sense, precisely, the peculiarly reduced and focused materials of Beckett's literary explorations are incontestably accessible at the very point of the *Vorhabe*. How could it be otherwise? And therefore, by the *non sequitur*, a critical reading of Beckett *can* proceed phenomenologically.

But that is still to confuse critical procedure and the plausible findings of *some* defensible procedure once applied. To the extent that Heidegger's account of *Dasein* has validity, that is, to the extent that it illuminates the existential features of human existence in some nonexclusively viable way, criticism may bring it to bear instructively on Beckett himself. But the validity of bringing it to bear in the way one does does not follow from the congruence of Beckett's work and the structure of *Dasein*, but from a theory of the relationship among texts and authors and interpretive communities and a contextually sensitized but corrigible proposal regarding the meaning of particular texts. The hypothesis involved is *not* Kaelin's – linking the anticipated existential sense of confronting a text and "its principle of closure."[25] It is rather a hypothesis about the very nature of reading – on the assumption (perhaps) that the validity of Heidegger's account of *Dasein* will (must) bear indiscriminately on the reading of anything. The trouble with the phenomenological theory is simply that it is too strong. In some way, granting its own claim, no human way of proceeding can fail to exhibit *Vor-Struktur*. But how it is to illuminate Beckett or Kafka or Proust or Eliot or Joyce or Flaubert or Balzac is a matter that cannot be formally deduced or somehow derived from the fore-structure of Being itself. To succeed in interpretation entails *Vor-Struktur*; fore-structure need not be differentially specified *in* an interpretation in order to insure the validity of the latter. In a certain sense, it cannot be. But that means that phenomenology may afford a theory of *why* interpretation succeeds, not of *how* it succeeds in the actual play of particular critical arguments. We are, therefore, left with the problem of the latter, in the very midst of our enlightenment. There may be specifically phenomenological forms of criticism: certainly, Heidegger and Merleau-Ponty thought so. But what their advantages and particular power may be can hardly be said to belong to the essential insight in virtue of which such criticism is first gathered together with that of *any* opposing practice. This is the sense as well – that most disarming sense – in which no competent human reader can possibly go wholly wrong in his reading. Still, to concede that much is to say absolutely nothing about the proprieties of discipline any critic's scrupulosity concedes in pursuing his favored disclosures of meaning. And that is what we need to know about, in gauging the special discipline of careful readers.

*Temple University*

## NOTES

1 Eugene F. Kaelin, *The Unhappy Consciousness: The Poetic Plight of Samuel Beckett*, Analecta Husserliana, vol. 13 (Dordrecht: D. Reidel, 1981).
2 Samuel Beckett, *Mercier and Camier*, trans. by the author (New York: Grove Press, 1974), p. 7.
3 Kaelin, p. 69.
4 Ibid., p. 78; cf. p. 295.
5 *Mercier and Camier*, p. 7.
6 Ibid., p. 10.
7 Ibid., 15.
8 Kaelin, p. 8.
9 Cf. the relevant and extremely perceptive observations about authors and narrators that Gérard Genette offers, *Narrative Discourse*, trans. Jane E. Lewin (Ithaca: Cornell University Press, 1980).
10 Kaelin, p. 7.
11 Ibid., p. 7.
12 I have explored the question in 'The Logic and Structures of Fictional Narrative,' presented at a conference on style in fiction, at Harvard University, Spring 1982.
13 Hugh Kenner, *Samuel Beckett* (Berkeley: University of California Press, 1961), p. 70.
14 Ibid., p. 95.
15 Kaelin, p. 300; but cf. also, p. 276.
16 Michel Foucault, 'What Is an Author?' in Josué V. Harari (ed.), *Textual Strategies* (Ithaca: Cornell University Press, 1979), p. 159.
17 Kaelin, p. 268.
18 Ibid., pp. 289–90.
19 Roland Barthes, 'From Work to Text,' in Harari, pp. 74–77.
20 Kaelin, p. 300.
21 Cf. ibid., p. 17.
22 Cf. ibid., pp. 173–78.
23 Ibid., p. 136.
24 Martin Heidegger, *Being and Time*, trans. John Macquarrie and Edward Robinson (New York: Harper and Row, 1962), pp. 193–95.
25 Kaelin, p. 136.

# INDEX OF NAMES


Adams, H. 508, 535
Adorno, T. W. 253
Albright, W. F. 185
Alvarez, A. 308
Arendt, H. 120
Aristotle 19, 164–66, 227, 242, 277, 308, 313, 325, 341–49, 352, 354, 356, 360, 393, 396, 432, 490
Arnold, M. 120–23, 125–27, 129–39
Auden, W. H. 144, 146, 542
Auerbach, E. 167, 176, 183, 185, 187–88, 192, 228, 397, 512–13, 518

Bachelard, G. 170
Barthes, R. 48, 59, 265, 269, 281, 292, 516, 565
Baudelaire, C. 104
Beckett, S. 356–59, 540, 559–68
Bender, H. 144, 147
Bergson, H. 262, 536, 541–43
Bettelheim, B. 62, 73
Birault, H. 460
Black, M. 243, 245, 247
Bloom, H. 85, 88
Bly, R. 152
Böll, H. 249–51, 258–59
Bonnefoy, Y. 145
Brinnin, J. 146
Brooke-Rose, C. 244, 248
Broughton, J. 145
Bruzina, R. 479
Büchner, G. 328–30
Burke, K. 179

Camus, A. 104
Casella, A. 295–96
Cassirer, E. 38
Celan, P. 147
Cézanne, P. 96, 262, 521
Chesnut, M. 387
Claudel, P. 273–93
Cogswell, F. 144
Coleridge, S. T. 506, 511
Conrad, J. 234–35, 237–38
Corneille, P. 409–15, 435–47
Creeley, R. 150
Croce, B. 42–44
Curtius, E. R. 163

Dante, 510–16
Davie, D. 55, 59
Deleuze, G. 531
Derrida, J. 49, 59, 121, 230, 242, 247
Dickey, J. 144, 151
Domin, H. 146–47
Dufrenne, B. 11
Duncan, R. 149

Eichendorff, J. v. 26–29, 31
Eliade, M. 215, 223–24
Eliot, T. S. 146

Farrell, J. T. 144, 154
Fergusson, F. 179, 183
Flaubert, G. 527–28
Foucault, M. 417–18, 423
Frye, N. 239, 507

Gadamer, H.-G. 167, 239
Garelli, J. 49–51, 53–59
Geiger, M. xii, 5–7, 9–12, 14–15, 20–21
Genette, G. 525–26, 529, 533
Gide, A. 527–28
Ginsberg, A. 150
Goethe, J. W. v. 133–35, 148–49, 320, 326–27, 332
Golden, L. 341–42
Gombrich, E. H. 190
Gomringer, E. 141

Grass, G. 252–54, 258–59
Gusdorf, G. 40, 44

Hall, D. 150
Hampton, C. 145
Hanke, P. 246–47, 254–58, 260
Hardy, T. 307–18
Harrison, J. 145
Hartmann, N. 11, 320–21, 324
Harvey, L. 414
Hauser, A. 261, 270
Hawkes, T. 244–45, 248
Hebbel, F. 328–29
Hegel, G. W. F. 253, 319–20, 323, 328, 354, 396
Heidegger, M. 4, 39, 41–42, 48, 85, 87–88, 99–100, 107, 113, 119, 139, 231, 278, 304, 319, 397, 459–60, 464–65, 471, 567
Heissenbüttel, H. 142–43
Hesiod, 121–23, 125, 131
Hobbes, T. 327, 418, 422
Hölderlin, F. 12, 29, 37, 102, 119, 121, 124–25, 138
Höllerer, W. 150
Hugo, R. 146
Hume, D. 358
Husserl, E. 13, 19, 40, 92, 96–97, 163, 263–64, 270, 499, 521, 523–24, 536, 568

Ignatow, D. 145
Ingarden, R. 4–6, 11, 168, 359, 503
Ionesco, E. 4

James, W. 536
Jaspers, K. 231–32, 240
Jones, A. R. 61, 72
Jones, J. W. 291
Joyce, J. 529
Justice, D. 144

Kaelin, E. 559–68
Kant, I. 10, 113, 350, 353–54, 495, 499, 503
Keene, D. 204, 208
Kierkegaard, S. 232
Kinnell, G. 144
Kirk, G. S. 167
Kirsch, S. 33
Koestler, A. 241, 247
Krieger, M. 319
Krolow, K. 251–52, 258–59
Kulpe, O. 10
Kumin, M. 150
Kurman, G. 169

Lattimore, R. 171, 173
Lawall, S. N. 73
Lessing, G. E. 135–36, 325–26, 332, 344–45, 350
Lewis, C. D. 147
Lewis, T. E. 52–53
Lipps, T. 10

McCullough, H. C. 202
McFarland, T. 127–28
Mach, E. 523–24, 533
Magliola, R. 62, 73
Mallarmé, S. 467–75, 477
Malory, T. 363–78
Man, P. de 505, 509–10, 516–17
Marcel, G. 530–32
Marx, K. 396–97
Matthews, W. 148
May, G. 409
Medina, A. 240–41
Merleau-Ponty, M. 49–50, 58–59, 81, 87–88, 95–97, 264, 270, 278, 292, 451–58, 472, 475–76, 479–93, 521, 523–27, 529, 532–33, 567
Meyer, C. F. 31
Milhaud, C. 273–93
Mills, R. J. 61, 72
Milosz, C. 91, 95, 97
Mon, F. 142–43
Montable, E. 151
Muller, H. J. 320, 331

Nietzsche, F. 115, 351–52, 355–56, 396–97
Novalis 123

Olson, C. 23

Ortega y Gasset, J. 428, 433

Paci, E. 40, 44
Parmenides 459–67, 471, 473
Pascal, B. 41
Patterson, R. R. 144
Paz, O. 146
Peirce, C. S. 157
Phillips, R. 61, 72
Plato 119–20, 313, 397, 462, 490
Plotinus 37
Pope, A. 105
Poulet, G. 73, 266
Pound, E. 53–57, 59
Proust, M. 97, 529, 531

Racine, J. B. 417–23
Reiss, T. 417–18, 422
Reznikoff, C. 145
Richards, I. A. 502
Ricoeur, P. 4, 240, 242, 244, 246–48
Riddel, J. 86, 88
Rilke, R. M. 30, 37, 91–93, 97, 148, 528
Robbe-Grillet, A. 264–65, 268–70, 401–6, 540
Rojas, F. de 426–28
Romilly, J. de 168
Rosenthal, M. L. 380–81
Ruch, B. 200, 208
Rühmkorf, P. 151
Ryle, G. 99

Sarraute, N. 265, 267–68
Sartre, J.-P. 195, 230–31, 234, 303
Saussure, F. de 547–48
Schiller, F. 131, 135, 149, 300, 324, 328, 330, 350
Schlegel, F. 123
Schneidau, H. N. 185
Schopenhauer, A. 351, 354
Schrag, C. O. 48, 58
Schwedhelm, K. 148
Sexton, A. 61–73
Shaffer, E. 129
Shakespeare, W. 380–99, 425–26, 429–32
Simpson, L. 145
Singleton, C. 510–11, 518
Snyder, G. 147
Sophocles 39, 41
Spenser, E. 91
Spiegelberg, H. 480
Stafford, W. 149
Steiner, G. 179, 183
Stevens, W. 75–88, 92, 97
Stoll, E. 430, 433
Strand, M. 147
Stryk, L. 144
Sukenick, R. 78, 88

Thom, R. 549, 554

Unamuno, M. de 115

Vattimo, G. 51–52, 59
Vonnegut, K. 63

Waehlen, A. de 479
Warren, A. 506–7, 517
Welch, L. 149
Wellek, R. 11, 506–7, 517
West, M. L. 121
Wiese, B. v. 319, 322, 326, 331–32
Williams, G. 144
Williams, W. C. 23, 489
Wordsworth, W. 93–95, 97, 127–29, 137
Woolf, V. 209–24, 241, 307, 316–17

# ANALECTA HUSSERLIANA

The Yearbook of Phenomenological Research

*Editor:*
ANNA-TERESA TYMIENIECKA
*The World Institute for Advanced Phenomenological Research and Learning*
*Belmont, Massachusetts*

I. *Analecta Husserliana.* 1971.
II. *The Later Husserl and the Idea of Phenomenology. Idealism-Realism, Historicity and Nature.* 1972.
III. *The Phenomenological Realism of the Possible Worlds. The 'A Priori', Activity and Passivity of Consciousness, Phenomenology and Nature.* 1974.
IV. *Ingardeniana. A Spectrum of Specialised Studies Establishing the Field of Research.* 1976.
V. *The Crisis of Culture. Steps to Re-Open the Phenomenological Investigation of Man.* 1976.
VI. *The Self and the Other. The Irreducible Element in Man, Part I: The 'Crisis of Man'.* 1977.
VII. *The Human Being in Action. The Irreducible Element in Man, Part II: Investigations at the Intersection of Philosophy and Psychiatry.* 1978.
VIII. Yoshihiro Nitta and Hirotaka Tatematsu (eds.), *Japanese Phenomenology. Phenomenology as the Trans-Cultural Philosophical Approach.* 1979.
IX. *The Teleologies in Husserlian Phenomenology. The Irreducible Element in Man, Part III: 'Telos' as the Pivotal Factor of Contextual Phenomenology.* 1979.
X. Karol Wojtyła, *The Acting Person.* 1979.
XI. Angela Ales Bello (ed.), *The Great Chain of Being* and *Italian Phenomenology.* 1981.
XII. *The Philosophical Reflection of Man in Literature.* 1982.
XIII. Eugene F. Kaelin, *The Unhappy Consciousness. The Poetic Plight of Samuel Beckett, An Inquiry at the Intersection of Phenomenology and Literature.* 1981.
XIV. *The Phenomenology of Man and of the Human Condition. Individualisation of Nature and the Human Being. I. Plotting the Territory for Interdisciplinary Communication.* 1983.
XV. *Foundations of Morality, Human Rights, and the Human Sciences. Phenomenology in a Foundational Dialogue with the Human Sciences.* 1983.
XVI. *Soul and Body in Husserlian Phenomenology. Man and Nature.* 1983.
XVII. *Phenomenology of Life in a Dialogue Between Chinese and Occidental Philosophy.* 1984.